68.00

C0-FPX-952

MUSCLE CONTRACTION

Its Regulatory Mechanisms

Proceedings of the Oji Seminar, Symposium on Regulatory Mechanisms of Muscle Contraction (Kumagai-Natori Symposium), held on October 28-30, 1979, sponsored by Japan Society for Promotion of Science and Fujihara Foundation of Science.

MUSCLE CONTRACTION

Its Regulatory Mechanisms

Edited by
SETSURO EBASHI, KOSCAK MARUYAMA, and MAKOTO ENDO

1980

JAPAN SCIENTIFIC SOCIETIES PRESS, Tokyo
SPRINGER-VERLAG, Berlin Heidelberg New York

1980

© Japan Scientific Societies Press, 1980

All rights reserved. No part of this publication may be reproduced or transmitted in any form or by any means, electronic or mechanical, including photocopy, recording, or any information storage and retrieval system, without permission in writing from the publisher.

JSSP No. 02392-1104

ISBN 3-540-10411-9 Springer-Verlag Berlin Heidelberg New York
ISBN 0-387-10411-9 Springer-Verlag New York Heidelberg Berlin

Sole distribution rights outside Japan granted to Springer-Verlag Berlin Heidelberg New York

Published jointly by
Japan Scientific Societies Press Tokyo
and
Springer-Verlag Berlin Heidelberg New York

Printed in Japan

591.1852
S989m
1979

Contents

Dynamic Aspects of Thin Filaments

Troponin-linked Regulation

Regulatory Mechanisms in Smooth Muscle

Myosin-linked Regulation

Excitation-Contraction Coupling

Cytoskeleton

Contributors

Adelstein, R. S.	NIH, Bethesda, Maryland 20014, U.S.A.
Aksoy, M. O.	Department of Physiology, University of Virginia School of Medicine, Charlottesville, Virginia 22901, U.S.A.
Amemiya, Y.	Engineering Research Institute, Faculty of Engineering, University of Tokyo, Tokyo 113, Japan
Anderson, K.	Department of Molecular, Cellular, and Developmental Biology, University of Colorado, Boulder Colorado 80302, U.S.A.
Arai, K.	Department of Food Science, Faculty of Fisheries, Hokkaido University, Hakodate 040, Japan
Arata, T.	Department of Biology, Faculty of Science, Osaka University, Toyonaka 560, Japan
Birnbaum, R. J.	Department of Physiology, Boston University School of Medicine, Boston, Massachusetts 02118, U.S.A.
Botts, J.	Cardiovascular Research Institute, University of California, San Francisco, California 94143, U.S.A.
Chandler, W. K.	Department of Physiology, Yale University School of Medicine, New Haven, Connecticut 06510, U.S.A.
Conti, M. A.	Section on Molecular Cardiology, Cardiology Branch, National Heart, Lung, and Blood Institute, Bethesda, Maryland 20205, U.S.A.
Couch, J.	Medical Research Council Cell Biophysics Unit, King's College, London W.C. 2, U.K.

Ebashi, S.	Department of Pharmacology, Faculty of Medicine, University of Tokyo, Tokyo 113, Japan
Endo, M.	Department of Pharmacology, Tohoku University School of Medicine, Sendai 980, Japan
Endoh, M.	Department of Pharmacology, Tohoku University School of Medicine, Sendai 980, Japan
Fujime, S.	Mitsubishi-Kasei Institute of Life Sciences, Tokyo 194, Japan
Gergely, J.	Department of Muscle Research, Boston Biomedical Research Institute, Boston, Massachusetts 02114, U.S.A.
Gonzalez-Serratos, H.	Department of Biophysics, School of Medicine, University of Maryland, Maryland, U.S.A.
Hartshorne, D. J.	Department of Nutrition and Food Science, University of Arizona, College of Agriculture, Tucson, Arizona 85721, U.S.A.
Hashizume, H.	Engineering Research Institute, Faculty of Engineering, University of Tokyo, Tokyo 113, Japan
Head, J. H.	Department of Physiology, Boston University School of Medicine, Boston, Massachusetts 02118, U.S.A.
Hirabayashi, T.	Institute of Biological Science, University of Tsukuba, Niihari, Ibaraki 300-31, Japan
Huxley, A. F.	Department of Physiology, University College London, WC1E 6BT, U.K.
Huxley, H. E.	MRC Laboratory of Molecular Biology, University Medical School, Cambridge, CB2 2QH, U.K.
Kaminer, B.	Department of Physiology, Boston University School of Medicine, Boston, Massachusetts 02118, U.S.A.
Katayama, E.	Department of Pharmacology, Faculty of Medicine, University of Tokyo, Tokyo 113, Japan
Kitazawa, T.	The Pennsylvania Muscle Institute, University of Pennsylvania School of Medicine, Philadelphia, Pennsylvania 19104, U.S.A.
Konno, K.	Department of Food Science, Faculty of Fisheries, Hokkaido University, Hakodate 040, Japan
Kumagai, H.	Faculty of Medicine, University of Tokyo, Tokyo 113, Japan
Kuroda, M.	Institute of Basic Medical Sciences, University of Tsukuba, Niihari, Ibaraki 300-31, Japan
Leavis, P. C.	Department of Muscle Research, Boston Biomedical Research Institute, Boston, Massachusetts 02114, U.S.A.

Maruyama, K. — Department of Biology, Faculty of Science, Chiba University, Chiba 260, Japan

Masaki, T. — Institute of Basic Medical Sciences, University of Tsukuba, Niihari, Ibaraki 300-31, Japan

Matsubara, I. — Department of Pharmacology, Tohoku University School of Medicine, Sendai 980, Japan

Matsuda, R. — The Laboratory for Contractile Proteins, The Tokyo Metropolitan Institute of Medical Science, Tokyo 113, Japan

Matsumoto, J. J. — Department of Chemistry, Faculty of Science and Technology, Sophia University, Tokyo 102, Japan

McClellan, G. — The Pennsylvania Muscle Institute, Department of Physiology and Pathology, University of Pennsylvania School of Medicine, Philadelphia, Pennsylvania 19104, U.S.A.

Mikawa, T. — Department of Pharmacology, Faculty of Medicine, University of Tokyo, Tokyo 113, Japan

Morales, M. F. — Cardiovascular Research Institute, University of California Medical Center, San Francisco, California 94143, U.S.A.

Morris, E. P. — Medical Research Council Cell Biophysics Unit, King's College, London W.C. 2, U.K.

Murray, J. M. — Department of Biochemistry, University of Pennsylvania School of Medicine, Philadelphia, Pennsylvania 19104, U.S.A.

Natori, R. — Jikei University, Tokyo 105, Japan

Nonomura, Y. — Department of Pharmacology, Faculty of Medicine, University of Tokyo, Tokyo 113, Japan

Obinata, T. — Department of Biology, Faculty of Science, Chiba University, Chiba 260, Japan

O'Brien, E. J. — Medical Research Council Cell Biophysics Unit, King's College, London W.C. 2, U.K.

Ohara, O. — Institute for Chemical Research, Kyoto University, Uji 611, Japan

Ohashi, K. — Department of Biology, Faculty of Science, Chiba University, Chiba 260, Japan

Ohtsuki, I. — Department of Pharmacology, Faculty of Medicine, Kyushu University, Fukuoka 812, Japan

Ooi, T. — Institute for Chemical Research, Kyoto University, Uji 611, Japan

Oosawa, F. — Department of Biophysical Engineering, Osaka University, Toyonaka 560, Japan

Pato, M. D.	Section on Molecular Cardiology, Cardiology Branch, National Heart, Lung, and Blood Institute, Bethesda, Maryland 20205, U.S.A.
Perry, S. V.	Department of Biochemistry, University of Birmingham, Birmingham, B15 2TT, U.K.
Podolsky, R. J.	Laboratory of Physical Biology, National Institute of Arthritis, Metabolism, and Digestive Diseases, NIH, Bethesda, Maryland 20205, U.S.A.
Porter, K. R.	Department of Molecular, Cellular and Developmental Biology, University of Colorado, Boulder, Colorado 80302, U.S.A.
Sakai, T.	Department of Physiology, Jikei University School of Medicine, Tokyo 105, Japan
Seymour, J.	Medical Research Council Cell Biophysics Unit, King's College, London W.C. 2, U.K.
Shuman, H.	The Pennsylvania Muscle Institute, Department of Physiology and Pathology, University of Pennsylvania School of Medicine, Philadelphia, Pennsylvania 19104, U.S.A.
Siemankowski, R. F.	Department of Nutrition and Food Science, Muscle Biology Group, University of Arizona, Tucson, Arizona 85721, U.S.A.
Somlyo, A. P.	The Pennsylvania Muscle Institute, Department of Physiology and Pathology, University of Pennsylvania School of Medicine, Philadelphia, Pennsylvania 19104, U.S.A.
Somlyo, A. V.	The Pennsylvania Muscle Institute, Department of Physiology and Pathology, University of Pennsylvania School of Medicine, Philadelphia, Pennsylvania 19104, U.S.A.
Sugi, H.	Department of Physiology, School of Medicine, Teikyo University, Tokyo 173, Japan
Szent-Györgyi, A. G.	Department of Biology, Brandeis University, Waltham, Massachusetts 02154, U.S.A.
Takahashi, S.	Institute for Chemical Research, Kyoto University, Uji 611, Japan
Tameyasu, T.	Department of Physiology, School of Medicine, Teikyo University, Tokyo 173, Japan
Tanaka, H.	Department of Physiology, School of Medicine, Teikyo University, Tokyo 173, Japan
Tawada, K.	Laboratory of Physical Biology, National Institute of

Arthritis, Metabolism, and Digestive Diseases, NIH, Bethesda, Maryland 20205, U.S.A.

Tonomura, Y. — Department of Biology, Faculty of Science, Osaka University, Toyonaka 560, Japan

Tsuchiya, T. — Department of Chemistry, Faculty of Science and Technology, Sophia University, Tokyo 102, Japan

Umazume, Y. — Department of Physiology, Jikei University School of Medicine, Tokyo 105, Japan

Wakabayashi, T. — Department of Physics, Faculty of Science, University of Tokyo, Tokyo 113, Japan

Watanabe, S. — Department of Chemistry, Faculty of Science, Tokyo Institute of Technology, Tokyo 152, Japan

Weber, A. — Department of Biochemistry, University of Pennsylvania School of Medicine, Philadelphia, Pennsylvania 19104, U.S.A.

Wegner, A. — Department of Biochemistry, University of Pennsylvania School of Medicine, Philadelphia, Pennsylvania 19104, U.S.A.

Yagi, N. — Department of Pharmacology, Tohoku University School of Medicine, Sendai 980, Japan

Yagi, S. — Department of Pharmacology, Tohoku University School of Medicine, Sendai 980, Japan

Yamada, K. — Department of Physiology, Medical College of Oita, Oita 879-56, Japan

Yamashita, T. — Laboratory of Physiological Chemistry, Juntendo University School of Medicine, Tokyo 113, Japan

Yanagida, T. — Department of Biophysical Engineering, Osaka University, Toyonaka 560, Japan

Yoshino, S. — Department of Physics, Nagoya University, Nagoya 464, Japan

Yoshioka, T. — Department of Physiology, Tokai University School of Medicine, Isehara 259-11, Japan

Preface

One of the aims of this Symposium was to celebrate the seventy-fifth birthday of Prof. Hiroshi Kumagai and the twenty-fifth anniversary of the discovery of 'skinned fibers' by Prof. Reiji Natori. This did not mean that the meeting would be only a ceremonial one, placing rather minor emphasis on scientific aspects. On the contrary, it was a result of the desire of Japanese muscle scientists to pursue the mechanism of muscle contraction, which has been inseparably linked with their admiration for the unparalleled contribution to muscle science of Profs. Kumagai and Natori.

We take the liberty of quoting a part of the 'Introduction' of a book dedicated to Prof. Kumagai (*Molecular Biology of Muscular Contraction*, Igaku Shoin, Tokyo, and Elsevier, Amsterdam, 1965):

> ". . . , even this somewhat encouraging event did not change his deep feeling that the best days of his career as a research scientist were over and that the best he could do then would be to bring up the younger generation to be full-fledged scientists on an international level. Since then he has performed his chosen task faithfully and enthusiatically, often at the expense of his private life."

The latter half of the 1940's was a most difficult time for Japanese scientists. However, this could not depress Prof. Natori, but stimulated him to create a new thing, 'Natori's fibers' or skinned fibers. His ardor for muscle

Professors H. Kumagai and R. Natori in front of the International House of Japan, Tokyo, where the 1957 Muscle Conference was held. June 28, 1980

science further compelled him to organize periodical muscle physiology meetings, 'Kin-seiri no Tsudoi', every two or three months, where young people could find a place to report their findings.

It was very timely that in 1955 Prof. Kumagai as the chairman of the research group on 'Chemistry of Muscular Contraction' invited Prof. Natori to join his group as a chief member. Thus links between muscle physiology and biochemistry began to bud in Japan, too.

We wish to quote again the above book:

> "He (Prof. Kumagai) moved further in organizing the 'Conference on the chemistry of muscular contraction', Tokyo, in 1957. To this conference, he invited many scientists to come from abroad. At that time, Japan was still experiencing difficult economic situation, and no financial support to the conference could be obtained from official sources. Nevertheless, the meeting was held, largely through his personal efforts. There were some difficulties in communication, but the conference was a

> great gain for the Japanese scientists, first because they could estimate their own level of achievement in the world, and second, because they made many friends from abroad."

It was our great pleasure to have Profs. M. F. Morales and J. Gergely, who played main roles in that Conference, in attendance at this Symposium. We were particularly delighted to welcome Prof. T. C. Tsao from Shanghai, who tried hard to attend the 1957 Conference but without success owing to unavoidable circumstances.

Our original plan was to have this Symposium in the same place where the 1957 Conference was held (see the photograph above) to refresh the nice memories of twenty-two years ago. However, Tokyo has changed very much in a direction unfavorable for such a meeting. So the meeting was held in a hotel near Mt. Fuji, where participants could enjoy the beautiful scenery and the courtesy of the hotel.

Taking this opportunity, we wish to express our cordial thanks to all participants, especially those from abroad. We are particularly indebted to Sir Andrew Huxley for his historical survey of the studies on regulatory mechanisms of muscle contraction, which has exempted us from explaining the scientific reason why we planned this Symposium.

We are very grateful to the officials of The Japan Society for the Promotion of Science and the Fujiwara Foundation of Science for their extending us cordial assistance in organizing this Symposium, especially to Mr. Y. Ichikawa, the director of the Fujiwara Foundation, for his sincere encouragement. Our thanks are also due to our colleagues, especially Dr. Y. Nonomura, for their efforts behind the curtain in managing the Symposium.

July 1980

S. Ebashi
K. Maruyama
M. Endo

October 18, 1979

My dear Dr. Ebashi:

You have made me very happy with your letter—a new expression of an old friendship. You have also deeply honored me by the generous invitation to the Kumagai-Natori Symposium on Regulatory Mechanisms of Muscle Contraction. It would have been a very great pleasure for me to attend the meeting and meet old friends. It would have given me a special pleasure to convey personally the expression of my admiration for Professor Kumagai, the father to Japanese muscle scientists, for the wonderful research he has conducted all his life, and the equally wonderful example and teaching which he has given to the younger Japanese generation devoted to science. During my long research I could never separate ethics, morals, goodwill and the search of new truth which formed one unit in Dr. Kumagai's life. I also would have loved to express personally my admiration for the wonderful scientific achievements of Professor R. Natori, inventor of the fiber carrying his name. To both of them I wanted to express also my best wishes for their further work, long life, health and happiness.

Very unfortunately my advanced age and many obligations prevent me from partaking in the Sypmosium in commemoration of Professor Kumagai's 75th anniversary, and the 25th anniversary of the publication of the skinned fiber which opened an entirely new road for research.

I could not have been a useful member, having left muscle research long ago, but I would have been proud of seeing the splendid research going on. If one had even a small part in planting a tree, one is always proud and glad to see that the tree grows and blossoms, even if one can do nothing for it any more.

With the warmest regards and the best wishes to all of you.

Yours,

Albert Szent-Györgyi

SPECIAL LECTURE

Muscle Contraction: Its Regulatory Mechanisms (Ebashi, S. et al., eds.), pp. 3–18,
Japan Sci. Soc. Press, Tokyo / Springer-Verlag, Berlin (1980)

1

The Variety of Muscle Activating Systems

A. F. HUXLEY

I am delighted to have the privilege of taking part in a meeting to do honour to two men who have made such original and far-reaching contributions to the understanding of muscle as those we associate with the names of Professor Kumagai and Professor Natori.

As far as I know the history, the key step in Professor Natori's contribution was the earlier. I believe it was around 1949 that he found that he could remove the sarcolemma from a single muscle fibre while retaining the ability of the fibre contents to give a contractile response to electrical stimulation. It was, however, a number of years before this discovery became widely known outside Japan. In self-defence, I think we can in this case excuse ourselves to some extent on the grounds of Professor Natori's modesty. He did publish a description of his experiments in 1951, but that was in his text-book of muscle physiology, which was in the Japanese language. His first publication of the method in a Western language was in 1954, in the first volume of the Jikeikai Medical Journal, the journal of the medical school where he then worked and of which he is now the President. This too was not widely read in the west, but in the meantime—as so often happens—personal contacts were more effective than formal publication in spreading the news. I was fortunate to be among the first in England to hear of his work, and this came about through a very informal channel. At that time both Victor Perry and I were fellows of Trinity College Cambridge, and Perry went to Japan in 1953. He visited Professor Natori's laboratory, and when he returned he told his friends who were interested in muscle about the remarkable experiments that he had seen there. That trip of Perry's to Japan, by the way, was not primarily a scientific visit. As well as being a biochemist of international repute, he was an international Rugby player, and he went to Japan as manager of a Cambridge Rugby team.

Professor D. R. Wilkie was also aware of Professor Natori's work through letters that they exchanged in 1953; he has shown me an account of the properties of the skinned-fibre preparation that was presented by Professor Natori at a conference of the Japanese Union of Physiological Sciences in 1952, and was sent to him at that time by Professor Natori.

The next contact with Professor Natori's work that I remember was when he came to England in 1960 or 1961, and visited University College London, to which I had just moved from Cambridge. He showed us films of some of his experiments, and very impressive they were. The most striking of his films—which I remember not from that occasion but from the Physiological Congress in Tokyo in 1965—is the one which shows a skinned muscle fibre which has been split longitudinally for a short part of its length, forming a ring of contractile material. Professor Natori stimulates this preparation electrically and succeeds in establishing a contraction which is conducted many times round and round this ring. After seeing that film, no one could possibly doubt the reality of all-or-none propagation in the skinned muscle fibre.

These propagated waves have the unfamiliar feature that they start opposite the anode, not the cathode. Professor Natori pointed out, however, that this is only to be expected since the electrodes are on the sarcoplasm side of the membranes which can be expected to be involved in the propagation—those of the transverse tubular system and of the sarcoplasmic reticulum proper. I think, however, that it is still not clear which of these membrane systems has the chief role in this type of propagation—whether it is the sarcoplasmic reticulum, which in amphibian muscle fibres runs throughout the length of each sarcomere and has occasional elements going past the Z line and uniting the reticulum of adjacent sarcomeres, or whether it is the transverse tubular system, which although—as its name implies—it runs chiefly in the transverse direction, nevertheless has a substantial number of longitudinal connections from the transverse network at the level of one Z line to that at the next (Eisenberg, 1971). This seems to me the more likely possibility since it has been clear for some time that in a normal intact fibre the transverse tubular system is capable of inward propagation (Costantin, 1970; Costantin and Taylor, 1971).

There was, however, another type of experiment performed by Professor Natori on his skinned fibres which was extremely puzzling for many years. These were experiments in which small volumes of solutions of various salts were applied to the preparation. The puzzling thing was that a contraction was produced whatever the composition of the solution—indeed, even if distilled water was used. The only explanation seemed to be the rather unlikely one that the relevant factor was the dilution of some material present in the sarcoplasm that kept the myofibrils in a relaxed state. The credit for resolving this paradox is, I believe, due chiefly to Richard Podolsky, who showed that contraction occurred only if the solution contained calcium, and that in cases where application of a nominally calcium-free solution had caused contraction, this was due to contamination with the trace quantities of calcium that are sufficient to turn on the contractile system (Podolsky and Hubert, 1961)—not the only case where scientists have been misled by the extraordinarily high sensitivity of the muscle regulatory proteins to calcium. Since that time, of course, the skinned-fibre preparation has become the preparation of choice for investigating the response of the complete contractile system to activation by calcium ions.

In Professor Natori's original skinned-fibre preparation, the operation of removing the sarcolemma was performed with the fibre under oil. As is all too well known to anyone who has dissected intact single muscle fibres in Ringer's solution—which contains a few millimolar ionized calcium—any damage to the surface membrane leads immediately to a violent local contraction followed by disintegration of the contractile material into a row of clots. The fact that nothing of this kind happens when the sarcolemma is removed under oil is satisfactorily explained by the very small volume of extra-cellular fluid which is present between the fibre surface and the oil. The calcium in this small volume of aqueous solution is rapidly taken up by the sarcoplasmic reticulum of the fibre after the sarcolemma has been removed, and the free calcium level is again reduced below the threshold for causing contraction. With the development of calcium buffer systems, however, it has become possible to skin a muscle fibre in an aqueous solution containing magnesium ATP and a calcium buffer in which the ionized calcium concentration is below the threshold for contraction. In this situation the contractile material survives in a remarkably normal condition, and is able to develop tension, and to give shortening speeds, comparable to those produced by intact fibres. This preparation is much more nearly normal than the glycerinated preparation of Szent-Györgyi, and is now being used in many laboratories for investigating the effects of changes of concentration of various relevant substances on the contractile characteristics of the actomyosin system. Even the mechanical transients associated with sudden stretch or release can now be investigated on this preparation. It is now the preparation which gives the best hope of linking purely biochemical investigations of isolated protein with observations that can be made on intact muscle fibres. It has largely superseded Szent-Györgyi's glycerinated preparation, and its position now in the investigation of muscle contraction can be likened to that of the giant nerve fibre of the squid in the investigation of nerve conduction.

I believe that the only actual publication by Professor Kumagai that has to do with muscle was a letter in Nature in 1955. The authors were H. Kumagai, S. Ebashi, and F. Takeda. Kumagai was then Head of the Pharmacology Department of Tokyo University; Ebashi was a Research Associate in the same department, who had taken his Ph. D. in the previous year. Takeda's name disappears from the scientific literature, but she herself reappears under the name of F. Ebashi, as she and our Chairman were married soon after that letter was published.

That letter reported the first piece of real progress in elucidating the mysterious "relaxing factor" discovered by Marsh in 1951, in the laboratory of Kenneth Bailey. Professor Kumagai and his colleagues showed that the relaxing action was due to a microsomal component which was identical with the "Kielley-Meyerhof enzyme," an ATPase, discovered in 1948, which had previously been one of the least understood components of muscle. I remember being completely puzzled why muscle should contain an enzyme which was capable of hydrolysing ATP at least as fast as actomyosin could, but which did not seem to be a part of the contractile system at all. Kumagai, Ebashi, and Takeda found a function for this enzyme, as well as taking the step which opened up the way for the series of discoveries—most of them by Professor Ebashi and his colleagues—which have led to the extraordinarily detailed understanding that we now have of the regulatory system of vertebrate striated muscle.

Important though this letter in Nature was, Professor Kumagai has exerted a much wider influence on the progress of muscle research. His own earlier research career, severely interrupted by World War II, was concerned chiefly with oxytocic drugs and with the pharmacology of the brain stem. But he made also a tremendous contribution which is little known outside Japan, and this was in the encouragement of the younger generation of muscle physiologists and biochemists. He organised in 1955 a Research Committee for muscle biochemistry, which included younger scientists as well as senior and long-established ones. In 1957 he arranged a successful international "Conference on the Chemistry of Muscular Contraction" in Tokyo—one of the first international scientific meetings in Japan after World War II. In 1960, he was one of the moving spirits in the formation of the Biophysical Society of Japan. In such ways, his unselfish activity has been a most important factor in stimulating the tremendous contributions to the physiology, biochemistry, and biophysics of muscle that have been made by Japanese scientists during the last quarter of a century.

I want to devote the rest of this lecture to a discussion of some of the obstacles that stood in the way of reaching the degree of understanding of the activation mechanism of muscle that has so far been achieved. One that I have already mentioned is the great sensitivity of the contractile system to calcium. We have all come to think of 10^{-5} molar free calcium as a rather high concentration, and it is easy to forget that without using calcium buffers it is very difficult to make solutions with a lower free calcium concentration than this. Impurity calcium comes partly from the glassware, and partly from the other salts used in the solution: after all, in a 0.1 molar solution of potassium chloride, it takes only a 0.01% impurity level of calcium chloride in the KCl to give 10^{-5} molar calcium in the solution. The complicated interactions of this impurity calcium with chelators such as ATP itself or EDTA, in the presence of competing magnesium ions, were analysed for the first time by Annemarie Weber (1959), who drew the important conclusion that maximal activation of the Mg-ATPase activity of a myofibril suspension required the presence of free calcium ions at a concentration of a few micromolar. It was a remarkable achievement of Professor Ebashi (1961) to demonstrate, without any use of calcium binders, an acceleration of superprecipitation of actomyosin caused by 1 micromolar added calcium. In his Croonian Lecture to the Royal Society in London this year, Ebashi (1979) said: "To carry out this experiment, all the apparatus and reagents had to be free of calcium ion. I appreciated the use of wares made of hard glass, which were not easily available in Japan in 1959; even so, it was the most tedious experiment in my life—so far." But from the time of Marsh's discovery of the relaxing factor in 1951 until the general acceptance of calcium as the activator (about 1962), the majority of muscle biochemists (excepting Annemarie Weber and Ebashi) did not realise that calcium concentrations due to impurities could be important, and attributed the effects of substances known to chelate calcium ions to direct actions on the contractile proteins rather than to the lowering of free calcium concentration.

Suggestions that calcium might be the activator had been around for many years. Ringer (1883) and Mines (1913) had shown that calcium was needed for excitation-contraction coupling in the heart: Chambers, the pioneer of microdissection, had found in 1932 that a local contraction occurred when a solution of calcium chloride was injected into a piece of a muscle cell, and this had been repeated by Heilbrunn (1940), who postulated a role for calcium in turning on many intracellular processes. Kenneth Bailey (1942) wrote: ". . . we suggest that the essential feature of excitation and contraction . . . is the liberation of the Ca ion in the neighbourhood of the ATPase grouping, which can thus by the almost instantaneous catalysis of ATP breakdown make available a large amount of energy. For it is legitimate to assume that the living cell in the resting state can provide a mechanism for the separation of enzyme and activator, until they are brought together as the result of excitation." Emil Bozler (1951) succeeded in making glycerinated muscle preparations (in the presence of Mg and a high ATP concentration) go through cycles of contraction and relaxation by adding and removing calcium, and showed that citrate caused relaxation when calcium was present. Marsh, in the full paper on his relaxing factor (1952), though not in his first report in 1951, mentions that the relaxing effect is abolished by 0.02% $CaCl_2$ (about 1.8 mM Ca), and this observation was strikingly confirmed and extended by Bendall (1953), who showed complete abolition on adding calcium to a concentration of 0.2 mM. Bozler (1954) showed that all the effects obtained by Bendall with the natural relaxing factor were closely imitated by the calcium chelator EDTA.

In spite of these numerous and powerful hints, I remember biochemists with whom I discussed these questions in the 'fifties (again excepting Weber and Ebashi) being very resistant to the idea that contraction might be regulated by free calcium concentration. I remember Professor Ebashi telling me that he found a similar resistance when he was in Lipmann's laboratory in New York in 1959, and that this was based largely on the conviction that so specific a function could not be performed by anything as simple as a calcium ion. I imagine that another major factor was the absence of a straightforward activating effect of added calcium on ATPase activity of actomyosin systems *in vitro*, due partly to the complicated interactions between the effects of calcium and magnesium, partly to the fact that the concentrations of Ca *etc.* used by Bailey and others were of the order of a few millimolar, similar to those known to produce activating effects in other enzyme systems, and partly to the ease with which the regulatory system was lost either through purification or by the passage of time.

At any rate, there was a widespread assumption that there must be a soluble "relaxing factor," and this role was attributed to a number of different substances—for example, creatine phosphokinase (Lorand, 1953), myokinase (Bendall, 1954), and cyclic AMP (Uchida and Mommaerts, 1963; these authors did recognise that calcium removal also plays an important part). I wonder whether there is a significant parallel between this search for a relaxing factor and the search for complex "high-energy intermediates" in oxidative phosphorylation, where the function in question was also being performed by small ions—in this case by a chemi-osmotic mechanism (Mitchell, 1961). Another parallel, in which the credit goes to the other side, is the resistance shown by electrophysiologists to chemical transmission in the 'twenties and 'thirties, well described by Feldberg (1977).

The next example I shall mention is one where by good fortune a misconception led to progress. In a paper published in 1949, A. V. Hill had made it clear that the whole of the cross-section of each muscle fibre was brought into action more rapidly than could be accounted for by inward diffusion of some "activating substance" from the surface membrane, which was clearly the place where the action potential was having its direct effect. One possibility suggested by Hill was a regenerative chemical or physico-chemical process; the "calcium-stimulated calcium release" which is now known to occur in certain circumstances is such a process. The only obvious alternative was that there might be some specialised structure capable of conducting an influence from the surface membrane inwards. It was natural to imagine that any such structure would be located at some definite position relative to the striations; furthermore, there were many suggestions from the old literature of light microscopy that there was a continuous transverse structure at the level of each Z line (see A. F. Huxley, 1957). This could be called a "Z-membrane," or "Krause's membrane," and consisted of the Z lines themselves within the myofibrils together with transverse structures connecting the Z lines of adjacent myofibrils to one another, and also running from the most superficial myofibrils to the surface membrane. Dr. R. E. Taylor and I (1955, 1958) devised a way of testing this hypothesis. We depolarised very small areas of surface membrane of isolated fibres from frog muscle by applying negative-going electric pulses to a small pipette (tip diameter about 1 μm) in contact with the fibre surface. We found what we were looking for: a contraction was produced only if the pipette overlapped the attachment of a Z line to the fibre surface. The phenomenon was very striking, though not quite as striking as we had hoped, since the contraction was graded with the strength of the applied electric pulse and did not spread across the fibre in an all-or-none manner. But the fact that we got a response only if the depolarised patch included the point of attachment of a Z-membrane was very clear, and we published the result in Nature in a letter entitled "Function of Krause's membrane."

Our first doubts about the interpretation of this result were raised when we presented the result in a communication to the Physiological Society. During the discussion of our paper, the electron microscopist J. D. Robertson produced from his pocket a slide (Robertson, 1956, Fig. 10) of a longitudinal section of muscle which very clearly showed transverse tubules in the spaces between the myofibrils at definite positions in the striation pattern. These were just the sort of structure that might be the basis for an inward conduction process but they were not at the level of the Z lines: there was a set of tubules near each side of each I band, *i.e.* associated with each Z there were two sets of tubules, one on either side.

Robertson made the very reasonable suggestion that perhaps our pipettes were not small enough to distinguish the two sets of tubules in each of these pairs. So we repeated our local-stimulation experiment on muscle fibres from a crab, with sarcomeres several times longer than those of vertebrate muscle. We found exactly what was to be expected on the basis of Robertson's observations: there was no response when the pipette was over a Z line or over the middle of an A band, but when the pipette was over an A-I boundary, there was a contraction involving only that half-sarcomere: the Z line was pulled into contact with the A band, the half-I-band being completely obliterated.

The next thing was to look afresh at frog muscle. The experiments I have so far described were done under a polarising microscope; in crab muscle this shows up the Z line clearly, as well as the A band, but in frog muscle the Z line is not sufficiently strongly birefringent to be seen. So we repeated the experiment under an interference microscope, and clearly saw that the phenomenon was different from what we had seen in fibres from crab muscle: the Z line always remained central in the I band, and we could not get one half of the I band to shorten more than the other.

Robertson's electron micrographs were of muscle from a lizard, so we tried the local-activation experiment on fibres from lizard muscle (A. F. Huxley and Straub, 1958). We found that each half-sarcomere could be activated separately, as in crab muscle but unlike what we found in the frog—not what one would have expected from the evolutionary relationships of these animals.

Meanwhile, Porter and Palade (1957) had shown with the electron microscope a considerable amount of detail in the tubular and vesicular structures between the myofibrils. "Triads" were seen at the level of the Z line in an amphibian (the same is true of frog muscle) and near each A-I boundary in mammalian muscle; we found a little later (A. F. Huxley, 1959) that the arrangement in the reptilian fibres we had used is similar to the mammalian layout. So both in frog and in lizard, "triads" were present at the centre of each region which could be activated separately by local depolarisation. But these "triads" did not appear to be continuous structures in the transverse direction. Robertson's micrograph that I have already mentioned does show continuous tubules at the positions of the centre of the triads, but the outer elements of the triads and the rest of the reticulum were not preserved. It was not until a couple of years later that micrographs were obtained showing a continuous central element in each triad (A. F. Huxley, 1959), and Andersson-Cedergren (1959) showed, by reconstruction from serial sections, that the "intermediary vesicles" seen by Porter and Palade at the centre of the triad were really sections through a highly convoluted tubule. With the advent of glutaraldehyde as a fixative, it became a matter of routine to obtain sections in which the middle element of the triad was obviously a "transverse tubule."

The inward spread of activation along the transverse tubules now seems extremely straightforward but there was a period of four years, from our discovery of the local-activation phenomenon in 1955 to the demonstration of a "transverse tubular system" in 1959 during which the phenomenon was very puzzling, partly because of unexpected differences in anatomical layout in the muscle fibres of different kinds of animals and partly because of fixation artifacts in electron micrographs. Not that all the problems were solved by 1959. Even now, surface openings of the tubules in adult frog muscle and in mammalian muscle have only seldom been clearly seen in electron micrographs, though continuity with the extracellular space was demonstrated by experimental methods in 1964 (Endo, 1964; Page, 1964; H. E. Huxley, 1964). The situation in crab muscle was also very puzzling for another five years. L. D. Peachey and I looked for transverse tubules with the electron microscope in 1958; we found them easily but they were in the wrong place—at the Z line level! However, their appearance was very different from that of the middle elements of the triads of vertebrate muscle, and when we looked again, with glutaraldehyde as a fixative (Peachey and A. F. Huxley, 1964) we found another set of tubules, near each A-I junction which formed contacts with large vesicles of the sarcoplasmic reticulum, similar to the triads exccpt that they are one-sided and are therefore referred to as "dyads." The function of the Z tubules in arthropod muscles is still completely unknown.

With the small pipettes necessary to restrict activation to one I band or half-sarcomere, we never did see an all-or-none inward propagation process, and we concluded that the inward conduction up the tubules was a passive process of electrotonic spread. This has turned out to be wrong: González-Serratos (1971) showed that the speed of inward spread of shortening in a twitch, and its temperature coefficient, suggested a spike-like process, and the all-or-none character of inward spread when the surface membrane potential was clamped was shown by Costantin (1970) and Costantin and Taylor (1971). Transverse propagation was also seen by Strickholm (1962) who used a local-stimulation technique similar to ours but with larger pipettes. Presumably an all-or-none action potential fails to be set up in the transverse tubules because the propagation mechanism has a low safety factor (Adrian and Peachey, 1973) and stimulation at a single point is unfavourable for setting up propagation in a two-dimensional network.

Apart from imperfect preservation of membrane structures in preparation for electron microscopy, the main factor which caused confusion in the interpretation of local-activation experiments was the unforeseen diversity of layout of tubule systems in different kinds of animals. The variety is even greater than what I have already spoken of. In the slow fibres of frogs the transverse tubular system is irregular and triads are few in number (Page, 1965), and local depolarisation may cause contraction wherever the pipette is placed in relation to the striation pattern (Peachey and A. F. Huxley, 1960); it may be that significant amounts of calcium enter through the surface membrane as well as being released intracellularly from the sarcoplasmic reticulum. A more extreme case is *Amphioxus* (Peachey, 1961), where the "fibres" are sheets only about 1 μm thick and there is no internal tubular system; presumably activation depends entirely on calcium entry through the surface membrane.

Very unexpected differences between different contractile tissues, and between different kinds of animals, have also turned up in the role of the calcium ion. By about 1962, it was generally accepted that the first step in excitation-contraction coupling in the familiar striated muscles was, as I have just been saying, the inward spread of depolarisation along the transverse tubules, which were invaginations of the surface membrane, and also that a subsequent step was the intracellular release of calcium ions, presumably from the lateral vesicles of the triads (or dyads). Nothing definite is known even now about the way in which the first of these steps causes the second, and in 1962 we were equally ignorant about all the subsequent steps in the processes by which calcium ions turn on the contractile material. Then in 1963 came what was perhaps the most unexpected, and the most spectacular, of Professor Ebashi's contributions to our understanding of muscle: he prepared extracts of muscle which conferred a sensitivity to calcium on purified actomyosin, or on myofibril preparations which had lost their calcium-sensitivity through ageing. Just as in the case of Kumagai, Ebashi, and Takeda's discovery of the identity of the relaxing factor with the Kielley-Meyerhof enzyme, this fresh breakthrough combined a major step forward in understanding the regulation of contraction with the discovery of a function for something that had been puzzling us for a very long time—in this case, Bailey's tropomyosin (1946, 1948). It is probably difficult for those who have entered the muscle field in the last ten or fifteen years, and probably take the tropomyosin-troponin system for granted, to realise how suddenly this whole field was clarified for us. Several things conspire together to fix in my memory the occasion when I first heard of this discovery: it was mentioned briefly by Dr. M. Endo at a Discussion Meeting at the Royal Society in London, attended by muscle experts from all over the world, on 22 November 1963; it was my birthday; it was the day my brother Aldous Huxley died; and it was the day that J. F. Kennedy was shot.

For some years, everyone—as far as I know—supposed that all types of muscle were regulated by calcium acting through troponin and tropomyosin. Another surprise was in store: in 1970, Kendrick-Jones, Lehman, and Andrew Szent-Györgyi found that the fast adductor of the fan-shell Pecten, and other muscles of molluscs, were regulated through an action of calcium on the myosin, and subsequently he and his colleagues found that striated muscles in many arthropods have regulatory systems on both thick and thin filaments, the latter being worked by troponin and tropomyosin by much the same process as in vertebrate muscle.

There is at present vigorous controversy about the mechanism of regulation in vertebrate smooth muscle; probably the only point on which everyone is agreed is that it is not closely similar to either the thin-filament or the thick-filament regulatory systems that are known in striated muscles. Perhaps stimulated by the discovery of Adelstein and Conti (1975) that phosphorylation of platelet myosin raises its ATPase activity, several groups (see references in Ebashi, 1979) found in 1976 that a similar result could be obtained with smooth muscle myosin, and concluded that calcium-activated phosphorylation of the myosin was the normal process by which vertebrate smooth muscle was brought from rest into contractile activity. Ebashi, however, has continuously maintained that he could extract a protein factor which conferred on the smooth muscle actomyosin a sensitivity to free calcium concentration which was not associated with changes in the degree of phosphorylation of the myosin. He and his colleagues have recently obtained very impressive evidence in favour of this view (Mikawa *et al.*, 1977). But in spite of a superficial resemblance to the troponin system of striated muscle, it seems to work in a totally different way: in the troponin system, calcium acts as a derepressor, releasing the actomyosin from an inhibition produced by troponin and tropomyosin, while in smooth muscle the regulatory protein (named "leiotonin" by Ebashi) is not an inhibitor of the actomyosin ATPase, and calcium in the presence of leiotonin acts as a genuine activator, raising the ATPase activity far above that of the unregulated actomyosin.

The "spasmoneme," or contractile thread, of Vorticella and related ciliates is also brought into contraction by a rise in the free calcium concentration, but the mechanism is totally unlike any of those I have been discussing. Hoffmann-Berling (1958) showed that the contraction did not require the presence of ATP or other organic substrate; his work has been greatly extended by Weis-Fogh and Amos (1972), who have shown clearly that the contraction is a physical change (plasticisation) in the rubbery contractile protein, brought about by binding of calcium and reversed by removal of the calcium, the work being provided by the calcium-removing system since the binding of the calcium is stronger the more the spasmoneme shortens. This mechanism is one that seemed an attractive explanation for muscle contraction (*e.g.*, Pryor, 1950) before the sliding filaments were discovered. The spasmoneme has for a long time been regarded by some (*e.g.*, Bernstein, 1901) as an almost ideal model for muscle contraction. It is perhaps fortunate that the mechanism by which it contracts was not fully elucidated till after sliding filaments were well established. If it had been, the sliding-filament idea might have met the sort of resistance that I mentioned earlier in this lecture in connection with the regulatory action of calcium, the chemi-osmotic mechanism in mitochondria, and the chemical theory of synaptic and neuro-effector transmission.

So we have the extraordinary situation that, in all the contractile systems in which the regulatory system is at least partially known, the normal activating event is a rise of the intracellular free calcium concentration, but there are at least four different ways in which the contractile proteins are influenced by the calcium. I do not suppose the range of mechanisms has yet been fully explored, and in any case there are numerous modulating effects—dependence of contractile force on previous activity, on muscle length, on hormone concentrations—that are still to be worked out—and we have no clear idea yet of the roles played by variation of calcium release, phosphorylation of various proteins, cyclic nucleotides, and no doubt other factors, in these modulating effects. The only conclusion we can safely draw from the examples I have gone through in this lecture is that it would be very rash to assume that similar-looking processes in different tissues, or in different kinds of animals, are operated by similar biochemical or biophysical processes. The furthest one can safely go is to use the knowledge gained by investigating one system as a clue in designing experiments to investigate another.

REFERENCES

Adelstein, R. S. and Conti, M. A. (1975). Phosphorylation of platelet myosin increases actin-activated myosin ATPase activity. *Nature* **256**, 597–598.

Adrian, R. H. and Peachey, L. D. (1973). Reconstruction of the action potential of frog sartorius muscle. *J. Physiol.* **235**, 103–131.

Andersson-Cedergren, E. (1959). Ultrastructure of motor end plate a sarcoplasmic components of mouse skeletal muscle fiber. *J. Ultrastruct. Res.* (*Suppl.* 1), 1-191.

Bailey, K. (1942). Myosin and adenosinetriphosphatase. *Biochem. J.* **36**, 121–139.

Bailey, K. (1946). Tropomyosin: a new asymmetric protein component of muscle. *Nature* **157**, 368–369.

Bailey, K. (1948). Tropomyosin: a new asymmetric protein component of the muscle fibril. *Biochem. J.* **43**, 271–279.

Bendall, J. R. (1953). Further observations on a factor (the 'Marsh' factor) effecting relaxation of ATP-shortened muscle-fibre models, and the effect of Ca and Mg ions upon it. *J. Physiol.* **121**, 232–254.

Bendall, J. R. (1954). The relaxing effect of myokinase on muscle fibres; its identity with the 'Marsh' factor. *Proc. Roy. Soc. Lond. B* **142**, 409–426.

Bernstein, J. (1901). Die Energie des Muskels als Oberflächenenergie. *Pflüg. Arch. Ges. Physiol.* **85**, 271–312.

Bozler, E. (1951). Mechanism of relaxation in extracted muscle fibers. *Am. J. Physiol.* **167**, 276–283.

Bozler, E. (1954). Relaxation in extracted muscle fibers. *J. Gen. Physiol.* **38**, 149–159.

Chambers, R. and Hale, H. P. (1932). The formation of ice in protoplasm. *Proc. Roy. Soc. Lond. B* **110**, 336–352.

Costantin, L. L. (1970). The role of sodium current in the radial spread of contraction in frog muscle fibers. *J. Gen. Physiol.* **55**, 703–715.

Costantin, L. L. and Taylor, S. R. (1971). Active and passive shortening in voltage-clamped frog muscle fibres. *J. Physiol.* **218**, 13–15P.

Ebashi, S. (1961). Calcium binding activity of vesicular relaxing factor. *J. Biochem.* **50**, 236–244.

Ebashi, S. (1979). Regulation of muscle contraction (The Croonian Lecture). *Proc. Roy. Soc. Lond. B* **207**, 259–286.

Eisenberg, B. (1971). The sarcotubular system of frog skeletal muscle. Ph. D. Thesis, University of London.

Endo, M. (1964). Entry of a dye into the sarcotubular system of muscle. *Nature* **202**, 1115–1116.

Feldberg, W. (1977). The early history of synaptic and neuromuscular transmission by acetylcholine: reminiscences of an eye witness. In *The Pursuit of Nature* (Hodgkin, A. L. *et al.*), Cambridge Univ. Press, London.

González-Serratos, H. (1971). Inward spread of activation in vertebrate muscle fibres. *J. Physiol.* **212**, 777–799.

Heilbrunn, L. V. (1940). The action of calcium on muscle protoplasm. *Physiol. Zool.* **13**, 88–94.

Hill, A. V. (1949). The abrupt transition from rest to activity in muscle. *Proc. Roy. Soc. Lond. B* **136**, 399–420.

Hoffmann-Berling, H. (1958). Der Mechanismus eines neuen, von der Muskelkontraktion verschiedenen Kontraktionszyklus. *Biochim. Biophys. Acta* **27**, 247–255.

Huxley, A. F. (1957). Muscle structure and theories of contraction. *Prog. Biophys. Chem.* **7**, 255–318.

Huxley, A. F. (1959). Local activation of muscle. *Ann. N.Y. Acad. Sci.* **81**, 446–452.

Huxley, A. F. and Straub, R. W. (1958). Local activation and interfibrillar structures in striated muscle. *J. Physiol.* **143**, 40–41P.

Huxley, A. F. and Taylor, R. E. (1955). Function of Krause's membrane. *Nature* **176**, 1068.

Huxley, A. F. and Taylor, R. E. (1958). Local activation of striated muscle fibres. *J. Physiol.* **144**, 426–441.

Huxley, H. E. (1964). Evidence for continuity between the central elements of the triads and the extracellular space in frog sartorius muscle. *Nature* **202**, 1067–1071.

Kendrick-Jones, J., Lehman, W., and Szent-Györgyi, A. G. (1970). Regulation in molluscan muscles. *J. Mol. Biol.* **54**, 313–326.

Kielley, W. W. and Meyerhof, O. (1948). Studies on adenosinetriphosphatase of muscle. II. A new magnesium-activated adenosinetriphosphatase. *J. Biol. Chem.* **176**, 591–601.

Kumagai, H. (ed.) (1958). *Conference on the Chemistry of Muscular Contraction, 1957*, Igaku-Shoin, Tokyo and Osaka.

Kumagai, H., Ebashi, S., and Takeda, F. (1955). Essential relaxing factor in muscle other than myokinase and creatine phosphokinase. *Nature* **176**, 166.

Lorand, L. (1953). Adenosinetriphosphate-creatine transphosphorylase as relaxing factor of muscle. *Nature* **172**, 1181.

Marsh, B. B. (1951). A factor modifying muscle fibre synaeresis. *Nature* **167**, 1065–1066.

Marsh, B. B. (1952). The effects of adenosine triphosphate on the fibre volume of a muscle homogenate. *Biochim. Biophys. Acta* **9**, 247–260.

Mikawa, T., Nonomura, Y., and Ebashi, S. (1977). Does phosphorylation of myosin light chain have direct relation to regulation in smooth muscle? *J. Biochem.* **82**, 1789–1791.

Mines, G. R. (1913). On functional analysis by the action of electrolytes. *J. Physiol.* **46**, 188–235.

Mitchell, P. (1961). Coupling of phosphorylation to electron and hydrogen transfer by a chemiosmotic type of mechanism. *Nature* **191**, 144–148.

Natori, R. (1954). The property and contraction process of isolated myofibrils. *Jikeikai Med. J.* **1**, 119–126.

Page, S. G. (1964). The organization of the sarcoplasmic reticulum in frog muscle. *J. Physiol.* **175**, 10–11P.

Page, S. G. (1965). A comparison of the fine structures of frog slow and twitch muscle fibres. *J. Cell Biol.* **26**, 477–497.

Peachey, L. D. (1961). Structure of the longitudinal body muscles of Amphioxus. *J. Biophys. Biochem. Cytol.* **10** (Suppl.), 159–176.

Peachey, L. D. and Huxley, A. F. (1960). Local activation and structure of slow striated muscle fibers of the frog. *Fed. Proc.* **19**, 257.

Peachey, L. D. and Huxley, A. F. (1964). Transverse tubules in crab muscle. *J. Cell Biol.* **23**, 70–71A.

Podolsky, R. J. and Hubert, C. E. (1961). Activation of the contractile mechanism in isolated myofibrils. *Fed. Proc.* **20**, 301.

Porter, K. R. and Palade, G. E. (1957). Studies on the endoplasmic reticulum. III. Its form and distribution in striated muscle cells. *J. Biophys. Biochem. Cytol.* **3**, 269–300.

Pryor, M.G.M. (1950). Mechanical properties of fibres and muscles. *Prog. Biophys. Chem.* **1**, 216–268.

Ringer, S. (1883). A further contribution regarding the influence of the different constituents of the blood on the contraction of the heart. *J. Physiol.* **4**, 29–42.

Robertson, J. D. (1956). Some features of the ultrastructure of reptilian skeletal muscle. *J. Biophys. Biochem. Cytol.* **2**, 369–380.

Strickholm, A. (1962). Excitation currents and impedence of a small electrically isolated area of the muscle cell surface. *J. Cell. Comp. Physiol.* **60**, 149–167.

Uchida, K. and Mommaerts, W.F.H.M. (1963). Modification of the contractile responses of actomyosin by cyclic adenosine 3′, 5′ phosphate. *Biochem. Biophys. Res. Commun.* **10**, 1–3.

Weber, A. (1959). On the role of calcium in the activity of adenosine 5′-triphosphate hydrolysis by actomyosin. *J. Biol. Chem.* **234**, 2764–2769.

Weis-Fogh, T. and Amos, W. B. (1972). Evidence for a new mechanism of cell motility. *Nature* **236**, 301–304.

Muscle Contraction: Its Regulatory Mechanisms (Ebashi, S. *et al.*, eds.), pp. 19–29,
Japan Sci. Soc. Press, Tokyo / Springer-Verlag, Berlin (1980)

2

Skinned Fiber, Past and Present

REIJI NATORI

To begin with, I should like to touch on the motive of this study. About 43 years ago, when I first became interested in muscle physiology, the so-called "Holism" of Smuts or Driesch's conception of "Ganzheitsbezogenheit" was the dominating current in the Physiological Society of Japan. This theory advocated that life was a whole phenomenon, and that the cell was the ultimate unit of life which could be isolated from living organisms from the stand-point of physiology. If the cell is decomposed, the characteristics of life is lost. In studies of muscle physiology, it was thought proper to consider the single muscle fiber as the functional unit of living muscle.

When we pursued the problem along this line, however, we faced a great limitation. For example, when the various numerical values of visco-elastic properties of a single muscle fiber were measured, it could not be determined to what structure of the cell the measured values were to be attributed.

It was then considered necessary to connect experimentally the sub-cellular findings with the molecular ones, in spite of the conceptual restric-

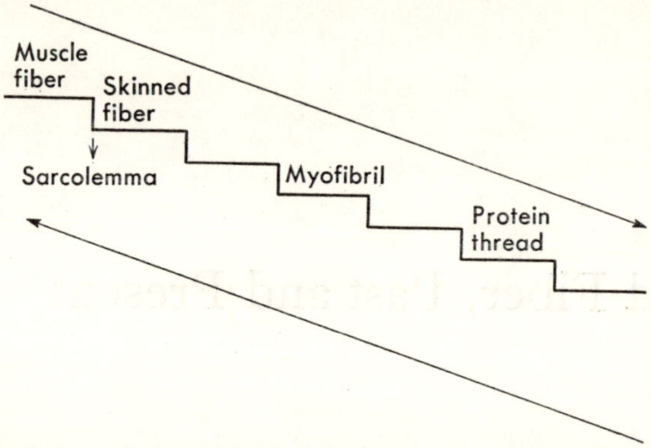

Fig. 1. Schematic illustration of "decomposition and reconstruction" of muscle fiber.

tion due to "Ganzheitsbezogenheit," in order to elucidate analytically the relationship between the structure and function of muscle.

If the muscle fiber could be decomposed with its living state maintained as far as possible and its properties were clearly discriminated at each step of gradual decomposition toward the final step of extracted substances as shown in Fig. 1, one would be able to find a way to reconstruct the physiological properties of a living muscle fiber from those of extracted substances. In other words, although the total could not be regarded simply as a sum of the parts, the accumulated findings with respect to "decomposition and reconstruction" were considered to enable us to understand the nature of a living muscle fiber.

About 30 years ago, with this motive, I tried to isolate bundles of myofibrils still retaining their *in situ* properties (Natori, 1949, 1950, 1951, 1952). In the beginning of this trial, on the basis of the chemical composition of the sarcoplasma, various solutions were applied for removing sarcolemma in them, but the mechanical removal of sarcolemma in the solutions induced irreversible contractures. In order to avoid this contracture, removal of sarcolemma was tried in oil, since oil was considered to inhibit diffusion into or out of fiber and the injury current. The application of oil immersion was found to be successful, and this marked the beginning of a series of experiments on skinned fibers (Natori, 1954, 1955, 1956; Natori and Sakai, 1957; Matsunaga, 1957; Natori and Isojima, 1958, 1959; Wachi, 1958).

The reasons why bundles of myofibrils could not be isolated in the solutions are now clearly explained, partly because of the localization of intracellular Ca^{2+} in sarcoplasmic reticulum (SR) and the role of Ca^{2+} as a trigger of contraction established by Professor Ebashi and others; later, Professor Endo prepared a solution in which sarcolemma could be removed from muscle fiber without causing any irreversible contracture, and his solution is now being widely used.

Returning to my previous experiments, we examined whether or not the application of a small amount of salt solution could induce a transient reversible contraction with skinned fibers isolated from fibers immersed in oil.

When the amount of solution applied was small, contraction was short in duration and was induced repeatedly many times. At that time it was supposed that some physico-chemical changes due to the applied solution triggered contraction.

The response of skinned fiber in oil to electrical stimulation was then examined. Contrary to my expectation, electrical stimulation was found to induce contractions, because, since the sarcolemma was then considered to be exclusively responsive to electrical stimulation, no contraction could be expected with skinned fibers. Actually, the application of square pulse induced contraction in the neighborhood of the anodic-stimulating electrode so that the above view needed to be abandoned. The electrical responsiveness was tentatively explained by the assumption that there would be some physico-chemical phase boundary around the myofibrils. The localization of electrically-induced contraction on the anodic side was later confirmed by experiments in which contractions were observed to be propagated, starting at the anodic portion in response to square pulse with skinned fibers.

About 30 years ago, no detailed information of the structure of internal membrane was available, but at present the morphological reality and physiological role of the internal membrane have been elucidated by electron microscopy and by biophysical and biochemical means. What I had imagined once has now been replaced by what is real. The following is a brief summary of my study concerned with the properties of internal membranes (Natori and Isojima, 1960, 1962; Higa, 1960; Natori, 1965, 1968, 1969, 1972, 1975; Natori and Morishita, 1972; Natori and Ishii, 1972; Morishita, 1972).

The skinned fiber in paraffine oil isolated from toad skeletal muscle responded to a square pulse with a twitch-like contraction. Repetitive stimulation evoked a tetanus-like contraction. The voltage-duration relation of the square pulse stimuli which were able to evoke the twitch-like contraction was similar to that of fiber. The frequency-threshold relation by alternating current was nearly linear in the range of 100–7,000 Hz.

The fact that the electrical excitability of skinned fiber is almost similar to that of muscle fibers suggests that within the skinned fiber there is some mechanism which plays a role similar to that of the plasma membrane in a muscle fiber. On the other hand, when the intensity of a square pulse was increased slightly above the threshold, local contraction was spatially augmented and was abruptly changed into a propagated contraction whose starting point was on the anodic side and was reversed by reversing the direction of the stimulating current.

It seems unlikely that the propagation of contraction in skinned fiber is a result of summation of isolated propagation along each myofibril, because, when a skinned fiber was partially longitudinally split into a shape of a ring of bundle of myofibrils, a contraction starting at the point of one of the half-circles was propagated to the other by transversal propagation through the portion connecting both half-circles.

If polarizable membranes surrounding each myofibril are connected with each other by formation of a network, there is a possibility that a change in the polarizable membrane is propagated transversely along the network of polarizable membranes, although postulation of the network of polarizable membranes is not always needed if the transverse propagation is due to diffusion of some substance; however, if the polarizable membranes are situated at the T tubules and SR, the postulation of the network-like structure is probable.

A skinned fiber isolated from a muscle fiber which had been depolarized by the application of isotonic KCl or K_2SO_4 solution did not respond to a square pulse of short duration, but responded to a direct current with an off contracture which was made to relax by making current flow again just like a depolarized muscle fiber. This result could also be understood by assuming a kind of polarizable membrane around the myofibrils.

On the basis of the findings mentioned above, being interested in the analysis of the physiological role and the mechanism of each morphologically discriminated component belonging to the internal membrane—that is, T tubule, triadic junction and SR—we attempted a series of experiments relevant to this problem. Since 1970 our studies have been focused on aspects of the properties of skinned fibers—for example, the recording of potential change with propagated contractions, the effect of pretreatment of fiber with various solutions on the contractility of skinned fibers, the galvanic surge-like contractions and the caffeine-induced periodic contractions.

However, as far as our physiological experiments are concerned, there remain many problems to be elucidated in order to distinguish the roles of the T tubule, the triadic junction, and the sarcoplasmic reticulum.

In connection with these experiments with skinned fibers, some recent experimental results with skinned fibers will be introduced, relating to the elastic property and structure of muscle fiber (Natori *et al.*, 1974, 1976; Umazume and Fujime, 1975; Isojima and Natori, 1976; Maruyama *et al.*, 1976, 1978, 1979). The properties of thin and thick filaments were examined in overstretched skinned fibers in a physiological state, and one of my interests was in the elastic structure of skinned fibers. This was motivated by Professor Maruyama's extraction of connectin as a new elastic protein (Maruyama *et al.*, 1976, 1978, 1979). This finding was considered to explain the visco-elastic properties of skinned fibers which had been thought to be attributable to some structures of unknown elastic substances. The length tension relation of skinned fibers was similar to that of muscle fibers, and the bundle of myofibrils did not sever even when stretched to nearly triple its resting length *in situ*. Disruption due to overstretching was considered to be avoided by some elastic structure within the skinned fibers.

When the stretching force was removed after skinned fibers were overstretched, a stretch residual was observed, but the resting length was gradually restored after a short time. The application of $CaCl_2$ solution in this state led to the occurrence of contractions nearly the same as those before stretching. Hence, it was inferred that overstretching of skinned fibers did not induce any decisive changes in the properties of thin and thick filaments respectively, but removed only the overlap between them. This was a starting point in examining the properties of thin and thick filaments, respectively, in the overstretched skinned fibers.

Umazume of our laboratory found that, when a voltage of 30–50 V/cm was applied to both ends of an over-stretched skinned fiber whose sarcomere length was more than 8 μm, the application of voltage induced a change in the intensity of light transmitted through the skinned fiber. The birefringence decreased at the half I band of one side of the Z line and increased at the half at the other side. This effect was almost the same in a goat skinned fiber in which myosin was extracted by immersion in KCl solution of a high concentration.

From these results and the analysis of diffraction pattern of overstretched skinned fiber, Umazume and Fujime supposed that the dipole property of an actin filament is similar to that of extracted actin (Umazume and Fujime, 1975). It seems likely that these results with thick and thin filaments of sarcomere give some evidence for the applicability of the properties of extracted myosin and actin to the study of properties of thick and thin filaments in a living state, taking into consideration my idea concerning "decomposition and reconstruction" which I have already mentioned here.

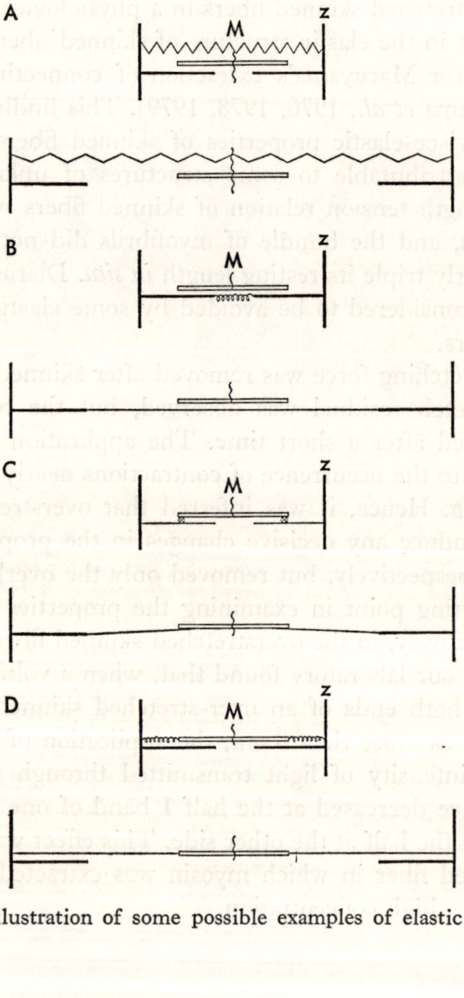

Fig. 2. Schematic illustration of some possible examples of elastic structure by connectin.

In connection with these experiments, I wondered whether the skinned fibers were over-stretched evenly without irregular change in the length of the sarcomeres and whether removal of stretching force leads to sliding of thin filaments between thick filaments. Regarding this matter, one aspect of our scheme proposed in the symposium "Cross-bridge mechanism in muscle contraction" held last year will be introduced here (Natori *et al.*, 1979).

Although the overlap between thin and thick filaments was lost when the sarcomere length was stretched to more than about 3.6 μm, the sarcomere did not sever in spite of stretching of more than 8 μm. There is a possibility that the internal membrane system plays a role of parallel elastic structure as a preventive factor of disruption, as shown in Fig. 2A. If this is the case, one would expect that overstretching of skinned fibers only decreases the overlap between the two kinds of filaments without causing any injuries to either kind of filament and that such overstretched skinned fibers could be used as specimens to examine the properties of thin and thick filaments. However, if scheme B, C or D is adopted, it follows that the internal elastic structure would function as series elastic components. Hence, changes in filament arrangement during overstretch-

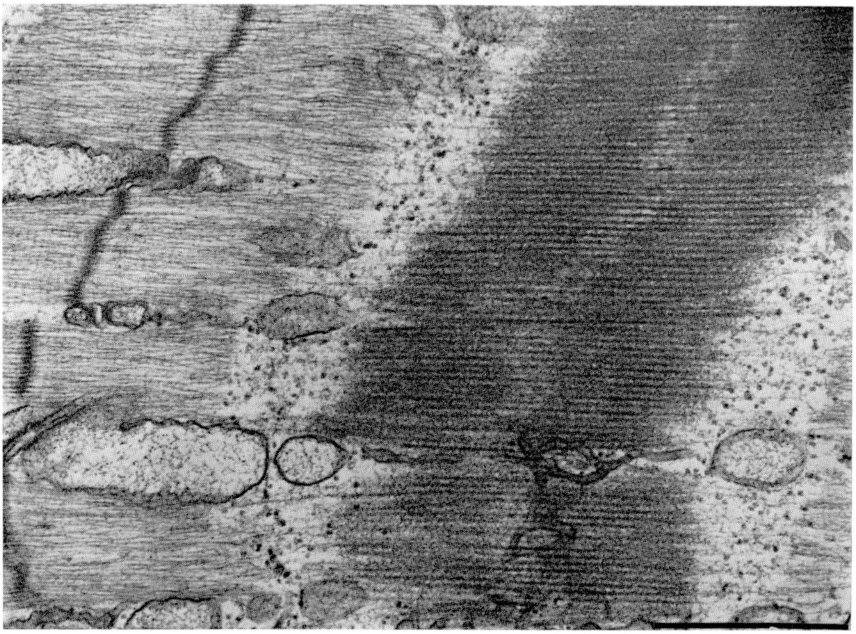

Fig. 3. Electron micrograph of overstretched muscle fiber. M. hyoglossus of the frog.

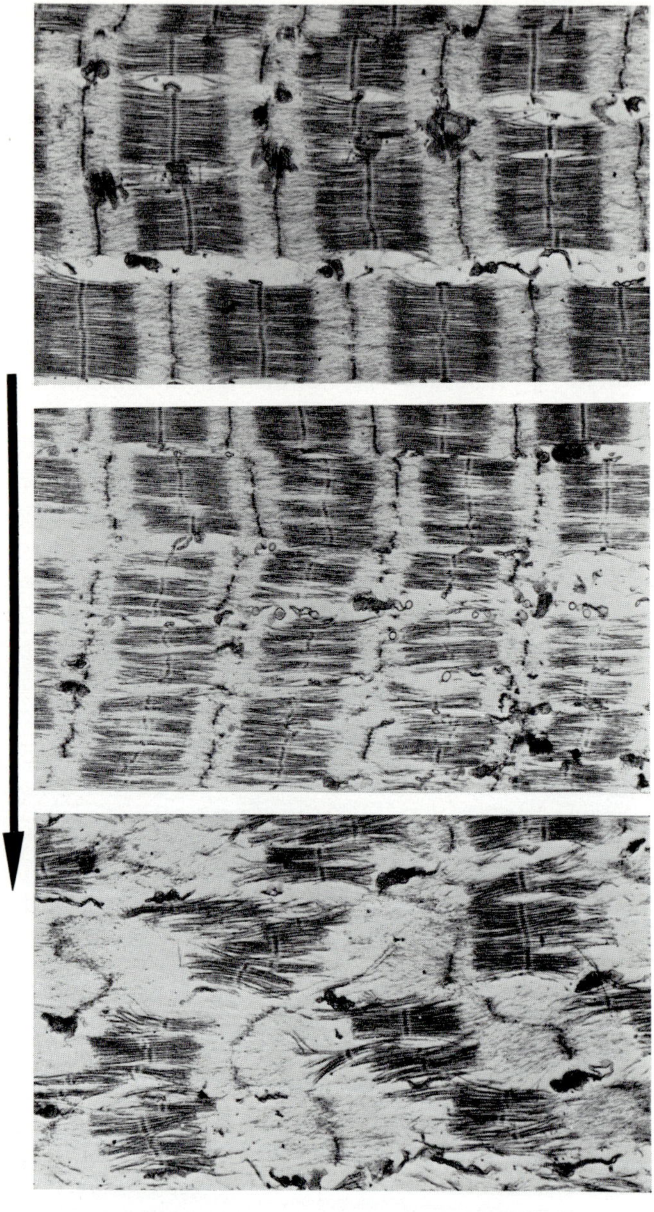

Fig. 4. Electron micrographs of skinned fiber following application of Endo's solution containing 1 μg/ml trypsin. M. sartorius of bullfrog. Thick and thin filaments showing gradually disordered arrangement with lapse of time (black arrow).

ing were examined. In the case of B, thin filaments are expected to be stretched; in the case of C, both kinds of filaments; and in the case of D, only thick filaments.

Electron micrographs were taken to look into this problem, with the cooperation of T. Yoshioka. As shown in Fig. 3, in overstretched frog, hyoglossus muscle thick and thin filaments are separated from each other and filament-like fibers which are thinner than thin filaments are observed in the space between separated thin and thick filaments.

It was considered that thick filaments were stretched in overstretched skinned fiber; in some cases the arrangement of thick filaments tended to demonstrate a zigzag figure, while thin filaments indicated a more or less relaxed state. Although it has not been decided whether these thinner filaments are connection (Maruyama) or not, when 0.75–1 μg/ml trypsin was applied to the skinned fiber, thin filaments relaxed markedly, and a few minutes after application of trypsin some thick filaments leaned to one side of the Z line, and represented random arrangement of thick filaments.

These results may indicated that the thinner filaments were dissolved or were broken by trypsin and may suggest a possibility that the situation of thick filaments located always in the center of sarcomere are sustained by these thinner filaments.

Although it is supposed for the present that the presence of a parallel elastic component may be most probable among the schemes mentioned above, but if scheme D is also the real structure in a living muscle fiber, it must be considered that thick filaments will be affected by a stretch force in stretched skinned fiber.

Therefore, there remain many points to be elucidated concerning our understanding of thc correlation between the function and structure of muscle fibers, although the visco-elastic properties of skinned fiber would have to be tentatively interpreted by introduction of various possible schemes of fine structure of muscle fibers.

REFERENCES

Higa, Y. (1960). Repetitive propagated contractions induced in depolarized muscle fibre and isolated myofibrils (in Japanese). *Jikei Idai Zasshi* **75**, 591–593.

Isojima, C. and Natori, R. (1976). On the break contraction in depolarized smooth muscle and skeletal muscle. *Jikeikai Med. J.* **23**, 215–222.

Maruyama, K., Matsubara, S., Natori, R., Nonomura, Y., Kimura, S., Ohashi, K., Murakami, F., Handa, S., and Eguchi, G. (1979). Connectin, an elastic protein of muscle, characterization and function. *J. Biochem.* **82**, 317–337.

Maruyama, K. and Natori, R. (1978). Connectin, an elastic protein from myofibrils and sarcolemma of frog skeletal muscle. *Zool. Mag.* **87**, 162–164.

Maruyama, K., Natori, R., and Nonomura, Y. (1976). New elastic protein from muscle. *Nature* **262**, 58–60.

Matsunaga, R. (1957). Structure of isolated myofibrils and sarcoplasm (in Japanese). *Jikei Idai Zasshi* **72**, 866–870.

Morishita, Y. (1972). A role of the internal membrane in the fatigue of muscle fibre (in Japanese). *Jikei Idai Zasshi* **87**, 671–677.

Natori, R. (1949). Studies on mechanism of muscle contraction (in Japanese). *Nippon Seirigaku Zasshi* **11**, 14–15.

Natori, R. (1949). Contraction mechanism of skeletal muscle (in Japanese). *Seitai no Kagaku* **1**, 187–190.

Natori R. (1950). Contraction mechanism of skeletal muscle (in Japanese). *Nippon Seirigaku Zasshi* **12**, 11.

Natori, R. (1951). The properties of myofibrils sarcoplasm and sarcolemma (in Japanese). *Nippon Seirigaku Zasshi* **13**, 6.

Natori, R. (1951). *Kin Seirigaku* (*Muscle Physiology*) (in Japanese). Maruzen, Tokyo.

Natori, R. (1952). The property and contraction process of myofibrils (in Japanese). *Seitai no Kagaku* **3**, 209–212.

Natori, R. (1954). The role of myofibrils, sarcoplasm and sarcolemma—(Reported at the 2nd Conference of the Japanese Union of Physiological Sciences, November 10, 1952). *Jikeikai Med. J.* **1**, 18–28.

Natori, R. (1954). Contraction process of myofibrils (in Japanese). *Denshi Kenbikyo* **3**, 90–94.

Natori, R. (1954). The property and contraction process of isolated myofibrils. *Jikeikai Med. J.* **1**, 119–126.

Natori, R. (1955). Repeated contraction and conductive contraction observed in isolated myofibrils. *Jikeikai Med. J.* **2**, 1–5.

Natori, R. (1956). Differences in physiological properties of myofibrils of small and large muscle fibres. *Jikeikai Med. J.* **3**, 36–42.

Natori, R. and Sakai, T. (1957). The properties of myofibrils. Conference on the chemistry of muscular contraction. In *The Committee of Muscle Chemistry of Japan*, pp. 125–130, Igakushoin, Tokyo.

Natori, R. (1965). Propagated contractions in isolated sarcolemma-free bundle of myofibrils. *Jikeikai Med. J.* **12**, 214–221.

Natori, R. (1968). On the propagated and periodic contractions in the sarcolemma-free bundle of myofibrils (in Japanese). *Symp. Cell. Chem.* **19**, 211–216.

Natori, R. (1969). On the excitation-contraction coupling of skeletal muscle fibres (in Japanese). *Japan. J. Smooth Muscle Res.* **5**, 127–134.

Natori, R. (1972). Some physiological aspects of internal membrane of skeletal muscle fibres. *Jikeikai Med. J.* **19**, 51–64.

Natori, R. (1975). The electric potential change of internal membrane during propagation of contraction in skinned fibre of toad skeletal muscle. *Japan. J. Physiol.* **25**, 51–63.

Natori, R. and Isojima, C. (1958). Restoration of denatured myofibrils (in Japanese). *Nippon Seirigaku Zasshi* **20**, 641.

Natori, R. and Isojima, C. (1959). Rhythmical contractions of isolated myofibrils (in Japanese). *Nippon Seirigaku Zasshi* **21**, 936–937.

Natori, R. and Isojima, C. (1960). Excitability of isolated myofibrils (in Japanese). *Nippon Seirigaku Zasshi* **22**, 731.

Natori, R. and Isojima, C. (1962). Excitability of isolated myofibrils. *Jikeikai Med. J.* **9**, 1–8.

Natori, R., Isojima, C., and Natori, Rb. (1976). On various types of periodic contractions in skinned fibers of skeletal muscle. *Jikeikai Med. J.* **23**, 255–270.

Natori, R. and Morishita, Y. (1972). On the break-contracture of sarcolemma free bundle of myofibrils immersed in hypertonic solution (in Japanese). *Jikei Idai Zasshi* **87**, 700–706.

Natori, R. and Ishii, R. (1972). On the direct current-induced surge-like contractions of sarcolemma-free bundle of myofibrils (in Japanese). *Jikei Idai Zasshi* **87**, 842–849.

Natori, R. and Natori, Rb. (1979). Contraction in skinned fibers. In *Cell Motility: Molecules and Organization* (Hatano, S., Ishikawa, H., and Sato, H., eds.), pp. 273–279, Univ. Tokyo Press, Tokyo.

Natori, R., Natori, Rb., and Umazume, Y. (1979). Contraction in overstretched skinned muscle fibers. In *Cross-bridge Mechanism in Muscle Contraction* (Sugi, H. and Pollack G. H., eds.), pp. 41–50, Univ. Tokyo Press, Tokyo.

Natori, R., Umazume, Y., and Yoshioka, T. (1974). Viscous elastic properties of internal membrane of skeletal muscle fibers. *Jikeikai Med. J.* **21**, 135–150.

Natori, Rb. and Natori, R. (1979). The contraction process of the skinned fiber of skeletal muscle following application of 1 M KCl solution. *Jikeikai Med. J.* **26**, 209–220.

Umazume, Y. and Fujime, S. (1975). Electro-optical property of extremely stretched skinned muscle fibers. *Biophys. J.* **15**, 163–180.

Wachi, K. (1958). Structure and contractility of isolated myofibrils (in Japanese). *Jikei Idai Zasshi* **73**, 2043–2047.

Yazaki, Y. and Natori, R. (1955). Studies on the skeletal muscle by fluoromicroscopy (1). The relation between contractility of muscle fibre and its fluorescence by acridine orange and alminium morine. *Jikeikai Med. J.* **3**, 60–68.

CROSS-BRIDGE MOVEMENT

Muscle Contraction: Its Regulatory Mechanisms (*Ebashi, S. et al., eds.*), *pp. 33–43,*
Japan Sci. Soc. Press, Tokyo / Springer-Verlag, Berlin (*1980*)

3

The Movement of Myosin Cross-bridges during Contraction

H. E. HUXLEY

Summary: It is now generally agreed that the sliding force between myosin and actin filament in contracting muscle is produced in some way by myosin cross-bridges which interact with specific sites on the actin filaments. However, the precise mechanism by which force is developed cannot be established without further experimental evidence. Time resolved low-angle X-ray diffraction studies on contracting muscles offer the possibility of monitoring both lateral and longitudinal cross-bridge movement during different phases of the contractile response, both under steady state and transient conditions. Such measurements have required considerable technological development to attain the necessary speed of recording, but this has now been achieved. The present status of this work will be reviewed.

In this paper, I will attempt to review very briefly what we know about the force producing mechanism in muscle, at what points our evidence is ambiguous or non-existent, and how some of the deficiencies in the evidence might be remedied.

I will take it for granted that the basic mechanism involves an active sliding process developed between filaments of actin and of myosin. I will also take it for granted that the responsible structural elements involved in tension development are the cross-bridges, the projecting ends of the myosin molecules containing the globular subfragment 1 (S-1) moieties which in turn contain the ATPase sites of the molecule. Furthermore, I will assume that all the cross-bridges in the region of overlap between actin and myosin filaments are potentially able to interact with actin and contribute to tension and shortening. I do not consider that any current controversies about the details of cross-bridge action provide in any way sufficient reasons to call into question these three basic tenets, which are founded on a very large body of well-established experimental evidence and which also seem to make a good deal of commonsense.

It seems to me therefore that the question which is really worth concerning ourselves with is the precise mechanism by which we suppose a myosin cross-bridge exerts a force on the actin filament with which it interacts. I do not believe it is possible to arrive at a solution to this problem simply by applying reason and logic to the existing experimental evidence, but such an analysis may at least show what are the crucial issues about which additional evidence is now required.

Probably the best starting point is the Tonomura-Taylor-Trentham scheme for the enzymatic splitting of ATP by actomyosin in solution, in which it has been very clearly demonstrated that the splitting takes place by a repetitive cyclical process in which the actin and myosin are physically associated together for one part of the cycle, and dissociated again during another part of the cycle. Since the maximum rate at which this cycle can take place is comparable with the rate at which ATP is split, on average, at each cross-bridge in a contracting muscle, it seems entirely feasible that the same cycle is employed in the two situations.

In any cross-bridge mechanism in which attachment to actin is involved in force generation, a cyclical process of attachment and detachment must necessarily take place in order to allow the actin and myosin filaments to slide past each other by the several thousands of angstrom units which can occur during a normal contraction. Whilst the biochemical results do not necessarily prove that this mechanism actually operates, only that it seems possible, it would be a rather remarkable coincidence if the similarity between the two cycles—the one proposed largely on structural grounds, the other discovered by studying reaction kinetics—turned out to be accidental.

Nevertheless, it is important to settle this question as decisively as possible, and we have tried to do this by X-ray diffraction studies on surviving muscles which enable us to get structural information about cross-bridges during normal contractions *in vivo*. These results have been described previously (H. E. Huxley and Brown, 1967; H. E. Huxley, 1968; Haselgrove and H. E. Huxley, 1973) so I will summarize them only very briefly here. We can recognise two basic types of equatorial X-ray diagram given by vertebrate striated muscles. Each one corresponds to the pattern given by a double hexagonal lattice of filaments, with essentially the same lattice spacing. The first type is given by resting muscle and has a strong (10) reflection (the inner reflection) and a weaker (11) reflection. The second type is given by muscle in rigor and has a weak (10) reflection and a strong (11) reflection. Fourier reconstructions show that this implies a shift of density away from the myosin filaments and towards the actin filaments. Electron-micrographs of muscles in the two states show that this is indeed what has taken place and that in the rigor muscles the cross-bridges can be seen attached to the actin filaments, (H. E. Huxley, 1957, 1968), thereby increasing the mass near the trigonal positions in the lattice.

Equatorial X-ray diagrams from isometrically contracting muscles near rest length are quite similar to those given by muscles in rigor at the same lengths, with a weak (10) and a strong (11), though the (11)/(10) ratio is slightly lower than it is in rigor. Whether this reflects a slightly smaller number of bridges attached, or a slightly different average angle of attachment is not known. However, the most straightforward interpretation of the data would be that the pattern arose in essentially the same way as it did in rigor, namely from cross-bridges attaching to actin, as indeed can be seen to be the case in rigor in the electron microscope. Nevertheless, one could wonder whether the change in pattern in contracting muscle could be due in whole or in part to some myosin-based control mechanism which allowed the cross-bridges to move outwards upon activation, so that the change did not necessarily reflect attachment to actin as such. We have investigated this possibility by recording the equatorial diagram from muscles stretched to sarcomere lengths at which thick and thin filaments no longer overlap. A strong (10) reflection is still produced by such muscles, but it does not change in intensity when the muscle is stimulated in a series of 1/2 sec or 1 sec tetani (H. E. Huxley, 1978). This is true even in muscles swollen in two-thirds strength Ringer solution. Thus the presence of actin alongside myosin is necessary for outwards movement of the average centre of gravity of the cross-bridges to occur. Since all the available evidence on the behaviour of the control proteins on the actin filaments indicate that they operate by producing changes in the myosin binding site which are equally effective either in a highly structured muscle or in a random suspension of thin filaments and S1 molecules, it would seem most reasonable to deduce that the average outward movement of the cross-bridges towards actin in an activated muscle really does represent physical attachment as such, mediated by Brownian movement. No doubt random fluctuations of bridge position also occur in resting muscles, but in that case the bridges which approach actin closely will not stick to it.

During the last few years we have made time-resolved measurements on these very characteristic changes in the equatorial X-ray diagram which show that they take place at least as fast as tension development, (apparently even faster) and that they revert to the resting state with approximately the same time course as the relaxation of tension. These results have been reported on a number of occasions previously (H. E. Huxley, 1975; H. E. Huxley and Haselgrove, 1976; H. E. Huxley, 1978) and I will not repeat them here. We have also shown that the changes occur in absolutely fresh and unfatigued muscles which have only undergone a second or two of contraction after dissection. The same changes have also been seen by a number of other groups (Podolsky *et al.*, 1976; Matsubara and Yagi, 1977; Sugi *et al.*, 1977).

I therefore feel a rather high degree of confidence that we are on firm ground in our supposition that the cross-bridges do move during contraction and that one component of that movement is a lateral one associated with their binding to actin. Since it has also been shown that the characteristics of the contracting pattern persist when the muscle is allowed to shorten, it follows that the attachment can be broken and reformed. Thus the evidence that the biochemical cycle seen in solution actually takes place in a muscle seems rather strong, and in addition there is good evidence that a substantial amount of cross-bridge movement is able to take place.

However, the movements with which we have been concerned so far clearly cannot themselves produce muscle shortening since they are lateral rather than longitudinal. Thus the crucial question we have to answer is whether there is evidence for longitudinal movement of the cross-bridges in a contracting muscle while they are attached to actin, and if so whether we can elucidate the detailed nature and the course of that movement. The kind of longitudinal movement we are concerned with is one that will, if it occurs in an attached bridge, bring about a translation of the actin filament towards the centre of the A band. Thus in principle it could be either a tilting movement of the myosin head (S-1) about its attachment site to actin, with a component of the motion which tended to move the region of the myosin head to which the S_2 is attached (*i.e.*, presumably the opposite end of the elongated head) away from the A-band centre: or it could be a rotation about the S_1S_2 hinge: or it could be a change in shape of the S-1 having essentially the same effect: or it could be a change in length of the S-2 region: or it could be some change within the backbone of the thick filament which produced a longitudinal cross-bridge movement. There may be good reasons for thinking that one or more of these types of movement is more likely than the others, but the first step must be to examine the quality of reasons and the evidence for supposing that any significant movement at all takes place.

The starting point for such consideration must be the fact that the axial periodicities of the subunits in the thick and thin filaments (143 Å and 27.3 Å) are incommensurate, so that irrespective of helical mismatching, the longitudinal positions of a cross-bridge and an actin monomer only repeat approximately the same relative spatial relationships once in eight cross-bridge repeats (H. E. Huxley and Brown, 1967). It is clear from micrographs of muscle in rigor that attached cross-bridges occur at least about every 400 Å between a given thick and thin filament (H. E. Huxley, 1957) and so some of these must have moved longitudinally from their positions in the resting muscle, assuming that they have to attach to a specific site on actin.

The next very important evidence derives from X-ray diffraction and electron microscope studies on insect flight muscle (Reedy *et al.*, 1965; Reedy, 1968). These showed very clearly that the tilt of the cross-bridges was different between the relaxed and rigor states. Not only was this shown by the micrographs, but there was decisive evidence from the X-ray diagrams which showed a large intensity decrease in the 143 Å meridional reflection. Since this meridional reflection arises from the repeating periodicity of the density associated with the myosin filaments and projected along a line parallel to the filament axis, large changes in its intensity, unaccompanied by any evidence of change in filament register, must necessarily reflect a longitudinal redistribution of density along the myosin filaments. And since it is clear that the major part of the 143 Å reflection arises from the cross-bridges, because of their contrast against the lower electron density of the sarcoplasm around them (rather than variation in density along the myosin filament backbone which are clearly much smaller), then the large decrease in the 143 Å intensity shows that longitudinal cross-bridge movement does take place, as indeed is apparent from the micrographs. Since the latter show a tilted configuration of the cross-bridges in rigor, which would spread out their axially projected density and lead to a weaker 143 Å reflection than that given by the more perpendicularly oriented bridges (visible in micrographs of the relaxed muscle), there are very strong reasons for taking the evidence of the micrographs at its face value. Thus we can conclude that a change in cross-bridge tilt does take place in the intact muscle before processing for electron microscopy.

There are a number of additional reasons for supposing that changes in cross-bridge orientation relative to the axis of the myosin filaments can indeed take place. Since muscles can operate over a range of side spacings (at different sarcomere lengths) it has been supposed that the S_2 portion of the myosin rod hinges out sideways by a variable amount to allow the myosin heads always to attach in an identical manner to actin. Such a mechanism necessitates a flexible link at either end of the S_2 moiety. The susceptibility of myosin to cleavage at these regions by a wide variety of proteolytic enzymes would accord very well with the concept of looser, more open regions of structure here. There is also good fluorescence and EPR evidence (Mendelson *et al.*, 1973; Thomas *et al.*, 1975) for flexibility in the attachment of the S_1 head units to the rest of the myosin molecule. And a similar impression of flexibility is given by very irregular arrangement of the projections on myosin filaments (when not attached to actin) seen in electron micrographs.

None of these are new arguments, but I think it is important to be clear about them and to appreciate that they do provide very strong reasons for believing that the myosin heads *can* undergo longitudinal tilting movements, relative to the backbone of the thick filaments. That being the case, it is of course very plausible to suggest such a tilting movement of attached bridges as the basis for force production. However, we still need to look for direct structural evidence concerning such movement, and it is that particular issue that I want to consider here. The question is then "What evidence is there that the myosin heads can be attached to actin in more than one configuration, with substantial structural differences between them, and that transitions between the states develop the force for contraction."

The best structural evidence relevant to this question comes from a comparison of the axial parts of the low angle X-ray diagram given by vertebrate striated muscle *in rigor* and that given by an isometrically contracting muscle. The former gives, in place of the characteristic 430 Å myosin layer line pattern of relaxed muscle, a pattern of layer line reflections which can be indexed on the actin helix, with a prominent 360–380 Å repeat. It is apparent that a substantial proportion of the bridges have attached to actin more or less similar orientations and are producing the 'labelled actin' pattern. The earlier pictures that we obtained (H. E. Huxley and Brown, 1967) of contracting muscles showed a great weakening of the myosin layer line pattern, but no sign of the appearance of any other layer line pattern to replace it. This provided *prima facie* evidence that the attached bridges (clearly present in substantial numbers because of the similarity of the contracting and the rigor equatorial X-ray diagrams) were attached in a much less regular way than in rigor, *i.e.*, that they were attached over a range of angles.

This earlier evidence, however, suffered from a number of weaknesses. The pattern was built up over a very long series of 1 or 2 sec tetani, and there was no evidence 1) how the change in pattern during a contraction corresponded temporally with the development of tension and its subsequent relaxation, 2) whether the change in pattern was present throughout the contraction series or whether it only developed as the muscle fatigued, 3) whether the progressive fading of the relaxed pattern which occurs during the long contraction series reflected changes in the muscle which were also making it more difficult to see evidence of a rigor pattern during contraction.

In view of the importance of the question, therefore, we have been attempting to obtain much better evidence about the behaviour of the layer line pattern during contraction, both as regards its behaviour during very short contraction series in unfatigued muscles, and also as regards its detailed time course relative to tension. Since the layer line patterns are weak ones, they cannot be recorded sufficiently quickly, using laboratory X-ray sources, even if electronic X-ray detectors are used, and for this reason we have carried out our experiments (H. E. Huxley *et al.*, 1980) with synchrotron radiation from the electron storage ring DORIS, using the facilities provided by the European Molecular Biology Outstation at DESY, Hamburg.

In our initial experiments we were concerned to establish whether or not the loss of layer line intensity was a genuine effect occuring during an unfatigued contraction in a freshly dissected muscle. We used a frog sartorius muscle, maintained in an oxygenated Ringer solution and stimulated directly by platinum electrodes lying along its length on either side. Tension was measured and recorded electronically. A focussed and monochromatized X-ray beam was produced from the incident radiation from the storage ring by a camera designed by G. Rosenbaum and A. Harmsen, comprising an eight element totally reflecting glass mirror of total length 160 cm and a germanium curved crystal monochromator. A good quality layer line pattern from muscle could be recorded on film in a few minutes with this apparatus, as compared to about 24 hr for a similar quality pattern using a high power rotating anode X-ray tube. In the actual experiments themselves, however, the pattern was recorded on an X-ray image intensifier-TV detector (Reynolds *et al.*, 1978), since this enabled good layer line data to be collected with a total exposure time of only 1 sec. When collecting time-resolved data, only the pattern corresponding to one particular 'time-slot' could be recorded in a given series of experiments because of read-out time limitations with this type of detector. To investigate the behaviour of the pattern during single twitches we recorded the pattern during a time interval of 150 msec and in each run of 6–10 twitches this 'slot' was set at the same time during contraction. A total exposure time of 1–1 1/2 sec was sufficient to record a good quality pattern. The run was then repeated for a different time slice. Within this limited time resolution it could be seen that the disappearance of the layer line pattern at the beginning of the twitch and its return during relaxation followed a time course very similar to tension development and its decay. This could be observed in quite short contraction series in which there was no doubt that the muscle was in good condition throughout and where there was none of the weakening of the resting layer line reflections that we have observed in longer series of tetanic contractions, especially at lower temperatures; nor did we see the delayed return of off-meridional layer line pattern after contraction apparent in very long contraction series (H. E. Huxley, 1976). Thus the experiments showed unambiguously that the loss of the layer line pattern is a specific and reversible accompaniment of normal contraction.

In order to compare the time course of the layer line changes more accurately with that of tension, a different system was used. This employed a position-sensitive one-dimensional proportional counter of the delay line type, designed and built by Dr. A. Gabriel of EMBL Heidelberg. This was positioned over the muscle pattern with its axis parallel to the meridian but displaced sideways from it so as to record the strongest parts of the set of myosin layer lines. The one-dimensional patterns were stored in 10 msec time bins using a bin switching device and a PDP 11/45 computor, the time slots being synchronized electronically with the stimulation of the muscle. With the storage ring operating at 3.185 GEV and 70–90 m/a (positrons only), total counting rates over one quadrant of the layer line pattern were in excess of 10^5 cps and adequate data could be collected in 10 msec time bins from a series of a few hundred single twitches spaced at 30 sec intervals. Thus a typical experimental run would last 2–3 hr.

The most striking change in the pattern was the large decrease in all the layer-line intensities during the time that the muscle was developing tension. The time course of the change in all the layer lines appeared to be identical and for simplicity we will consider only the first and strongest of these, that at 429 Å. In a typical run at 10°C (300 twitches) it could be seen that the reflection begins to decrease in intensity within 20 msec of stimulation, falls almost to zero intensity (in some cases) by about 70 msec and then returns to its resting value (or even slightly more) after about 200–250 msec.

The relationship between change in intensity and tension could be seen much more clearly when the change in each of them was plotted as a percentage of the maximum change occurring. It could be seen that both during the onset of activity and during relaxation, isometric tension and change in X-ray intensity follow a very similar time course indeed; this was the case in all the records we obtained (six good runs so far) though in one or two a very slight lag in the X-ray pattern (by about 20 msec or less) was noticed during relaxation. We do not have sufficient data yet, however, to establish whether or not the intensity change leads the tension change during the rising phase of tension by a few milliseconds at this temperature, as it does in the case of equatorial reflections (H. E. Huxley, 1976).

We have also carried out some preliminary experiments on the behaviour of the layer line pattern in frog semitendinosus muscles stimulated at very stretched lengths where overlap of thin and thick filaments is either completely abolished or very greatly reduced. In all our experiments so far we have found a very large reduction in the extent of decrease of the layer line intensities from very stretched muscles during stimulation as compared to that seen in similar muscles at normal sarcomere lengths, and in most of the experiments no decrease at all in intensity was observed. Thus we conclude provisionally that interaction with actin is necessary for helical disordering of the myosin cross-bridges to occur following stimulation. This accords with recent observations showing an absence of radial cross-bridge movement at no-overlap sarcomere lengths, as judged by the absence of change in the equatorial X-ray diagram (H. E. Huxley, 1978).

However, the principle finding whose significance we want to consider here is the large decrease in intensity of the layer lines at resting sarcomere lengths unaccompanied by any sign of a new layer line pattern. It may be that a weak pattern will show up with further improvements in technique, but it is significant that the substantial number of attached bridges, indicated by the substantially changed equatorial X-ray diagram, gives so little of a diffraction pattern in an axial direction. This indicates that the bridges are not all attached to actin in the same orientation as they are in rigor. Thus the evidence does appear to show that bridges can be attached to actin in more than one orientation, and that they move from their regular helical arrangement around the myosin filaments and occupy these different orientations attached to actin during contraction.

The experiments therefore provide very strong evidence that some type of longitudinal cross-bridge movement occurs during contraction. However, they do not in themselves prove that the bridges are already attached to actin when they undergo that movement, and that is the issue which still needs more direct experimental evidence. There is some possibility that evidence on that crucial remaining question can be obtained from time resolved X-ray studies on muscles during rapid length transients, and we are now pursuing such experiments on the storage ring at Hamburg.

REFERENCES

Haselgrove, J. C. and Huxley, H. E. (1973). *J. Mol. Biol.* **77**, 549–568.

Huxley, H. E. (1957). *J. Biophys. Biochem. Cytol.* **3**, 631–648.

Huxley, H. E. (1968). *J. Mol. Biol.* **37**, 507–520.

Huxley, H. E. (1975). *Acta Anat. Nippon* **50**, 310–325.

Huxley, H. E. (1976). In *Cell Motility* (Goldman, R. *et al.* eds.), pp. 115–126, Cold Spring Harbor Laboratory.

Huxley, H. E. (1978). In *Cross-bridge Mechanism in Muscle Contraction* (Sugi, H. and Pollack, G. H., eds.), pp. 391–401, Univ. Tokyo Press, Tokyo.

Huxley, H. E. and Brown, W. (1967). *J. Mol. Biol.* **30**, 383–434.

Huxley, H. E. and Haselgrove, J. C. (1976). In *Myocardial Failure* (Riecker, G., Weber, A., and Goodwin, J., eds.), pp. 4–15, Springer-Verlag, Berlin.

Huxley, H. E., Faruqi, A. R., Bordas, J., Koch, M., and Milch, J. (1980). *Nature* **284**, 140–143.

Matsubara, I. and Yagi, N. (1978). *J. Physiol.* **287**, 297–307.

Mendelson, R. A., Morales, M. F., and Botts, J. (1973). *Biochemistry* **12**, 2250–2255.

Podolsky, R. J., St. Onge, R., Yu, L., and Lymn, R. W. (1976). *Proc. Natl. Acad. Sci. U.S.* **73**, 813–817.

Reedy, M. K. (1968). *J. Mol. Biol.* **31**, 155.

Reedy, M. K., Holmes, K. C., and Tregear, R. (1965). *Nature* **207**, 1276.
Reynolds, G. T., Milch, J. R., and Gruner, S. M. (1978). *Rev. Sci. Instr.* **49**, 1241–1249.
Sugi, H., Amemiya, Y., and Hashizume, H. (1977). *Proc. Japan Acad.* **53**, 178–182.
Thomas, D. D., Seidel, J. C., Gergely, J., and Hyde, J. S. (1975). *Proc. Natl. Acad. Sci. U.S.* **72**, 1729–1733.

Muscle Contraction: Its Regulatory Mechanisms (Ebashi, S. et al., eds.), pp. 45–52,
Japan Sci. Soc. Press, Tokyo / Springer-Verlag, Berlin (1980)

4

The States of Myosin Heads in Heart Muscle during Systolic and Diastolic Phases*

ICHIRO MATSUBARA, NAOTO YAGI, AND
MASAO ENDOH

Summary: The molecular events underlying the staircase phenomenon in mammalian heart muscle were studied by the X-ray diffraction method. When peak systolic tension was increased by increasing the frequency of contraction, the number of myosin heads transferred to the vicinity of the thin filament during the systolic phase increased. This suggests that the inotropic effect is caused by an increase in the number of cross-bridges at the peak of contraction.

In cyclically contracting heart muscle, the array of myosin heads did not go into the fully rested state during the diastolic phase. A considerable number of myosin heads remained near the thin filament during the diastolic phase, but they did not seem to produce significant force since a change in their number did not affect the diastolic tension. However, an increase in their number appeared to be associated with an increase in the rate of tension development and in the peak tension of the subsequent

* The work was presented also at 5th Workshop of Contractile Behaviour of the Heart, Antwerp.

contraction. This suggests that the myosin heads remaining near the thin filament during the diastolic phase may produce tension in the subsequent contraction more readily than those which have returned to the vicinity of the thick filament.

The contractile apparatus of vertebrate heart muscle is similar to that of vertebrate skeletal muscle when viewed under electron microscope. The sarcomere contains regular overlapping arrays of thick and thin filaments arranged in a hexagonal lattice, and the filament lengths and the sarcomere lengths are about the same as those of skeletal muscle. X-ray diffraction studies on mammalian heart muscle have shown that the molecular packing of myosin and actin in the myofilaments is the same as in skeletal muscle (Matsubara and Millman, 1974). The changes in the diffraction pattern occurring when heart muscle goes into rigor are almost the same as those in skeletal muscle. These findings suggest that a cross-bridge mechanism, similar to that in skeletal muscle, operates in heart muscle. Namely, the relative sliding force between the thick and thin filaments appears to be produced by myosin heads forming cross-bridges between the two types of filaments (A. F. Huxley, 1957; H. E. Huxley, 1969; A. F. Huxley and Simmons, 1971). Recent advances in the X-ray diffraction technique have made it possible to study certain movements of myosin heads in cyclically contracting heart muscle (Matsubara *et al.*, 1977, 1979). In this paper, we present an analysis of the movements of myosin heads in relation to the systolic and diastolic tensions. Some of the experimental results have already been published (Matsubara *et al.*, 1978).

MATERIAL AND METHODS

1. *Specimen Preparation*

A papillary muscle, together with the ventricular septum, was isolated from the right ventricle of a canine heart and was cross-perfused with the arterial blood of a donor dog according to the method described by Endoh and Hashimoto (1970); the anterior septal artery was cannulated and connected to the carotid artery of the donor dog. A part of the arterial tubing was immersed in a temperature-controlled bath to keep the perfusing blood at 38°C. The venous outflow from the preparation was returned to the jugular vein of the donor dog.

The ventricular septum was set in a chamber surrounded by a warm bath (38°C), and the papillary muscle was pulled out of the chamber through a close fitting hole. The base of the muscle was stitched to the edge of the hole, and the tendinous end was connected to an isometric tension transducer. The muscle was stimulated with square pulses of 5 msec duration through a pair of platinum electrodes inserted into the ventricular septum. The frequency of the stimulation was varied to obtain diffraction patterns at different inotropic states. At the beginning of each experiment, the muscle length was adjusted to L_{max}, where the developed tension was greatest.

2. *Recording of Equatorial Diffraction Patterns*

A low-angle X-ray camera of the mirror-monochromator type (H. E. Huxley and Brown, 1967) was used. The X-ray source was a high-power rotating anode generator (Rigaku type FR) operated at 50 kV with a tube current of 80 mA (nominal focal size 1 mm$\times$0.1 mm). The muscle was mounted vertically in the X-ray camera and the equatorial diffraction pattern was recorded on X-ray film with a specimen-to-film distance of 45 cm. An electromagnetic shutter was coupled with the tension transducer and passed the X-ray beam only in a specified phase of the cardiac cycle. To obtain the systolic pattern, the shutter was opened as the tension exceeded two thirds of the peak systolic tension and closed as the tension fell below half of the peak tension (Matsubara *et al.*, 1977). To obtain the diastolic pattern, the shutter was opened and closed while the tension was at the diatolic level. The cardiac cycle was repeated approximately 10,000 times to record one diffraction pattern.

3. *Analysis of Diffraction Data*

Each diffraction pattern showed the 1,0 and 1,1 equatorial reflections from the hexagonal array of the myofilaments. The integrated intensities of the two reflections were measured and the ratio ($I_{1,0}/I_{1,1}$) was obtained. Using the ratio, the electron density distribution in the myofilament lattice projected onto the plane perpendicular to the muscle axis was calculated (H. E. Huxley, 1968). The cross-sections of the thick and thin filaments were clearly visible in each density-distribution diagram; from these, the apparent mass of the thick and thin filaments were calculated following the method described by Haselgrove and H. E. Huxley (1973).

An increase in the apparent mass of the thin filament, accompanied by a decrease in that of the thick filament, was interpreted as a result of radial transfer of myosin heads from the vicinity of the thick filament to that of the thin filament (H. E. Huxley, 1968; Haselgrove and Huxley, 1973;

Haselgrove *et al.*, 1976). Based on this interpretation, the number of myosin heads in the vicinity of the thin filament was calculated using the following assumptions: (i) all the heads are in the vicinity of the thick filament in quiescent heart muscle which gives an intensity ratio of 3.07; (ii) all the heads are in the vicinity of the thin filament in muscle in rigor which gives an intensity ratio of 0.37 (Matsubara *et al.*, 1977).

RESULTS

1. *Systolic Phase*

The diffraction pattern during the systolic phase was recorded at contraction frequencies of 60, 120, and 180 min^{-1}. At 120 min^{-1}, which was chosen as the "control" frequency, the peak systolic tension was 287 ± 26 g/cm^2 (mean$\pm$SEM, $n=10$) and the equatorial intensity ratio ($I_{1,0}/I_{1,1}$) was 1.34 ± 0.07 ($n=10$). The tensions and the intensity ratios at other frequencies are shown in Table I as percentages of the values at the control frequency. The peak tension was larger, and the intensity ratio was smaller, at higher frequencies of contraction.

The number of myosin heads in the vicinity of the thin filament was calculated from the average intensity ratio at each contraction frequency, and was plotted against the peak systolic tension (Fig. 1). The open circle in the middle indicates the value obtained at the control frequency of 120 min^{-1}, and the open circles on its left and right indicate the values at 60 and 180 min^{-1} respectively. These results indicate that a rise in peak systolic tension caused by increasing the frequency of contraction is accompanied by an increase in the number of myosin heads in the vicinity of the thin filament during the systolic phase.

The two filled circles in Fig. 1 indicate the results obtained from different types of preparation producing different systolic tensions: the one

TABLE I
Changes in the Equatorial Intensity Ratio during the Staircase Phenomenon
A decrease in the intensity ratio indicates an increase in the number of myosin heads transferred to the vicinity of the thin filament during the systolic phase (see text).

Contractions per min	Peak systolic tension (%)	Intensity ratio $I_{1,0}/I_{1,1}$ (%)
60	74 ± 13	130 ± 18
120	100	100
180	126 ± 6	81 ± 10

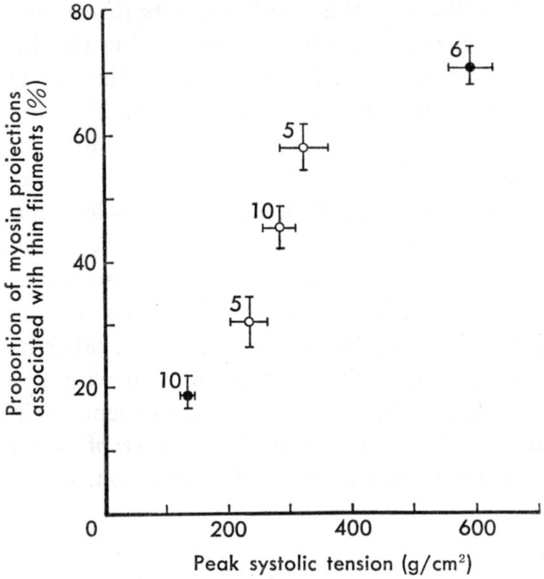

Fig. 1. The proportion of myosin heads found in the vicinity of the thin filament during the systolic phase plotted against the peak systolic tension. The mass transferred from thick to thin filaments when a quiescent muscle goes into rigor is assumed to represent the total mass of myosin heads (see text). The open and filled circles represent mean values (±SEM) obtained from three different types of heart muscle preparations, including one type contracting at three different frequencies (○). The numbers beside the circles indicate the number of observations.

with the smallest peak tension indicates the results from isolated papillary muscles perfused with Tyrode solution with 1% blood added (12–20 contractions per min), and the one with the greatest peak tension indicates the results from paipllary muscles attached to whole hearts which were cross-circulated (115 contractions per min) (Matsubara *et al.*, 1977). These results suggest that a rise in the peak tension, caused by improving the quality of preparation, is also accompanied by an increase in the number of myosin heads in the thin filament vicinity.

2. *Diastolic Phase*

Preliminary observations during the staircase phenomenon have shown that the intensity ratio during the diastolic phase decreases when the frequency of contraction is changed from 60 to 180 min^{-1}. As a decrease in the intensity ratio indicates an increase in the number of heads in the vicinity of the thin filament, this result suggests that a greater number of myosin heads remain near the thin filament during the diastolic phase at a higher contraction frequency. In contrast, the diastolic tension was not affected the change in the contraction frequency. Therefore, the number of myosin heads in the thin filament vicinity seems to be unrelated to the diastolic tension.

The experiments on different types of heart muscle preparation (Matsubara *et al.*, 1977) have shown more clearly that the number of heads is unrelated to the diastolic tension. In papillary muscles attached to cross-circulated whole hearts, 51–52% of myosin heads remained in the thin filament vicinity during the diastolic phase. On the other hand, in isolated papillary muscles perfused with Tyrode solution with 1% blood added, only 7–9% of heads remained near the thin filament during the diastolic phase. In spite of this difference in the number of heads, the diastolic tension was the same in the two types of preparation.

DISCUSSION

The present study has shown that the number of myosin heads in the vicinity of the thin filament during the systolic phase changes with the systolic tension, whereas the number during the diastolic phase is not related to the diastolic tension. The simplest interpretation of these results is that the number affects the tension only when the thin filaments are activated. One possibility is that the tension is roughly proportional to the product of the number of heads and the degree of thin filament activation;

$$\text{Tension} \propto (\text{number of myosin heads in the vicinity of thin filaments}) \times (\text{number of active actin sites})$$

A preparation with a greater number of heads transferred to the vicinity of the thin filament would then produce a greater systolic tension, as can be seen in Fig. 1. Here, according to the above idea, the tension may not be linearly related to the number of heads since tension would also depend on the extent of thin filament activation; the latter may be different at various systolic tensions. The diastolic tension would be unrelated to the number of heads remaining near the thin filament, since most thin filaments would be inactive during the diastolic phase.

The result suggests that changes in the number of heads induced by alteration of the contraction frequency are responsible for the staircase phenomenon. Two mechanisms can be proposed as to how an increase in the frequency leads to an increased transfer of heads to the vicinity of the thin filament: (i) Allen and Blinks (1978) have studied the transient rise of the intracellular calcium concentration of heart muscle using the calcium sensitive photoprotein aequorin. They have shown that, when the contraction frequency is raised, the amplitude of the transient rise of the calcium level increases. This is likely to lead to a greater number of myosin heads attached to actin. (ii) A fast X-ray diffraction study on heart muscle (Matsubara *et al.*, 1979) has shown that the return of myosin heads towards the thick filament after the peak of contraction is rather slow, and continues even after the tension has fallen to the steady diastolic level. At a higher frequency, the diastolic interval is shorter so that a greater number of heads remain near the thin filament at the end of the diastolic phase. As a result of this, the number of heads available for attachment in the next systolic phase would be increased.

Although the myosin heads found in the vicinity of the thin filament during the diastolic phase are apparently not producing tension, they may be primed to produce tension on activation of the thin filaments more readily than other heads. The rate of tension development and the peak systolic tension may be greater when the thin filaments are activated in the presence of heads in the "ready" state. This idea has been supported by (i) comparing different types of heart muscle preparation with different systolic tensions (Matsubara *et al.*, 1977), and (ii) examining a given preparation under different patterns of stimulation, either with single pulses at various frequencies or with paired pulses (Matsubara *et al.*, 1979). In all these experiments, the rate of tension development and the peak systolic tension were greater when a larger number of heads remained near the thin filaments during the diastolic phase. Thus, changes in the distribution of myosin heads in the diastolic phase may contribute to physiological alterations of cardiac contractility.

REFERENCES

Allen, D. G. and Blinks, J. R. (1978). *Nature* **273**, 509–513.

Endoh, M. and Hashimoto, K. (1970). *Japan. J. Physiol.* **20**, 320–331.

Haselgrove, J. C. and Huxley, H. E. (1973). *J. Mol. Biol.* **77**, 549–568.

Haselgrove, J. C., Stewart, M., and Huxley, H. E. (1976). *Nature* **261**, 606–608.

Huxley, A. F. (1957). *Prog. Biophys. Biophys. Chem.* **7**, 255–318.

Huxley, A. F. and Simmons, R. M. (1971). *Nature* **233**, 533–538.

Huxley, H. E. (1968). *J. Mol. Biol.* **37**, 507–520.

Huxley, H. E. (1969). *Science* **164**, 1356–1366.

Huxley, H. E. and Brown, W. (1967). *J. Mol. Biol.* **30**, 383–434.
Matsubara, I., Kamiyama, A., and Suga, H. (1977). *J. Mol. Biol.* **111**, 121–128.
Matsubara, I. and Millman, B. M. (1974). *J. Mol. Biol.* **82**, 527–536.
Matsubara, I., Suga, H., and Yagi, N. (1977). *J. Physiol.* **270**, 311–320.
Matsubara, I., Yagi, N., and Endoh, M. (1978). *Nature* **273**, 67.
Matsubara, I., Yagi, N., and Endoh, M. (1979). *Nature* **278**, 474–476.

Muscle Contraction: Its Regulatory Mechanisms (Ebashi, S. et al., eds.), pp. 53–63,
Japan Sci. Soc. Press, Tokyo / *Springer-Verlag*, Berlin (1980)

5

Kinetic Properties of the Cross-bridges in Frog Skeletal Muscle as Studied by Time-resolved X-ray Diffraction

HARUO SUGI, YOSHIYUKI AMEMIYA,
HIROO HASHIZUME, TSUKASA TAMEYASU,
AND HIDEHIRO TANAKA

Summary: 1) The change in the intensity ratio of two equatorial reflections, $I_{1,0}/I_{1,1}$, during a single twitch in frog skeletal muscle at low temperatures (2–4°C) was studied with the time-resolved X-ray diffraction technique (time resolution, 25 msec). 2) The minimum value of the intensity ratio attained during isotonic twitch under a small load (about $0.03P_0$) was 0.8–1.0, being significantly larger than that during an isometric twitch (0.5–0.6). 3) During an isotonic twitch under a moderate load (about $0.3P_0$), the minimum value of the intensity ratio was similar to that during an isometric twitch. 4) When a muscle was made to shorten against an inertial lever, the tension first rose to a peak and then fell quickly when the lever was maximally accelerated. The change in the intensity ratio was a mirror image of the tension change, showing an early reduction immediately followed by a subsequent recovery. 5) When a quick release of an appropriate magnitude was applied during the rising phase of an isometric twitch, the tension fell to zero and then started to rise again. The intensity ratio showed, however, a time course similar to that during an isometric twitch without quick release. 6) These results are discussed in connection with kinetic properties of the cross-bridges during contraction.

Although it has been well established that the contraction in striated muscle results from a relative sliding motion between the thick and thin filaments, the molecular mechanism which produces the sliding force still remains a matter for debate and speculation. In spite of a number of detailed kinetic studies on actin-myosin-ATP interactions in solution, no direct evidence has hitherto been obtained about the actual behavior of the cross-bridges which are generally believed to cause the sliding between the filaments.

Low angle X-ray diffraction technique is expected to be effective in obtaining information about the structural changes in actin and myosin molecules during contraction in living muscles. The equatorial X-ray diffraction pattern arises from the double hexagonal array of the thick and thin filaments. When vertebrate skeletal muscle contracts isometrically, the intensity of 1,0 equatorial reflection ($I_{1,0}$) decreases while that of 1,1 reflection ($I_{1,1}$) increases, each by a factor of about 2 (Elliott *et al.*, 1967; H. E. Huxley and Brown, 1967; H. E. Huxley, 1968; Haselgrove and H. E. Huxley, 1973). This marked effect has been generally interpreted as being due to the radial movement of the cross-bridges from the vicinity of the thick filaments towards the thin filaments (H. E. Huxley and Brown, 1967; Haselgrove and H. E. Huxley, 1973). As a matter of fact, the change in the intensity ratio between 1,0 and 1,1 reflections ($I_{1,0}/I_{1,1}$) takes place roughly in parallel with the development of isometric tension (H. E. Huxley, 1975, 1979; Matsubara and Yagi, 1978).

Thus, it is of interest to investigate how the intensity ratio between the two equatorial reflections behaves under various experimental conditions, while it should be kept in mind that the change in the intensity ratio can also be explained in various ways other than the radial movement of the cross-bridges during contraction (*e.g.*, Lymn, 1975). The experiments to be described in this paper were undertaken to examine the time course of change in the intensity ratio during single twitches of frog skeletal muscle under various mechanical conditions by use of the time-resolved X-ray diffraction technique.

METHODS

The sartorius muscle dissected from the bullfrog *Rana catesbeiana* was mounted vertically in an experimental chamber with two Mylar windows and a multielectrode assembly; the pelvic end was clamped while the tibial end was connected to an isotonic lever, the movement of which was sensed with a differential transformer. The initial sarcomere length was adjusted to 2.2–2.4 μm by the diffraction pattern of He-Ne laser light. The muscle was continuously perfused with oxygenated Ringer solution (2–4°C) and stimulated with single supramaximal current pulses to produce afterloaded twitches. The low angle equatorial X-ray diffraction patterns (specimen to focus distance, 36 cm) during twitches were recorded with a linear position-sensitive proportional counter, the output of which was fed into a data collection system (Hashizume *et al.*, 1978; Amemiya *et al.*, 1979). The period of a twitch was divided into 40 phases of 25 msec duration, and the data in each phase were stored in the corresponding separate memory segment of the data collection system to be displayed on a chart recorder for analysis. To obtain reasonable photon statistics for each phase, twitches were repeated 120–150 times at intervals of 10–40 sec. X-rays were obtained from a rotating anode fine-focus generator (Rigaku Denki, RU-200PL) operated with a Cu target at 40 kV and 30 mA, and collimated by a mirror-monochrometer focusing optical system (Amemiya *et al.*, 1979). Further details of the experimental apparatus have been described elsewhere (Amemiya *et al.*, 1979).

RESULTS AND DISCUSSION

1. *Time Course of Change in the Intensity Ratio, $I_{1,0}/I_{1,1}$, during an Isotonic Twitch under a Small Load*

A typical example of the time course of change in the intensity ratio ($I_{1,0}/I_{1,1}$) during an afterloaded isotonic twitch (initial sarcomere length, 2.4 μm) under a small load (0.02–0.03P_0) is shown in Fig. 1. It can be seen that the intensity ratio started to decrease on stimulation to reach a minimum value of 0.8–1.0 within the first 20–30% of the shortening phase, and maintained this value until the beginning of relaxation (Sugi *et al.*, 1978). The intensity ratio returned to the resting value during the relaxation phase. The time course of change in the intensity ratio was also examined during an isometric twitch (Fig. 2) with the tibial end connected to a strain gauge. In agreement with previous reports (H. E. Huxley, 1975; Matsubara and Yagi, 1978), the intensity ratio decreased to a minimum value of 0.5–0.6 during the rising phase of isometric tension, and started to return to the resting value after the beginning of relaxation (Sugi *et al.*,

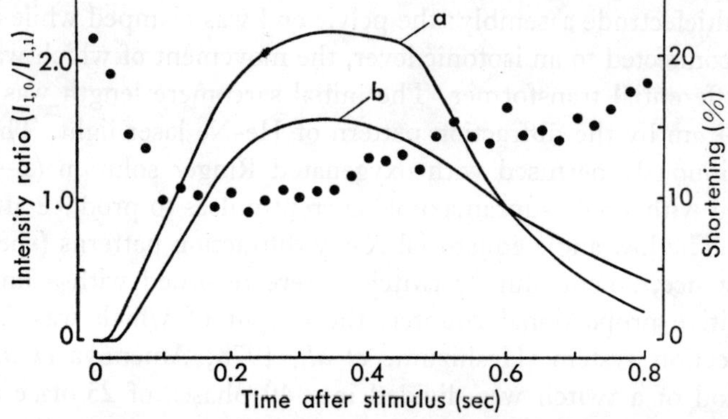

Fig. 1. Time course of change in the intensity ratio $I_{1,0}/I_{1,1}$ during an isotonic twitch under a small load ($0.025P_0$). The initial sarcomere length was 2.4 μm. The degree of shortening is expressed in percentage of the initial length. In this and subsequent figures, curves a and b show the first and the 120th twitches respectively. (Sugi *et al.*, 1978).

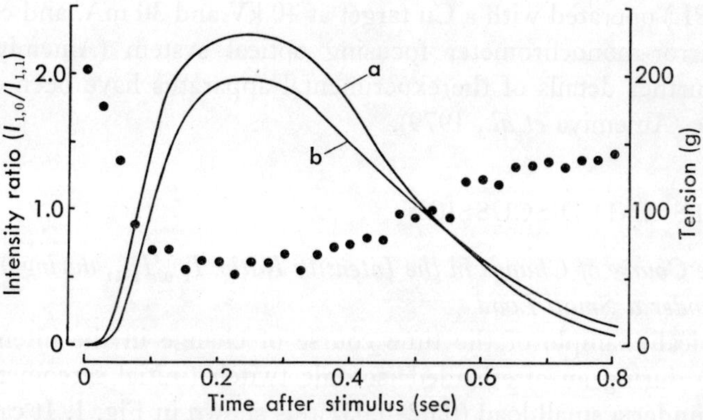

Fig. 2. Time course of change in the intensity ratio $I_{1,0}/I_{1,1}$ during an isometric twitch. The initial sarcomere length was 2.4 μm.

1978). In both isotonic and isometric twitches, the decrease in the intensity ratio resulted from both a decrease in the 1,0 intensity and an increase in the 1,1 intensity (Haselgrove and H. E. Huxley, 1973; Podolsky *et al.*, 1976; Matsubara and Yagi, 1978).

During isometric twitches, the amount of overlap between the thick and thin filaments is kept nearly constant except for some degree of internal shortening against the compliance of the tendons and the recording system

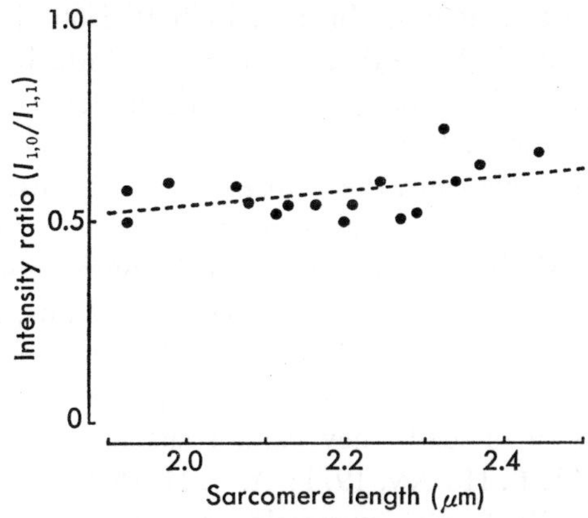

Fig. 3. The intensity ratio $I_{1,0}/I_{1,1}$ for isometrically tetanized muscles plotted as a function of sarcomere length. The data points were obtained from different preparations. Each preparation was tetanized ten times with 1.5 sec train of pulses at 20 Hz with an interval of 1–2 min, so that the total exposure time during the period of steady isometric force was 10 sec or more. The line was fitted to the data points by the method of least squares. (Sugi *et al.*, 1978).

(less than 1.5% of muscle length), while during isotonic twitches the overlap increases considerably as the muscle length shortens by 10–20%. It seems, therefore, necessary to examine the effect of sarcomere length on the intensity ratio attained during isometric contraction, before comparing the data on isotonic twitches with those on isometric ones. In Fig. 3, the values of the intensity ratio attained during isometric tetani are plotted as a function of sarcomere length, though the dependence of the intensity ratio on sarcomere length was much less marked (Sugi *et al.*, 1978) compared with previous reports (Elliott *et al.*, 1963; Haselgrove and H. E. Huxley, 1973). Similar results have also been obtained by Podolsky *et al.* (1976). The steep relation between the intensity ratio and sarcomere length reported by the previous authors may be due to fatigue or damage to the specimen caused by a long exposure time to X-rays, resulting in a decrease in the degree of activation.

Thus, the result that the minimum value of the intensity ratio attained during an isotonic twitch (Fig. 1) is definitely larger than that attained during an isometric one (Fig. 2) is opposite to the effect expected from the relation between the intensity ratio and sarcomere length (Fig. 3), since both types of twitch started from the same sarcomere length. Then, if it is assumed that the minimum value of the intensity ratio serves as a measure of the number of attached cross-bridges in both isotonic and isometric conditions, these results indicate that the number of attached cross-bridges is definitely smaller during isotonic shortening under a small load then under isometric conditions.

In accordance with this view, the stiffness of intact muscle fibers, which is also generally regarded as a measure of the number of attached cross-bridges, decreases with decreasing isotonic load, approaching a finite minimum value when the load tends to zero (Julian and Sollins, 1975; Tsuchiya *et al.*, 1979; Sugi and Tsuchiya, 1979). In addition, when the load on muscle fibers is suddenly increased to about P_0 after a period of isotonic shortening under a small load, they cannot sustain the new load to be lengthened rapidly (A. F. Huxley, 1971; A. F. Huxley and Simmons, 1973; Sugi and Tsuchiya, 1979), also suggesting a decrease in the number of attached cross-bridges during shortening. Thus, both the X-ray and mechanical data described above are qualitatively consistent with the sliding filament model of A. F. Huxley (1957), in which the attachment of the cross-bridges to the thin filaments has a moderate rate constant so that the proportion of attached cross-bridges decreases as the velocity of shortening increases.

Since the amount of overlap between the filaments increases during isotonic shortening, the reflecting mass in the 1,1 plane increase while that in the 1,0 plane decreases as a result of shortening. Consequently, the intensity ratio, $I_{1,0}/I_{1,1}$, may be expected to decrease rather sharply with decreasing sarcomere length. From this point of view, it seems somewhat strange that the intensity ratio stays almost unchanged during the period of most isotonic shortening (Fig. 1). A possible explanation for this is that the 1,0 lattice becomes more regular in the overlap zone than in the non-overlap zone, so that the effect of increase in the reflecting mass in the 1,1 plane is canceled out. The relatively small dependence of the intensity ratio during isometric contraction on sarcomere length can also be accounted for on this basis.

2. *Time Course of Change in the Intensity Ratio during a Twitch under a Moderate Isotonic or Inertial Load*

It seems possible that, when the thick and thin filaments slide past each other, the configuration of the cross-bridges differs from that during isometric contraction, so that the intensity ratio no longer serves as a measure of the number of attached cross-bridges. In order to make this point clear, the following experiments were performed.

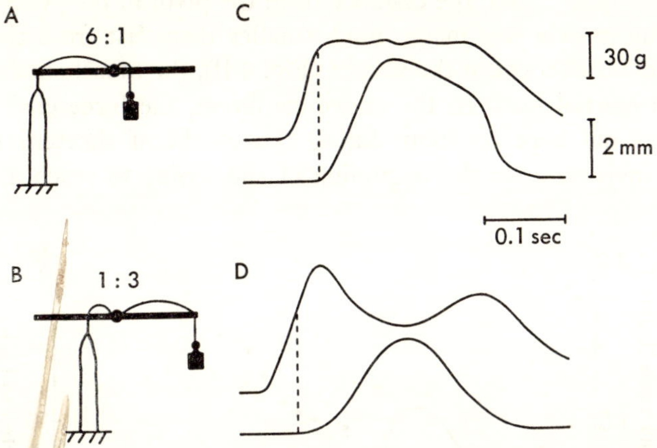

Fig. 4. Diagrams showing the points of attachment of the preparation and the weight to the lever. In A, the tibial end of the preparation was attached to one arm at a point 6 cm from the pivot, while the weight was attached to the opposite arm at a point 1 cm from the pivot to serve as an isotonic load. In B, the preparation was attached to one arm at a point 1 cm from the pivot, while the weight was attached to the opposite arm at a point 3 cm from the pivot to serve as an inertial load. C and D: length and force changes during twitches under isotonic (C) and inertial (D) loads.

Figure 4 illustrates the experimental conditions. A moderate weight was attached to one arm of the lever, while the muscle was connected to the opposite arm. When the distance from the pivot to the point of attachment of the muscle was made several times longer than that from the pivot to the point of attachment of the weight (A), the muscle shortened isotonically on stimulation. This was proved to actually be the case by simultaneously recording the length and tension changes during a twitch (C). The muscle first contracted isometrically, and when the force exerted by it reached the value of isotonic load, the muscle started to shorten with almost a constant velocity while the force was kept almost constant. Figure 5 shows the time course of change in the intensity ratio during an isotonic twitch under a moderate load of about $0.3P_0$. The intensity ratio started to fall on stimulation, reached a minimum value of 0.5–0.6 at the early part of the shortening phase, and maintained this value until the beginning of relaxation. Thus, the minimum value of the intensity ratio attained during an isotonic twitch under a moderate load did not differ appreciably from that attained during an isometric twitch (Fig. 1). Similar results have also been obtained by Podolsky *et al.* (1976) and H. E. Huxley (1979).

If, on the other hand, the distance from the pivot to the point of attachment of the muscle was made much smaller than that from the pivot to the point of attachment of the weight (Fig. 4 B), the muscle had to shorten against an inertial load. In the above condition, the force exerted by the muscle was not kept constant during the course of shortening; it continued to increase after the beginning of shortening to reach a peak, and

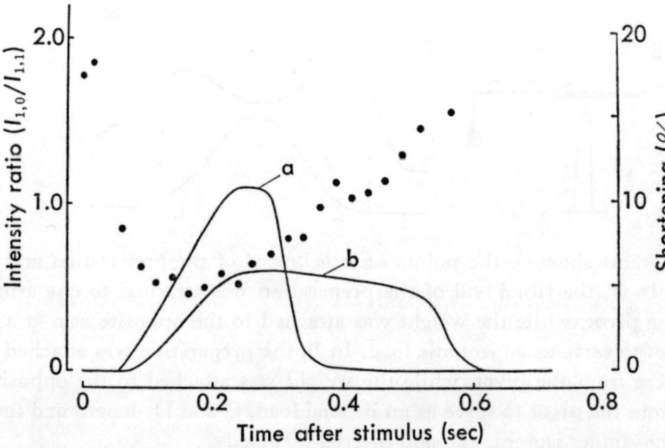

Fig. 5. Time course of change in the intensity ratio $I_{1,0}/I_{1,1}$ during an isotonic twitch under a moderate load (about $0.3P_o$). The initial sarcomere length was 2.4 μm.

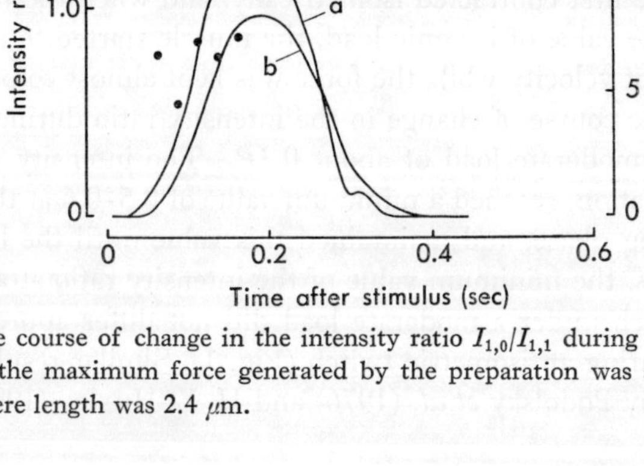

Fig. 6. Time course of change in the intensity ratio $I_{1,0}/I_{1,1}$ during a twitch under an inertial load (the maximum force generated by the preparation was about $0.4P_o$). The initial sarcomere length was 2.4 μm.

then started to decay while the shortening was still in progress (Fig. 4 D). Figure 6 shows the time course of change in the intensity ratio during a twitch against an inertial load, the magnitude of peak tension exerted by the muscle being about $0.4P_0$. It can be seen that the intensity ratio also fell to a minimum value of about 0.5 at the early part of the shortening phase, but the minimum value was not maintained until the beginning of relaxation, starting to return towards the resting value while the shortening was still in progress. Thus, the change in the intensity ratio was roughly a mirror image of the change in force exerted by the muscle (Fig. 4D).

These results seem to indicate that, during shortening under isotonic or inertial loads, the intensity ratio, $I_{1,0}/I_{1,1}$ reflects the process of force generation rather than the process of sliding between the thick and thin filaments. This implies that the contribution of the change in configuration of the cross-bridges to the intensity ratio may not be very large.

3. Effect of Quick Release Applied during an Isometric Twitch on the Intensity Ratio

The effect of quick release applied during an isometric twitch is shown in Fig. 7. When a quick release (3–4% of muscle length) was applied to the muscle during the rising phase of an isometric twitch, the force fell to zero and then started to rise again to a peak appreciably smaller than

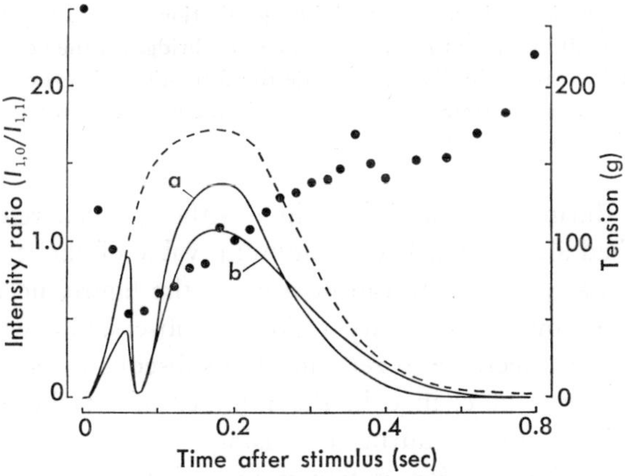

Fig. 7. Time course of change in the intensity ratio $I_{1,0}/I_{1,1}$ during an isometric twitch with a quick release applied at 50 msec from the onset of stimulation. The amount of release was 3–4% of the length of the preparation, causing a transient fall of tension to zero followed by the subsequent redevelopment of isometric force. The dotted line shows an isometric twitch without quick release.

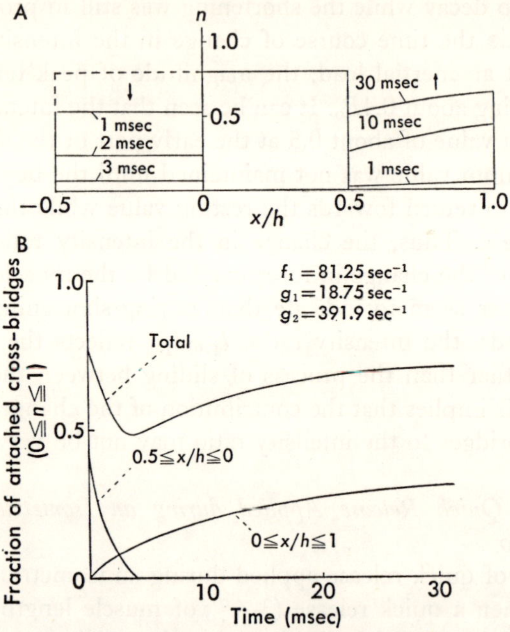

Fig. 8. Effect of quick release applied during the rising phase of an isometric twitch. The transient nature of activation during a twitch (Julian, 1969) was taken into consideration. A: change in the distribution of attached cross-bridges after a quick release. A quick release results in a displacement of the distribution immediately before release to the left to cause a fall in force to zero. Then, the cross-bridges at the negative force region begin to detach from the thin filaments, while the formation of new cross-links starts at the positive force region. B: time course of change in the number of attached cross-bridges relative to that immediately before release.

that reached during an isometric twitch without quick release. The intensity ratio had already reached a minimum value of about 0.5 before the quick release was applied. As can be seen in the figure, no distinct effect of quick release on the intensity ratio was observable, while the force showed a transient decrease to zero and the subsequent redevelopment.

The time course of change in the number of attached cross-bridges after quick release applied during the rising phase of an isometric twitch was calculated using the contraction model of A. F. Huxley (1957), and the results are shown in Fig. 8. Since only the ratios of the rate constants for making and breaking of the cross-links (f and g respectively) are given in Huxley's paper, we used appropriate values chosen by Julian (1969): f=81.25/sec, g_1=18.75/sec, and g_2=391.9/sec. The time course of activation during a twitch was also taken into consideration (Julian, 1969). Immediately after the completion of quick release, the cross-bridges start to detach from the thin filament in the negative force region, while they start to attach to the thin filament in the positive force region (Fig. 8A). As a result the total number of attached cross-bridges reaches a minimum (about 50% of the initial value) at about 10 msec after the release, and then returns towards the initial value (Fig. 8B). Thus, the time resolution of the present experiments (25 msec) is not sufficient to decide whether the effect of quick release on the intensity ratio is consistent with the Huxley model or not. In order to clarify the behavior of the cross-bridges during and after quick release, it is necessary to record the changes in the intensity ratio and in the muscle stiffness with a much better time resolution.

REFERENCES

Amemiya, Y., Sugi, H., and Hashizume, H. (1979). *Cross-bridge Mechanism in Muscle Contraction* (Sugi, H. and Pollack, G. H., eds.), pp. 425–443, Univ. Tokyo Press, Tokyo.

Elliott, G. F., Lowy, J., and Millman, B. M. (1967). *J. Mol. Biol.* **25**, 31–42.

Elliott, G. F., Lowy, J., and Worthington, C. R. (1963). *J. Mol. Biol.* **6**, 295–305.

Haselgrove, J. C. and Huxley, H. E. (1973). *J. Mol. Biol.* **77**, 549–568.

Hashizume, H., Mase, K., Amemiya, Y., and Kohra, K. (1978). *Nucl. Instr. Method* **152**, 199–203.

Huxley, A. F. (1957). *Prog. Biophys. Biophys. Chem.* **7**, 255–318.

Huxley, A. F. (1971). *Proc. Roy. Soc. Lond. B* **178**, 1–27.

Huxley, A. F. and Simmons, R. M. (1973). *Cold Spring Harbor Symp. Quant. Biol.* **37**, 669–680.

Huxley, H. E. (1968). *J. Mol. Biol.* **37**, 507–520.

Huxley, H. E. (1975). 5th Int. Biophys. Cong., Copenhagen, S53.

Huxley, H. E. (1979). In *Cross-bridge Mechanism in Muscle Contraction* (Sugi, H. and Pollack, G. H., eds.), pp. 391–405, Univ. Tokyo Press, Tokyo.

Huxley, H. E. and Brown, W. (1967). *J. Mol. Biol.* **30**, 383–434.

Julian, F. J. (1969). *Biophys. J.* **9**, 547.

Julian, F. J. and Sollins, M. R. (1975). *J. Gen. Physiol.* **66**, 287.

Lymn, R. W. (1975). *Nature* **258**, 770.

Matsubara, I. and Yagi, N. (1978). *J. Physiol.* **278**, 297–307.

Podolsky, R. J., Onge, R. St., Yu, L., and Lymn, R. W. (1976). *Proc. Natl. Acad. Sci. U.S.* **73**, 813–817.

Sugi, H., Amemiya, Y., and Hashizume, H. (1978). *Proc. Japan Acad.* **54**, 559–564.

Sugi, H. and Tsuchiya, T. (1979) (submitted for publication).

Tsuchiya, T., Sugi, H., and Kometani, K. (1979). In *Cross-bridge Mechanism in Muscle Contraction* (Sugi, H. and Pollack, G. H., eds.), pp. 225–240, Univ. Tokyo Press, Tokyo.

Muscle Contraction: Its Regulatory Mechanisms (Ebashi, S. et al., eds.), pp. 65–78, Japan Sci. Soc. Press, Tokyo / Springer-Verlag, Berlin (1980)

6

A Rate-limiting Step in Muscle Contraction

RICHARD J. PODOLSKY AND
KATSUHISA TAWADA

Summary: The contractile response of calcium-activated skinned frog muscle fibers has been studied in solutions of various ionic strengths, and in the presence and absence of an ATP regenerating system, which presumably affects the ADP level in the fiber during contraction. These factors were found to change the isotonic transients in the motion that follows a step change in load from P_0, the steady isometric force, to $P<P_0$, but not the steady speed of shortening at a given relative load, P/P_0. The influence of ionic strength and ADP on the transient response of the fiber indicates that these factors modulate the interaction of actin and myosin in the fiber as well as in solution. The insensitivity of the steady motion to these factors is the first actual evidence that the cross-bridge cycle in the fiber contains a single rate limiting step. The nature of this step is not known, but one possibility is a slow intramolecular transition in the "unattached" myosin molecule, as has been suggested by Eisenberg and Kielley (1973).

It is indeed a pleasure to participate in this symposium, and to acknowledge our scientific debt to Professors Kumagai and Natori. In 1954, Professor Natori published his famous paper on skinned fibers (Natori, 1954). At that time the fibers were prepared and studied in oil. One type of experiment was to apply small droplets of various solutions to the surface of the fiber with a micropipette and to monitor the contractile response microscopically. A great deal was learned in this way about the internal membrane system, the importance of which was recognized early on by Professor Kumagai (Kumagai *et al.*, 1955), and about the contractile mechanism. However the responses were difficult to score quantitatively, and it seemed worthwhile to develop recording methods for skinned fibers that were similar to those used for intact fibers. Since the outer membrane was no longer present, the preparation had to be activated with a perfusion technique rather than by electrical stimulation. This became a feasible maneuver when Annemarie Weber (Weber *et al.*, 1963) showed us how to use EGTA to control the calcium ion concentration of solutions in the micromolar range, which is the range required to modulate the contractile activity of a skinned fiber (Ebashi and Endo, 1968; Hellam and Podolsky, 1969). After various technical problems were solved, it became possible to study the force and motion of such preparations under a wide range of conditions.

We would like to review some of these experiments because, with many results now in hand, an interesting generalization seems to emerge regarding the rate limiting step in the cross-bridge cycle.

EFFECT OF CALCIUM AND IONIC STRENGTH ON THE STEADY MOTION OF SKINNED FIBERS

At the outset, only steady state force-velocity relations were measured; the response of the displacement transducer was not yet quick enough to record the pre-steady motion. An obvious question was whether the degree of activation of the fibers affects the contraction kinetics. To answer this, preparations were activated with low levels of calcium, producing isometric forces ranging from 20–75% of that reached at saturating calcium levels. At each calcium level, the contraction velocity was measured with different loads. As shown in Fig. 1, at a given relative load (that is, the load relative to the isometric force at that calcium level), the velocity was the same at all calcium ion concentrations (Podolsky and Teichholz, 1970). This result is consistent with the idea that calcium acts as a switch on the actin sites involved in cross-bridge formation, and that the kinetic pro-

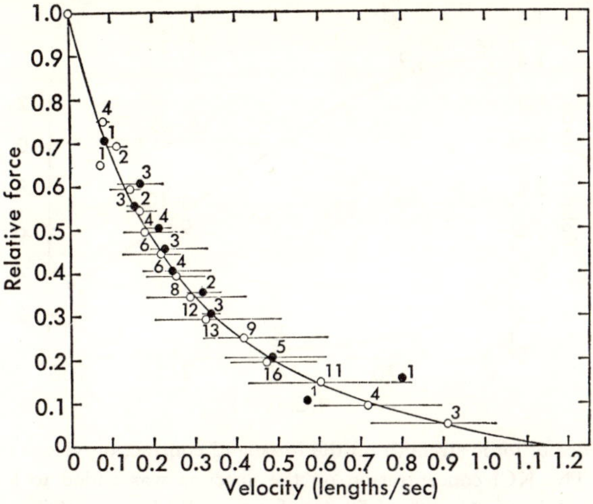

Fig. 1. Force-velocity relation for skinned frog muscle fibers at different calcium ion concentrations. The open circles are from fibers at high calcium levels (pCa 5.0 or 5.7) where maximum force was generated; the filled circles are from fibers at lower calcium levels (pCa 6.5 or 6.75) where the steady force was 0 to 60% or 50 to 80%, respectively, of the maximum force. Temperature, 5–7°C. Note that all data points fall on the same curve. (From Podolsky and Teichholz, 1970).

perties of the cross-bridge cycle are unaffected by calcium ion concentration.

Similar experiments were carried out when the isometric force was changed by increasing the ionic strength of the bathing solution. We knew from biochemical data that ionic strength affected the association of actin and myosin during ATP hydrolysis in solution (Rizzino *et al.*, 1970), and therefore expected to see a change in the cross-bridge rate constants, and a corresponding change in contraction velocity, during motion of the fiber. However, the velocity at a given relative load was the same even when the ionic strength was increased considerably (Thames *et al.*, 1974). This is seen in the right side of Fig. 2, where the upper solid curve is the steady isometric force and the interrupted curve is the velocity at a given relative load. When the KCl concentration in the bathing solution was increased from 140 mM to 280 mM, the force decreased three fold, but the contraction velocity was unchanged. This result was unexpected and is difficult to explain if the contraction velocity is controlled primarily by the rate constants for making and breaking cross-bridges, as is the case,

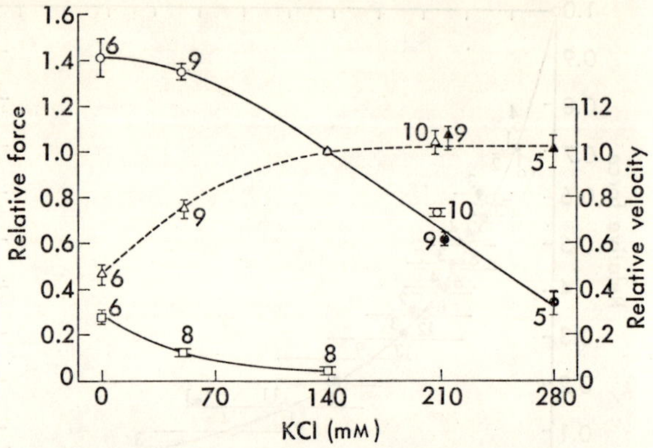

Fig. 2. Influence of ionic strength on the force and contraction velocity of skinned frog muscle fibers. The KCl concentration on the abscissa was added to a simple bathing solution having an ionic strength of 50 mM. Upper solid line is calcium-activated force; lower solid line is resting tension, and interrupted line is velocity. Calcium-activated force and velocity are expressed as a fraction of that obtained in the same fiber in 140 mM KCl at the same relative load. Temperature, 5–7°C. Note that when the KCl concentration is increased above 140 mM, the calcium-activated force decreases while the speed of shortening is unchanged. (From Thames *et al.*, 1974).

for example, in the model put forward by A. F. Huxley in 1957 (Huxley, 1957). (The lower solid line in Fig. 2 is irreversible resting tension. This force appears when fibers are activated at low ionic strength and is associated with a decrease in contraction velocity; this is another story which has been discussed elsewhere (Thames *et al.*, 1974)).

TRANSIENT RESPONSES

After the calcium and ionic strength experiments were completed, the response time of the displacement transducer was improved and velocity transients following step changes in load were recorded (Podolsky *et al.*, 1974). The panels in Fig. 3 show a typical experiment at 3°C. Displacement is given by the upper trace, force in the middle, and zero force in the lower trace. The general pattern of the pre-steady motion is the same as that found in electrically stimulated intact fibers: the velocity is first greater than, and then less than, the steady value. The duration of the transient is measured by the time τ when the actual motion intersects the back extrapolation of the steady motion. The time scale here is 20 msec; the

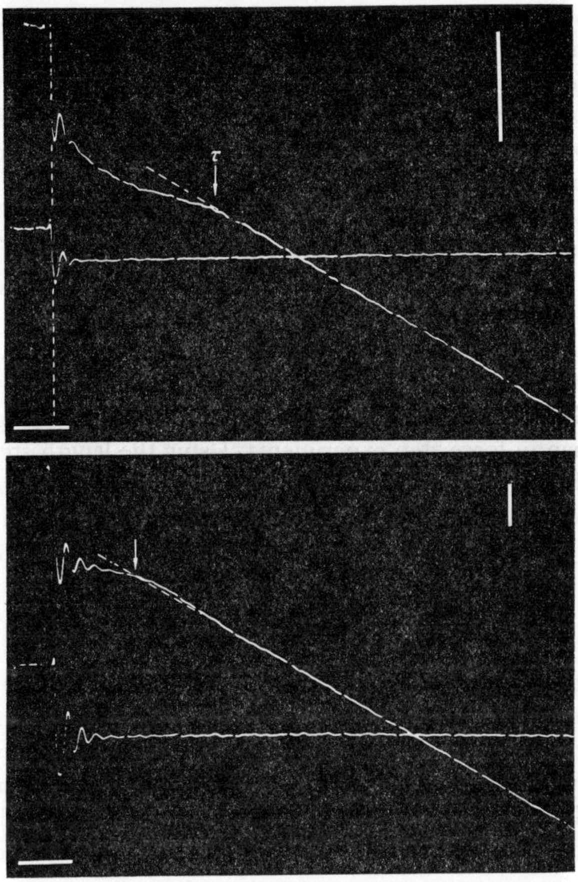

Fig. 3. Response of skinned muscle fiber in simple bathing solution to step changes in load at 3°C. Top trace, displacement; middle trace, force; bottom trace, zero of force. The downward pointing arrow marks the null time, τ, where the actual motion intersects the back extrapolation of the steady motion. Vertical scale bar, 50 Å/half sarcomere; horizontal scale bar, 20 msec. (From Podolsky *et al.*, 1974).

transient in the upper panel lasts about 60 msec, while in the lower panel, where the force step is greater, the duration is 30 msec.

These transients contain information about processes that occur during the cross-bridge cycle. They probably reflect the shift in cross-bridge distribution from that set up during isometric contraction to that formed during the steady motion. Some evidence for this shift comes from X-ray diffraction studies which show that the equatorial diffraction pattern is nearly the same in the two states (Podolsky *et al.*, 1976; H. E. Huxley, 1979; Amemiya *et al.*, 1979). Although the interpretation of this experiment is not completely settled (Lymn, 1978; Yu *et al.*, 1979), the observation suggests that the number of cross-bridges is nearly the same during isometric contraction and during the steady motion, in which case the force difference between the two physiological states would be due mainly to a shift in the configuration of a given number of cross-bridges. Since cross-bridges presumably turn over during this shift, the rate functions for making and breaking cross-bridges would be expected to affect the properties of the transient.

Similar transients are seen in length step experiments (A. F. Huxley and Simmons, 1971). For technical reasons, the time resolution in the first few milliseconds after the step is generally better in length step experiments than in force step experiments. Using the length step technique with intact fibers, Huxley and Simmons (1971) found a very rapid force transient, lasting only a few milliseconds, which they identified with the transition of an attached cross-bridge from a low force to a high force state. They suggested that a myosin projection normally attaches to an actin site in the low force state, and that force is generated by a transition to the high force state. They also suggested, on the basis of data from double length step experiments (Ford *et al.*, 1974), that the attachment and detachment processes are much slower than the force generating transition. Since the force steps in Fig. 3 take about 3 msec to complete, these ideas imply that the force generating transitions of attached cross-bridges equilibrate during the step in our experiments, and that the relatively slow transient seen after the step is probably caused mainly by cross-bridge turnover.

COMPARISON OF SKINNED AND INTACT FIBER RESPONSE

With these ideas in mind, it is interesting to compare the motion of skinned and intact fibers in more detail. The bathing solution in the experiment shown in Fig. 3 was a simple salt solution: 140 mM KCl, 5 mM ATP, 1 mM $MgCl_2$, 10 mM imidazole, and 3 mM (EGTA+CaEGTA), pCa 5.0, pH 7.0. The steady state force-velocity relation was the same as that for an intact fiber at the same temperature. However, as shown in Fig. 4, the duration of the transient is about twice as long in skinned fibers as in intact fibers (Podolsky *et al.*, 1974).

The question arises as to whether this prolongation of the isotonic

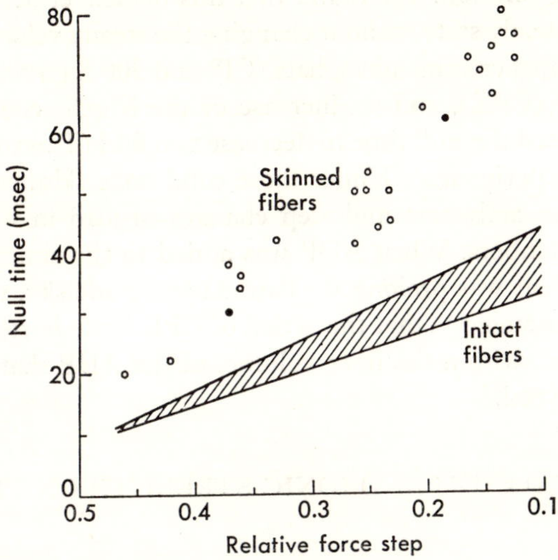

Fig. 4. Influence of force step on the null time in intact fibers (hatched area) and in skinned fibers in a simple bathing solution (○). The filled circles are data from the traces in Fig. 3. Note that the null time for the motion of the skinned fibers is about twice as long as that for intact fibers. (From Podolsky *et al.*, 1974).

transient, with no change in the steady motion, can be obtained by a straightforward modification of the two state cross-bridge model that describes the motion of intact fiber (Podolsky and Nolan, 1971). The answer is that there is no simple way to do so. All three of the model parameters, the rate functions for making and breaking cross-bridges and the cross-bridge force function, must be changed to simulate the prolongation, which is possible perhaps but not too likely. This result, as well as the ionic strength effect, suggested that factors in addition to the making and breaking of cross-bridges and a rapid force generating process must be involved in the cross-bridge cycle.

EFFECT OF AN ATP REGENERATING SYSTEM

What causes the prolongation of the transient? Perfused skinned fibers lose soluble enzymes and metabolites to the bathing solution. In addition, the concentration of ionized magnesium is probably much higher in intact fibers than in the simple salt solution used in our experiments. Therefore we studied the effect of adding a potent ATP regenerating system to the simple bathing solution. We found that this maneuver accelerated the approach to the steady state without changing the steady velocity. Specifically, addition of 10 mM creatine phosphate (CP) and 300 Sigma units/ml creatine phosphokinase (CPK), and an increase of the $MgCl_2$ concentration from 1 to 5 mM, caused the null time to decrease two fold (Tawada and Podolsky, unpublished experiments). Under these conditions skinned fibers respond to both calcium activation and step changes in load in ways that closely resemble intact fibers. When ADP was added to the simple bathing solution, the transient was prolonged (Tawada and Podolsky, unpublished experiment). This suggests that the action of CPK that shortens the transient is the reduction, within the fiber volume, of the ADP that accumulates as ATP is hydrolyzed.

EFFECT OF CALCIUM AND IONIC STRENGTH ON THE ISOTONIC TRANSIENT

As already mentioned, early experiments showed that the relative force-velocity relation was unaffected when isometric force was decreased either by decreasing calcium ion concentration or by increasing ionic strength. The calcium effect was easy to understand, but the ionic strength effect was unexpected. To gain more insight into the ionic strength effect, these experiments were repeated with improved time resolution, so that we could study the transient as well as the steady motion (Gulati and Podolsky, 1978).

The ionic strength results are shown in Fig. 5. The left hand panel is the control response, in a simple bathing solution containing 140 mM KCl, and the right hand panel is the response in a solution containing 210 mM KCl. As before, the increase in ionic strength causes the steady isometric force to decrease, but the steady motion at a given relative load does not change. However, ionic strength does change the transient response: the null time is the same at both ionic strengths, but the amplitude at the higher ionic strength is twice that in the control solution. Therefore ionic strength appears to change certain parameters in the cross-bridge cycle without affecting the steady velocity.

With regard to the calcium effect, both the transient and the steady motion were the same, at the same relative load, at pCa 5 and pCa 6.4, where the steady isometric force was reduced two fold. This is good evidence that calcium controls the number of active sites without changing the kinetic properties of an active site.

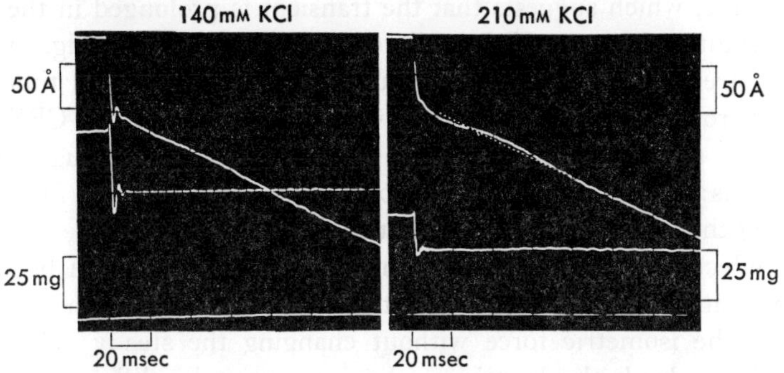

Fig. 5. Influence of ionic strength on the response of a skinned fiber to step changes in load. Temperature, 5–6°C. Note that the isometric force in 210 mM KCl is about half that in 140 mM KCl, while the amplitude of the transient at the same relative load is about two times greater. (From Gulati and Podolsky, 1978).

SUMMARY OF THE EXPERIMENTAL RESULTS

This is given in Table I. For skinned fibers under various conditions, the transient null time and amplitude, and the steady state force and velocity, are referred to the corresponding values for intact fibers. The main result, shown in the last column of Table I, is that the steady state contraction velocity remains constant under a variety of conditions that cause the transient to change markedly.

The increased null time seen in skinned fibers in simple bathing solutions (row *b*), relative to intact fibers (row *a*), is probably due to the loss of CPK activity in the skinned fibers, since the null time can be brought back to the intact fiber value by adding back CPK, CP, and $MgCl_2$ (row *c*). CPK appears to act by reducing the concentration of ADP within the

TABLE I

Influence of Various Parameters on the Force, Velocity, and Transient Response of Skinned Muscle Fibers

	Fiber	(mM)	pCa	CPK	Transient		Steady state	
					τ	Amp	P_0	V
a	Intact	190	5	+	1	1	1	1
b	Skinned	190	5	−	2	1	1	1
c	Skinned	190	5	+	1	1	1	1
d	Skinned	260	5	−	2	2	0.5	1
e	Skinned	190	6.4	−	2	1	0.5	1

fiber volume, which suggests that the transient is prolonged in the simple bathing solution because ADP inhibits a step in the cross-bridge cycle.

Ionic strength changes the amplitude rather than the null time of the transient (row *d*), which is qualitatively different from the CPK effect. This is not surprising since the two factors clearly act through different mechanisms: CPK has little effect on the steady isometric force, while increasing the ionic strength reduces this force.

The effect of changing the calcium ion concentration can be thought of as a control for the ionic strength effect. Although both parameters modulate the isometric force without changing the steady velocity at a given relative load, the kinetic mechanisms must be different, since increasing the ionic strength changes the transient (row *d*), while lowering the calcium level does not affect the transient (row *e*).

THE CROSS-BRIDGE CYCLE

These results have implications regarding the principal steps in the cross-bridge cycle. The diagram on the left of Fig. 6 shows the two state cycle.* The myosin molecule, M, is either attached to the actin filament,

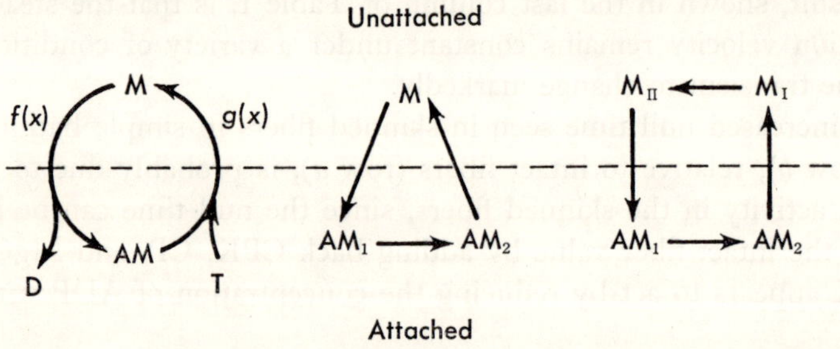

Fig. 6. Diagrams of cross-bridge cycles with different numbers of states. In a two state cycle (left) myosin M is either attached to actin A, forming a cross-bridge AM, or unattached. The three state cycle (center) has two attached states, AM_1 and AM_2; the four state cycle has, in addition, two unattached states, M_I and M_{II}, of which only M_{II} can form a cross-bridge. $f(x)$ and $g(x)$ are rate functions for making and breaking cross-bridges, and T and D are ATP and ADP.

* The cycles in Fig. 6 are drawn as though myosin makes transitions between states that are either unattached or attached to actin. However, the arguments that follow would be the same if the interrupted line separated states of myosin that are either weakly or strongly bound to actin, and the major force producing state is one of the latter (Stein *et al.*, 1979).

forming a force generating cross-bridge, AM, or unattached. The rate of making the bridge, *f*, is a function of *x*, the relative position of the myosin molecule and the actin site (A. F. Huxley, 1957). The breaking rate, *g*, is also a function of *x*. The ATP molecule (T) provides energy by binding to AM breaking the cross-bridge. ATP is split into ADP (D) and phosphate while it is bound to M; the reaction products are released after A binds to MD.

Calculations show that in this scheme both the velocity for a given force and the time course of the isotonic transient, depend on the magnitudes and spatial distribution of *f* and *g*. The sensitivity of the velocity to *f* and *g* may depend on the load, but, in general, there will be certain loads for which the velocity will vary when *f* and/or *g* are varied (see, for example, Simmons and Jewell, 1974). Therefore, if we take the effects of CPK and ionic strength on the null time and amplitude of the transient as evidence that *f* and *g* depend on both the ADP concentration and ionic strength of the bathing solution, the insensitivity of the steady velocity to these parameters cannot be explained easily by a two state cycle.

In the center diagram of Fig. 6, the activity cycle contains two attached states, AM_1 and AM_2, and one unattached state. AM_1 represents a low force configuration of the cross-bridge, and $AM_1 \rightarrow AM_2$ represents a transition to a high force state. Huxley and Simmons (1971) associated this transition with the very rapid force transient seen on the 2–4 msec time scale after a quick length step was imposed on the activated fiber. The cross-bridge making ($M \rightarrow AM_1$) and breaking ($AM_2 \rightarrow M$) processes are supposed to take place on a time scale that is considerably slower than the force generating transition. Therefore for experiments that take place on the 10 msec time scale, the center diagram tends to collapse into a two state cycle which, as mentioned above, cannot easily account for the response pattern given in Table I.

The diagram on the right of Fig. 6 is the next stage of complexity for the cross-bridge cycle. In this four state cycle (Eisenberg and Hill, 1978), only one of the unattached states, M_{II}, can react with A to form a cross-bridge. Eisenberg and Kielley (1973) pointed out that if the transition M_I to M_{II} is the slowest step in the cycle, it could limit the turnover rate in the organized system and therefore the contraction velocity. In this case changes in the rates and/or spatial characteristics of the other transitions could produce changes in the transients, but would not be expected to change the force-velocity relation. This is consistent with the response pattern summarized in Table I.

RELATION TO THE MYOSIN ATPase IN SOLUTION

These ideas concerning the rate limiting step in the physiological activity cycle are supported by certain properties of myosin ATPase in solution. The ATPase activity of myosin is activated by Ca^{2+} as well as by actin. Barany (1967) measured both activities in myosins isolated from many different muscles, and found that both were proportional to the contraction velocity of the muscles; that is, the greater the ATPase activity the greater the contraction velocity (Fig. 7). The fact that this correlation was found in the actin-free, Ca^{2+}-activated system as well as in the actin-activated system shows that information about contraction velocity is encoded in the myosin molecule and, in addition, that this information can be expressed in the absence of actin.

The connection between the Ca^{2+} ATPase in solution and contraction velocity in the fiber is not known, but it could be that the rate of both

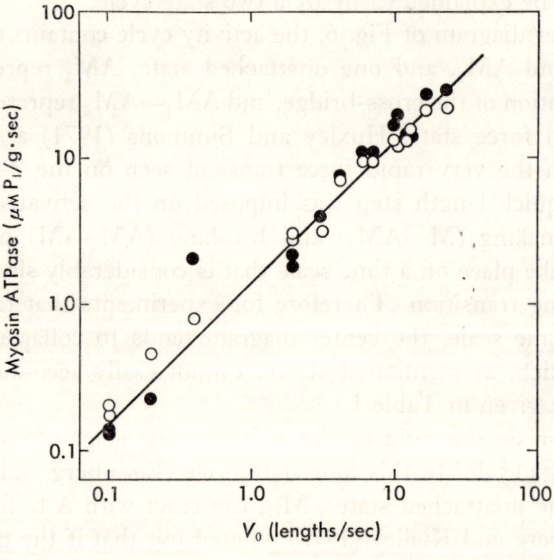

Fig. 7. Relation between maximum speed of shortening and the myosin ATPase activity of various muscles. Actin-activated ATPase measured in 30 mM KCl, 20 mM Tris Cl, 1 mM $MgSO_4$, 0.1 mM $CaCl_2$, 1 mM ATP, 0.4 mg F-actin/mg myosin, pH 7.4. Ca^{2+} ATPase measured in 30 mM KCl, 20 mM TrisCl, 10 mM $CaCl_2$, 1 mM ATP, pH 7.4. The temperature is the same as that used to measure the speed of shortening, which ranges from 0 to 37°C. ○ actin-activated; ● Ca^{2+}-activated. (Data taken from Barany, 1967).

processes is limited by an intramolecular transition in the myosin molecule. Evidence that myosin undergoes such a transition, and that the transition can be relatively slow, comes from studies of actin-activated hydrolysis of ATP by heavy meromyosin (HMM) and subfragment 1 (S-1) at low temperature and low ionic strength. Under these conditions the myosin fragments appear to exist in two states, one which does not bind actin and another which does, and the ATPase rate reflects the transition rate between these states (Eisenberg and Kielley, 1973). The similarity of the actin-activated and Ca^{2+}-activated rates, shown in Fig. 7, suggests that both hydrolysis processes are limited by the same step, and the close correlation of both with contraction velocity can be taken as evidence that it too is limited by the same transition.

CONCLUSION

To sum up, we have studied the motion of skinned muscle fibers in various solutions, and have found that the steady motion is unaffected by factors that change the isotonic transients. These factors are known to modulate the interaction of actin and myosin in solution, and their influence on the transients indicates that this happens in the fiber as well. The insensitivity of the steady motion to these factors is probably the first evidence that the cross-bridge cycle in a muscle fiber contains a single rate limiting step. The nature of this step is not known, but one possibility is a slow intramolecular transition in the myosin molecule, as was suggested some years ago by Eisenberg and Kielley (1973).

REFERENCES

Amemiya, Y., Sugi, H., and Hashizume, H. (1979). In *Cross-bridge Mechanism in Muscle Contraction* (Sugi, H. and Pollack, G. H., eds.), pp. 425–440, Univ. Tokyo Press, Tokyo.

Barany, M. (1967). *J. Gen. Physiol.* **50**, 197–218.

Ebashi, S. and Endo, M. (1968). *Prog. Biophys. Mol. Biol.* **18**, 123–183.

Eisenberg, E. and Hill, T. L. (1978). *Prog. Biophys. Mol. Biol.* **33**, 55–82.

Eisenberg, E. and Kielley, W. W. (1973). *Cold Spring Harbor Symp. Quant. Biol.* **37**, 145–152.

Ford, L. E., Huxley, A. F., and Simmons, R. M. (1974). *J. Physiol.* **240**, 42–43P.

Gulati, J. and Podolsky, R. J. (1978). *J. Gen. Physiol.* **72**, 701–716.

Hellam, D. C. and Podolsky, R. J. (1969). *J. Physiol.* **200**, 807–819.

Huxley, A. F. (1957). *Prog. Biophys. Chem.* **7**, 255–318.

Huxley, A. F. and Simmons, R. M. (1971). *Nature* **233**, 533–538.

Huxley, H. E. (1979). In *Cross-bridge Mechanism in Muscle Contraction* (Sugi, H. and Pollack, G. H., eds.), pp. 391–400, Univ. Tokyo Press, Tokyo.

Kumagai, H., Ebashi, S., and Takeda, F. (1955). *Nature* **176**, 166.

Lymn, R. W. (1978). *Biophys. J.* **21**, 93–98.

Natori, R. (1954). *Jikeikai Med. J.* **1**, 119–126.

Podolsky, R. J., Gulati, J., and Nolan, A. C. (1974). *Proc. Natl. Acad. Sci. U.S.* **71**, 1516–1519.

Podolsky, R. J. and Nolan, A. C. (1971). In *Contractility of Muscle Cells* (Podolsky, R. J., ed.), pp. 247–260, Prentice-Hall, Englewood Cliffs.

Podolsky, R. J., St. Onge, R., Yu, L., and Lymn, R. W. (1976). *Proc. Natl. Acad. Sci. U.S.* **73**, 813–817.

Podolsky, R. J. and Teichholz, L. E. (1970). *J. Physiol.* **211**, 19–35.

Rizzino, A. A., Barouch, W. W., Eisenberg, E., and Moos, C. (1970). *Biochemistry* **9**, 2402–2408.

Simmons, R. and Jewell, B. (1974). *Recent Adv. Physiol.* **9**, 87–147.

Stein, L. A., Schwarz, R. P., Chock, P. B., and Eisenberg, E. (1979). *Biochemistry* **18**, 3895–3909.

Thames, M. D., Teichholz, L. E., and Podolsky, R. J. (1974). *J. Gen. Physiol.* **63**, 509–530.

Weber, A., Herz, R., and Reiss, J. (1963). *J. Gen. Physiol.* **46**, 679.

Yu, L. C., Hartt, J. E., and Podolsky, R. J. (1979). *J. Mol. Biol.* **132**, 53–67.

Muscle Contraction: Its Regulatory Mechanisms (*Ebashi, S. et al., eds.*), *pp. 79–97, Japan Sci. Soc. Press, Tokyo / Springer-Verlag, Berlin* (*1980*)

7

Three-dimensional Image Analysis of the Complex of Thin Filaments and Myosin Molecules Which Form Skeletal Muscle

TAKEYUKI WAKABAYASHI

Summary: A three-dimensional image of the "rigor" complex of actin and myosin subfragment 1 (S-1) was reconstituted.

The S-1 molecules were attached to the actin filament in a configuration in which they were tilted by only 15° from the plane normal to the helix axis.

It was also shown that morphological unit of actin-S-1 complex consists of five domains. The possible ways to assign actain to one of these domains are discussed.

In skeletal muscle fibres, actin-containing thin filaments and myosin-containing thick ones are arrayed interdigitatingly (H. E. Huxley, 1953), and the relative sliding of two kind of filaments (A. F. Huxley and Niedergerke, 1954; H. E. Huxley and Hanson, 1954) induced by the interaction of actin and myosin in the presence of ATP is the basic step in muscle contraction (Szent-Györgyi, 1951).

Muscle contraction is regulated by the intracellular concentration of calcium ions, of which the target protein is troponin (Ebashi and Kodama,

1963). In the absence of calcium ions, troponin and tropomyosin inhibit the interaction between actin and myosin heads. Activation of muscle contraction is triggered by the binding of calcium ions, as they are released from the sarcoplasmic reticulum, which was first isolated as the relaxing factor (March, 1951; Kumagai *et al.*, 1955), to troponin, which then removes the inhibition (Ebashi and Endo, 1968).

Since there is only one troponin molecule for every seven actin monomers (Ohtsuki and Wakabayashi, 1972; Spudich *et al.*, 1972; Bremel and Weber, 1972), and since actin interacts more strongly with tropomyosin (Laki *et al.*, 1962) than with troponin, it has been presumed that the change in the physical state of troponin induced by calcium ions (Wakabayashi and Ebashi, 1968) is transmitted to actin *via* tropomyosin (Ebashi *et al.*, 1969).

It is important to clarify the structure of thin filaments in active or inhibited states and also the structure of the thin filament which is interacting with part of the myosin molecule (*e.g.* myosin subframent 1 (S-1) or heavy meromyosin), in order for such structural information to be helpful in elucidating the regulatory as well as basic mechanism of muscle contraction on the molecular level.

Earlier structural studies by X-ray diffraction showed that the gross structure of the thin fialments and the structural parameters of the actin helix were rather insensitive to the state of the muscle (Elliott *et al.*, 1967; H. E. Huxley and Brown, 1967).

It was shown, however, that the second "actin" layer line can be observed only in the X-ray diffraction pattern from contracting frog muscle (H. E. Huxley, 1971a, b). Similar changes were observed in *Taenia coli* of the guinea pig (Vibert *et al.*, 1972a, b) *i.e.*, the intensity of the second "actin" layer line increases and that of the third one decreases when the muscle contracts. These changes were interpreted as being the consequence of the shift of the tropomyosin in the grooves of the actin helix (H. E. Huxley 1972; Haselgrove, 1972; Parry and Squire, 1973). They also related the shift of tropomyosin with the regulation of muscle contraction; *i.e.*, the tropomyosin in the shifted position would sterically block the stable binding of the myosin heads, whose binding mode to actin was described by Moore *et al.* (1970). When they related the position of the shifted tropomyoisn and that of bound S-1, it was necessary to assume that the shifted tropomyosin is situated at the right side of the actin groove when one looks at the thin filament in such a way that the Z band is at the bottom and the tip of the thin filament points upwards.

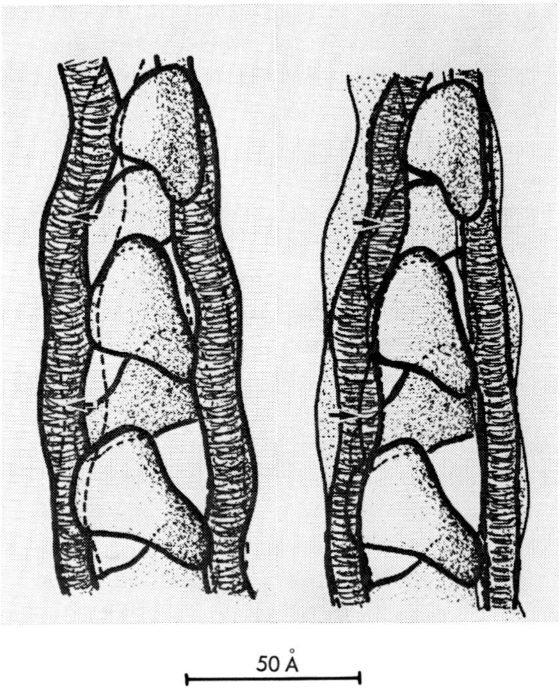

Fig. 1. Schematic presentation of the three-dimensional structure of actin-tropomyosin (left) and actin-tropomyosin-troponin (T+I) (right) based on Wakabayashi *et al.* (1975). The 10 Å shift of tropomyosin is indicated by arrows.

By reconstructing the three-dimensional structure of actin-tropomyosin which represents the thin filaments in the active state and that of actin-tropomyosin-troponin (T+I) which represents the inhibited filaments, Wakabayashi *et al.* (1975) obtained detailed and more direct structural evidence for the mode of the interaction between tropomyosin and actin. They showed that one tropomyosin molecule interacts loosely with 14 actin molecules in active thin filaments; in inhibited ones the interaction between tropomyosin and actin becomes conspicuously tighter, and one tropomyosin binds only 7 actin molecules which belong to the homo-strand. Wakabayashi *et al.* also showed that the tropomyosin shifts to the homostrand actin by 10 Å in the inhibited state (Fig. 1), supporting the interpretation of the change in the intensities of "actin" layer lines by H. E. Huxley (1972), Haselgrove (1972), and Parry and Squire (1973). Since the time when this image analysis of thin filaments in active and inhibited states was done, an effort has been made to correlate the position of the shifted tropomyosin and that of the S-1 bound to actin by analyzing the image of the natural thin filaments which are still bound to the Z bands. However, unambiguous determination is not yet possible, since the quality of our electron micrographs of the thin filaments in I-segments or I-z-I brushes is limited. Therefore, it is not possible, for the moment, to prove or disprove the assumption about the correlation of the position of tropomyosin and that of S-1 bound to actin.

We are reconstituting, therefore, the three-dimensional image of the complex consisting of thin filaments and parts of myosin molecules, *e.g.* S-1, since it is the more direct way to determine the relative positions of actin, tropomyosin, and S-1. Before going into the complicated system which contains actin, tropomyosin, troponin, and S-1, we analyzed the simplest system, actin-S-1 complex. The image of actin-S-1 on carbon film is less contrasty and specimens are more severely flattened in comparison with images taken over the holes of microgrids. Usually stain sheets over holes break when they are illuminated by an electron beam, and the specimens embedded in them are distorted anisotropically. Sometimes, however, it was possible to prepare microgrids with holes over which the stain sheets do not break.

The images of the actin-S-1 photographed over holes were thinner in diameter, and their optical diffraction patterns showed more symmetrical layer lines across the meridian. The computed Fourier transforms showed extremely good correlation in phase angles across the meridian, and it was possible to reconstitute the three-dimensional image with effective resolution of 20 Å radially and 26 Å axially.

The three-dimensional model of actin-S-1 thus reconstituted revealed that the mode of morphological interaction between actin and S-1 is different from that described by Moore *et al.* (1970). It was also shown that the angle between the main axis of the main part of S-1 and the plane perpendicular to the helix axis of the thin filament (tilt angle) is about 15°, which is far smaller than that described previously (Reedy *et al.*, 1965; Moore *et al.*, 1970).

Therefore, the type of steric block mechanism which has been proposed for the regulatory mechanism has to be reexamined in the light of the improved three-dimensional image of the actin-S-1 (Toyoshima and Wakabayashi, 1979). The conventional type of tilting mechanism of the movement of myosin heads also needs to be modified to accommodate the non-tilting angle of the major axis of the main part of S-1. Possible alternatives for the mode of movement of the myosin heads are discussed, and the reason why the tilt angle of the major axis of S-1 has been described to be about 45° is explained.

MATERIALS AND METHODS

Actin was prepared from acetone powder of rabbit skeletal muscle according to the procedure described by Spudich and Watt (1971). Myosin was prepared from rabbit back muscle essentially according to the method of Szent-Györgyi (1951), and was stored in 50% glycerol (Merck) containing 0.3 M KCl and 0.15 mM $NaHCO_3$. Chymotryptic S-1 (Yagi and Ohtani, 1974) was prepared and purified by the procedure described by Weeds and Taylor (1975). Though S-1 could be separated into two fractions, their mixture was used as the S-1 preparation. The rigor complex of actin and myosin was formed by mixing S-1 and F-actin in 50 mM KCl, 10 mM potassium phosphate buffer (pH 7.6) 0.01 mM $MgCl_2$, and 0.01 mM ATP, and leaving the mixture overnight in the cold room. The specimens were negatively stained with 1% uranyl acetate or 0.5% uranyl oxalate (pH 7, adjusted by adding 1 M Tris) over a carbon coated perforated microgrid (Fukami and Adachi, 1965) at room temperature and examined in a JEM-100B electron microscope with an accelerating voltage of 80 kV using an anticontamination device and cold trap. The calibration of magnification was done by measuring the 395 Å periodicity in magnessium tactoids of tropomyosin (Caspar *et al.*, 1969).

Three-dimensional image reconstruction was done following the procedure described by Shirakihara and Wakabayashi (1979) according to the method established by DeRosier and Moore (1970). Each image that gave a clear transform was digitized using a computer-controlled microdensitometer (Wakabayashi and Nagasawa, unpublished work), and its Fourier transform and three-dimensional density were computed on the HITAC 8700/8800 at the Computer Centre of University of Tokyo. The density map of the projection of the three-dimensional image was plotted with a dot printer controlled by PDP-11/03.

To average the reconsituted images from different images, a fitting procedure was carried out as described previously (Wakabayashi *et al.*, 1975; Amos and Klug, 1975).

The horizontal sections across the helix axis (z axis) and the vertical sections which are parallel to the helix axis and normal to the x axis were computed. Equidensity lines with four levels of density were drawn.

Contour maps were copied onto the acetate film which were then attached to a Perspex plate (1 mm thick) and were stacked using spacers of appropriate thickness. This transparent model was built in order to examine and interpret the three-dimensional image. The solid model was made for easier inspection and presentation of the structure after choosing the optimal cut-off level to represent the main features which are recognized during the time when the transparent models were examined.

RESULTS

Several hundred electron micrographs of the "rigor" complex of actin-S-1 were collected. The image of the "rigor" complex is straight and contrasty when the specimen is embedded in the *unbroken* stain sheets over holes in the microgrids. Examples of such photographs are presented in Fig. 2 a, b, c. Though arrowhead appearance is evident, the polarity of the structure is less distinct than that of acto-heavy meromyosin. Figure 2 d shows an optical diffraction pattern of the filament shown in Fig. 2b. The layer lines of this particular optical transform could be indexed with the selection rule

$$l = 7n + 13m\,,$$

and the true pitch was about 360 Å. The first, fourth, fifth, and sixth layer lines were strong, but the seventh and eighth lines were weak. Though not reproduced in Fig. 2 d, the meridional reflection on the 13th layer line could be observed in the optical and computed transform.

It should be pointed out that the intensity distribution of the optical transforms of actin-S-1 is similar to that of the X-ray diffraction patterns from crustacean muscle in rigor state (Wray *et al.*, 1978). This probably means that the preservation of the structure of actin-S-1 was good.

Though the intensity distribution of the optical transforms was constant from filament to filament, the exact selection rule with which the layer lines are indexed was found to be variable. For determining the

→ Fig. 2. a, b, c: Electron micrographs of the complex of actin and myosin S-1 negatively stained with 1% uranyl acetate. d: optical transform of the micrograph shown in b. Two copies of the original optical transform were superimposed after one of them was turned upside-down to form a composite negative, which was then printed on paper. This was done to average the intensities of the diffraction spots which are related by Friedel's law. e: a density plot of the projection of the three-dimensional image, a solid model of which is shown in Fig. 3. The projection was done to show that the model accounts for the arrowhead pattern seen in the micrograph presented in a, b, and c.

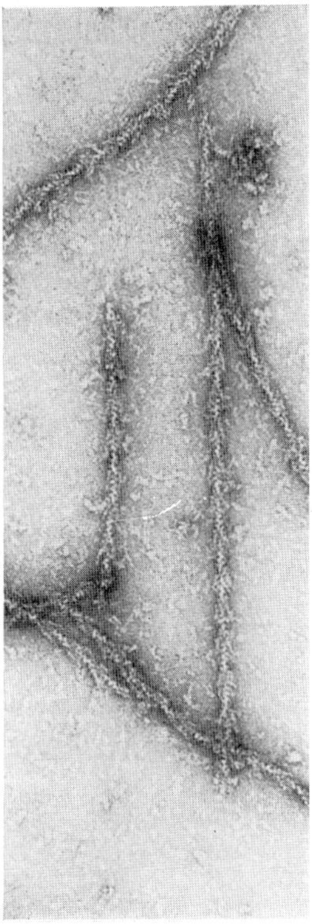

a

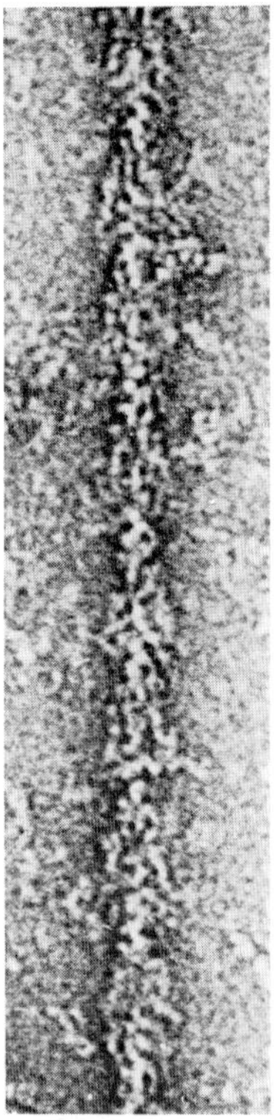

b

c

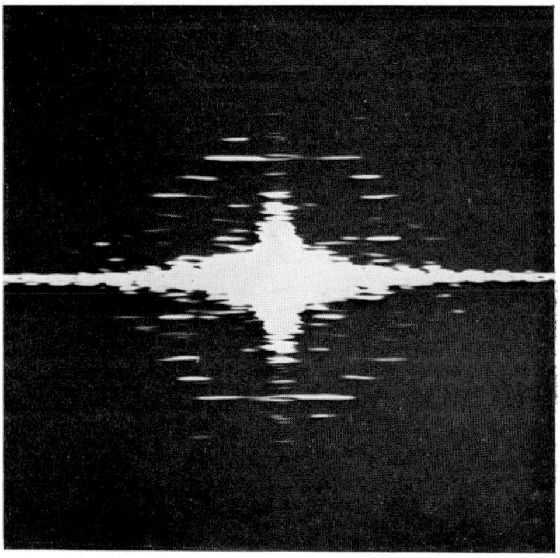

d

e

selection rule it was important to measure the spacing of the "first" and "sixth" layer lines. We obtained the former by measuring the distance between the crossover points in a real space, since the position of the "first" layer line is near the equator and relative error of measurement is large in Fourier space. The majority of transforms could be indexed with the conventional selection rule of $l=7n+13m$, but the rest of them are indexed with either $l=22n+41m$, $l=15n+28m$, or $l=8n+15m$. It should be pointed out that these latter selection rules can be obtained when the helix which obeys the conventional selection rule is untwisted by a small amount of angle.

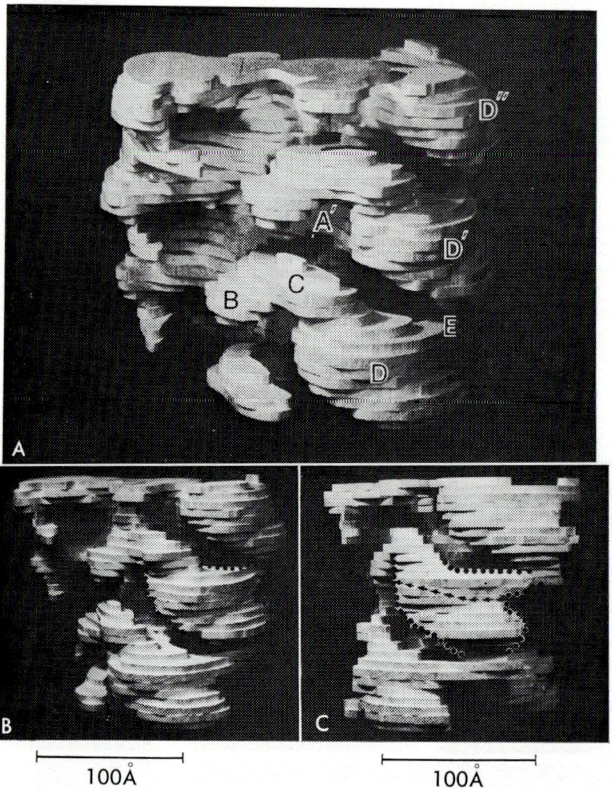

Fig. 3. Oblique views (A, B) and a side view (C) of a model of a portion of the actin-S-1 complex reconstituted from two filaments including the one shown in Fig. 2 b. The outline of the main part of S-1 is enclosed by a dotted line and the major axes are indicated by arrows in C. The sections of balsa wood used to build the model are 5.4 Å thick.

On the first layer line there was sometimes a meridional spot, which is not allowed by any of the selection rules described above. This meridional spot, which was weaker in the optical transform of the filament in the side-by-side aggregates and may originate from the periodic variation of thickness of the staining reagent, was not included in the data to reconstitute the three-dimensional image.

Transparent models of the three-dimensional images reconstituted both from individual filaments and from the averaged data were examined, and the solid model was made after the appropriate cut-off level was determined.

Figure 3 shows oblique and side views of a model of a portion of the three-dimensional image reconstituted from the averaged data from two filaments of actin-S-1, including the one shown in Fig. 2b. The resolution of the averaged data went out to 20 Å radially and 26 Å axially.

Figure 2e shows the density plot (protein is white to simulate the print of the electron micrograph) of the projection of the three-dimensional image, of which a solid model is shown in Fig. 3.

We projected a three-dimensional image onto a plane to show that the model accounts for the arrowhead pattern seen in the original electron micrograph.

Using the three-dimensional image of the actin-tropomyosin complex reconstituted by Wakabayashi *et al.* (1975) as a guide, the location of the actin and the S-1 was examined. However, unambiguous identification of actin and S-1 at radii smaller than 30 Å was difficult, since the model of decorated filament was rather complicated at these radii. Therefore, the size and shape of whole S-1 could not be determined uniquely at present time. Yet the shape of the S-1 at higher radii can be identified unambiguously, since actin molecule does not exist at these radii. The outline of the identified part of S-1 is enclosed by a dotted line in Fig. 2 b, c. We call this region the "main" part of S-1, since it is the highest in integrated density and the largest in size.

The main part of S-1 appears curved and swallow-tailed. We define tilt angle as the angle between the major axis of the main part of S-1 and the horizontal plane perpendicular to the helix axis, and the slew angle as the angle between the plane which contains both helix axis and the radius from the helix axis to the proximal end of the major axis and the vertical plane (P_1 or P_2) which contains major axis (Fig. 4). We found that the "main" part of S-1 is about 80 Å in length and can be represented by two major axes (u_1, u_2 in Fig. 4). The slew angles of u_1 and u_2 were about

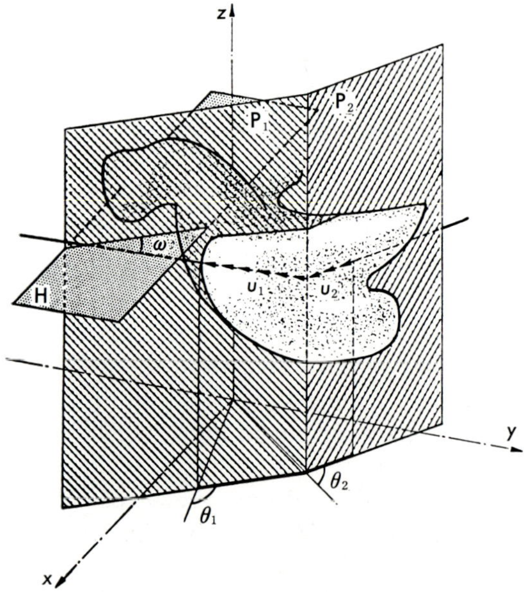

Fig. 4. Diagrammatic representation of an oblique view of the "main" part of myosin S-1 corresponding to Fig. 3B. We define the tilt angle (ω) as the angle between the major axis of S-1 and the horizontal plane (H) normal to the helix axis (z), and the slew angle (θ) as the angle between the plane which contains the helix axis and the radius from the helix axis to the proximal end of the major axis and the vertical plane (P_1, P_2) containing the major axis (u_1 or u_2, respectively).

120° (θ_1) and 105° (θ_2) respectively. The tilt angles of u_1 and u_2 were about 15° (ω_1) and 0° (ω_2) respectively. This result means that S-1 is attached to the actin filament approximately perpendicular to the helix axis, or that S-1 binds actin filament in non-tilted configuration.

It should be pointed out that S-1 molecules need not be tilted to give rise to the arrowhead pattern and that the arrowhead appearance arises as a result of the slewing and curvature of the S-1 molecules, as discussed by Moore *et al.* (1970).

The three-dimensional image of the actin-S-1 complex is complicated at radii smaller than 35 Å, as described above, and it is difficult to examine both transparent and solid models. We *untwisted* the transparent model so that the ($n=2$, $l=1$) helices (convention according to Klug *et al.*, 1958) with the spacing of about 360 Å become parallel to the axis of the helical symmetry. After this operation, the right-handed ($n=2$,

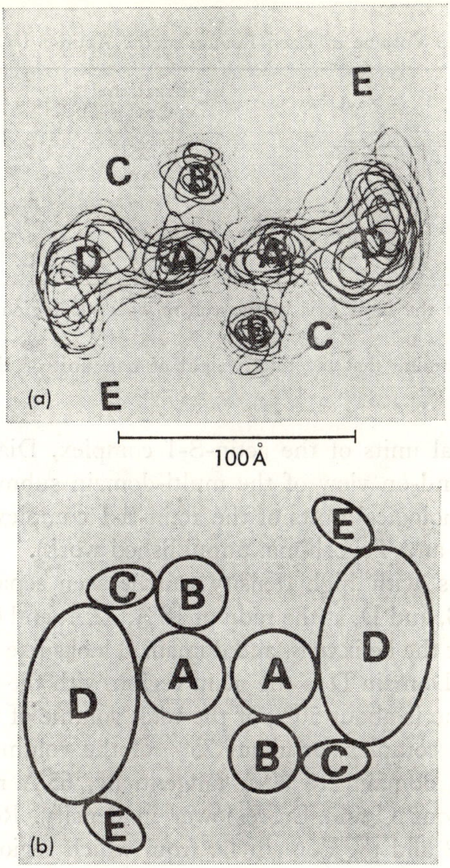

Fig. 5. An end-on view (a) and its schematic diagram showing the multi-domain structure of the *untwisted* transparent model of a portion of the three-dimensional image (b). The depth of the horizontal slices shown in (a) is about 60 Å in the axial direction. The three-dimensional model was *untwisted* so that double strands of actin helix become parallel to the helix axis.

$l=1$) helices become the ($n=2$, $l=0$) helices which are straight lines and the ($n=1$, $l=7$) helices become the ($n=1$, $l=6.5$) helices. This means that the true pitch of the structure becomes twice as that of the original structure.

Figure 5a shows the end-on view of the *untwisted* transparent model of a portion of the three-dimensional image of the actin-S-1 complex. The thickness of the model is about 60 Å, which covers approximately

TABLE I
The Position and the Volume of Each Region of the Actin-S-1 Complex

Domain	r (Å)[a]	Fractional volume (%)	Weight (dalton)[b]
A	18	25	39,000
B	32	12	19,000
C	50	7	11,000
D	62	53	84,000
E	75	3	4,000

[a] r: the radius from the helix axis to the centre of each region which is shown in Fig. 5b.

[b] Calculated by assuming that the total weight of one unit of the actin-S-1 complex is 157,000 daltons.

two morphological units of the actin-S-1 complex. Diagrammatic representation of an end-on view of the multi-domain submolecular structure of the two morphological units of the actin-S-1 complex is shown in Fig. 5b (Wakabayashi and Toyoshima, unpublished work).

Three domains with high density can be seen separately in Fig. 5a: *i.e.*, domains A, B, and D, at the radii of 18 Å, 32 Å, and 62 Å, respectively. Domain A is near the helix *axis*, and domain B looks like a *bulge* protruding from domain A. Domain D is the main region with the highest integrated *density;* its volume is about 50% of the total volume of the morphological unit of actin-S-1 complex, or about 75% of the volume of S-1. The dimensions of this domain are 80 Å tangentially, 65 Å radially, and 60 Å axially. The regions C and E are lower in density. Region C seems to *connect* domain B and the domain D, from which region E *extends*. The radius, volume, and calculated weight of each of these five domains are listed in Table I. Figure 3A shows the identification of each of the five domains in the solid model which is not untwisted.

It is not surprising that the morphological unit of actin-S-1, whose weight is about 157,000 daltons, shows the multi-domain structure, since the multi-domain submolecular structure can be observed in the three-dimensional image of a flagellin molecule (mol.wt. about 40,000) which is the morphological unit of bacterial flagella (Shirakihara and Wakabayashi, 1979). Total volume per unit of the solid model shown in Fig. 3 is about 179,000 Å^3, which is about 95% of the volume calculated by assuming that the partial specific volume is 0.73 (Lowey *et al.*, 1969) and that the molecular weight of S-1 and actin are 115,000 (Lowey *et al.*, 1969) and 41,785 (Elzinga *et al.*, 1973) respectively.

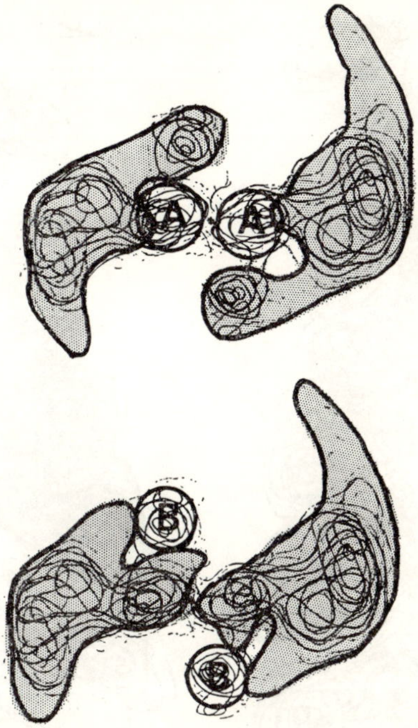

Fig. 6. An end-on view of the three-dimensional model showing the candidates for the S-1 as shaded area in the type A assignment (upper) and the type B assignment (lower). Note that the morphological contacts occur at multiple sites in either type of assignment.

Since domains C, D, and E exist at higher radii than 35 Å, one cannot assign actin to these domains. Indeed, the "main" part of S-1 described above is equivalent to domains D–E. Therefore, we are obliged to assign actin either to domain A (type A assignment) or to domain B (type B assignment). The S-1 molecule is represented by the domains B–C–D–E in the type A assignment and by the domains A–C–D–E in the latter assignment.

Figure 6 and 7 show the end-on and frontal views of S-1 in type A (Figs. 6 (upper) and 7 (left)) and type B (Figs. 6 (lower) and 7 (right)) assignment. It should be pointed out that Fig. 7 is the view along the major axes of S-1 rather than the side view of the outline of S-1. The side view of S-1 can be seen at the region near the crossover point where many contour lines overlap. It is interesting to note that in both types of assignment the morphological contact between actin and S-1 is multi-sited; *i.e.*, in type A assignment, actin and S-1 seem to interact A–B and

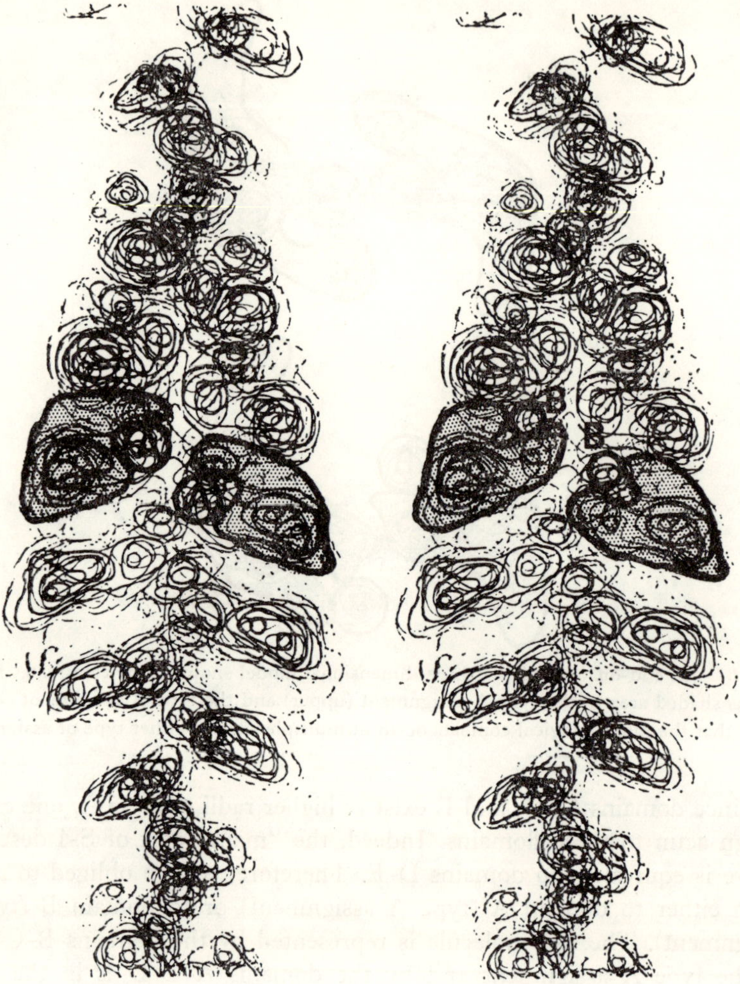

Fig. 7. A side view of the three-dimensional model showing the candidates for the S-1 as shaded area in the type A assignment (left) and the type B assignment (right). Note that the morphological contacts occur at multiple sites in either type of assignment. It should be pointed out that the shaded area is the frontal rather than the side view of S-1 and corresponds to the S-1 which situates higher than the D″ domain shown in Fig. 3A.

A–D interfaces and in type B assignment they seem to interact at A–B and B–C interfaces. It is also a constant feature of the end-on view of S-1 that it is two-pronged or y-shaped (especially when viewed from the Z band side or the bottom of the illustration shown in Fig. 3).

DISCUSSION

1. *The Shift of Tropomyosin*

Before the shift of tropomyosin was shown directly by three-dimensional image reconstruction of the active and inhibited thin filaments (Wakabayashi *et al.*, 1975), it was proposed by H. E. Huxley (1971a, b, 1972), Haselgrove (1972), and Parry and Squire (1973). They proposed the shift of tropomyosin on the basis of the changes in intensity of the "actin" layer lines of the X-ray diffraction pattern when the resting muscle goes into the rigor state. When they interpreted the change in intensity, they presumed that the change is caused solely by the shift of tropomyosin.

However, it is clear from the three-dimensional image shown in Fig. 5a that the intensity of the second "actin" layer line can be increased not only by the shift of tropomyosin but also by the attachment of S-1 heads to actin filaments. This is independent of the type of assignment of actin (Fig. 6), for the relative azimuthal angle between domain A and domain B is about 90°. Since the amount of the mass of S-1 which occupies the actin "grooves" is much more than that of tropomyosin, one should be extremely careful when one deduces the position of tropomyosin from the relative intensities of the "actin" layer lines in X-ray diffraction pattern of non-resting muscle.

The X-ray diffraction pattern from the crustacean muscle in rigor state showed the far stronger second "actin" layer line (Wray *et al.*, 1978) than that calculated on the basis of the tropomyosin shift only.

Vibert *et al.* (1972 a, b) carefully examined the possibility of the contribution of the attachment of S-1 to the second layer line by taking the rigor pattern of the frog semitendinosus muscle which had been strectched to a sarcomere length above 3.7 μm, so that the thin and thick filaments cannot overlap in theory. They observed that the second layer line is still there and has the same radial position as that in the pattern from a rest length rigor muscle, and excluded the possibility of the contribution of the S-1 attachment in frog striated muscles. However, as this type of experiment is not possible using smooth vertebrate muscle (*Taenia coli*) or *Mytilus* (ABRM), it is not possible to exclude the possibility of cross-bridge attachment.

2. *Polarity of Actin Filaments*

Wakabayashi *et al.* (1975) showed that the actin molecule in the actin-tropomyosin complex and the actin-tropomyosin-troponin (T+I) looks like a "chick" whose side view is triangular and is polar in structure. Though we have been making an effort to determine the direction of the actin molecule by analyzing the thin filament in the I-z-I brushes, it is not yet determined whether the actin "triangle" points to the tip of the thin filament or in the Z-band direction.

In addition to the image analysis of actin-S-1 described above, we are now finishing the three-dimensional image reconstruction of actin-tropomyosin-S-1. The resolution of the latter structure is better than the former. If we can reach the resolution of 15 Å, unambiguous identification of actin in actin-S-1 complex, determination of the direction of the actin "triangle" and correlation of the positions of tropomyosin and S-1 will become possible.

3. Tilt Angle of Myosin Subfragment-1 in the Rigor Complex

Reedy *et al.* (1965) reported that the angle between the "cross-bridge" and the actin filament is about 45° in the rigor state but about 90° (or non-tilted) in relaxed muscle. They obtained these values by analyzing some excellent electron micrographs of thin section samples. However, it should be noted that the whole S-1 is not a simple rod, and there could be other minor axes which connect the "main" part and the other parts of S-1 at smaller radii and that the "main" part is wedge-shaped when it is viewed along the major axis. The centre line through the wedge was tilted about 25°. It is difficult to determine the tilt angle of S-1 in micrographs which are two-dimensional projections of three-dimensional speciments. Therefore, the tilt angle of the "cross-bridge" and that of the S-1 major axis are not necessarily equal. Moore *et al.* (1970) also reported that the S-1 is tilted by 40°. But the boundaries between the S-1 molecules are not clear in their three-dimensional model of the decorated actin filaments; and in addition their original electron micrograph was distorted anisotropically, and the exact angle could not be obtained.

It did not escape our attention that the axis which connects domain C and domain D is tilted about 40°. This angle is very similar to that obtained by Moore *et al.* (1970). However, it should be pointed out that domain C is rather low in density and its radial dimension in a solid model is as thin as 10 Å, and that domain C is not so bulky as it looks. Therefore, the axis connecting domain C and domain D cannot represent the tilt angle of S-1.

There are at least two ways to reconcile the non-tilted S-1 configuration in the rigor complex obtained by us with the sliding theory. First, one can still retain the tilting mechanism by assuming the tilt angle to be about −40°, for instance, when myosin binds actin. The second possibility is to introduce a pivoting mechanism. In this mechanism myosin heads bind actin at a non-tilted angle and rotate so that the slew angle increases. It is interesting to note that this pivoting mechanism can also explain the change in the intensity ratio of two major equatorial X-ray diffraction spots induced by activation of the muscle. A similar model has also been proposed by Lymn (1978).

4. Ca-sensitivity of Actin-subfragment-1 Binding in the Absence of ATP

It is well known that the interaction of regulated thin filament and myosin becomes less sensitive to calcium when the concentration of ATP is decreased. This suggests that the actin-S-1 interaction in the absence of ATP is insensitive to calcium concentration. Therefore one can argue that the rigor link should be insensitive to the regulation by the calcium ion and that it may not be appropriate to compare the S-1-binding site(s) of actin in rigor complex and the binding site of the shifted tropomyosin. Though it is possible that the stability of rigor complex decreases at the lower concentration of calcium, it is more plausible that the interaction between the regulated in filament and nucleotide-loaded myosin is much more sensitive to the calcium ion.

This means that it is important to solve the structure of the complex of thin filament and nucleotide-loaded myosin in order to elucidate the molecular mechanism of the regulation of muscle contraction. Since myosin is probably loaded with nucleotide when the initial complex of actin and myosin is formed, it is important to solve this structure to clarify the basic mechanism of muscle contraction.

5. Arrowhead Appearance of Actin-subfragment-1 Complex

It was an unexpected feature of the three-dimensional image of actin-S-1 that the major axis of the "main" part of S-1 is not tilted even though the original micrographs show a polar arrowhead pattern. The three-dimensional image and its end-on view show that outline of S-1 can be approximated by a half turn of the spiral. When this horizontal spiral is projected along the line perpendicular to the helix axis, the part of the spiral which is tangential to the projection line and therefore gives the densest spots in projected image varies when the direction of the projection varies. Since S-1 is approximated by about a *half* turn of the spiral, the radius to the densest spot increases gradually from zero to the maximum and then abruptly decreases to zero. This abrupt decrement of the radius is the main factor in producing the arrowhead pattern; S-1 itself need not be tilted.

Acknowledgments

I would like to thank Professor Ebashi for his interest and continuous encouragement. I am thankful to Mr. C. Toyoshima who collaborated with me. The constructive criticism and helpful suggestions of Dr. E. Katayama and the advice of Dr. Y. Shirakihara are acknowledged. I also thank Ms. I. Masuo for her expert technical assistance and Mr. M. Suzuki, who helped to build the balsa wood model.

This work was supported in part by a grant from the Ministry of Education, Science and Culture.

REFERENCES

Amos, L. A. and Klug, A. (1975). *J. Mol. Biol.* **99**, 51–64.
Bremel, R. D. and Weber, A. (1972). *Nature New Biol.* **238**, 97–101.
Caspar, D.L.D., Cohen, C., and Longley, W. (1969). *J. Mol. Biol.* **41**, 87–107.
DeRosier, D. and Moore, P. B. (1970). *J. Mol. Biol.* **52**, 355–369.
Ebashi, S. and Kodama, A. (1965). *J. Biochem.* **58**, 107–108.
Ebashi, S. and Endo, M. (1968). *Prog. Biophys. Mol. Biol.* **18**, 123–183.
Ebashi, S. Endo, M., and Ohtsuki, I. (1969). *Q. Rev. Biophys.* **2**, 351–384.
Elliott, G. F., Lowy, J., and Millman, B. F. (1967). *J. Mol. Biol.* **25**, 31–45.
Elzinga, M., Collins, J. H., Kuehl, W. M., and Adelstein, S. (1973). *Proc. Natl. Acad. Sci. U.S.* **70**, 2687–2691.
Fukami, A. and Adachi, K. (1965). *J. Electr. Microsc.* **14**, 112–118.
Haselgrove, J. (1972). *Cold Spring Harbor Symp. Quant. Biol.* **37**, 341–352.
Huxley, A. F. and Niedergerke, R. (1954). *Nature* **173**, 971–973.
Huxley, H. E. (1953). *Biochim. Biophys. Acta* **12**, 387–394.
Huxley, H. E. and Hanson, J. (1954). *Nature* **173**, 973–976.
Huxley, H. E. and Brown, W. (1967). *J. Mol. Biol.* **30**, 383–434.
Huxley, H. E. (1971a). *Am. Biophys.* abstr. p. 235.
Huxley, H. E. (1971b). *Biochem. J.* **125**, 85 p.
Huxley, H. E. (1972). *Cold Spring Harbor Symp. Quant. Biol.* **37**, 361–376.
Klug, A., Crick, F.H.C., and Wyckoff, H. W. (1958). *Acta Crystallogr.* **11**, 199–213.
Kumagai, H., Ebashi, S., and Takeda, F. (1955). *Nature* **176**, 116.
Laki, K., Maruyama, K., and Kominz, D.R. (1962). *Arch. Biochem. Biophys.* **98**, 323–330.
Lowey, S., Slayter, H. S., Weeds, A. G., and Baker, H. (1969). *J. Mol. Biol.* **42**, 1–20.
Lymn, R. W. (1978). *Biophys. J.* **21**, 93-98.
Marsh, B. B. (1951). *Nature* **167**, 1065.
Moore, P. B., Huxley, H. E., and DeRosier, D. J. (1970). *J. Mol. Biol.* **50**, 279–295.

Ohtsuki, I. and Wakabayashi, T. (1972). *J. Biochem.* **72**, 369–377.

Parry, D.A.D. and Squire, J. M. (1973). *J. Mol. Biol.* **75**, 33–55.

Reedy, M. K., Holmes, K. C., and Tregear, R. T. (1965). *Nature* **207**, 1276–1280.

Shirakihara, Y. and Wakabayashi, T. (1979). *J. Mol. Biol.* **131**, 485–507.

Spudich, J. A. and Watt, S. (1971). *J. Biol. Chem.* **246**, 4866–4871.

Spudich, J. A., Huxley, H. E., and Finch, J. T. (1972). *J. Mol. Biol.* **72**, 619–632.

Szent-Györgyi, A. (1951). *Chemistry of Muscular Contraction*, 2nd ed., Academic Press, New York.

Toyoshima, C. and Wakabayashi, T. (1979). *J. Biochem.* **86**, 1887–1890.

Vibert, P. J., Haselgrove, J. C., Lowy, J., and Poulsen, F. R. (1972a). *Nature New Biol.* **71**, 757–767.

Vibert, P. J., Haselgrove, J. C., Lowy, J., and Poulsen, F. R. (1972b). *J. Mol. Biol.* **71**, 757–767.

Wakabayashi, T. and Ebashi, S. (1968). *J. Biochem.* **64**, 731–732.

Wakabayashi, T., Huxley, H. E., Amos, L. A., and Klug, A. (1975). *J. Mol. Biol.* **93**, 477–497.

Weeds, A. G. and Taylor, R. S. (1975). *Nature* **257**, 54–56.

Wray, J., Vibert, P., and Cohen, C. (1978). *J. Mol. Biol.* **124**, 501–521.

Yagi, K. and Ohtani, F. (1974). *J. Biochem.* **76**, 365–373.

Muscle Contraction: Its Regulatory Mechanisms (*Ebashi, S. et al., eds.*), *pp. 99–114, Japan Sci. Soc. Press, Tokyo / Springer-Verlag, Berlin* (*1980*

8

Relationship of Cross-bridge Movement to ATP Splitting in Glycerol-treated Muscle Fibers

TOSHIAKI ARATA AND YUJI TONOMURA

Summary: The reaction mechanism of acto-H-meromyosin (acto-HMM) or acto-subfragment 1 (acto-S-1) ATPase in solution was analyzed by measuring its activity in the steady state (Δv_0), the rate of recombination of the reaction intermediate myosin-phosphate-ADP complex (M_P^{ADP}) with F-actin (v_{recomb}), and the extent of binding of a myosin head with F-actin in the steady state ($1-\alpha$). In the presence of a relatively high concentration of KCl and at low temperature, the rate of the acto-HMM or acto-S-1 ATPase reaction was limited by the recombination steps of M_P^{ADP} with F-actin, and almost all myosin heads dissociated from F-actin. On the other hand, in the presence of a low concentration of KCl and at room temperature, the steady-state rate of the acto-HMM or acto-S-1 ATPase reaction was much higher than the rate of recombination of M_P^{ADP} with F-actin, and was proportional to the extent of binding of myosin heads with F-actin. In general, Δv_0 was given by $(1-\alpha)\cdot\Delta V_0+\alpha\cdot v_{recomb}$, where ΔV_0 is Δv_0 at an infinitively high concentration of F-actin. Therefore, it was concluded that the two cycles for ATP hydrolysis, *i.e.*, the dissociation and recombination cycle of the actin-myosin complex induced by ATP and the cycle *via* direct decomposition of an undissociated complex of M_P^{ADP} with F-actin (AM_P^{ADP}), coexist in solution. It is reasonable to assume that the cycle *via* direct decomposition of AM_P^{ADP} induces the rotation of myosin heads on the thin filament, which generates force. Then, to induce sliding of thick filaments past thin ones, the two cycles must occur alternately during the cyclic action of cross-bridges. In other words, the probability of dissociation of cross-bridges by ATP depends on their angular position, and we can expect that after quick release of a slow stretch of muscle fibers, the cross-bridges can rotate to a more acute angle then hydrolyze 1 mol of ATP per mol of myosin, accompanied by their rapid dissociation by ATP and subsequent recombination.

Glycerol-treated muscle fibers at rest length were stretched slowly by about 2%, released rapidly, then fixed at the rest length for α sec. When the length change cycle was repeated, the ATPase activity increased markedly, especially when α was about 2 sec at 3–4°C. From the dependence of the extent of acceleration on α, the time course of ADP liberation after the quick release of the slow stretch was calculated. After the quick release, about one mol of ADP per mol of myosin was rapidly liberated, then the ADP liberation was interrupted and finally its rate approached that of static fibers. All the results obtained were consistent with the assumption that the probability of dissociation of cross-bridges by ATP increases greatly after their rotation to a more acute angle.

Analyses of the elementary steps in the acto-H-meromyosin (acto-HMM) or acto-myosin subfragment 1 (acto-S-1) ATPase reaction in solution have advanced greatly in recent years. Two types of reaction mechanism have already been proposed. One is the Lymn-Taylor mechanism (Lymn and Taylor, 1971; Taylor, 1973, 1979) and the other is the two-route mechanism proposed by us (Tonomura, 1972; Tonomura and Inoue, 1975; Inoue *et al.*, 1973, 1979). In the Lymn-Taylor mechanism, the actomyosin ATPase reaction occurs *via* dissociation and recombination steps and its rate is limited by the recombination steps of the myosin-phosphate-ADP complex (M_P^{ADP}) with F-actin. In the two-route mechanism, two pathways coexist in solution: one is that proposed by Lymn and Taylor, and the other is the direct decomposition of an undissociated complex of M_P^{ADP} with F-actin (AM_P^{ADP}). In this article, we will first review our recent kinetic studies on acto-HMM or acto-S-1 ATPase in solution, which have shown clearly that the two-route mechanism operates in the actomyosin ATPase reaction in solution.

When applied to myofibrils, the Lymn-Taylor mechanism provides a very simple model in which force is generated by the rotation between two attached states, from AM_P^{ADP} to AM, following the dissociation and recombination of actin-myosin (Taylor, 1973, 1979). If we adopt the two-route mechanism, we must assume a flip-flop model to explain cross-bridge movement in which a myosin head alternately uses the two different routes for its movement on the thin filament (Tonomura, 1972; Tonomura and Inoue, 1975; Inoue *et al.*, 1979). In this model, the probability of dissociation of cross-bridges by ATP should depend on their angular position and increase markedly after the rolling of myosin heads to a more acute angle. Therefore, if this model is correct, we can expect that after quick release of a slow stretch of muscle fibers, the cross-bridges will rotate to a more acute angle then hydrolyze 1 mol of ATP per mol of myosin, accompanied by their rapid dissociation by ATP and subsequent recombination.

We slowly stretched the glycerol-treated muscle fibers at rest length by about 2%, released them rapidly, then fixed them at the rest length for α sec. When the length change cycle was repeated, the ATPase activity increased markedly, especially when α was about 2 sec at 3–4°C. From the dependence of the acceleration on α, we calculated the time course of ADP liberation after the quick release of the slow stretch, and found that about 1 mol of ADP per mol of myosin was rapidly liberated, then the ADP liberation was interrupted and finally its rate approached that of static fibers. All the results obtained were consistent with the assumption that the probability of dissociation of cross-bridges by ATP increases greatly by their rotation to a more acute angle.

ACTOMYOSIN ATPase IN SOLUTION

To clarify the relationship of ATP hydrolysis to cross-bridge movement in myofibrils, we must know how the actomyosin ATPase reaction proceeds in solution. It is well established that decomposition of the key intermediate of the myosin ATPase reaction, M_P^{ADP}, is accelerated by F-actin (Tonomura, 1972). The simplest mechanism for this acceleration has been proposed by Lymn and Taylor (1971). According to them (Lymn and Taylor, 1971; Taylor, 1973, 1979), the dissociation of actomyosin is induced very rapidly by ATP binding, and the rate of the actomyosin ATPase reaction is limited by the recombination steps of M_P^{ADP} with F-actin. Their mechanism can explain the ATPase reaction of acto-HMM or acto-S-1 only in the presence of a relatively high concentration of KCl

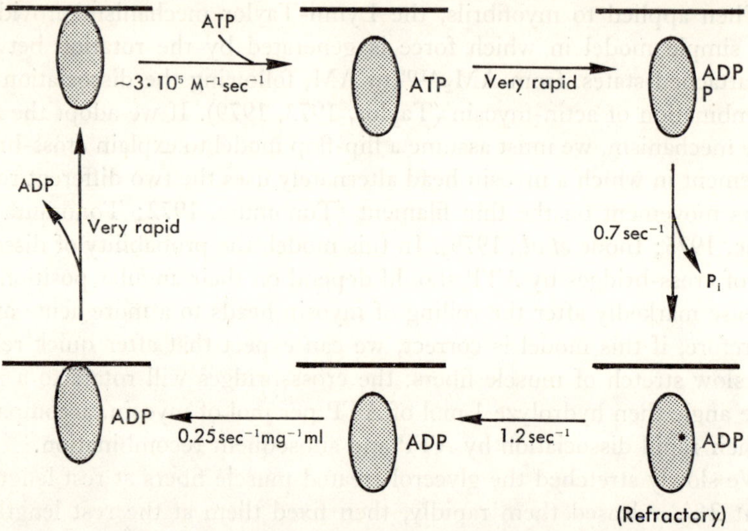

Fig. 1. Reaction sequence of Mn^{2+}-ATPase of acto-HMM in 0.1 M KCl at 5°C. The rate constants of various elementary steps, which were separately measured by Ikebe *et al.* (1980), are given in the figure, where the dissociation-recombination cycle of myosin head with F-actin is shown. The rates of the over-all reaction under various conditions are quantitatively expressed by the rate constants given here. In other words, under the conditions used in this experiment, the actomyosin ATPase reaction was satisfactorily explained by the Lymn-Taylor mechanism.

at low temperature. Under these conditions, almost all myosin heads dissociate from F-actin during the steady-state ATP hydrolysis. We (Ikebe *et al.*, 1980) have recently determined rate constants of various elementary steps in the Mn^{2+}-ATPase reaction of acto-HMM in 0.1 M KCl at 5°C. Over wide ranges of F-actin and ATP concentrations, the ATPase activity in the steady state could be explained quantitatively by the rate constants given in Fig. 1. In other words, under these conditions the actomyosin ATPase reaction was satisfactorily explained by the Lymn-Taylor mechanism.

However, in the presence of a low concentration of KCl and at room temperature, the rate of the acto-HMM or acto-S-1 ATPase reaction, Δv_0, was found to be much higher than that of recombination of M_P^{ADP} with F-actin, v_{recomb}, as shown in Fig. 2A (Inoue *et al.*, 1973; Inoue *et al.*, 1980). This result is contradictory to the Lymn-Taylor mechanism. Furthermore, the rate of ATP hydrolysis, Δv_0, was proportional to the extent of binding of the myosin head with F-actin, $1-\alpha$, as shown in Fig. 2B.

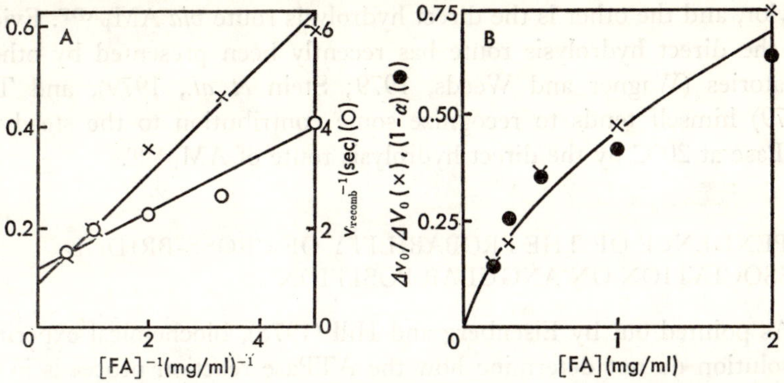

Fig. 2. Properties of Mg^{2+}-ATPase reaction of acto-S-1 in the absence of KCl at 20°C (Inoue *et al.*, 1980). A: double reciprocal plots of the rate of the acto-S-1 ATPase reaction in the steady state, Δv_0 (×), and the rate of recombination of M_P^{ADP} of S-1 with F-actin, v_{recomb} (O), against the F-actin concentration. Note that Δv_0 is about 10-fold higher than v_{recomb}. B: dependence on the F-actin concentration of the rate of acto-S-1 ATPase, $\Delta v_0/\Delta V_0$ (×), and the extent of binding of S-1 with F-actin, $(1-\alpha)$ (●). ΔV_0 is the ATPase rate at an infinitively high concentration of F-actin.

In 1973, we (Inoue *et al.*, 1973) reported that the rates of ATP hydrolysis under various conditions are given by the equation: $\Delta v_0=(1-\alpha)\cdot\Delta V_0+\alpha\cdot v_{recomb}$, where ΔV_0 is the rate of ATP hydrolysis at an infinitively high concentration of F-actin. We (Inoue *et al.*, 1980) reconfirmed this relation with more accurate measurements of Δv_0, ΔV_0, v_{recomb}, and α. Therefore, we proposed a two-route mechanism for the actomyosin-ATPase reaction (Tonomura and Inoue, 1975; Inoue *et al.*, 1970; Ikebe *et al.*, 1979):

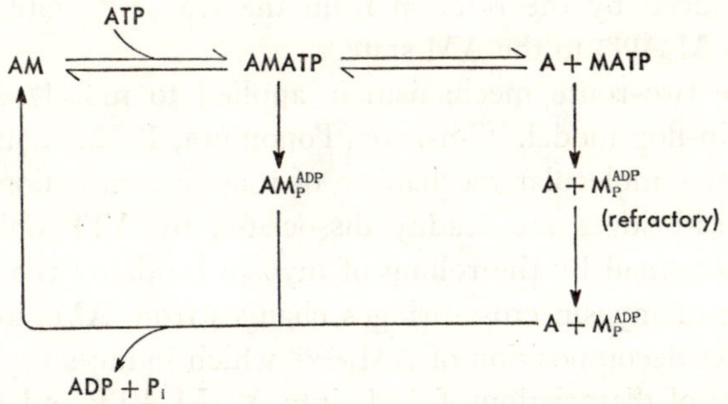

One route is the hydrolysis with dissociation and recombination of the actin-myosin complex and is the same as that proposed by Lymn and

Taylor, and the other is the direct hydrolysis route *via* AM_P^{ADP}. Evidence for the direct hydrolysis route has recently been presented by other laboratories (Wagner and Weeds, 1979; Stein *et al.*, 1979), and Taylor (1979) himself tends to recognize some contribution to the steady-state ATPase at 20°C by the direct hydrolysis route of AM_P^{ADP}.

DEPENDENCE OF THE PROBABILITY OF CROSS-BRIDGE DISSOCIATION ON ANGULAR POSITION

As pointed out by Eisenberg and Hill (1978), biochemical experiments in solution do not determine how the ATPase reaction proceeds in myofibrils, since the actomyosin systems in solution are always essentially at their minimum free energies. Also, the two routes of ATP hydrolysis by actomyosin coexist in solution, as mentioned above. It is generally accepted (A. F. Huxley, 1974) that the cross-bridges in myofibrils assume at least three different states, *i.e.*, a detached state and two attached states with different angular positions, *i.e.*, AM_1 and AM_2. In myofibrils, the free energy level of cross-bridges should depend on their angular position, and the probability of dissociation of cross-bridges by ATP depends on their angle, as predicted by A. F. Huxley (A. F. Huxley and Simmons, 1971; A. F. Huxley, 1974). When applied to actual muscle contraction, the Lymn-Taylor mechanism provides a very simple model by assuming that the AM_P^{ADP} complex has a minimum free energy at a vertical angle and the AM state has a minimum at an acute angle (Taylor, 1973, 1979). The consequence of their model have been discussed by Eisenberg and Hill (1978). In the model, the probability of dissociation of two attached states, AM_P^{ADP} and AM, depends on the angular position of each complex, and force is produced by the rotation from the AM_P^{ADP} state (which is reformed from M_P^{ADP}) to the AM state.

When the two-route mechanism is applied to muscle contraction, it provides a flip-flop model. Thus, we (Tonomura, 1972; Inoue *et al.*, 1979) proposed a new molecular mechanism of muscle contraction by assuming that the cross-bridges are readily dissociated by ATP only when their angle becomes small by the rolling of myosin heads on the thin filament, *i.e.*, the state of myosin cross-bridges changes from AM_2 into AM_1. The route of direct decomposition of AM_P^{ADP} which induces the head rotation and the route of dissociation of AM_1 into $A+M_1ATP$ and recombination into AM_2 then occur alternately in the cycle. Then, myofibrils contract with high efficiency in coupling with ATP hydrolysis:

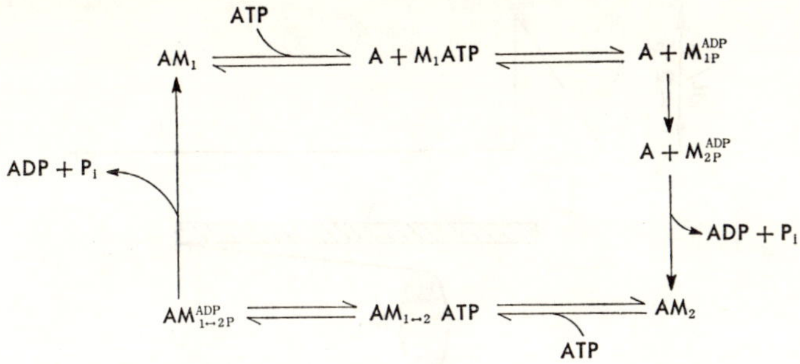

Figure 3 shows our experimental design to demonstrate that the dissociation of cross-bridges by ATP is initiated after the transition of myosin cross-bridges from AM_2 to AM_1. In isometrically contracted muscle, most of the myosin heads become attached to thin filaments to develop a maximum tension under our conditions. When the muscle fibers at rest length are slowly stretched, the cross-bridges slip on thin filaments. If they are then released quickly, the cross-bridges rotate to a more acute angle. In other words, they are transformed from AM_2 into AM_1. Therefore, if our assumption is correct, we can expect that after the quick release, 1 mol of ATP per mol is hydrolyzed, accompanied by dissociation of cross-bridges by ATP and subsequent re-formation.

ACCELERATION OF ATPase ACTIVITY BY REPETITIVE JAGGED STRETCH-RELEASE FOLLOWED BY RETURN TO REST LENGTH

The ATPase activity could be measured readily and exactly by continuously monitoring the decrease in NADH absorbance linked with ADP production in the presence of pyruvate kinase and lactate dehydrogenase (Arata *et al.*, 1979). We measured the steady-state ATPase activity of fibers subjected to the repetitive stretch-release cycles to avoid diffusional artifacts, and calculated the time course of ADP liberation from fibers subjected to a single cycle of the length change. We used glycerol-treated muscle fibers to fix the concentrations of ATP and Ca^{2+} ions within the fibers.

The fibers at rest length were stretched slowly by 2.5%, then released quickly to be held isometrically at rest length for α sec. This length change was repeated several tens of times (see Fig. 4 inset). ADP was liberated linearly with time, and the rate of ADP liberation was markedly increased by applying the repetitive length change, especially when α was around

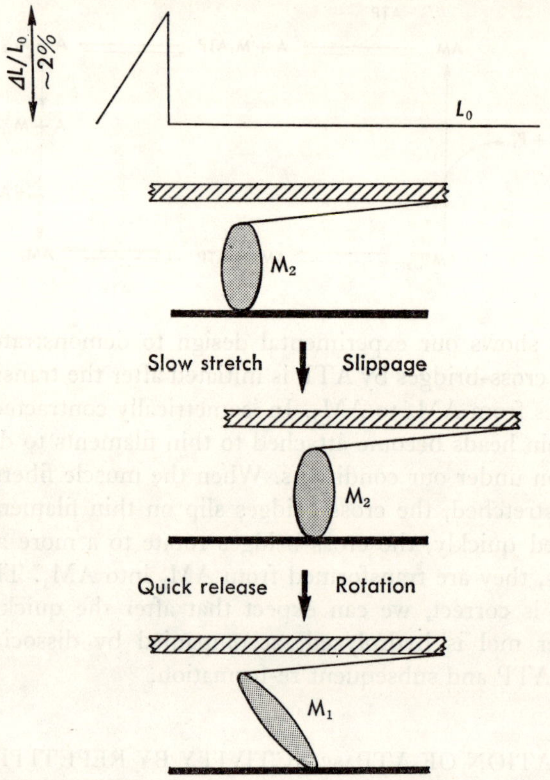

Fig. 3. Experimental design to show that the probability of dissociation of cross-bridges by ATP depends on their angular position. When muscle fibers at rest length (L_0) are slowly stretched as shown by the top trace, the cross-bridges slip on thin filaments. If the stretch is then released quickly, the cross-bridges rotate to a more acute angle and the state of the myosin heads is transformed from M_2 into M_1. Thus, we can expect that after the quick release, the M_1 cross-bridges liberate 1 mol ADP per mol of myosin, accompanied by rapid detachment and reattachment of cross-bridges (see text for details).

2 sec. At the end of the experiment, we confirmed that the rate of the static fibers had returned to the original level (Arata *et al.*, 1979).

Figure 4 is a plot of the ATPase activity of oscillated fibers against the duration of the isometric phase following the sudden release, α. At $\alpha=0$, *i.e.*, when the fibers were subjected to the simple slow jagged wave, the activity was almost the same as that of static fibers. This suggests that the state of the cross-bridges does not essentially change during the slow stretch. The activity increased with an increase in α to reach the maximum

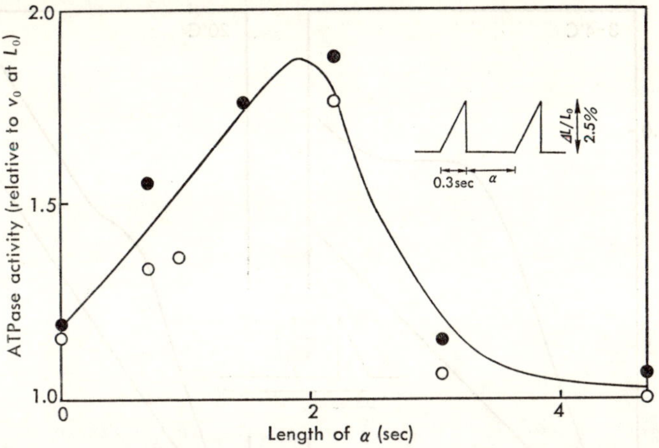

Fig. 4. Mg^{2+}-ATPase activity of muscle fibers subjected to repetitive jagged stretch-release followed by return to rest length (L_0), as a function of the duration of the isometric phase at rest length, α (Arata *et al.*, 1979). The fiber bundles were stretched linearly with time for 0.3 sec by 2.5% of L_0, released suddenly, then fixed at L_0 for α sec. The cycle was repeated (see inset). The ATPase activity of the oscillated fibers was normalized to that of static fibers (v_0 at L_0). The activating solution contained 0.25 mM ATP, 1.5 mg/ml pyruvate kinase, 0.75 mg/ml lactate dehydrogenase, 10 mM phosphoenolpyruvate, 0.3 mM NADH, 50 mM KCl, 2 mM $MgCl_2$, 10 μM $CaCl_2$, and 10 mM Tris-HCl (pH 7.8) at 3–4°C. The symbols indicate two different series of experiments with different fiber preparations.

of about twice that of static fibers at $\alpha \approx 2$ sec at 3–4°C, then decreased with further increase in α to reach a value almost equal to that of static fibers.

The extent of acceleration reached its limit at an amplitude of about 1.5% of the rest length, and decreased with an increase in the sarcomere length beyond its rest length (Arata *et al.*, 1978). Furthermore, no activity acceleration was observed on applying the repetitive length change when we used 8-bromo-ATP as a substrate (Arata *et al.*, 1979); it is a good substrate for hydrolysis by myosin, but this hydrolysis is not activated by F-actin (Takenaka *et al.*, 1976, 1978). These results indicate that the enzymic activity of active cross-bridges is accelerated by the repetitive stretch-release cycles.

To calculate the time course of ADP liberation after quick release of the slow stretch, we assumed that the acceleration of ATPase activity is turned off and on by the stretch and release, respectively, and that the state of the cross-bridges immediately after the stretch-release is independent

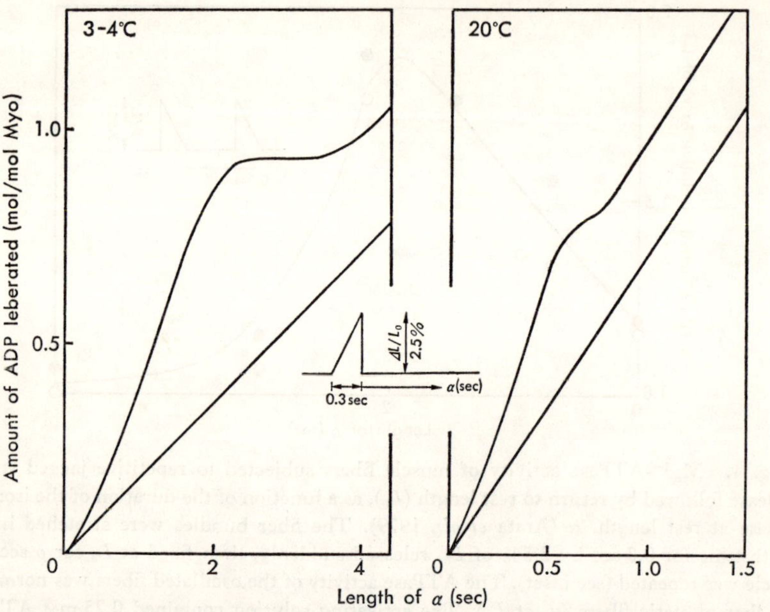

Fig. 5. Time courses of ADP liberation at 3–4°C and 20°C by muscle fibers after a stretch-release. The time courses were calculated from the ATPase activity of the summed duration of the isometric phase, α, assuming that the ATPase acceleration was turned on and off by the release and stretch, respectively, and that the state of the cross-bridges immediately after the stretch-release was independent of α of the cycle. The straight line indicates the time course of ADP liberation of static fibers.

of α of the cycle. As shown in the left of Fig. 5, at 3–4°C, after the quick release, about 1 mol of ADP per mol of myosin was rapidly liberated, then the ADP liberation was interrupted and finally its rate approached that of static fibers. As shown in the right of Fig. 5, on raising the temperature from 3–4 to 20°C, the time required for the series of changes in the rate of ADP liberation decreased to one-fourth. On the other hand, the ATPase activity of static fibers was increased 4-fold by raising the temperature. Thus, at 20°C about 1 mol of ADP also was liberated during the cycle of changes in the rate of ADP liberation after the quick release. These results are consistent with our assumption that when the myosin cross-bridges are transformed from AM_2 into AM_1, their detachment by ATP is rapidly initiated and the detachment-reattachment of cross-bridges occurs. This reaction of cross-bridges is accompanied by ATP splitting of 1 mol per mol of myosin.

The dependence of ATPase activity on the time interval α became broad and the time required for maximal acceleration increased 2-fold by lowering the concentration of Ca^{2+} ions, suggesting that the rate of ATP splitting by the detachment and reattachment of cross-bridges decreases as the Ca^{2+} concentration is reduced (Arata *et al.*, 1978). We also found that the extent of acceleration by the length change showed a marked substrate inhibition (Arata *et al.*, 1979).

The assumption that the probability of detachment of cross-bridges by ATP is determined by their angular position is also supported by the following findings. The ATPase activity of myofibrils during shortening was much higher than that of those after completion of shortening or of glycerol-treated muscle fibers fixed in an isometrically contracted state (Arata *et al.*, 1977). During shortening, myosin heads rotate on thin filaments, then the detachment and reattachment of cross-bridges occur. Therefore, the ATPase activity during shortening becomes much higher than that in an isometrically contracted state.

ACCELERATION OF ATPase ACTIVITY BY INTERMITTENT STRETCH-RELEASE

Flitney and Hirst (1978) recently observed nonlinear changes of sarcomere length during the stretching of muscle fibers, and suggested that when muscle fibers are stretched linearly, the cross-bridges rotate and subsequently become detached. Therefore, we examined the effect of applying intermittent stretch-release on the ATPase activity of glycerol-treated muscle fibers. The fibers were stretched quickly by 2.5% of their rest length, held at the plateau length, then re-stretched immediately after a sudden release to rest length. The cycle was repeated (see Fig. 6 inset). We found two types of acceleration of ATPase activity. The activity increased with an increase in the length of one cycle, β, to reach the first maximal level at β of 50 msec, then decreased to a value almost equal to that of static fibers. The activity increased again with further increase in β to reach the second maximal level at β of about 1 sec. We found that a simple repetitive jagged stretch-release produced the first type of acceleration on intermittent stretch-release, as shown by the dashed line in Fig. 6. All the properties of the first acceleration could be explained by the mechanism that it is caused by rotation of the cross-bridge and is abolished by its subsequent detachment (Arata *et al.*, 1979).

As summarized in Table I, the acceleration of ATPase produced by

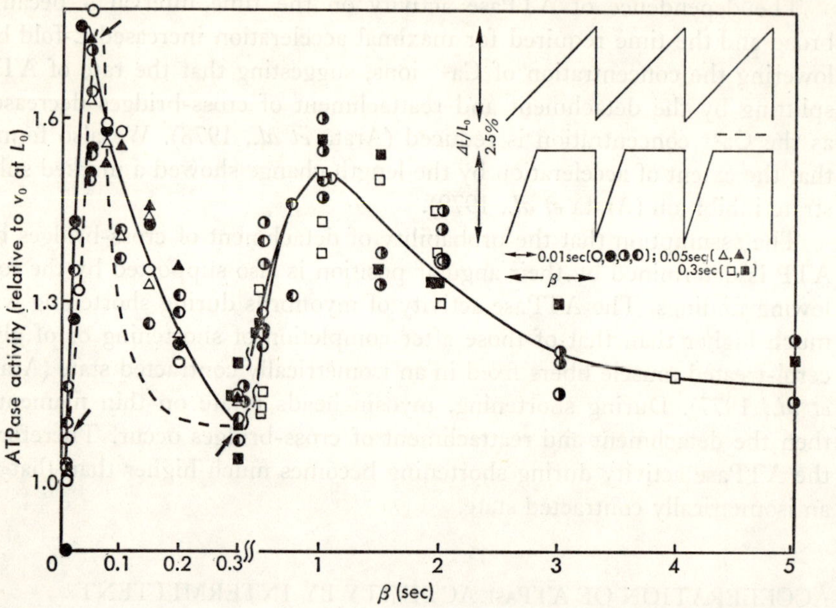

Fig. 6. Mg^{2+}-ATPase activity of muscle fibers subjected to intermittent stretch-release, as a function of the length of one cycle, β (Arata *et al.*, 1979). The fibers were stretched linearly with time for 0.01, 0.05, or 0.3 sec by 2.5% of the rest length (L_0), held at the plateau length, then stretched immediately after a sudden release to L_0. The cycle was repeated (see inset). Other experimental conditions are as described for Fig. 4. The dashed line indicates the result obtained by applying a simple repetitive jagged stretch-release (see inset).

applying a repetitive jagged stretch-release followed by return to rest length and the second acceleration produced by applying an intermittent stretch-release have common properties, which are very different from those of the first acceleration by an intermittent stretch-release and the acceleration by a simple repetitive jagged stretch-release. In the former two cases, the time for maximum acceleration at 3–4°C was at 1–2 sec. The extents of acceleration reached their limits when the stretch reached about 2% of the rest length. Both types of ATPase acceleration showed a marked substrate inhibition. On raising the temperature from 3–4° to 20°C, the time required for the maximum acceleration decreased to one-fourth. When Ca^{2+} concentration was reduced to 0.1 μM, the time for maximum acceleration increased 2-fold. When ATP was replaced with

TABLE I
Properties of Acceleration of the Mg^{2+}-ATPase Activity of Glycerol-treated Muscle Fibers by Applying Two Types of Stretch-release Cycles

Experimental observations	Repetitive jagged stretch-release followed by return to rest length	Intermittent stretch-release (2nd acceleration)
Time for maximum acceleration (τ_{max})	2 sec	1 sec
Extent of stretch	Maximum at $\Delta L/L_0 \geqq 2\%$	Maximum at $\Delta L/L_0 \geqq 2\%$
ATP concentration	Maximum acceleration at 0.25 mM ATP; substrate inhibition	Maximum acceleration at 0.25 mM ATP; substrate inhibition
Increase in temperature to 20°C	τ_{max} decreases to 0.5 sec	τ_{max} decreases to 0.25 sec
Decrease in Ca^{2+} ions to 0.1 μM	τ_{max} increases to 4–5 sec	τ_{max} increases to 5 sec
PCMB-DTT treatment	τ_{max} decreases to 0.3 sec; extent of acceleration is decreased	τ_{max} decreases to 0.5 sec; extent of acceleration is decreased
Replacement of ATP with ITP	Markedly decreased	Markedly decreased
Replacement of Mg^{2+} with Mn^{2+}	Markedly decreased	Markedly decreased

(Arata *et al.*, 1979).

ITP or Mg^{2+} with Mn^{2+}, both types of acceleration were observed only slightly.

The acceleration of ATPase activity produced by applying a repetitive jagged stretch-release followed by return to rest length could be explained if ATP splitting of 1 mol per mol of myosin was induced by the length change, as we mentioned above. Furthermore, the second acceleration on applying an intermittent stretch-release was observed after the inactivation process of the first acceleration, which was attributed to the detachment of cross-bridges. We previously showed that the dissociation of actomyosin by ATP is inhibited by modification of one specific SH group in the myosin molecule with *p*-chloromercuribenzoate (Shibata-Sekiya and Tonomura, 1976). As shown in Table I, this modification of fibers markedly inhibited the acceleration of ATPase activity upon applying the repetitive jagged stretch-release followed by return to rest length and the second acceleration upon applying the intermittent stretch-release (Arata *et al.*, 1979). This observation also supports the mechanism that the acceleration is due to the initiation of cross-bridge detachment by applying the length changes. Therefore, we concluded that ATP splitting of 1 mol per mol of myosin is initiated by the detachment of cross-bridges, either after quick release of a slow stretch or after a quick stretch followed by maintenance of the plateau length.

RELATIONSHIP OF ATP SPLITTING TO CROSS-BRIDGE MOVEMENT AND A MOLECULAR MECHANISM OF MUSCLE CONTRACTION

Figure 7 shows a molecular mechanism of the relationship of ATP splitting to cross-bridge movement, deduced from kinetic analyses of actomyosin ATPase in solution and also in myofibrils. This model is depicted on the basis of a three-state model of cross-bridge movement (H. E. Huxley, 1969; A. F. Huxley, 1974). We assumed that the myosin head during isometric contraction took the form of the M_P^{ADP} complex attached to F-actin, because the ATPase activity at isometric contraction was much smaller than that during shortening and the tension development was almost unaffected by changing the temperature under the conditions used here (Arata *et al.*, 1977; Arata and Tonomura, 1979). The important point

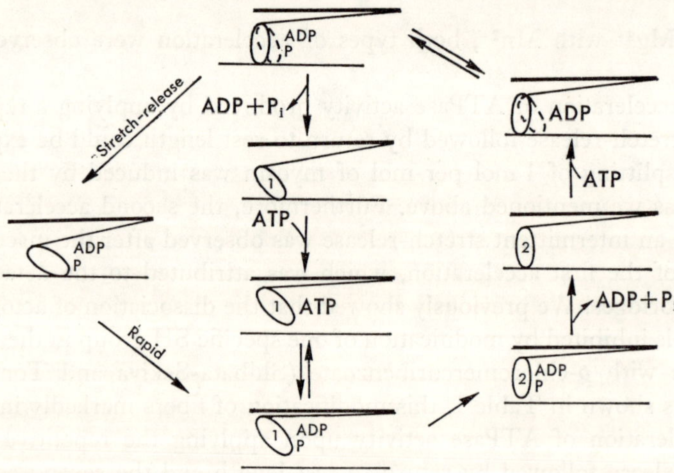

Fig. 7. A molecular mechanism for the relationship of cross-bridge movement to ATP splitting based on the actomyosin ATPase cycle in solution and in myofibrils. The probability of dissociation of cross-bridges, AM_P^{ADP}, depends on their angle. ATP splitting is rapidly initiated by the dissociation of cross-bridges after the transformation of their state from 2 to 1, which is induced by the sliding process or by quick release of a slow stretch.

in this model is that the probability of dissociation of cross-bridges depends on their angle. Thus, ATP splitting is initiated by cross-bridge dissociation after transformation of the state of myosin cross-bridges from 2 to 1, which is induced by the sliding process or by quick release of a slow stretch, as described in this paper.

In constructing the molecular model of muscle contraction shown in Fig. 7, we assumed that the myosin head in the M_1 state binds with the thin filament at an angle of 45° and in the M_2 state at 90°. However, note that the results mentioned here can be explained well only by assuming three cross-bridge states, and our conclusions do not depend on the angular positions of cross-bridges corresponding to the states AM_1 and AM_2 adopted in this model.

Acknowledgment

This work was supported by grants from the Ministry of Education, Science and Culture of Japan and the Muscular Dystrophy Association, Inc.

REFERENCES

Arata, T., Mukohata, Y., and Tonomura, Y. (1977). *J. Biochem.* **82**, 801–812.
Arata, T., Mukohata, Y., and Tonomura, Y. (1978). *J. Biochem.* **84**, 751–761.
Arata, T., Mukohata, Y., and Tonomura, Y. (1979). *J. Biochem.* **86**, 525–542.
Arata, T. and Tonomura, Y. (1979). *J. Biochem.* **86**, 543–551.
Eisenberg, E. and Hill, T. L. (1978). *Prog. Biophys. Mol. Biol.* **33**, 55–82.
Flitney, F. W. and Hirst, D. G. (1978). *J. Physiol.* **276**, 467–479.
Huxley, A. F. (1974). *J. Physiol.* **243**, 1–43.
Huxley, A. F. and Simmons, R. M. (1971). *Nature* **233**, 533–538.
Huxley, H. E. (1969). *Science* **202**, 1356–1366.
Ikebe, M., Inoue, A., and Tonomura, Y. (1980). *J. Biochem.* (in press).
Inoue, A., Shigekawa, M., and Tonomura, Y. (1973). *J. Biochem.* **74**, 923–934.
Inoue, A., Takenaka, H., Arata, T., and Tonomura, Y. (1979). *Adv. Biophys.* **13**, 1–194.
Inoue, A., Ikebe, M., and Tonomura, Y. (1980). *J. Biochem.* (in press).
Lymn, R. W. and Taylor, E. W. (1971). *Biochemistry* **10**, 4617–4624.
Shibata-Sekiya, K. and Tonomura, Y. (1976). *J. Biochem.* **77**, 543–557.
Stein, L. A., Schwarz, R. P., Jr., Chock, P. B., and Eisenberg, E. (1979). *Biochemistry* **18**, 3895–3909.
Takenaka, H., Ikehara, M., and Tonomura, Y. (1976). *J. Biochem.* **80**, 1381–1392.
Takenaka, H., Ikehara, M., and Tonomura, Y. (1978). *Proc. Natl. Acad. Sci. U.S.* **75**, 4229–4233.
Taylor, E. W. (1973). In *Current Topics in Bioenergetics* (Sanadi, O. R. and Packer, L., eds.), vol. 5, pp. 201–231, Academic Press, New York.

Taylor, E. W. (1979). *CRC Crit. Rev. Biochem.* **6**, 103–164.
Tonomura, Y. (1972). *Muscle Proteins, Muscle Contraction and Cation Transport,* Japan Sci. Soc. Press and Univ. Park Press, Tokyo and Baltimore.
Tomomura, Y. and Inoue, A. (1975). In *Energy Transducing Mechanisms,* MTP Int. Rev. Sci. Biochem. Series One (Racker, E., ed.), vol. 3, pp. 121–160, Butterworth, London.
Wagner, P. D. and Weeds, A. G. (1979). *Biochemistry* **18**, 2260–2266.

Muscle Contraction: Its Regulatory Mechanisms (*Ebashi, S. et al., eds.*), *pp. 115–131, Japan Sci. Soc. Press, Tokyo / Springer-Verlag, Berlin (1980)*

9

The Role of the Sulfhydryl Group of L-Meromyosin in the Actin-Myosin Interaction

TATSUHISA YAMASHITA

Summary: The reactivity of the sulfhydryl groups in myosin B to N-ethylmaleimide (NEM) was examined under various conditions. Under the conditions where actin and myosin associate, *i.e.*, at low ionic strength, only Mg^{2+}-ATPase activity was markedly activated by NEM modification. NEM-modified myosin B exhibited a shift of the KCl dependence curve of ATPase activity to high concentration, a shift of the maximum activation of ATPase activity to high Mg ion concentration, a suppression of substrate inhibition of ATPase activity, a marked enhancement of superprecipitation even at high concentration of Mg^{2+} and substrate and greater resistance to clearing by ATP. Electron microscopic observations showed that myosin and actin filaments of modified myosin B in the presence of Mg^{2+}-ATP gathered in parallel while in the case of unmodified myosin B they dispersed. These findings all indicate that the blocking by NEM of supecific sulfhydryl group (S_a), the sulfhydryl group responsible for the activation of Mg^{2+}-ATPase of myosin B, brings about an increase in the affinity of myosin and actin in the myosin B system.

When Mg^{2+}-ATPase activity was maximally activated, SDS-gel electrophoresis of [^{14}C]-NEM-modified myosin B indicated that NEM was incorporated into myosin heavy chain and actin, but not into tropomyosin and troponin. Mg^{2+}-ATPase activities of actomyosins reconstituted from myosins and actins prepared from S_a-blocked and intact myosins B revealed that myosin modification was essential for activation. ActoH-meromyosin, composed of heavy meromyosin (HMM) prepared from S_a-blocked myosin B and intact actin, did not retain the effect of S_a modification of Mg^{2+}-ATPase activity. Radioactivity was detected in light meromyosin (LMM) obtained from [^{14}C]-NEM-modified myosin B, but not in HMM, indicating that S_a is located in the LMM region of myosin.

Based on these results, a possible mechanism of the participation of S_a in the myosin-actin interaction is discussed.

We have studied the role of the sulfhydryl groups of myosin in muscular contraction and have already reported that two specific sufhydryl groups are involved in the active site of myosin A: one (S_1) is responsible for EDTA-ATPase activity, and the other (S_2) is necessary for Ca^{2+}-ATPase activity elevated by blocking S_1 with N-ethylmaleimide (NEM) (Sekine *et al.*, 1962; Yamaguchi and Sekine, 1966; Yamashita *et al.*, 1974). It has also been reported that some sulfhydryl groups located in the myosin molecule are essential for the Ca^{2+} sensitivity of natural actomyosin (Daniel and Hartshorne, 1972, 1973). Therefore, further studies on the role of the sulfhydryl groups in muscular contraction were carried out using myosin B, the primary mechanochemical energy-transducing system in molecular terms.

The reactivity of the sulfhydryl groups in myosin B to NEM under various conditions and properties of resulting NEM-modified myosin B were examined. The results were obtained that under the conditions where actin and myosin associate, *i.e.*, at low ionic strength, a specific sulfhydryl group (S_a) located in the light meromyosin (LMM) region of myosin was modified, resulting in the increase in the affinity of actin and myosin (Yamashita *et al.*, 1975; Yamashita and Horigome, 1977; Horigome and Yamashita, 1977, 1978, 1979a, 1979b).

METHODS

S_a-blocked myosin B was prepared by incubating myosin (7 mg/ml) with 0.24 mM NEM (3.3 mol per 10^5 *g* of protein) in 0.04 M Tris-maleate buffer (pH 7.0) at 0°C while stirring, in the absence of KCl, for 90 min and by stopping the reaction on addition of a 50-fold excess of β-mercapto-ethanol unless otherwise noted. The reaction mixtures were diluted with an appropriate KCl solution to give a final KCl concentration of 0.45 M, and aliquots were used for assays. Unmodified myosin B, used as a control, was treated under the same conditions as for NEM-blocking, except that NEM was omitted.

RESULTS

1. *Modification of Myosin B with NEM*

The effect of KCl concentration in the reaction medium on the chemical modification of myosin B with NEM is shown in Fig. 1. The lower the

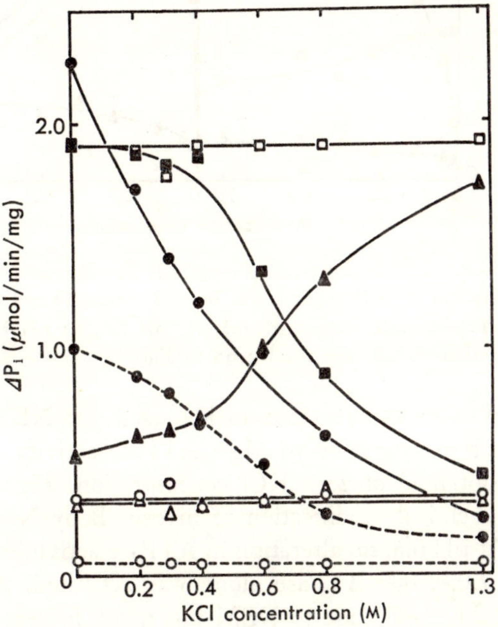

Fig. 1. Effect of ionic strength on the modification of myosin B with NEM. After the incubation of myosin B with NEM in the indicated KCl concentrations for 15 min at pH 8 (40 mM Tris-maleate), each incubation mxiture was diluted with the appropriate KCl solution with excess β-mercaptoethanol to a final concentration of 0.45 M, and used for ATPase assay. ATPase activity was measured for 5 min at 37°C either in 0.5 M KCl, pH 7.6, in the presence of 5 mM $CaCl_2$ (△, ▲) or 1 mM EDTA (□, ■) or in 0.04 M KCl, pH 7.0, in the presence of 1 mM $MgCl_2$ (○, ●). The concentration of ATP was 2 mM. Open and solid symbols reperesent unmodified and NEM-modified myosin B, respectively. Mg^{2+}-ATPase activity in the presence of 1 mM EGTA is shown by dotted lines.

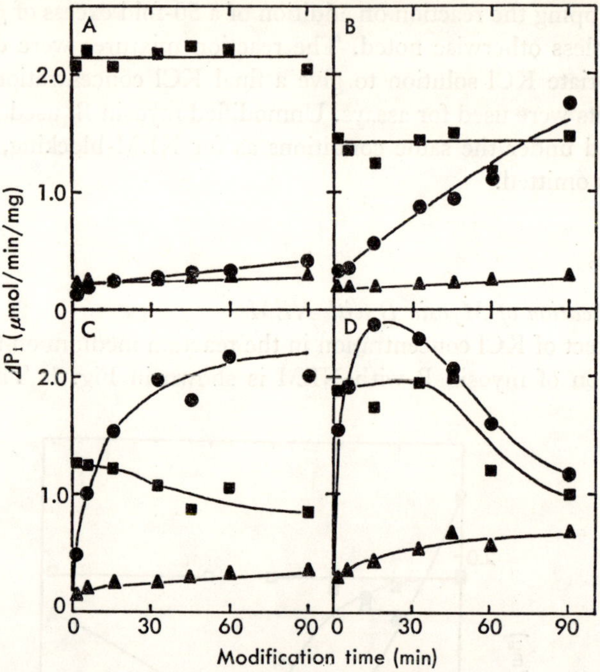

Fig. 2. Changes in the ATPase activities of myosin B during reaction with NEM in the absence of KCl. Myosin B (3.18 mg/ml) was reacted with 3.14 mol of NEM per $10^5 g$ protein in 40 mM Tris-malcate buffer (pH 6.5, A; pH 7.0, B; pH 7.5, C; pH 8.0, D) with stirring. For other details, see the legend to Fig. 1.

KCl concentration in the reaction meida used for NEM-blocking, the higher the degree of activation of Mg^{2+}-ATPase activity. The maximum activation was obtained at zero KCl concentration. Figure 2 shows the time course of chemical modification of myosin B by NEM at zero KCl concentration. At pH 6.5, no alteration in ATPase activities were observed, whereas at pH 7.0, Mg^{2+}-ATPase alone was activated. At alkaline pH, the reactivity of Mg^{2+}-ATPase to NEM was much increased with a slight modification of S_1. The above results indicate that S_a, the SH group responsible for Mg^{2+}-ATPase activity, became more specifically reactive to NEM at low ionic strength and neutral pH.

2. *Properties of NEM-modified Myosin B*

ATPase activities of S_a-blocked and unmodified myosins B are summarized in Table I. As seen in Fig. 3, the Mg^{2+}-ATPase activity of NEM-

TABLE I
Comparison of the ATPase Activities of S_a-blocked Myosin B and Myosin A Prepared from S_a-blocked Myosin B
Myosin B was modified under the same conditions as F in Table II. Modified myosin A was separated from S_a-blocked myosin B by centrifugation in the presence of Mg^{2+}-ATP, and then purified by ammonium sulfate fractionation (40–50% saturation). Unmodified myosin A was prepared from intact myosin B by the procedure used for the preparation of modified myosin A.

	ATPase activity (P_i μmol/min/mg)			
	Ca^{2+}-	EDTA-	Mg^{2+}-	Mg^{2+}-EGTA-
Unmodified myosin B	0.42	4.45	0.35	0.14
Modified myosin B	0.68	4.34	1.07	0.48
Unmodified myosin A	1.18	6.83	0.04	—
Modified myosin A	1.54	6.60	0.06	—

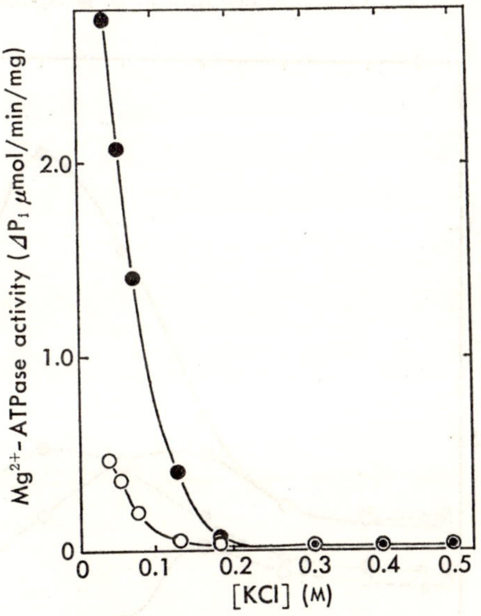

Fig. 3. Dependence of the Mg^{2+}-ATPase activity of S_a-blocked myosin B on KCl concentration. The ATPase activity was measured in 0.04 M Tris-maleate buffer (pH 7.0) at various KCl concentrations in the presence of 1 mM $MgCl_2$ and 2 mM ATP. ○ unmodified myosin B; ● S_a-blocked myosin B.

modified myosin B was the same as that of unmodified material at higher KCl concentrations, but was activated 6–7-fold at lower KCl concentra-

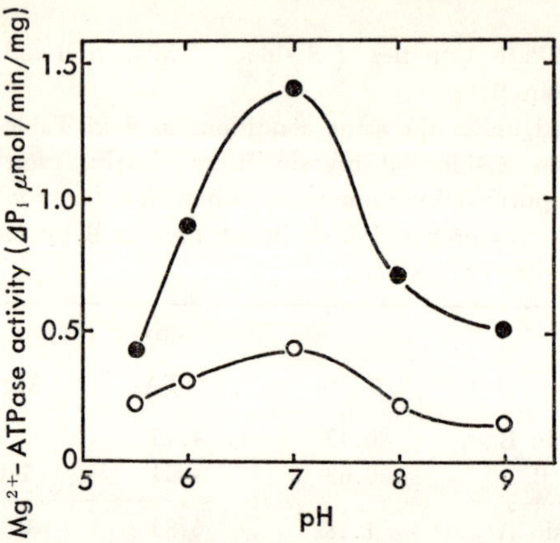

Fig. 4. pH dependence of the Mg^{2+}-ATPase activity of S_a-blocked myosin B. The ATPase activity of modified and unmodified myosins B was measured in 0.04 M KCl, 2 mM $MgCl_2$, and 2 mM ATP; ○ unmodified myosin B. ● S_a-blocked myosin B.

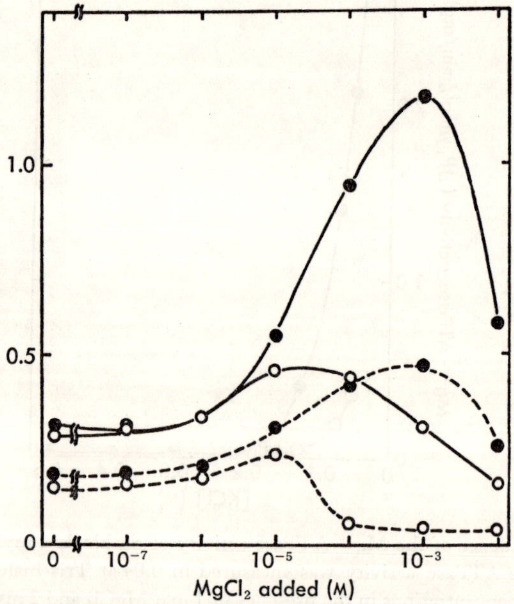

Fig. 5. Effect of Mg^{2+} concentration on the ATPase activity of NEM-modified myosin B. Myosin B was modified with NEM in 0.2 M KCl and 40 mM Tris-maleate buffer, pH 7.5, for 15 min at 0°C. ATPase activities were determined in 0.02 M KCl, pH 7.0, in the presence of $MgCl_2$ (——) or $MgCl_2$ plus 1 mM EGTA (---). ○ unmodified myosin B; ● modified myosin B.

tions. The maximum activation of Mg^{2+}-ATPase was observed around pH 7.0 (Fig. 4). The dependence of myosin B ATPase activity on Mg ion concentration showed that the Mg^{2+}-ATPase activity of control myosin B was slightly dependent on Mg concentration, with a maximum between 10^{-5} and 10^{-4} M followed by the ATPase inhibition, whereas addition of Mg ion to the S_a-blocked myosin B system above 10^{-5} M brought about a marked activation of ATPase with maximum activation at 10^{-3} M (Fig. 5). As can be seen in Fig. 6, a Lineweaver-Burk plot of Mg^{2+}-ATPase of unmodified myosin B showed remarkable substrate inhibition at higher substrate concentrations. The Mg^{2+}-ATPase of S_a-blocked myosin B was much higher than that of unmodified material over all the substrate concentrations examined; in particular, the substrate inhibition at higher concentrations was markedly suppressed. These results suggest that dissociation of myosin and actin is suppressed by S_a-blocking. If myosin B were entrapped in 9% agarose gel, even though myosin and actin might be

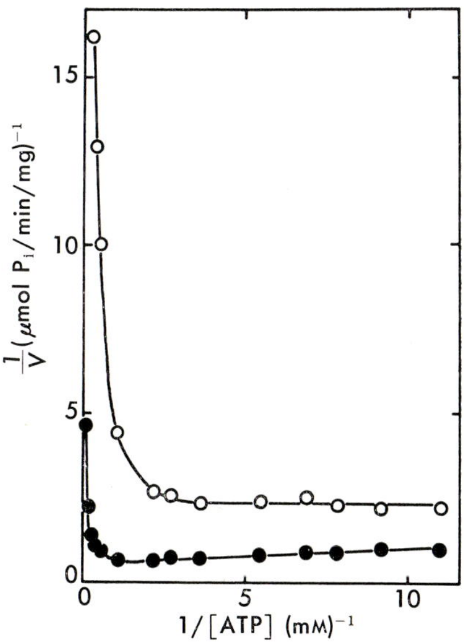

Fig. 6. Effect of ATP concentration on the Mg^{2+}-ATPase activity of NEM-modified myosin B. The initial rate of P_i production is plotted against ATP concentration in the Lineweaver-Burk form. NEM-modified myosin B was prepared by incubating myosin B with NEM in 50 mM Tris-maleate buffer, pH 7.0, for 90 min at 0°C. ATPase assay was carried out in 0.04 M KCl, pH 7.0, in the presence of 1 mM $MgCl_2$. ○ unmodified myosin B; ● modified myosin B.

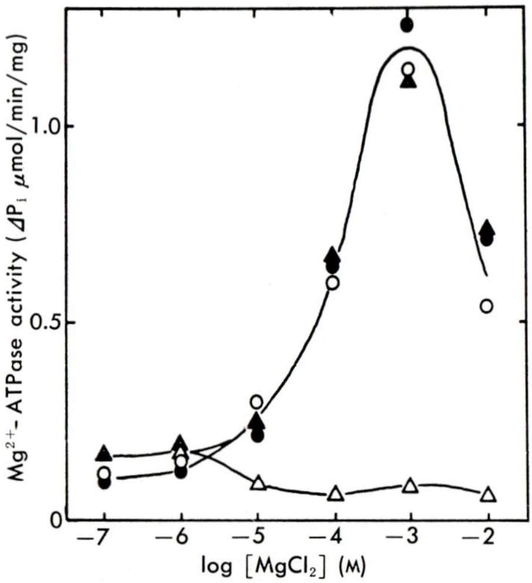

Fig. 7. Effect of Mg^{2+} concentration on the ATPase activity of myosin B entrapped in agarose gel. ATPase activity was determined by measuring P_i relased in 20 min in 0.04 M KCl, pH 7.0, at various $MgCl_2$ concentrations at 37°C with shaking: unmodified (○) and S_a-blocked (●) myosins B entrapped in 9% agarose gel, unmodified (△) and S_a-blocked (▲) myosins B mobilized during the homogenization of the entrapped samples.

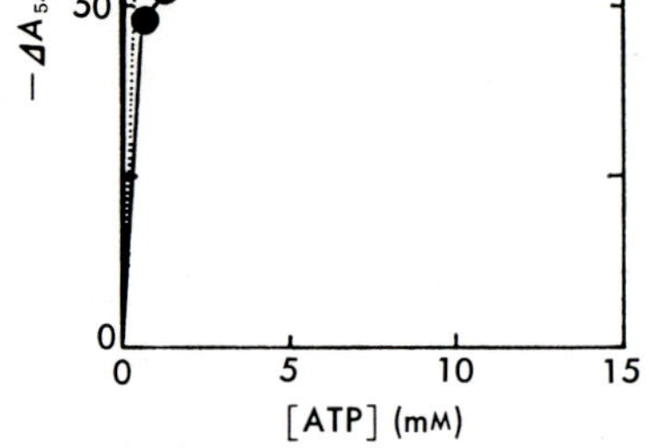

Fig. 8. Clearing response of S_a-blocked myosin B to successive additions of ATP. Clearing response was tested at 25°C. The reaction mixture contained 0.1 M KCl, 5 mM $MgCl_2$, 40 mM Tris-maleate, pH 7.0, and 2 mM EGTA with 2 mM $CaCl_2$ (——) and without $CaCl_2$ (······). ● S_a-blocked myosin B; ○ unmodified myosin B. Protein concentrations used were 0.363 mg (●) and 0.373 mg (○) per ml.

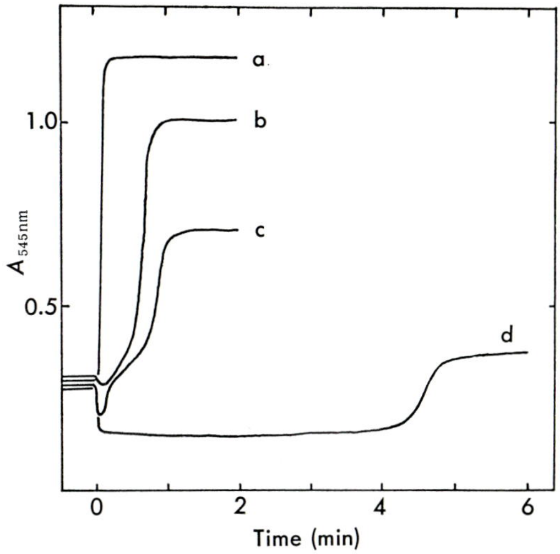

Fig. 9. Time course of superprecipitation of S_a-blocked myosin B. Superprecipitation was measured by recording the increase in the absorbance of myosin B suspensions at 545 nm in a Hitachi 200-20 recording spectrophotometer under continuous stirring. The reaction mixture contained 0.36 mg/ml myosin B in the presence of 0.1 M KCl, 40 mM Tris-maleate buffer (pH 7.0), 5 mM $MgCl_2$, 2 mM EGTA (b, d), and 2 mM $CaCl_2$ (a, c). The reaction was started at 25°C by adding 0.01 ml of 10 mM ATP to 3.0 ml of reaction mixture. a and b, S_a-blocked myosin B; c and d, unmodified myosin B treated under the same conditions as for S_a-blocking as control.

dissociated by Mg^{2+}-ATP, they could not disperse as freely as in the soluble state. Thus, the Mg^{2+}-ATPase activities of unmodified and S_a-blocked myosins B in gel were examined. As shown in Fig. 7, the Mg concentration dependence of Mg^{2+}-ATPase of the immobilized myosin B, irrespective of S_a modification, was similar to that of soluble S_a-blocked myosin B. These results support our interpretation of the activation of Mg^{2+}-ATPase by S_a-blocking. Clearing response to ATP of S_a-blocked and unmodified myosins B is shown in Fig. 8. The turbidity of control myosin fell rapidly and sharply even at the minimum concentration used here, so that Ca sensitivity was not observed under our conditions. S_a-blocked gel showed greater resistance to clearing than unmodified gel. A 50% clearing of S_a-blocked myosin gel occurred at 1 mM ATP in the presence of Ca^{2+}, and 100% clearing was obtained on addition of 10 mM ATP. The time courses of superprecipitation of unmodified and S_a-blocked myosins B are shown in Fig. 9. The blocking of S_a markedly accelerated

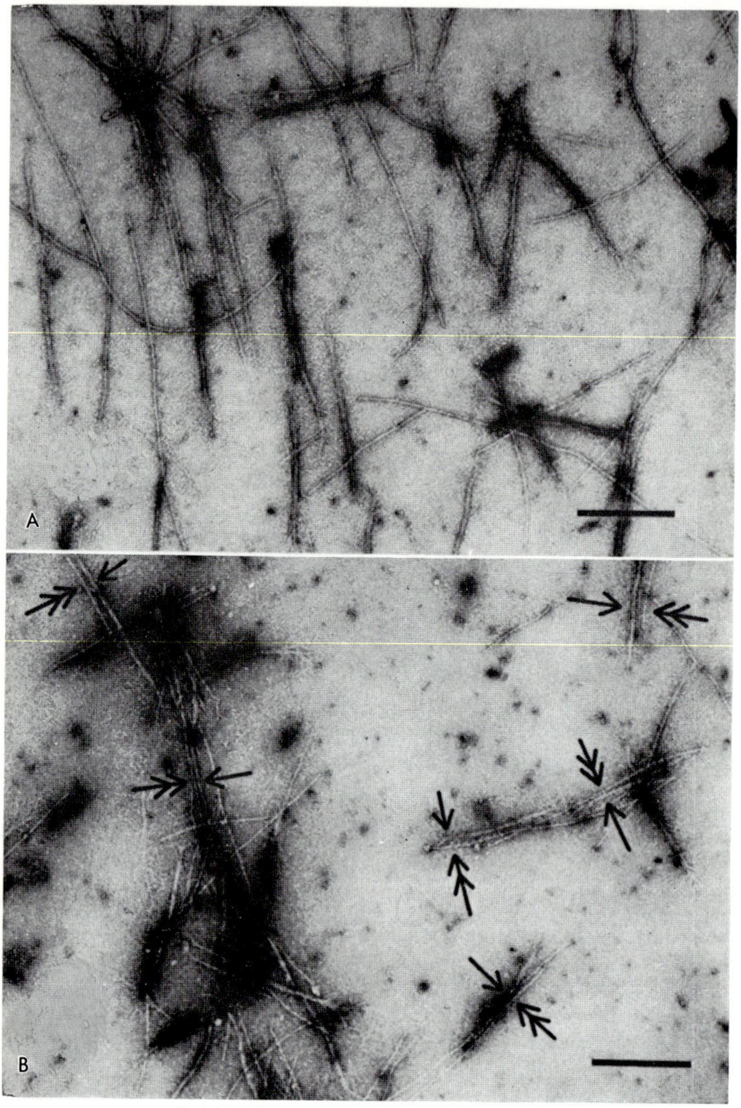

Fig. 10. Electron micrographs of unmodified and S_a-blocked myosins B in the presence of Mg^{2+}-ATP. One-tenth ml of myosin B solution (0.5 mg/ml, 0.42 M KCl) was mixed with 1 ml of a solution containing 40 mM Tris-maleate (pH 7.0), 2 mM ATP, and 1 mM $MgCl_2$ at 25°C. One drop of the mixture was taken out whitin 1 min and negatively stained with uranyl acetate. A: unmodified myosin B. B: S_a-blocked myosin B. Arrows and double-arrows indicate myosin and actin filaments gathered in parallel, respectively. The bar represents 0.3 μm.

superprecipitation; *i.e.*, S_a-blocked myosin B showed a much larger value of ΔA and a much smaller value of $t_{1/2}$ than the unmodified myosin B, consequently superprecipitating at a faster rate than the control myosin B. However, no changes in the sensitivity of superprecipitation to Ca^{2+} were observed on blocking S_a; *i.e.*, myosin B superprecipitated rapidly with Ca^{2+} and very slowly without Ca^{2+}. The activation of Mg^{2+}-ATPase and the increase of resistance to clearing on S_a-blocking suggested that dissociation of myosin and actin filaments by Mg^{2+}-ATP was suppressed by the modification. Thus, the interaction of myosin and actin filaments in the presence of Mg^{2+}-ATP was checked directly by electron microscopy. As shown in Fig. 10, myosin and actin filaments of S_a-blocked myosin B gathered in parallel, whereas in the case of unmodified myosin B they dispersed. These results indicate clearly that dissociation of myosin and

TABLE II
The Relationship between the Activation of Mg^{2+}-ATPase and the Incorporation of Radioactive NEM into Components of Myosin B
The modification of myosin B (3 mg/ml) was carried out in 0.04 M Tris-maleate buffer at 0°C under the conditions shown in the table. Electrophoresis was carried out using 7.5% polyacrylamide gel in the presence of 0.1% sodium dodecyl sulfate and the gel was stained with amidoblack 10 B. Stained protein bands were cut out and their radioactivities were measured in a liquid scintillation counter. The protein concentration on the gel was calculated from the amounts of charged protein and the area ratio of each band to the total bands on the electrophoretogram scanned at 660 nm. The molecular weights of actin and the heavy chain of myosin were taken as 45,000 and 210,000, respectively, and the molar ratios of LC_1, LC_2, and LC_3 to two heavy chains were assumed to be 1, 2, and 1, respectively.

	Modification conditions			Mg^{2+}-ATPase activity (P_i μmol/min/mg)	[^{14}C]-NEM incorporated (mol/mol)				
					Myosin				Actin
	pH	NEM (mM)	Time (min)		Heavy chain[b]	LC_1	LC_2[b]	LC_3	
A	Unmodified			0.189 (1.0)[a]	0	0	0	0	0
B	7.0	0.1	5	0.299 (1.6)	0.24	0	0.10	0	0.24
C	8.0	0.05	6	0.663 (3.5)	0.46	0.10	0.08	0.04	0.34
D	7.5	0.1	15	0.946 (5.0)	0.70	0	0	0	0.45
E	6.8	0.2	90	1.08 (5.7)	0.70	0.17	0.08	0.04	0.67
F	7.0	0.1	90	1.35 (7.1)	0.84	0	0.10	0.07	0.60
G	7.0	0.1	180	1.84 (9.7)	1.0	0	0.14	0.11	0.79

[a] Figures in parentheses represent the ratio of each modified myosin B ATPase activity to that of the unmodified material.
[b] The values of NEM incorporated are expressed per 2 mol.

actin filaments with Mg^{2+}-ATP at low ionic strength is suppressed by S_a-blocking.

3. Submolecular Localization of S_a Thiol Group

When myosin B was incubated with [^{14}C]-NEM under various conditions, the degree of incorporation of radioactive NEM into components of myosin B which were separated by SDS-gel electrophoresis was examined. As indicated in Table II, radioactivity was detected at the bands corresponding to the heavy chain, actin and light chains, but not detected at the bands of tropomyosin and troponin. Mg^{2+}-ATPase activity was activated in parallel with increasing amount of NEM incorporated into myosin heavy chain and actin and reached almost the maximum level on incorporation of about 1 mol of NEM per mol of each protein. As the labeling of light chains by NEM was small and moreover not proportional to the activation of ATPase, light chains are probably not involved in ATPase activation. In order to determine which component, myosin A

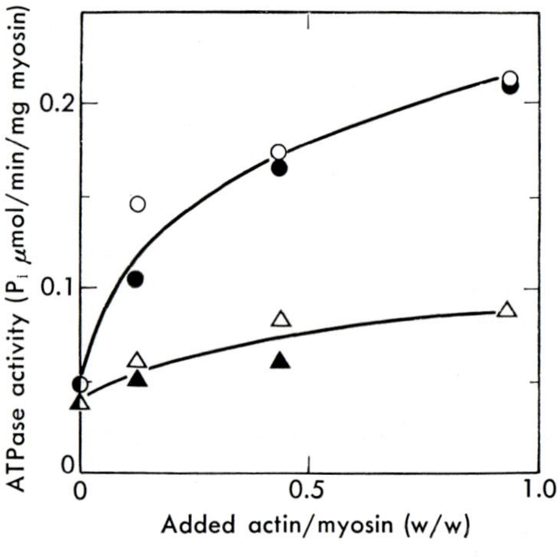

Fig. 11. Effect of S_a modification on the activation of Mg^{2+}-ATPase of myosin A by actin. Mg^{2+}-ATPase activity of actomyosins was assayed by mixing the indicated number of parts of modified or unmodified actin with 1 part of modified or unmodified myosin A by weight in 0.38 M KCl and allowing the mixture to stand for 6 hr at 0°C; ○ modified myosin A plus modified actin. ● modified myosin A plus unmodified actin; △ unmodified myosin A plus modified actin; ▲ unmodified myosin A plus unmodified actin. For preparations of modified and unmodified myosin A, see the legend to Table I. Modified actin was prepared from the same S_a-blocked myosin B used for the preparation of modified myosin A.

or actin, is responsible for the activation of myosin B ATPase, modified myosin A and actin were separated from S_a-blocked myosin B. As shown in Fig. 11, the Mg^{2+}-ATPase of modified myosin A was markedly activated not only by modified actin but also by unmodified actin, the degree of activation by both actins being the same. On the other hand, with unmodified myosin A the modified actin produced a less marked activation of ATPase than did modified myosin A, its ability to activate ATPase being the same as that of unmodified actin. Thus, it can be concluded that the modification of myosin heavy chain is essential for the activation of Mg^{2+}-ATPase of myosin B. The activation by actin of the Mg^{2+}-ATPase activity of heavy meromyosin's (HMM's) prepared from S_a-blocked and unmodified myosins B was studied. As shown in Fig. 12, the activation curves of Mg^{2+}-ATPase activity of HMM's by various amounts of Ca^{2+}-

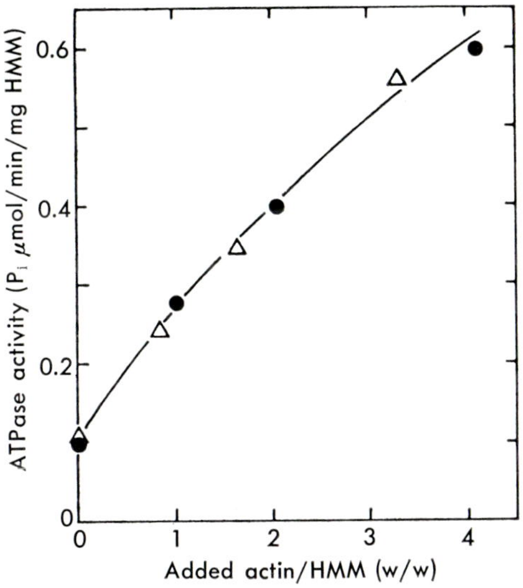

Fig. 12. Mg^{2+}-ATPase activities of actoHMM's constituted from HMM's prepared from S_a-blocked and unmodified myosin B. Modified HMM was prepared by incubation of myosin A separated from S_a-blocked myosin B with trypsin (1/200 w/w) at pH 6.5 and 25°C for 7 min followed by ammonium sulfate fractionation (50–60% saturation) of the digest. Unmodified HMM was prepared from intact myosin B by the same procedure. ActoHMM's were prepared by mixing various amounts of Ca^{2+}-sensitive actin with HMM in 0.3 M KCl and allowing the mixture to stand at 0°C for 6 hr, and their Mg^{2+}-ATPase activities were measured. ● HMM prepared from S_a-blocked myosin B plus Ca^{2+}-sensitive actin; △ HMM prepared from unmodified myosin B plus Ca^{2+}-sensitive actin.

sensitive actin were identical in both cases, indicating that the effect of S_a modification seen in actomyosin synthesized from modified myosin A was lost in the actoHMM system. This result suggests that the S_a site is located in the LMM moiety of the myosin molecule. To elucidate how many sulfhydryl groups S_a consists of, the chromatographic pattern of the radioactive tryptic peptides of modified myosin A prepared from [^{14}C]-NEM-modified myosin B was examined. As shown in Fig. 13A, a major portion of the radioactivity was concentrated in one spot and the rest was dis-

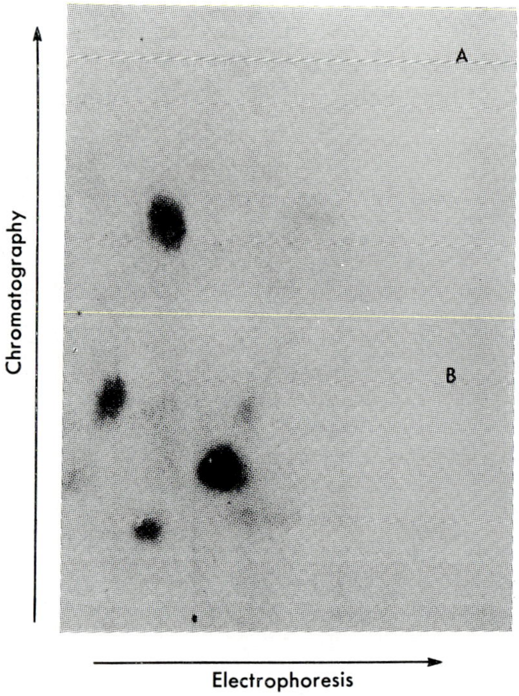

Fig. 13. Fingerprints of the tryptic digest of [^{14}C]-NEM-labeled myosins A. S_a and S_1 modifications were cariried out by incubating 3 mg/ml myosin B with 0.1 mM [^{14}C]-NEM in 0.04 M Tris-maleate buffer (pH 7.5) for 15 min at 0°C with stirring and by reacting 10 mg/ml of myosin A with 0.1 mM [^{14}C]-NEM in 0.04 M Tris-maleate buffer pH 7.9) at 0°C for 30 min, respectively. The modified proteins were digested with trypsin and the specified amount of the digest was run in one dimension by chromatography in pyridine-butanol-1 mM EDTA (3: 3: 4) and in the second dimension by electrophoresis at pH 8.8. A: 1.8 mg of tryptic digest of S_a-blocked myosin A. B: 0.42 mg of tryptic digest of S_1-blocked myosin A. A small dot on the bottom left of the photograph indicates the origin.

tributed in several spots. Radioactivity corresponding to about 0.7 mol of [^{14}C]-NEM per 450,000 *g* of myosin was present in the major spot (named S_a-peptide), whereas less than 0.1 mol of [^{14}C]-NEM/mol of myosin was located in other spots. Which part of the myosin molecule contains the S_a-peptide, HMM or LMM, was examined using high-voltage paper electrophoresis. Figure 14 shows that a radioactive pattern identical to that of S_a-peptide was obtained with LMM digest prepared from [^{14}C]-NEM-modified myosin B but not with HMM digest, showing that S_a-peptide [Thr(or Ser)–Asn–Ala–Ala–Cys–Ala–Ala–Leu–Asp–Lys–Lys] is involved in the LMM moiety of the myosin molecule. Next, the possible location of S_a in LMM was studied using a specific chemical cleavage method at the amino peptide bonds of cysteine residue, followed by SDS-

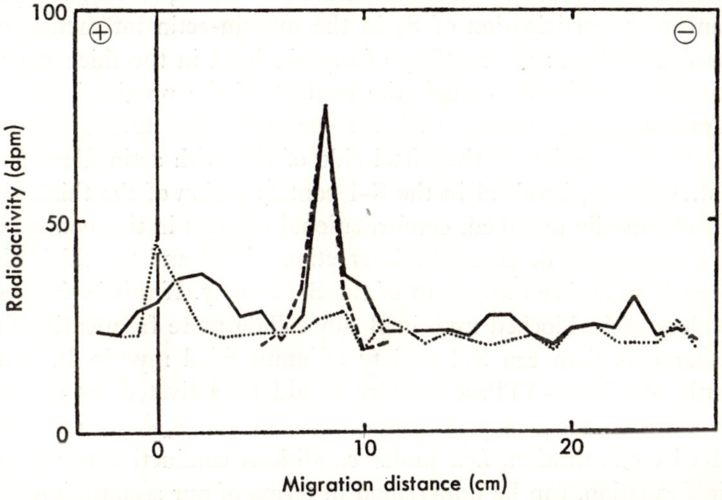

Fig. 14. High-voltage paper electrophoresis of S_a-peptide and the tryptic digests of modified-HMM and -LMM prepared from [^{14}C]-NEM-modified myosin B. [^{14}C]-NEM-modified myosin A was separated from S_a-blocked myosin B modified with radioactive NEM under the same conditions as F in Table II. From modified myosin A, modified HMM and LMM fraction-1 were prepared by digestion with trypsin as described in the legend to Fig. 12. After the modification of their remaining sulfhydryl groups with cold NEM, HMM and LMM were digested completely with TPCK-trypsin. These tryptic digests and S_a-peptide were electrophoresed at room temperature (2,000 V for 3.5 hr), using pyridine-acetic acid-water (1: 10: 289 by volume, pH 3.7). After the paper had dried, it was cut into 1 cm widths and the radioactivity of the pieces was counted in a liquid scintillation spectrometer. Solid, dotted, and broken lines indicate LMM and HMM digests and S_a-peptide, respectively.

gel electrophoresis. The S_a of LMM was preferentially modified with DTNB and the cleavage at the S_a cysteine residue produced a fragment with a molecular weight of 75,000 from LMM (mol. wt. 79,000). This result and the arrangement of the peptide fragments obtained by specific chemical cleavage of myosin rods suggested that S_a was located at a position of 4,000 in terms of molecular weight from the amino terminus of LMM.

DISCUSSION

The mechanism of the effect of modification of S_a in LMM on the interaction of myosin with actin is unclear. However, based on our results with S_a-blocking and on the reports of Sutoh and Harrington (1977), Mendelson *et al.* (1973), and Mendelson and Cheung (1976), one possible mechanism of participation of S_a in the myosin-actin interaction can be proposed as follows: the motility of myosin head in the thick filament is considerably restricted through the contact of the proximal side of the subfragment 1 (S-1) moiety with the surface of the thick filament core, so that the interaction of the distal side of S-1 with actin filament is restricted. If the S_a involved in the S-1 contact region of the thick filament core, is chemically modified, conformational changes in this contact region occur which would decrease the interaction of S-1 and the thick filament core, resulting in free movement of the S-1 moiety. If this is the case, the S-1 moiety of S_a-blocked myosin B can collide more frequently with the actin filaments than can S-1 moiety of unmodified myosin B, and consequently the Mg^{2+}-ATPase activity would be activated by an increase in the collision frequency. Disappearance of the S_a-blocking effects at high KCl concentration, *i.e.*, under conditions conductive to monomeric myosin formation, can be understood in terms of our assumption that the conformational change in the LMM region produced by S_a-blocking affects the ATPase-active head of the adjoining molecule.

REFERENCES

Daniel, J. L. and Hartshorne, D. J. (1972). *Biochim. Biophys. Acta* **278**, 567–576.
Daniel, J. L. and Hartshorne, D. J. (1973). *Biochem. Biophys. Res. Commun.* **51**, 125–131.
Horigome, T. and Yamashita, T. (1977). *J. Biochem.* **82**, 1085–1092.
Horigome, T. and Yamashita, T. (1978). *J. Biochem.* **83**, 49–56.
Horigome, T. and Yamashita, T. (1979a). *J. Biochem.* **85**, 221–228.
Horigome, T. and Yamashita, T. (1979b). *J. Biochem.* **85**, 229–237.

Mendelson, R. A., Morales, M. F., and Botts, J. (1973). *Biochemistry* **12**, 2250–2255.
Mendelson, R. A. and Cheung, P. (1976). *Science* **194**, 190–192.
Sekine, T., Barnett, L. M., and Kielley, W. W. (1962). *J. Biol. Chem.* **237**, 2769–2772.
Sutoh, K. and Harrington, W. F. (1977). *Biochemistry* **16**, 2441–2449.
Yamaguchi, M. and Sekine, T. (1966). *J. Biochem.* **59**, 24–33.
Yamashita, T., Soma, Y., Kobayashi, S., and Sekine, T. (1974). *J. Biochem.* **75**, 447–453.
Yamashita, T., Kobayashi, M., and Horigome, T. (1975). *J. Biochem.* **77**, 1037–1046.
Yamashita, T. and Horigome, T. (1977). *J. Biochem.* **81**, 933–939.

Muscle Contraction: Its Regulatory Mechanisms (*Ebashi, S. et al., eds.*), *pp. 133–143, Japan Sci. Soc. Press, Tokyo / Springer-Verlag, Berlin (1980)*

10

Commentary on Certain Problems in the Use of Molecular Probes in Muscle Protein Research

MANUEL F. MORALES AND JEAN BOTTS

Summary: This paper discusses two questions that commonly arise in using molecular probes. One is the question of whether an extrinsic probe being used to measure the affinity between two macromolecules is itself affecting the affinity. A test is formulated analytically, and is then applied illustratively to the measurement of acto-(subfragment 1) affinity. Numerical analysis shows that the test is capable of detecting probe effects if these are causing changes in affinity by factors of two or more. The other question is whether there is a logical contradiction between the assumptions underlying two different applications of probes. [In time-resolved anisotropy decay work rotation of probe optic axes is used to infer rotation of the macromolecule; in energy-transfer work the probe axes are taken to be randomized (orientational factor $\sim 2/3$)]. The question is not answered quantitatively, but arguments (based on two models for probe motion) are advanced to show that no contradiction *need* exist.

As is generally known, small fluorescent or paramagnetic molecules ("probes") artifically attached to the macromolecules of muscle can sometimes "report" useful information about the static properties of the system and about its dynamic behavior. Thus far the majority of studies with probes has centered on the contractile system, but there is a growing number of studies on the control system. Therefore it may be *a propos* in this symposium to comment on two general questions about probes that have arisen in studies of the contractile system.

The nemesis of all probe work is the danger that the attached probe may significantly affect the very phenomenon that it is intended to "observe." Sometimes there are ways of ascertaining whether this is so. Many observations have been made for example, by attaching SH-directed probes to the reactive (SH_1) thiols of the myosin subfragment 1 (S-1) moieties of muscle fibers. In those cases investigators were usually able to show that, say, the tetanic tensions developed by such fibers were not significantly affected by the attached probes. In studies of isolated proteins the situation is less definite. Probes have been used to sense rotational motion, to estimate point-to-point distances, and to measure protein-protein affinities in solution in which the probe method was often the only available method, and no parallel independent check seemed possible. Since often myosin has been one of the proteins in the system, investigators have resorted to asking whether attaching probes affects myosin ATPase. While this is somewhat illogical,* there has seemed to be no other way. As our first comment we want to say that there is another way, as illustrated by studies of protein-protein affinity.

Time-resolved fluorescence anisotropy decay (TRFAD) apparatus enables one to measure the rotational correlation time, Φ, of a labelled protein. If the labelled protein is in a binding equilibrium with another protein, the TRFAD curve reflects the fact that anisotropy decay is accomplished by the Brownian rotation of two populations of fluorophores (those on protein that is bound to the second protein and those on protein that is not bound), characterized by two rotational correlation times. Curve-fitting to find the mole fraction of, say, unbound fluorophores, $1-y$, is particularly easy when the rotational correlation time of the bound fluorophore is effectively infinite (*e.g.*, labelled S-1 bound to massive long strands

* Since there is no guaranteed correspondence between the measurement achieved by the probe (*e.g.*, measuring a distance between two points) and overall ATPase catalysis, neither observing of an effect nor observing no effect guarantees that the measurement made with the probe is or isn't affected by the probe. In the acto-myosin system the situation is even murkier, since attaching probes at SH_1 doesn't affect acto-myosin ATPase but does affect acto-S-1 ATPase.

of F-actin). If the mole fraction of unbound unlabelled protein, $(1-x)$, is exactly the same—*i.e.*, if labelling does not affect the measurement—then $x=y$, and the affinity constant calculated is representative of all S-1 molecules. However, if we assume that labelling does affect affinity we must proceed a little differently. Suppose that the molar concentration of actin is A, and the molar concentration of all forms of S-1 is B. The fraction of S-1 that is labelled, p, is under the experimenter's control (or at least it is ascertainable, either by using a radioactive probe or by denaturing the labelled protein and measuring label absorbance), so the concentrations of unlabelled and labelled S-1, $(1-p)B$ and pB, are known. If x and y are the mole fractions of bound unlabelled and labelled S-1, then the sum of the corresponding concentrations is $(1-p)Bx+pBy$, or,

$$GB \equiv [(1-p)x+py]B \tag{1}$$

and the affinity constants for actin of unlabelled and labelled S-1 (now assumed to be different) are,

$$K=[x/(1-x)](A-GB)^{-1} \tag{2}$$

$$J=[y/(1-y)](A-GB)^{-1} \tag{3}$$

If $x=y$, G is independent of p (Eq. 1), but if $x\neq y$, G is a function of p, and so therefore are x and y (Eqs. 2 and 3). Thus if from a TRFAD experiment one finds $y=y_1$ for a given A, B, and p_1, one can ask whether keeping A and B the same but changing p_1 to p_2 one can change y to y_2. In the strict mathematical sense $y_1=y_2$ means that $x=y$, $K=J$, and labelling has no effect on affinity, and conversely if $y_1\neq y_2$. At the suggestion of one of us (J.B.) this test was applied to acto-(S-1) affinity measurements by Highsmith *et al.* (1976). They found that in the range, $0.18\leqslant p\leqslant 0.80$, $y_1=y_2$ within their experimental accuracy, so they concluded that their measurements were unaffected by labelling; their conclusion was strengthened by reasonable agreement with the results of others using non-labelling techniques (Marston and Weber, 1975; Margossian and Lowey, 1978). We have thought it of interest to examine the *sensitivity* of the "varying p" test, granting that in principle it offers a means of deciding the issue of label effect.

What we have done on a computer is to use Eqs. 1, 2, and 3 to calculate values of K and J that would be obtained for experimentally plausible values of A and B and a median value of p. To do this we have assumed y's in the range of measurements made in our laboratory (Highsmith *et al.*, 1976; Mendelson *et al.*, 1975; Wadzinsky *et al.*, 1979), and for each y an x in the range, $y\leqslant x\leqslant 1$ (*i.e.*, we have assumed that native S-1 always

TABLE I
Representative Computer Calculations Showing the Percent Change in the Measured Parameter (y, the mole fraction of bound, labelled S-1) When the Degree of Labelling, p, is Changed from 0.1 to 0.9
For definition of other symbols see text.

A^a	B^a	K^b	J^b	$(\Delta y/y)\cdot 100$
0.1	0.1	2	1	7.4
0.1	0.1	45	7	45.0
0.1	0.1	45	6	47.0
0.1	0.1	100	8	65.0
0.5	0.5	7	5	3.8
0.5	0.5	9	1.4	45.0
0.5	0.5	65	6	52.7
0.5	0.5	75	5	69.7
1	1	1	0.5	12.5
1	1	120	20	15.9
1	1	12	0.7	88.2
1	1	30	3	50.0

[a] Concentrations of actin (monomer) and of S-1 are expressed in micromolarity.
[b] Affinity constants are multiples of $10^6\ \text{M}^{-1}$.

binds more tightly than modified S-1 (Botts *et al.*, 1979). The K, J pairs have then been taken to be "unknown situations" conceivably met in practice. Now solving the equations in the reverse direction (for x and y) we have asked (Table I) what percent change would be observed in the measured variable, y, as p is changed from 0.1 to 0.9. Table I shows that within the framework of our calculations our "varying p" test can only marginally detect situations in which labelling halves the binding affinity, but can reliably detect effects greater than that. While at first sight this sensitivity may seem unremarkable one must remember that with the present "state of the art" protein-protein affinities of the order of $10^6\ \text{M}^{-1}$ are seldom measured with an accuracy better than within a factor of two; also, because free energies are logarithmic functions of affinities, the test has a very respectable sensitivity if the question is whether labelling affects the measurement of free energy of binding.

To summarize our first comment: when probe methods are used to determine the affinity of a labelled protein to another protein, variation of the fraction of labelled protein bound with variation in the degree of labelling is diagnostic of whether labelling is affecting the affinity. But the test is reliable only if the affinities of labelled and unlabelled protein differ by a factor of at least two.

Our second comment concerns the motion of a molecular probe relative to the macromolecule to which it is attached. Measurements from our laboratory have caused some confusion about the consequences of this effect, and therefore a comment is timely. Some years ago we introduced TRFAD as a means of studying the rotation of the S-1 moiety of myosin (Mendelson *et al.*, 1973). We found that a dye, 1,5-N-iodoacetyl-N′-(1-sulfo-5-naphthyl)-ethylene diamine (IAEDANS), attached to the SH_1 thiol of free S-1 had a rotational correlation time Φ of 150–200 nsec (depending on solvent composition, temperature, *etc.*), and that Φ lengthened greatly when S-1 attached to F-actin. We could rationalize these various observations by assuming that the dye molecule was rigidly attached to an S-1 that was roughly a prolate ellipsoid of revolution, in such a way that the optic axes of the dye were approximately parallel to the major axis of the ellipsoid. Our conclusions were strengthened by the harmonious independent findings of Gergely and his colleagues, Thomas, Seidel, and Hyde (1975); and also later by our own X-ray scattering studies of S-1, showing that to a rough approximation S-1 could be modelled as an ellipsoid similar to that envisioned in the fluorescence work (Kretzschmar *et al.*, 1978). Five or six years have passed, and our laboratory has turned its attention to fluorescence resonance energy transfer in an effort to find geometrical relations in the acto-(S-1) complex. Assumptions made in the interpretation of this latter work may seem at first sight inconsistent with our earlier work.

In a paper soon to appear (Takashi, 1979), Reiji Takashi has shown that a fraction of the excitation energy absorbed by 1,5-IAEDANS attached to cysteine 373 of actin is transferred to fluoresceine attached to SH_1 of S-1 when the two molecules are in the rigor complex. Energy transfer of this kind is interpreted according to the treatment of von Förster. von Förster's treatment is quantum mechanical, but the plausibility of his result can be seen from a classical argument about dipole-dipole interaction. A dipole, $\vec{p}_1$ parallel to the z axis at the origin generates an electric field at $\vec{r}$,

$$\vec{E}_1(\vec{r}) = -\frac{1}{r^3}\vec{p}_1 + \frac{3(\vec{p}_1\cdot\vec{r})}{r^5}\vec{r} \tag{4}$$

If at $\vec{r}$ we place a second dipole, $\vec{p}_2$, then the force that $\vec{E}_1$ exerts along $\vec{p}_2$ is,

$$\vec{p}_2\cdot\vec{E}_1 = -\frac{1}{r^3}\vec{p}_1\cdot\vec{p}_2 + \frac{3(\vec{p}_1\cdot\vec{r})(\vec{p}_2\cdot\vec{r})}{r^5} \tag{5}$$

or setting $\vec{p}_i = q_i d_i \hat{p}_i$ and rewriting in terms of unit vectors,

$$\vec{p}_2 \cdot \vec{E}_1 = (q_1 d_1)(q_2 d_2) \frac{[3(\hat{p}_1 \cdot \hat{r})(\hat{p}_2 \cdot \hat{r}) - (\hat{p}_1 \cdot \hat{p}_2)]}{r^3} \tag{6}$$

The rhs of Eq. 6 is symmetrical, and would also be the expression for $\vec{p}_1 \cdot \vec{E}_2$. We can surmise that the probability of resonance is proportional to both $\vec{p}_2 \cdot \vec{E}_1$ and to $\vec{p}_1 \cdot \vec{E}_2$, therefore to,

$$\frac{[\hat{p} \cdot \hat{p} - 3(\hat{p} \cdot \hat{r})(\hat{p} \cdot \hat{r})]^2}{r^6} \tag{7}$$

Thus the probability of transfer should be proportional to r^{-6}, and also to a factor, κ^2, which with the aid of Fig. 1 can be interpreted as,

$$\kappa^2 = [\cos \alpha - 3 \cos \beta \cos \gamma]^2 \tag{8}$$

The probability of transfer can be measured, but it cannot immediately be used to measure r because ordinarily one does not know the angles α, β, γ, relating the optic axes of the donor and acceptor dyes, and κ^2 can assume any value between 0 and 4 depending on these angles. von Förster showed that if either $\vec{p}_1$ or $\vec{p}_2$ are randomly directed in space, $\kappa^2 \rightarrow 2/3$. If κ^2 can be assigned such a value then indeed the energy transfer method becomes, as Lubert Stryer calls it, a "spectroscopic ruler." As a matter of fact the method remains important even if κ^2 is only near to 2/3, since estimation of r requires taking the sixth root of an experimental expression. Of course if κ^2 were exactly zero (dipoles perpendicular to $\vec{r}$ and to each other) then the fraction, ε, of the excitation energy transferred would be zero even if $\vec{r}$ were vanishingly small. Because of the theoretical benefits of having κ^2 turn out to be 2/3 investigators have looked for evidence that the attached fluorophores either are in random motion relative to the macromolecule to which they are attached or else that they have multiple non-paralleled transition moments (*e.g.*, spherically symmetrical ions). At wavelengths greater than *ca.* 400 nm our much-used 1,5-IAEDANS dye appears to have a single congruent transition moment for absorption and emission. We have inquired whether the single axis of this dye when the

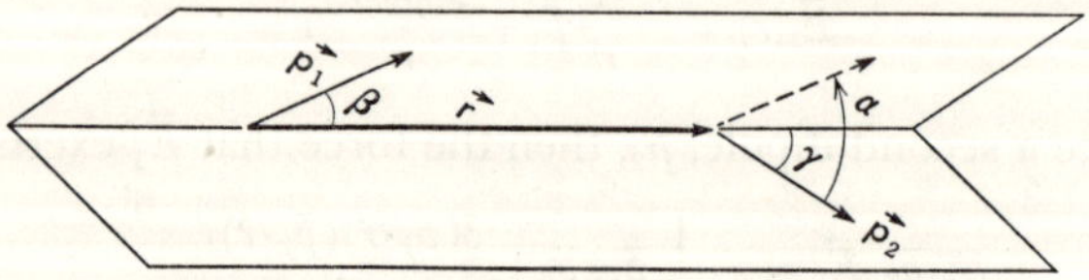

Fig. 1. The angles, α, β, and γ, that define the orientational relation between two dipoles, and whose cosines go to make up the "orientation factor," κ^2.

dye is attached moves relative to the macromolecule. Several observations suggest that it does. The first observation is one of dye interchange. Because of the excellent spectral overlap between them we routinely use 1,5-IAEDANS as the donor and iodoacetyl fluoresceine (IAF) as the acceptor. Although both dyes have iodoacetyl reactivity toward proteins their ring systems are very different in size and composition. *A priori* one expects the atomic neighborhoods of actin Cys-373 and of myosin SH_1 also to be quite different. Therefore it would be an outrage if accidentally the *static* attitudes of $\vec{p}_1$ and $\vec{p}_2$ have the same relationship to one another when IAEDANS is on actin and IAF is on myosin as when IAF is on actin and IAEDANS is on myosin. When Takashi made this interchange in the acto-(S-1) complex he observed *no* effect on ε. The second observation had to do with disrupting surface structure by applying certain ions. It is well known that ions such as SCN^- and ClO_4^- disrupt the structure of proteins and other macromolecules. So one might think that if $\vec{p}_1$ and $\vec{p}_2$ are holding some fixed relation to one another solutions of ions capable of wrecking ATPase might "thaw out" the fixed relation and thereby change ε. However, Takashi found again that ε remained unchanged. A rationalization of both of these observations is that the attached probes were already in random motion before both maneuvers (interchange and exposure to ions), in which case it would have been appropriate to use $\kappa^2=2/3$. The third observation concerns the behavior of the function of the polarized intensities known as the anisotropy, $r=(I_{\parallel}-I_{\perp})/(I_{\parallel}+2I_{\perp})$. It is possible to calculate r mathematically for an ensemble of randomly oriented *immobilized* fluorophores. If the moments are congruent (*e.g.*, IAEDANS at $\lambda \geqslant 400$ nm) the result is 2/5. Hudson and Weber (1973) froze IAEDANS at solid CO_2 temperatures, and measured $r(\lambda)$. As the top curve of Fig. 2 shows, the measured anisotropy of this compound when frozen approximates the theoretical 2/5 at higher wavelengths. We made the same measurement* on IAEDANS co-valently attached to S-1 particles in solution, but we added increasing concentrations of sucrose (0% to 40%) in order to slow down the diffusion of the particles. At each λ we measured the value of r extrapolated to $T/\eta \to 0$, where T is the absolute temperature and η is the viscosity. The results in the lower curve of Fig. 2 show that at all λ the emission of the attached dye is less polarized than that of the truly immobilized dye. We attribute this difference to the motion of the attached dye. On learning of these recent observations and

* And, through the courtesy of Professor G. Weber, on the same fluorometer.

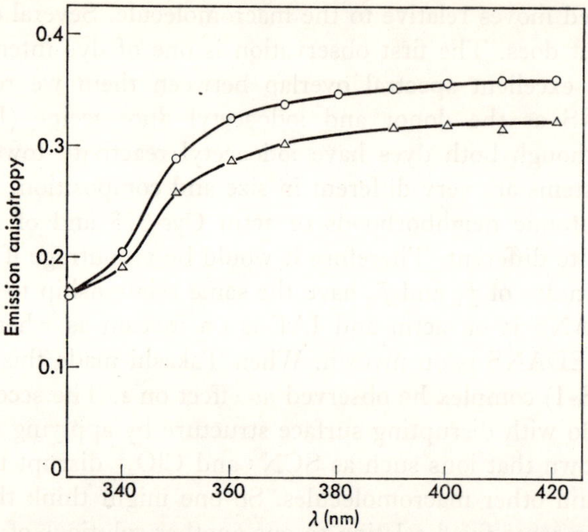

Fig. 2. Emission anisotropy of the dye, 1,5-IAEDANS as a function of wavelength. The top curve is the data of Hudson and Weber (1973), and refers to the dye frozen and practically immobilized. The bottom curve is our data for dye bound covalently to S-1 that is prevented from rotating by the viscosity of the medium.

interpretations it has occurred to some of our friends as well as to our referees that we may be on the horns of a dilemma. While making deductions about S-1 from TRFAD work we seem to claim that an attached dye such as IAEDANS on SH_1 is rigidly bound to S-1, whereas when we seek to make distance estimates from energy transfer measurements we say that the same dye, attached to the same thiol, is in random motion. We want to make it plausible to you that both statements are true; the reconciliation lies in the difference in speed of the two motions.

In TRFAD experiments we are accustomed to plots like that of Fig. 3A. We usually consider the data before some limiting time, t_1, as meaningless because the equation, $r = r(0) \exp(-t/\Phi)$, is derived on the assumption that there is only one relaxation process initiated by a lamp flash that is truly a "delta function" (Fig. 3B). In practice it is not a delta function, but more like the dotted curve of the figure, with a width of, say, 3 nsec. If really we were observing the relaxation of only one process, then the ln r-intercept obtained by extrapolation should give us an $r(0)$ that is the anisotropy of an immobilized randomly-oriented ensemble; if $\lambda \geqslant 400$ nm it should give us 2/5. But in actual practice this extrapolation always gives an apparent $r(0)$ less than 2/5. As we have explained, our apparatus cannot

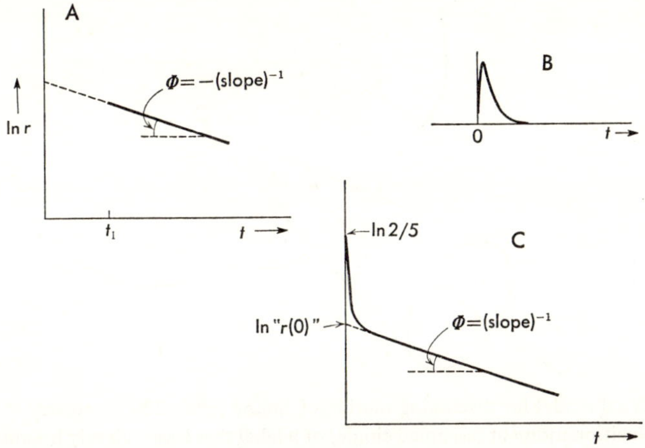

Fig. 3. Effect of label motion on TRFAD plots, ln $r(t)$ *vs*. t. A: schema of a plot as conventionally used to calculate Φ; early data (up to t_1) is ignored. B: schemas of lamp flash contours. Interpretative analysis assumes that it is a spike (delta function) at $t=0$; in practice it is usually an unsymmetrical curve. C: conception of a TRFAD plot under conditions in which meaningful data is collected from $t=0$ onwards. The initial value of anisotropy is the theoretical 2/5, but extrapolation from the portion of the curve that generates Φ leads to the calculation of an apparent anisotropy less than 2/5 due to rapid label motion.

yet assess precisely how much less, but the difference is not unlike that found using a steady state fluorometer (Fig. 2). So it seems likely that if we could observe TRFAD accurately between 0 and t_1 the trace would look like Fig. 3C. In other words, besides the depolarization on the nsec time scale that we have ascribed to the rotational Brownian motion of S-1 (and from which we have made deductions about S-1) there is a depolarization due to the superimposed motion of the label—the latter being very much faster (psec time scale) and thus manifesting itself on the TRFAD plot as merely a displaced ln r-intercept. If S-1 is taken to be rigid, and if we are conducting an energy transfer experiment between two sites on S-1* then the motion on the nsec scale would not affect the relation between the attitudes of the moments of the two dyes, but the motion on the psec scale would tend to randomize the relation. Even if

* Of course no protein is truly rigid; breathing movements illustrate the point. Also, if we are dealing with a molecular complex A-M—one site on protein A and the other site on protein M—there is sure to be some rocking motion at the interface. In both of these situations the relation between donor and acceptor movements would be blurred.

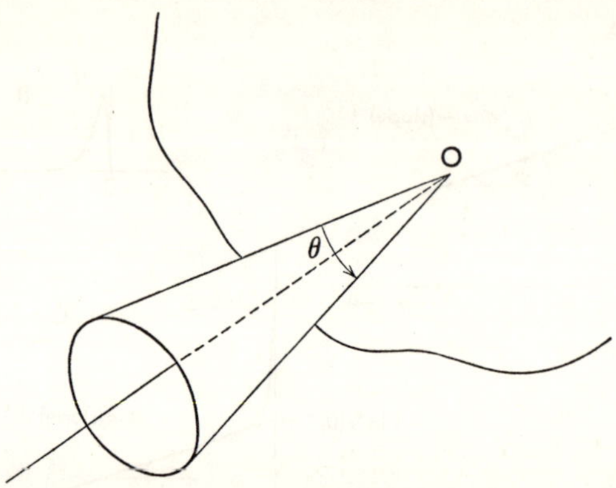

Fig. 4. Usual model for discussing motion of bound label. The symmetry axis coincides with the transition moment (assumed single) of a label that is co-valently bound to a macromolecule.

the relation called for $\kappa^2=0$ in the truly frozen state the motion of the labels would make it different from zero at room temperatures. The foregoing account of how things might be has not yet been quantitatively established, but it does show that there need not be a *logical* conflict between current interpretations of TRFAD and of energy transfer experiments.

In closing our second comment it is of interest to consider models that simulate both the slow and the rapid motions of the label. In the usual treatment illustrated in Fig. 4 the optic axis of the dye (assumed to be single for simplicity) is taken as a line segment which at its root end is fixed to the macromolecule. The distal tip is assumed to move randomly on the surface of the unit sphere, but to stay within a restricted area of surface (taken as circular for mathematical simplicity). Thus the line segment rapidly and randomly moves within a conical volume. The symmetry axis of the cone, representing a time-average position of the line segment is, of course, *rigidly* oriented relative to the macromolecule, but as the entire macromolecule rotates the symmetry axis rotates relative to the "laboratory framework" of coordinates. One may think of the rapid motion within the cone as producing the almost zero-time displacement in $r(0)$, and of the slow motion of the symmetry axis as generating the Φ-value that we ascribe to S-1 in TRFAD experiments. There may be a somewhat different way of looking at the same physical situation. We can imagine that in an ensemble of labelled S-1 molecules the labels can exist in one of two states—fixed to the S-1 absolutely rigidly, or considerably free to rotate (say in 2π steradians). We can suppose further that there is a very rapidly attained equilibrium between the two states. This model assumes in effect that there are two kinds of labels, one generating the vertical displacement in $r(0)$ and the other generating the Φ of S-1. If this model were correct it would facilitate energy transfer calculations, for then $\Phi = 2/3$ would be assigned only to the "free" labels. At present we cannot distinguish experimentally between the two views, but either provides a picture of how a probe could report on the rotation of a macromolecule and yet have a motion that makes $\kappa^2 \sim 2/3$ in energy transfer work.

Acknowledgment

The authors are privileged to contribute to a symposium honoring Professors Kumagai and Natori, and also express their indebtedness to the research support provided by the USPHS (HL-16683 and HL-06285), by the USNSF (PCM 75-22698), and by the AHA (CI-8).

REFERENCES

Botts, J., Ue, K., Hozumi, T., and Samet, J. (1979). *Biochemistry* (in press).

Highsmith, S., Mendelson, R. A., and Morales, M. F. (1976). *Proc. Natl. Acad. Sci. U.S.* **73**, 133.

Hudson, E. and Weber, G. (1973). *Biochemistry* **12**, 4154.

Kretzschmar, M. K., Mendelson, R. A., and Morales, M. F. (1978). *Biochemistry* **17**, 2314.

Margossian, S. and Lowey, S. (1978). *Biochemistry* **17**, 5431.

Marston, S. and Weber, A. (1975). *Biochemistry* **14**, 3868.

Mendelson, R. A., Morales, M. F., and Botts, M. F. (1973). *Biochemistry* **12**, 2250.

Mendelson, R. A., Putnam, S. V., and Morales, M. F. (1975). *J. Supramol. Struct.* **3**, 162.

Takashi, R. (1979). *Biochemistry* (in press).

Thomas, D. D., Seidel, J. C., Hyde, J. S., and Gergely, J. (1975). *Proc. Natl. Acad. Sci. U.S.* **72**, 1729.

Wadzinsky, L., Botts, J., Wang, A., Woodard, J., and Highsmith, S. (1979). *Arch. Biochem. Biophys.* (in press).

DYNAMIC ASPECTS OF THIN FILAMENTS

Muscle Contraction: Its Regulatory Mechanisms (Ebashi, S. et al., eds.), pp. 147–164,
Japan Sci. Soc. Press, Tokyo / Springer-Verlag, Berlin (1980)

11

Structure and Activity of the Thin Filament

E. J. O'BRIEN, E. P. MORRIS, J. SEYMOUR, AND J. COUCH

Summary: Paracrystals formed from actin+tropomyosin filaments are of two types, one ribbon-like in which adjacent filaments are antiparallel and the other more compact with the cross-over points of the long-pitch actin helices in adjacent filaments aligned. In addition, the actin helical symmetry is slightly different in each type. When the filaments contain also troponin-I (TN-I), the two paracrystal types are again formed, but the position of tropomyosin in the long-pitch helical grooves of actin is changed, being closer to the edge of each groove. Since TN-I causes inhibition of the ability of the filaments to activate myosin ATPase, this inhibition may be correlated with the altered tropomyosin position but not with a particular paracrystal type or actin symmetry.

In paracrystals containing actin+tropomyosin+TN-I+troponin-C, the tropomyosin occupies a position at the edge of each actin groove, although in this case the filament is not inhibitory. This suggests that the movement of tropomyosin in the groove may not be the only important structural change inducing regulation.

The hypothesis that the tropomyosin movement takes place between positions either permitting or obstructing myosin attachment, depends on the assumption that the attachment site is on the same side of the actin groove as the tropomyosin. Structural analysis of thin filaments still attached to the Z line establishes the polarity of the filament with respect to the characteristic arrowhead appearance of subfragment 1 decoration, and indicates that this assumption is false, unless the shape or attachment site of myosin to actin is revised.

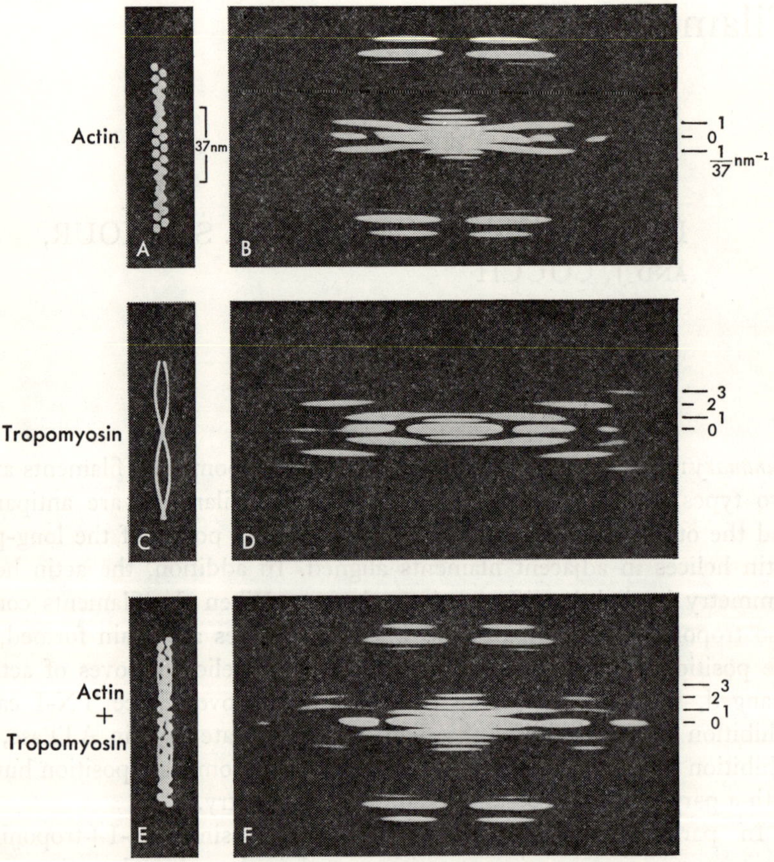

Fig. 1. A, C, and E are computer-drawn models of actin, tropomyosin, and actin+tropomyosin respectively, and B, D, and F are the corresponding optical diffraction patterns obtained from these models. In E the relative azimuthal angle of actin and tropomyosin is 70°, as in Fig. 7 A.

THE THIN FILAMENT AND ITS DIFFRACTION PATTERN

The thin filament of vertebrate skeletal muscle comprises three proteins: actin, tropomyosin, and troponin. Actin has the appearance of two long-pitch helical strands of globular subunits wound round each other, and in each of the grooves running along the filament between these strands are tropomyosin molecules joined end to end, with troponin molecules attached to the filament at axial intervals of 38.5 nm. This is the structure as first proposed by Ebashi *et al.* (1969) and since confirmed by a wealth of evidence principally from electron microscopy and X-ray diffraction studies.

The essential features of the low-angle diffraction pattern of actin+tropomyosin are illustrated in Fig. 1, where the optical diffraction patterns obtained from computer-drawn models of actin, tropomyosin and actin+tropomyosin are compared. The subunit structure of actin gives rise to reflections near the meridian at an axial spacing of about 6 nm, but the smoothly helical tropomyosin contributes only to the first few layer-lines. When the two proteins are combined, tropomyosin contributes significantly to the second layer-line, which is very weak in the diffraction pattern from pure actin. Moreover, azimuthal movement of the tropomyosin in the long-pitch helical grooves of actin modulates the intensity on the second layer-line (Hanson *et al.*, 1972; Haselgrove, 1972; H. E. Huxley, 1972; Parry and Squire, 1973), so the behaviour of this layer-line is of key importance in monitoring the tropomyosin position. Although not shown in Fig. 1 the principal effect of adding troponin to the filament is to introduce in the diffraction pattern meridional reflections at orders of 38.5 nm, of which the first is the strongest.

EFFECT OF Ca^{2+}

Gillis and O'Brien (1975) and O'Brien *et al.* (1975) studied the structure of reconstituted thin filaments in paracrystals induced by Mg^{2+}, and found differences according to whether the ambient Ca^{2+} concentration was "high" ($>10^{-5}$ M) or "low" ($<10^{-6}$ M). In electron micrographs of paracrystals at high Ca^{2+} the filaments were relatively thicker and widely spaced whereas at low Ca^{2+} they were thinner and closely spaced. In optical diffraction patterns of the micrographs, the second layer-line was strong when the Ca^{2+} level was high and weak when low, indicating a movement of tropomyosin from near the centre of the actin groove (as in Fig. 1E)

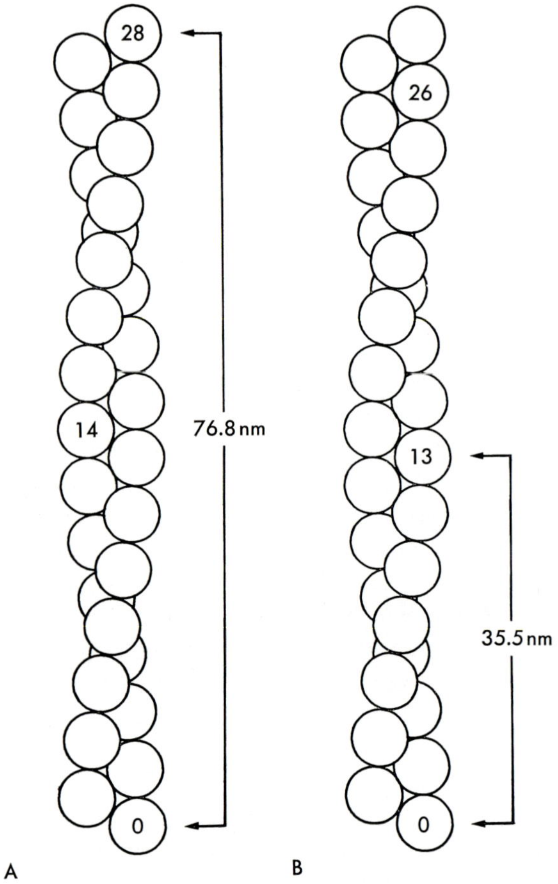

Fig. 2. A: actin with 28 subunits in 13 turns of the genetic helix, as present in ribbon-like paracrystals with antiparallel filaments (see Figs. 3 A, 5 A, 6, and 7). B: actin with 13 subunits in 6 turns of the genetic helix, as present in compact paracrystals with the cross-over points of the long-pitch helices in adjacent filaments aligned (see Fig. 4 A). The measurements marked indicate in each case the distance between subunits with identical orientation, *i.e.*, the crystallographic repeat.

to near the edge as the Ca^{2+} decreased. Measurements of the axial positions of the layer-lines in the diffraction patterns indicated a difference in the helical symmetry of the actin according to the paracrystal type. In the high Ca^{2+} paracrystals the helix had 28 subunits in 13 turns of the genetic helix, whereas in the low Ca^{2+} paracrystals there were 13 subunits in 6 turns, that is, a slightly tighter conformation (2.167 *vs.* 2.154 subunits/

turn). The two helices are shown in Fig. 2. For the 13/6 helix the axial crystallographic repeat is equal to the cross-over repeat of the two long-pitch helices, whereas for the 28/13 helix it is twice this repeat. From the positions of row-lines in the diffraction patterns and from the addition of heavy meromyosin to "decorate" the paracrystals, these authors also showed that in high Ca^{2+} paracrystals adjacent filaments were antiparallel and in the low Ca^{2+} paracrystals probably parallel.

The X-ray diffraction studies of Haselgrove (1972) and H. E. Huxley (1972) on frog muscle and of Vibert *et al.* (1972) on molluscan muscle have indicated a movement of tropomyosin on activation, but there is little evidence for a change of actin helix (H. E. Huxley and Brown, 1967). The measurement of layer-line positions in the pattern from contracting muscle is not very accurate, however. In the X-ray diffraction pattern from crab striated muscle, Maeda, Matsubara, and Yagi (1979) found some change when the muscle passed into rigor, a change which was in the same direction as that observed by Gillis and O'Brien (1975), but smaller in magnitude. Our experiments have been designed to further evaluate the significance of the helix change and of the tropomyosin movement observed in paracrystals.

ACTIN+TROPOMYOSIN PARACRYSTALS

Two types of Mg^{2+}-induced paracrystals of actin+tropomyosin are observed, in which the filament packing resembles the two forms observed in thin filament paracrystals at high and low Ca^{2+}. In Fig. 3A the filaments of actin+tropomyosin are relatively thick and widely separated and the paracrystals have a flat ribbon-like appearance, whereas in Fig. 4A the paracrystals are more compact with the filaments closer together, and the cross-over points of the long-pitch actin helices in adjacent filaments are aligned. In all these features the paracrystals resemble those formed from reconstituted thin filaments at high and low Ca^{2+}, apart from lacking the cross striation due to troponin that is observed in high Ca^{2+} paracrystals (Gillis and O'Brien, 1975).

In the optical diffraction pattern obtained from the ribbon-like paracrystals (Fig. 3B) the row-lines are spaced at 1/16 nm^{-1}, so that, since the filament width is about 8 nm, the unit cell contains two filaments. The row-line spacing in the pattern from the aligned paracrystals (Fig. 4B) corresponds to the interfilament spacing. The intensity of the second layer-line is relatively strong in both diffraction patterns. The layer-line posi-

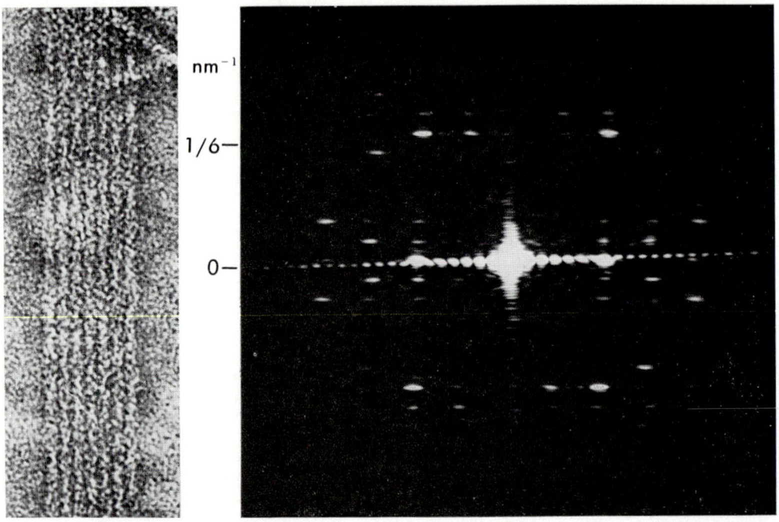

Fig. 3. Electron micrograph of a ribbon-like paracrystal of actin+tropomyosin, negatively stained (×270,000) (A) and its optical diffraction pattern (B). The transverse crystallographic repeat spans two filaments, and row-lines in the diffraction pattern are therefore at 1/16 nm^{-1}. Note the strong second layer-line attributable to the presence of tropomyosin in the long-pitch helical grooves of the actin (as in Fig. 1 E and F).

tions index on a helix with 28/13 symmetry in Fig. 3B and 13/6 in Fig. 4B, confirming the similarity of the paracrystals to the two types observed for reconstituted thin filaments.

The ribbon-like paracrystals of actin+tropomyosin have been described by O'Brien *et al.* (1975) and a three-dimensional reconstruction of the filament in these paracrystals (O'Brien and Couch, 1976) shows that the tropomyosin is located at a radius of 4.0 nm from the filament axis and at an azimuthal angle of 70°. The azimuthal angle is defined as the angle between the line joining the two actin strands and that joining the two tropomyosin strands. A reconstruction of the filament present in the aligned type of actin+tropomyosin paracrystal has been carried out by Wakabayashi *et al.* (1975), who found a similar angle (75°) for the tropomyosin position but a smaller radius (about 2.0 nm). The description of only one of the two paracrystal types by the different groups of authors implies an element of selection, since we observe the two types together in roughly equal proportions, but it is possible that under some conditions one or other form is preferred.

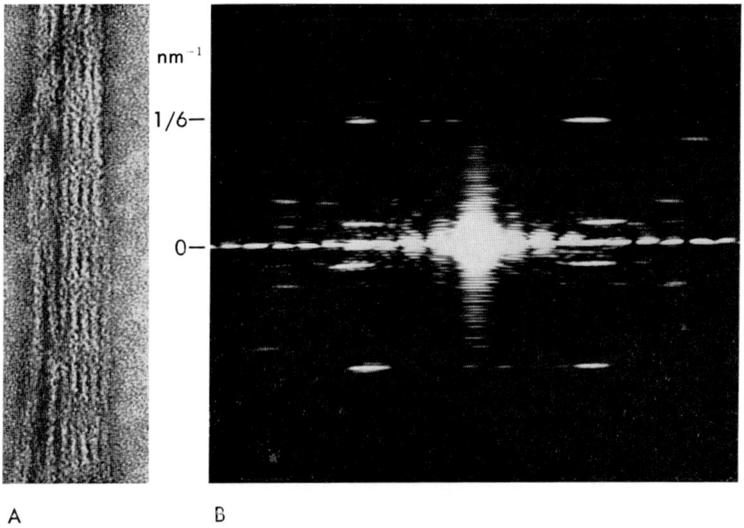

Fig. 4. Electron micrograph of an aligned type of paracrystal of actin+tropomyosin, negatively stained (×270,000) (A) and the corresponding optical diffraction pattern (B). The main row-line positions are at orders of the interfilament spacing, 6 nm, but because the transverse alignment of cross-over points of the long-pitch actin helices in adjacent filaments is not perfect there is some intermediate sampling, *e.g.* on the second layer-line. The second layer-line is strong as in Fig. 3 B.

ACTIN+TROPOMYOSIN+TN-I

When troponin-I (TN-I) is present the filaments are "inhibited," that is their ability to activate myosin ATPase is reduced to a very low level, as for fully reconstituted thin filaments containing whole troponin at low Ca^{2+}. Analysis of the structure of actin+tropomyosin+TN-I should therefore elucidate the role of the inhibitory troponin component in thin filament activity.

Both ribbon-type and aligned-type paracrystals are formed from filaments of actin+tropomyosin+TN-I. The ribbon-type is shown in Fig. 5A. A close examination of the micrograph indicates that although it superficially resembles the paracrystal of actin+tropomyosin shown in Fig. 3A, the white protein of the filaments is spaced by dark lines of stain that are more prominent, and give the paracrystal a banded appearance as indicated by the lines drawn at the side of Fig. 5A. In the optical diffraction pattern (Fig. 5B), the second layer-line is much weaker than that in the pattern from actin+tropomyosin (Fig. 3B), indicating a change in the tropomyosin position which should be correlated with the increased prominence of the

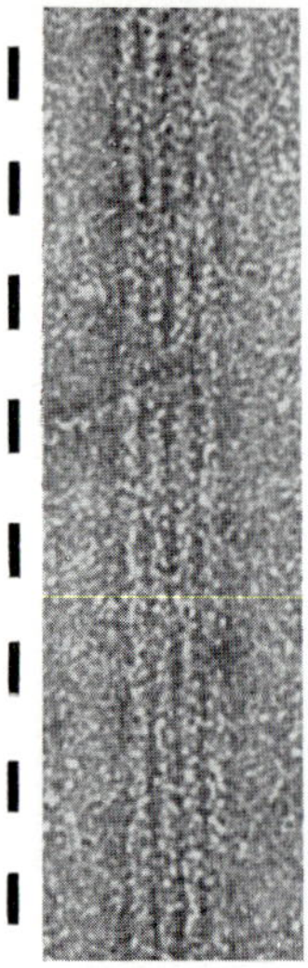

A

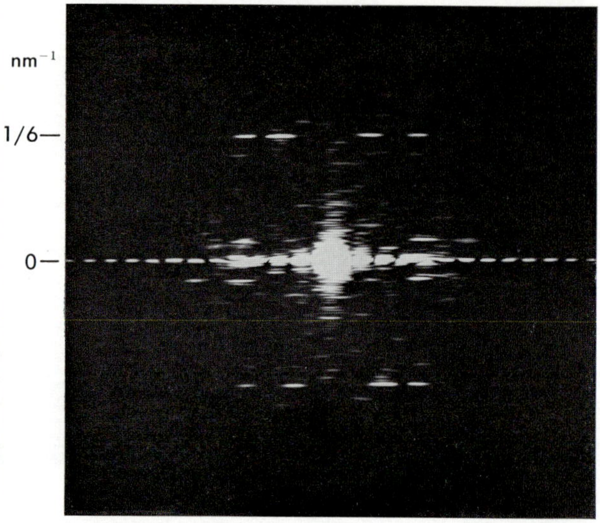

B

Fig. 5. A: electron micrograph of ribbon-type paracrystal of actin+tropomyosin+TN-I negatively stained (×270,000). When compared with ribbon-type paracrystals of actin+tropomyosin alone (Fig. 3 A), the dark lines of stain between filaments appear more prominent, indicating the presence of periodic gaps as indicated by the lines drawn at the side. In the optical diffraction pattern, in B, the second layer-line is weak, compared with Fig. 3 B or 4 B.

stain lines in the micrograph. Measurement of the axial positions of the layer-lines in Fig. 5B confirm that the helix has 28/13 symmetry, as is characteristic of the ribbon-type of paracrystal.

The comparison between paracrystals of actin+tropomyosin with and without TN-I is facilitated when the micrographs are processed by computer filtering. Figure 6 shows the filtered images, and the marked increase in the extent of the stain-filled regions between the filaments in the inhibited case can be easily discerned. Tropomyosin has moved to the edge of the actin groove under the influence of TN-I, and is therefore more closely integrated with the actin strands, so that the filaments look more like actin on its own. This is illustrated in the computer-drawn model paracrystals shown in Fig. 7. For the actin+tropomyosin model (Fig. 7A) the filaments have 28/13 actin helical symmetry, and the tropomyosin is at an angle of 70° as shown in the view along the filament shown underneath. In the model for the inhibited filament, troponin is omitted, but tropomyosin is at an angle of 50°. In both model paracrystals adjacent

Fig. 6. Computer-filtered images of electron micrographs of paracrystals of actin+tropomyosin (a) and actin+tropomyosin+TN-I (negative contrast, ×470,000) (b). The micrographs (Figs. 3 A and 5 A) were densitometered and converted into digital form, their Fourier transforms calculated and, after subtracting noise from the transforms, the images recalculated. (a) shows clearly the antiparallel packing of adjacent filaments in ribbon-type paracrystals. In (a) relatively little stain penetrates into gaps between the filaments, but in (b) the gaps are longer and more pronounced. The lines at the sides of the paracrystals mark the positions of the gaps.

filaments are antiparallel with an axial stagger between adjacent filaments of about 9 nm as deduced by O'Brien *et al.* (1975). Whereas in Fig. 7A the tropomyosin tends to fill in the spaces between actin filaments so that the overall density distribution is uniform, in 7B there are conspicuous gaps between filaments. In the aligned type of paracrystal of actin+tropomyosin+TN-I the tropomyosin is similarly moved towards the edge of the actin groove, though the movement is less clearly demonstrated in micrographs and filtered images.

The results of Gillis and O'Brien (1975) indicated that the change of filament activity and the movement of tropomyosin induced by Ca^{2+} in reconstituted thin filaments was accompanied by a change of helical sym-

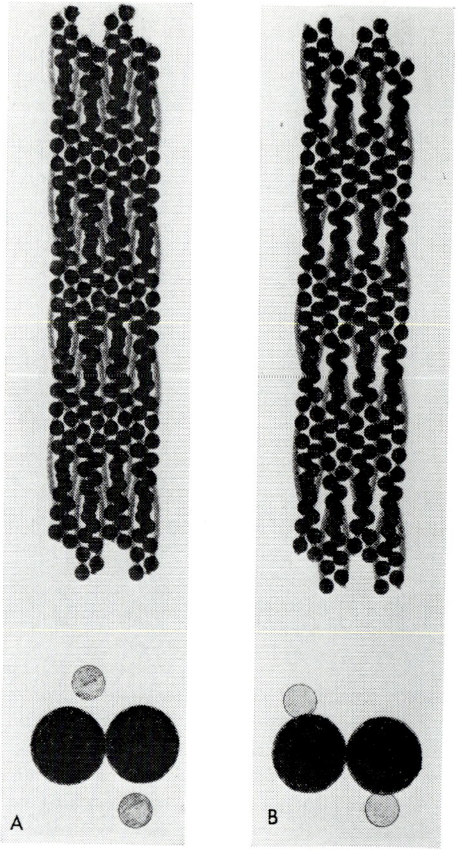

Fig. 7. Computer-drawn models of actin+tropomyosin in ribbon-type paracrystals. In A the azimuthal position of tropomyosin is at 70° as illustrated in the end-on view of a filament shown below, and in B this angle is 50°. At the smaller angle, the gaps between filaments are more pronounced, and these would produce the dark, stain-filled regions seen in electron micrographs of negatively stained paracrystals of actin+tropomyosin+TN-I (Figs. 5 A and 6 b).

metry and packing of the filaments. Our observations of the effect of TN-I show that within the *same* paracrystal type filament activity and the tropomyosin position may change, so that the helix change may be relatively unimportant. This result complements that of Wakabayashi *et al.* (1975) who showed that the combination of TN-I and troponin-T (TN-T) caused tropomyosin to move towards the edge of the actin groove in paracrystals of the aligned typed with 13/6 helical symmetry. Our observation that filaments of actin+tropomyosin or of actin+tropomyosin+TN-I may form either the aligned or the ribbon-type paracrystals indicate that the change of paracrystal type may take place without a change of filament activity, again implying that the helix change is unimportant. Stability of the helix in one form or the other seems to depend on the presence of TN-T, since we find that when this component is present only one paracrystal type is formed. Thus, actin+tropomyosin with TN-T, with TN-T+TN-I, or with TN-T+TN-I+troponin-C (TN-C) at low Ca^{2+}, form aligned type paracrystals with 13/6 symmetry and with TN-T+TN-C at high Ca^{2+}, or TN-T+TN-I+TN-C at high Ca^{2+} form ribbon-type with 28/13 symmetry.

ACTIN+TROPOMYOSIN+TN-I+TN-C

This combination is only stable at low Ca^{2+} concentration (Hitchcock *et al.*, 1973) and under this condition forms both ribbon and aligned types

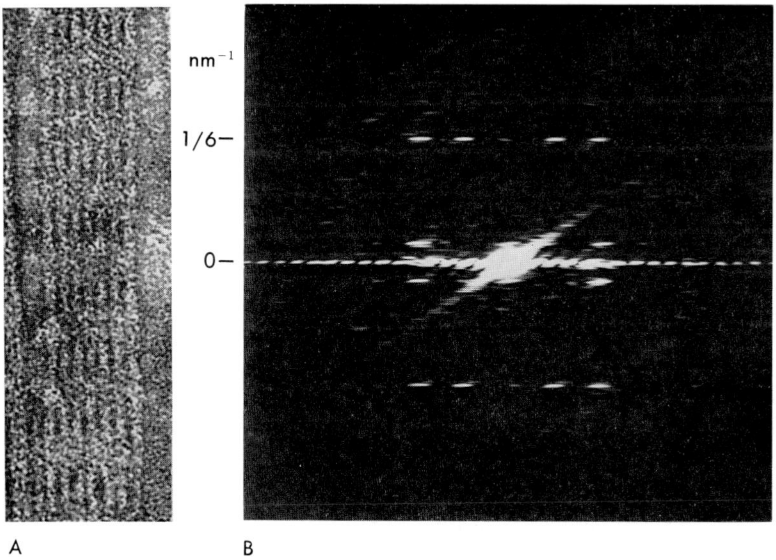

Fig. 8. A: electron micrograph of ribbon-type paracrystal of actin+tropomyosin+TN-I+TN-C, negatively stained (×270,000). The stain-filled gaps between filaments, illustrated in Fig. 6 b, are evident in this micrograph. In the optical diffraction pattern (B) the second layer-line is very weak indicating that the tropomyosin is at the edge of the actin groove, as in Fig. 7 B. On the first layer-line there is a meridional reflection arising from the axial repeat of the troponin components.

of paracrystal. The ribbon type paracrystal shows very clearly the banded appearance and the stain-filled gaps between the filaments (Fig. 8A), and indicates that the tropomyosin is near the edge of the actin groove. This is confirmed by the very weak intensity on the second layer-line in the optical diffraction pattern (Fig. 8B). The strong meridional reflection on the first layer-line arises from the repeat of the TN-I+TN-C along the filament, but it is difficult to identify the precise position of these components in relation to the stain-filled gaps in the micrograph. This reflection is considerably clearer than that observed when only TN-I is present (Fig. 5B), presumably because of the additional density introduced by TN-C, and because the paracrystal contains more filaments.

Although the tropomyosin in these paracrystals appears to occupy the inhibitory position, the filaments themselves have the same activity as actin+tropomyosin (Greaser and Gergely, 1971), that is, they are not fully inhibited as in the complex of actin+tropomyosin+TN-I. It is possible that tropomyosin may be only weakly held in the inhibitory position in these filaments, and that it may be displaced when myosin heads attach to actin. In this situation the movement of tropomyosin would be a consequence of actomyosin interaction rather than a regulator of it. This casts some doubt on the overriding importance of the tropomyosin movement, and suggests that a further factor, such as a change of the actin subunit conformation, may be involved. The observation that TN-I in a molar ratio of 1: 7 TN-I to actin produces inhibition of actin+tropomyosin filaments, indicates there is a joint binding site for TN-I to both actin and tropomyosin, and it is possible that TN-I may influence the actin conformation directly, with this influence communicated to all actin subunits by the tropomyosin. Any change of actin structure must be small, however, since it is not detected in optical diffraction patterns of electron micrographs. A change of actin structure might be expected to lead to a change of helical symmetry. Although we have shown that the change of helical symmetry from 13/6 to 28/13 does not necessarily accompany a change of filament activity when troponin components other than the full complement are bound, the observation remains that in the fully reconstituted thin filaments it does. Also our criterion of activity has been with respect to the actomyosin ATPase of filaments in solution. It is possible that in muscle the production of tension involves other factors, in particular the exact stereochemical relationship of myosin heads and their binding sites on actin, and a change of helical symmetry of the thin filament, possibly of a smaller extent than that observed in the paracrystal studies cannot be entirely ruled out.

THE POSITION OF TROPOMYOSIN IN RELATION TO THE ACTOMYOSIN BINDING SITE

In general, the movement of tropomyosin is established as an important and possibly the most important structural change in the thin filament taking place on activation and relaxation of the muscle. Does the tropomyosin move to either block or allow the attachment of myosin heads to actin? This steric blocking mechanism has been proposed by a number of authors (Hanson *et al.*, 1972; Haselgrove, 1972; H. E. Huxley, 1972; Parry and Squire, 1973) and is based largely on evidence from X-ray diffraction studies and electron microscopy.

The scheme presented by H. E. Huxley (1972) is shown in Fig. 9A, where the shape and attachment site of subfragment-1 (S-1) is based on the three-dimensional reconstruction of electron micrographs of the S-1 decorated thin filament calculated by Moore *et al.* (1970). This reconstruction does not have sufficient resolution to reveal tropomyosin, and although, as we have described above, its azimuthal position is established, there is an ambiguity whether it is on the same or the opposite side to the S-1 attachment site. The alternative arrangement, of which proponents of the steric blocking model were well aware (H. E. Huxley, 1972; Wakabayashi *et al.*, 1975) is shown in Fig. 9B. To resolve the ambiguity it is necessary to superimpose, with the correct relative orientation, reconstructions of decorated and undecorated thin filaments, since in the latter tropomyosin can usually be well resolved. If the reconstructions are carried out on I-segments, the assemblies of thin filaments still attached to the Z line, the correct orientation is known, since the characteristic arrowhead appearance obtained by adding S-1 to thin filaments always points away from the Z line (H. E. Huxley, 1963).

An I-segment from a crude muscle homogenate obtained at a low Ca^{2+} concentration is shown in Fig. 10. Some homogenates at a high Ca^{2+} level were studied, but acto-myosin precipitation made these experiments difficult. At low Ca^{2+} concentrations, all preparations produced up to 80% inhibition of the acto-myosin ATPase rate, confirming the presence of regulatory proteins. Furthermore, troponin gave rise to strong meridional reflections at $1/38.5\ nm^{-1}$ in the Fourier transforms of many micrographs

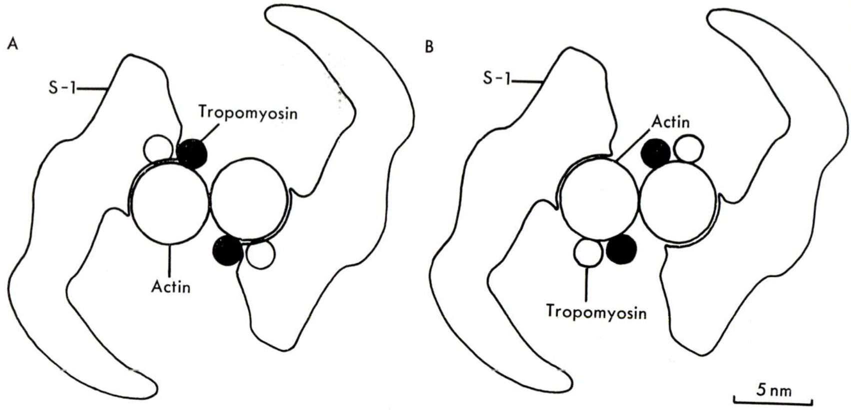

Fig. 9. The two ways of superimposing reconstructions from decorated (*i.e.*, complexed with S-1) and undecorated thin filaments. A is based on Huxley (1972), B has the same S-1 attachment scheme but the thin filament has the opposite orientation. The S-1 arrowheads are pointing out of the plane of the paper away from the Z line. The position of tropomyosin in relaxed and contracting muscle is shown by open and filled-in circles respectively.

and also to corresponding periodic stripes visible in large I-segments. Side-by-side aggregation of filaments into paracrystals was induced by raising the Mg^{2+} concentration, but single-layered paracrystals of more than two or three filaments were rare because movement of the filaments was restricted by their Z line attachment and tangling readily occurred. Filaments within paracrystals are usually better for image analysis studies because they are straighter, more evenly stained and less flattened (De Rosier and Moore, 1970).

Only filaments that appeared connected to a Z line and terminated at the end of the I-segment "brush" were considered for analysis. Optical diffraction patterns of electron micrographs of many hundreds of such filaments were examined, but only 20 patterns clearly showed at least three actin layer-lines. The density distribution of these micrographs was converted into digital form and their Fourier transforms calculated (Fig. 11). The quality of the phase and amplitude information justified three-dimensional reconstruction in eight cases where, for example, the phase residual was always less than 37° and had an average value of 28°. The phase residual is a measure of the amount by which the helical symmetry deviates from that expected for actin (Moore *et al.*, 1970), and was calculated for all non-equatorial layer-line data of medium or strong intensity. A high

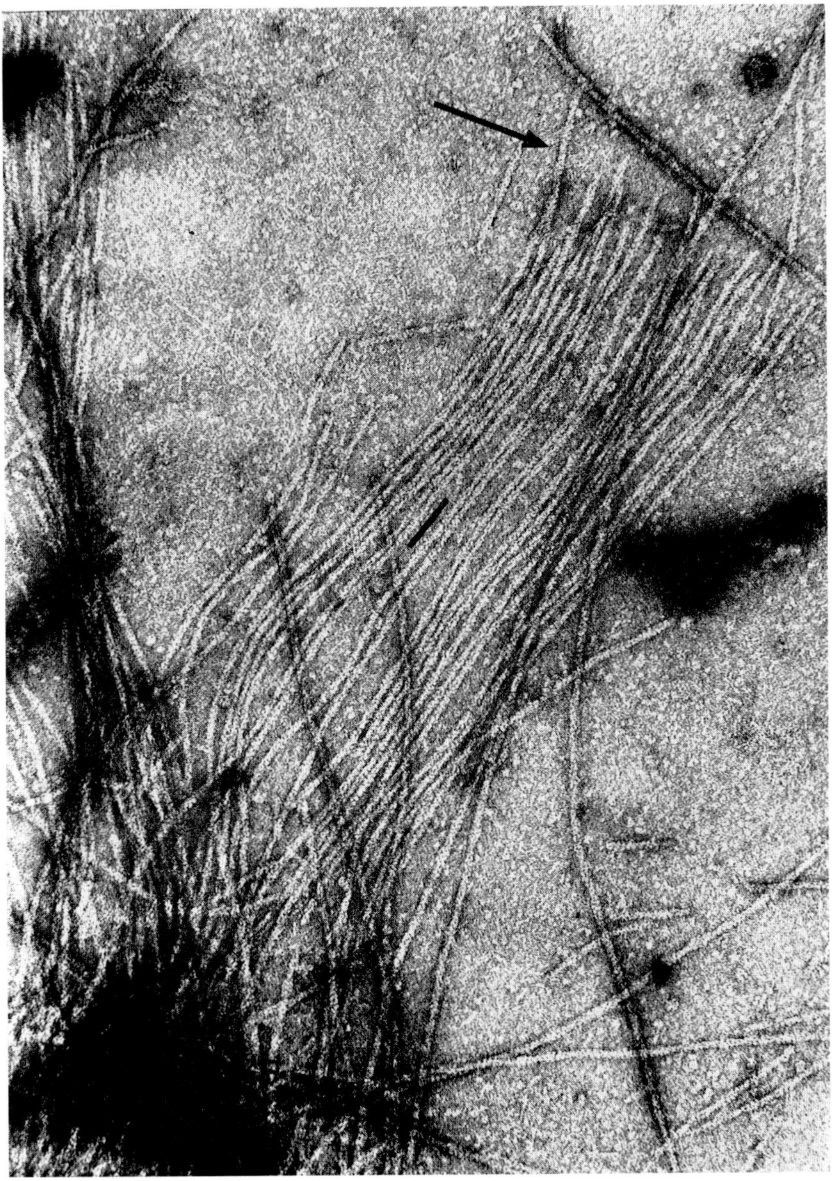

Fig. 10. Electron micrograph of an I-segment in a muscle homogenate prepared in a relaxing medium (negatively stained, ×170,000). The arrow indicates the end of the I-segment "brush" and the bar is set to the left of one of the filaments analysed.

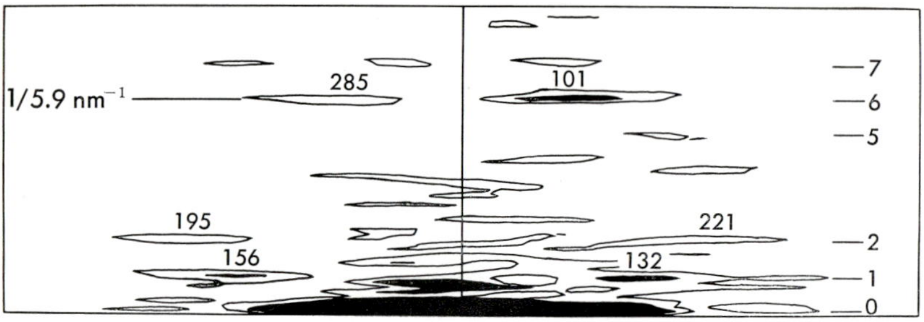

Fig. 11. Fourier transform of the filament marked in Fig. 10. The contours represent amplitude and the numbers above layer-line peaks indicate phases.

phase residual was associated with asymmetries between the upper and lower sides of the filaments and the relatively poor signal: noise ratio, arising from additional proteins in the homogenate and from filament curvature restricting the length of filaments processed. Straight filaments not in paracrystals sometimes gave low phase residuals, but in most cases it was better to mask off a filament within a paracrystal or to deconvolute the Fourier transform of the paracrystal to give the transform of an average filament. Six reconstructions clearly resolved the relative positions of actin and tropomyosin; in one case the Ca^{2+} level was high. By orientating each reconstruction with respect to the Z line, tropomyosin was consistently found to occupy positions shown in Fig. 9B, that is on the opposite side of the actin groove to that of the postulated S-1 binding site.

Two reconstructions can be calculated from any set of Fourier transform data, using layer-line information arising from either the upper or lower side of the filament (Klug and De Rosier, 1966). Five sets of data from three filaments were averaged by the methods of Amos and Klug (1975) and Fig. 12A shows the resulting helical projection. The averaging produced an improvement in the signal: noise ratio of the data, particularly on weaker layer-lines. Great care was taken to preserve the known orientation of each filament and the effect of each mathematical operation was monitored on a test object, which was an S-1 decorated thin filament having an obvious polarity. The helical projection shows on each side a well-resolved actin peak with, extending to higher radius, a smaller region of lower density attributable to tropomyosin. This assignment is confirmed by the three-dimensional analysis (Fig. 12B), where the size and shape of the actin subunit is consistent with that observed in other reconstructions (Wakabayashi *et al.*, 1975). In common with earlier reconstructions

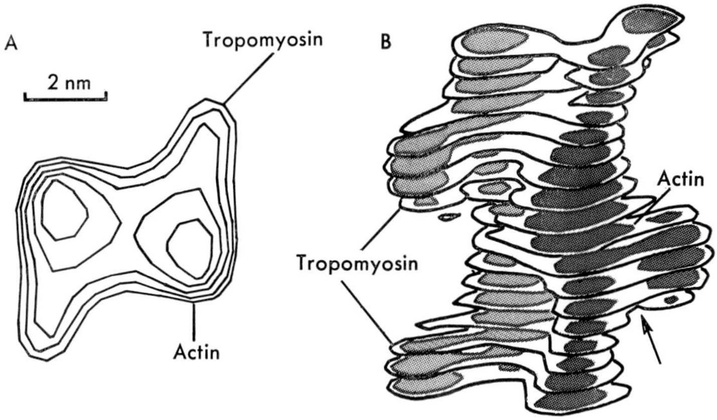

Fig. 12. A: contour map of the helical projection derived from averaged layer-line data. The helical projection is formed by projecting the structure down the long-pitch actin helices looking towards the Z line. B: three-dimensional reconstruction from averaged layer-line data, with the Z line at the bottom. The different shadings indicate the long-pitch actin and tropomyosin strands on opposite sides of the helical axis. The position of tropomyosin derived from both A and B indicates that the arrangement shown in Fig. 9 B is correct.

(Spudich *et al.*, 1972), tropomyosin appears to be discontinuous, which may be due to inadequate resolution and the periodic merging of tropomyosin with part of the actin subunit. Where tropomyosin diverges from one end of the actin subunit a V-shaped cleft is formed (arrow, Fig. 12B). Furthermore, each actin subunit makes a connection to each of the two adjacent subunits in the other long-pitch strand and these are asymmetrically disposed with respect to the subunit's centre of mass. These features are present in the reconstructions of Wakabayashi *et al.* (1975) and assist in assigning filament polarity.

These reconstructions remove the ambiguity concerning which side of the actin groove tropomyosin lies, and indicate that the steric blocking mechanism as presented in Fig. 9A was based on an incorrect assumption. In this figure there is no obvious ways in which tropomyosin can interfere with myosin attachment to actin, unless the shape of S-1 or its attachment site is revised. We understand that Taylor and Amos (personal communication) are currently re-examining the structure of decorated thin filaments and their results will be of particular interest.

CONCLUSION

The movement of tropomyosin in the long-pitch helical grooves of actin is established as a structural change taking place when various combinations of troponin components are present and in nearly all cases it accompanies a change of the filament activity. On the other hand, a change of actin helical symmetry is relatively unimportant, at least in the *in vitro* situation. Any changes of actin subunit structure that occur when the tropomyosin moves are likely to be very small, and the idea that tropomyosin might physically block myosin attachment to produce inhibition is made less attractive by the identification of its position on the opposite side of the actin groove.

REFERENCES

Amos, L. A. and Klug, A. (1975). *J. Mol. Biol.* **99**, 51–73.

DeRosier, D. J. and Moore, P. B. (1970). *J. Mol. Biol.* **52**, 355–369.

Ebashi, S., Endo, M., and Ohtsuki, I. (1969). *Q. Rev. Biophys.* **2**, 351–384.

Gillis, J. M. and O'Brien, E. J. (1975). *J. Mol. Biol.* **99**, 445–459.

Greaser, M. L. and Gergely, J. (1971). *J. Biol. Chem.* **246**, 4226–4233.

Hanson, J., Lednev, V., O'Brien, E. J., and Bennett, P. M. (1972). *Cold Spring Harbor Symp. Quant. Biol.* **37**, 311–318.

Haselgrove, J. C. (1972). *Cold Spring Harbor Symp. Quant. Biol.* **37**, 341–352.

Hitchcock, S. E., Huxley, H. E., and Szent-Györgyi, A. G. (1973). *J. Mol. Biol.* **80**, 825–836.

Huxley, H. E. (1963). *J. Mol. Biol.* **7**, 281–308.

Huxley, H. E. (1972). *Cold Spring Harbor Symp. Quant. Biol.* **37**, 361–376.

Huxley, H. E. and Brown, W. (1967). *J. Mol. Biol.* **30**, 383–434.

Klug, A. and DeRosier, D. J. (1966). *Nature* **212**, 29–32.

Maeda, Y., Matsubara, I., and Yagi, N. (1979). *J. Mol. Biol.* **127**, 191–201.

Moore, P. B., Huxley, H. E., and DeRosier, D. J. (1970). *J. Mol. Biol.* **50**, 279–295.

O'Brien, E. J. and Couch, J. (1976). *Proceedings of the Sixth European Congress on Electron Microscopy* **2**, 153–155.

O'Brien, E. J., Gillis, J. M., and Couch, J. (1975). *J. Mol. Biol.* **99**, 461–475.

Parry, D.A.D. and Squire, J. M. (1973). *J. Mol. Biol.* **75**, 33–55.

Spudich, J. A., Huxley, H. E., and Finch, J. T. (1972). *J. Mol. Biol.* **72**, 619–632.

Vibert, P. J., Haselgrove, J. C., Lowy, J., and Poulson, F. R. (1972). *J. Mol. Biol.* **71**, 757–767.

Wakabayashi, T., Huxley, H. E., Amos, L. A., and Klug, A. (1975). *J. Mol. Biol.* **93**, 477–497.

Muscl Contraction Its Regulatory Mechanisms (Ebashi, S. et al., eds.), pp. 165–172, Japan Sci. Soc. Press, Tokyo / Springer-Verlag, Berlin (1980)

12

Dynamics of Actin Filament

FUMIO OOSAWA

Summary: Experimental data on the flexibility of F-actin and the thin filament *in vitro* and *in vivo* are summarized. In the absence of Ca^{2+}, tropomyosin-troponin changes the orientation of actin monomers in F-actin and makes F-actin rigid. The binding of heavy meromyosin to F-actin changes the orientation of actin monomers and during a cyclic reaction with heavy meromyosin in the presence of ATP, F-actin assumes a conformation different from pure F-actin or F-actin keeping bound heavy meromyosin.

Actin is in a state of monomer, G-actin, in a salt-free solvent and forms a helical polymer, F-actin, with the addition of neutral salts. In a given environmental condition, a polymerization equilibrium is established between G-actin and F-actin (Oosawa and Asakura, 1975). The concentration of G-actin coexisting in equilibrium with F-actin is determined by the salt concentration, the pH, and the temperature, independently of the amount of F-actin (Oosawa *et al.*, 1959). This concentration of G-actin gives the critical concentration of actin, only above which can F-actin stably be formed. Such a specific feature of the polymerization of actin as a kind of condensation phenomena is understandable on the basis of the helical structure of F-actin (Oosawa and Kasai, 1962).

Macroscopic equilibrium is supported by the microscopic fluctuation. Each actin molecule undergoes a cycle of polymerization and depolymerization. If G-actin has bound ATP, the polymerization reaction is accompanied by the splitting of ATP. The repetition of the polymerization / depolymerization cycle by actin molecules in the presence of ATP results in a steady splitting of ATP, which was actually found under the condition where G-actin and F-actin coexist (Asakura and Oosawa, 1960).

Thus, the structure of F-actin cannot be permanently fixed. At the end of F-actin, the actin molecule repeats binding and detachment. Not only at the end, but also in the middle of F-actin, the bond between monomers is weakened and strengthened by the exchange of thermal energy with the environment. The monomer-monomer distance and the monomer conformation in F-actin must be fluctuating. Such a bond fluctuation has been suggested by the activation of ATP splitting in F-actin solutions under sonic vibration (Asakura, 1961) and at high temperatures (Asai and Tawada, 1966).

As a polymer made by the secondary bonding of globular protein molecules, F-actin is not completely rigid but flexible.

FLEXIBILITY OF F-ACTIN *IN VITRO*

The flexibility of F-actin was first estimated by Fujime using the quasi-elastic light scattering method (Fujime, 1970). The frequency shift of the light scattered from F-actin in solution is related to the relaxation time of the Brownian motion of F-actin. The Brownian motion consists of translation, rotation, and bending. By the measurement of scattering at different angles, the contribution of the bending motion to the frequency shift was separated from other factors. The relaxation time of the bending motion was found to be about 10 msec. From this value, the elastic modulus for bending (the flexural rigidity) was calculated to be about 1.7×10^{-17} dyn cm^2. This corresponds to the persistence length of F-actin, as a semiflexible filament, of about 6 μm.

In electron micrographs, negatively stained F-actin filaments are not always straight but more often are curved. If the curved form is due to the thermal bending motion of F-actin in solution, the flexibility of F-actin can be estimated from the relation between the end-to-end distance

TABLE I
Flexibility of F-actin

Methods	Flexibility (flexural rigidity) ($\times 10^{-17}$ dyn cm^2)
In vitro (in solution)	
Light scattering	1.7
Electron microscopy	1.7
Optical microscopy[a]	7.2
In vivo (in fiber)	
Electrooptic effect[b]	3.1
Polarized fluorescence	5.3

[a] F-actin with HMM at saturation.
[b] Thin filament ($+Ca^{2+}$).

and the contour length. The statistical analysis of a large number of F-actin filaments in electron micrographs gave a value of the elastic modulus for bending which was in good agreement with the vlaue obtained by the light scattering method (Takebayashi *et al.*, 1977).

Recently, it was found by Nagashima that when decorated with myosin or heavy meromyosin (HMM), single filaments of F-actin can be seen directly by dark field optical microscopy (Nagashima and Asakura, 1980). The thermal bending motion of the filaments can be observed with the eye. Now, it is certain that F-actin is flexible. The value of the elastic modulus for bending estimated from the shape of moving F-actin with bound HMM at saturation was about 7×10^{-17} dyn cm^2, which was a little larger than, but of the same order as, the values obtained previously. The data are summarized in Table I.

FLEXIBILITY OF THE THIN FILAMENT *IN VITRO* AND *IN VIVO*

In the striated muscle, F-actin binds tropomyosin and troponin to form the thin filament (Ebashi and Endo, 1968). When Ca^{2+} is bound to troponin, F-actin in the thin filament begins to interact with myosin cross-bridges to produce a sliding force. The quasielastic light scattering experiment by Ishiwata and Fujime showed that the thin filament reconstituted *in vitro* from F-actin, tropomyosin and troponin changes its flexibility at a micromolar concentration of Ca^{2+} (Ishiwata and Fujime, 1972). In the absence of Ca^{2+}, tropomyosin and troponin make F-actin rigid. After the addition of Ca^{2+}, the complex of F-actin, tropomyosin and troponin has a flexibility similar to that of pure F-actin.

To see if a flexibility change occurs in the thin filament *in vivo*, the measurement of ultraviolet linear dichroism has been very useful. It was found by Higashi that F-actin oriented by flow in solution gave negative peaks of dichroism at 260 nm and 297 nm (Higashi *et al.*, 1963). The former peak was due to a regular arrangement of ADP bound to F-actin, the absorption dipole being nearly perpendicular to the filament axis of F-actin (Miki and Mihashi, 1977). The dichroism of a glycerinated muscle fiber was investigated by Yanagida. The I band gave a similar negative peak at 260 nm (Yanagida *et al.*, 1974). Upon the addition of Ca^{2+} to the fiber, the height of the peak decreased by about 30%. This dichroism decrease was considered to correspond to the flexibility increase observed *in vitro*.

An electrooptic effect discovered by Umazume in a stretched muscle fiber has provided another method of estimating the flexibility of thin filaments *in vivo* (Umazume and Fujime, 1975). An electric field applied along the fiber induced better parallel orientation of thin filaments in I band on one side of the Z line and disorientation on the other side. The degree of orientation showed oscillation under an alternating field. The phase difference between the field and the orientation was related to the flexibility or the elastic modulus for bending of thin filaments (Yoshino *et al.*, 1978). The value of the modulus obtained by Yoshino using this method was in the same order of magnitude as those previously obtained *in vitro*, as listed in Table I. According to this method, the difference in the rigidity or flexibility of this filaments between the absence and presence of Ca^{2+} was very large.

A change in the helical symmetry of F-actin in thin filaments by Ca^{2+} was found in X-ray diffraction study and in the electron-microscopic analysis (Maeda, *et al.*, 1979; Gillis and O'Brien, 1975). It may correspond to the flexibility change described above.

FLEXIBILITY AND MONOMER ORIENTATION

The measurement of polarized fluorescence is more powerful than that of simple dichroism in the sense that the orientation of chromophores or actin monomers in thin filaments and the orientation of thin filaments in a muscle fiber can be separately determined. A fluorescent analogue of ADP, ε-ADP, is useful for this purpose, because ADP in F-actin can be

TABLE II
Angles of the Absorption and Emission Dipoles of ε-ADP with the Thin Filament Axis and the Average Angle of the Thin Filament Axis with the Fiber Axis

	ϕ_A	ϕ_E	$\sin^2 \theta$
F-actin	75.7	75.1	0.062
Thin filament ($-Ca^{2+}$)	75.9	76.4	0.057
Thin filament ($+Ca^{2+}$)	77.1	76.5	0.067

replaced with ε-ADP, keeping the regular arrangement, as shown by Miki (Miki and Mihashi, 1977).

A glycerinated muscle fiber, from which myosin and tropomyosin-troponin had been removed, was incubated with a high concentration of ε-ATP at 36°C for 20 min, and ADP in F-actin were partially replaced with ε-ADP (Yanagida and Oosawa, 1978). With exciting lights polarized parallel and perpendicular to the fiber axis, the intensities of emitted lights polarized parallel and perpendicular were measured. From four components of polarized fluorescence thus obtained, the angle of the absorption dipole of ε-ADP with the thin filament axis, ϕ_A, and that of the emission dipole, ϕ_E, were determined. To get the best fit to the experimental data of four components, the fluctuation of the thin filament axis from the fiber axis had to be taken into consideration. The average angle, θ, between the filament axis and the fiber axis was estimated. From this angle, the elastic modulus for bending of F-actin was calculated.

As shown in Table II, the angles ϕ_A and ϕ_E were nearly equal and the adenine plane of ε-ADP was nearly perpendicular to the filament axis. The value of the elastic modulus of F-actin obtained by this method was the same order as those previously obtained (Table I).

After replacement of ADP with ε-ADP, thin filaments were reconstituted again in the fiber by adding tropomyosin and troponin. Then the polarized fluorescence was measured in the absence and presence of Ca^{2+} (Yanagida and Oosawa, 1980). The angles ϕ_A and ϕ_E changed with the addition of Ca^{2+}. In the absence of Ca^{2+} ϕ_A was a little smaller than ϕ_E, whereas in its presence ϕ_A was a little larger than ϕ_E. The adenine plane of ε-ADP was rotated.

The average angle θ was larger in the presence of Ca^{2+} than in its absence. The flexibility of thin filaments was made larger by Ca^{2+}. Thus, the binding of Ca^{2+} to troponin on the thin filament brings about changes both in the overall flexibility of the filament and in the monomer orientation or monomer conformation in F-actin.

EFFECT OF HEAVY MEROMYOSIN AND ATP ON F-ACTIN AND THIN FILAMENT

After removal of myosin and replacement of ADP with ε-ADP, a glycerinated muscle fiber was incubated in a solution of HMM. Heavy meromyosin were bound to F-actin in the fiber within 10 min. Then the polarized fluorescence from ε-ADP was measured at a molar ratio of bound HMM to actin of about 1: 3 (Yanagida and Oosawa, 1978). As shown in Table III, both angles ϕ_A and ϕ_E were decreased by HMM. The adenine plane of ε-ADP was rotated in F-actin. The direction of rotation was different from that induced by binding of tropomyosin and troponin without Ca^{2+} (Table II). The average angle of fluctuation of the filament axis, θ, was increased by HMM, suggesting an increase in the flexibility of F-actin.

The increase in the flexibility of F-actin by binding of a small amount of HMM was previously reported in the quasielastic light scattering experiment and in the electron-microscopic analysis of the shape of F-actin (Takebayashi *et al.*, 1977; Fujime and Ishiwata, 1971). The measurement of simple dichroism of a muscle fiber at different overlapping lengths of thin and thick filaments gave similar results (Yanagida and Oosawa, 1975).

Since the absorption and emission of ε-ADP do not interfere with the absorption of ATP or ADP, the polarized fluorescence from ε-ADP in the fiber can be measured in the presence of ATP. The fiber is incubated with HMM and a few millimollar of Mg-ATP. In this condition, F-actin in the fiber must have a cyclic interaction with HMM, coupled with the splitting of ATP. The results of analysis of polarized fluorescence are shown in Table III (Yanagida and Oosawa, 1980). The angles ϕ_A and ϕ_E were not very different from those in the absence of HMM. However, the average angle of fluctuation θ was much larger than that in the absence

TABLE III
Angles of the Absorption and Emission Dipoles of ε-ADP with the F-actin Axis and the Average Angle of the F-actin Axis with the Fiber Axis

	ϕ_A	ϕ_E	$\sin^2 \theta$
F-actin	75.5	75.0	0.060
F-actin+HMM[a]	73.0	73.5	0.079
F-actin+HMM+Mg-ATP	76.5	76.0	0.099

[a] Bound HMM : actin=1 : 3.

of HMM and even larger than that in the presence of HMM without Mg-ATP. That is, during interaction with HMM having ATP or ADP-P, the monomer orientation or the monomer conformation in F-actin is, on the average, similar to that in pure F-actin, but the overall conformation of F-actin is very different from that of pure F-actin. The conformation of F-actin in this condition is not simply intermediate between those of pure F-actin and F-actin having bound HMM.

The binding of HMM having ATP or ADP-P causes a change in the conformation of F-actin. The effect of this change may be spread along the F-actin filament. In the presence of Mg-ATP, the lifetime of the bound state of HMM is rather short. Nevertheless, a large conformational change may be left in F-actin.

It must be noted here that a quasielastic light scattering experiment indicated, several years ago, that the relaxation time of bending motion of F-actin in solution was made longer by the interaction with HMM in the presence of a few millimollar of Mg-ATP (Oosawa *et al.*, 1973). Apparently, the flexibility of F-actin was increased. Two experiments, using completely different methods in solution and in fiber, showed good agreement.

When reconstituted thin filament was used instead of F-actin, the specific conformation of F-actin in the presence of HMM and a few millimollar of Mg-ATP was not found in the absence of Ca^{2+} but was found in the presence of Ca^{2+}.

CONCLUSION

F-actin is flexible *in vitro* and *in vivo*. The estimated value of the persistence length of F-actin is around 10 μm. In the resting state of muscle, the thin filament is more rigid. Upon activation of muscle, the monomer orientation changes and the thin filament becomes flexible. A cyclic interaction with HMM in the presence of Mg-ATP induces a large change in the overall conformation of F-actin or of the thin filament.

Spatial and temporal directionality of the conformational change must be important in the process of force generation and movement. F-actin and the thin filament have a structural polarity. The conformational change in F-actin induced by the binding of myosin is expected to affect the two sides of F-actin unequally. The time sequence of conformational changes which take place in F-actin and in myosin cross-bridges must be analyzed further.

REFERENCES

Asai, H. and Tawada, K. (1966). *J. Mol. Biol.* **20**, 403-417.
Asakura, S. and Oosawa, F. (1960). *Arch. Biochem. Biophys.* **87**, 273-285.
Asakura, S. (1961). *Biochim. Biophys. Acta* **52**, 65-75.
Ebashi, S. and Endo, M. (1968). *Prog. Biophys. Mol. Biol.* **18**, 123-183.
Fujime, S. (1970). *J. Phys. Soc. Japan* **29**, 751-759.
Fujime, S. and Ishiwata, S. (1971). *J. Mol. Biol.* **62**, 251-265.
Higashi, S., Kasai, M., Oosawa, F., and Wada, A. (1963). *J. Mol. Biol.* **7**, 421-430.
Gillis, J. and O'Brien, E. (1975). *J. Mol. Biol.* **99**, 445-459.
Ishiwata, S. and Fujime, S. (1972). *J. Mol. Biol.* **68**, 511-522.
Maeda, Y., Yagi, N., and Matsubara, I. (1979). *J. Mol., Biol.* **127**, 191-201.
Miki, M. and Mihashi, K. (1977). *Biophys. Chem.* **6**, 101-106.
Nagashima, H. and Asakura, S. (1980). *J. Mol. Biol.* **136**, 169-182.
Oosawa, F. and Asakura, S. (1975). *Thermodynamics of the Polymerization of Protein*, Academic Press, New York.
Oosawa, F., Asakura, S., Ooi, T., Hotta, K., and Imai, N. (1959). *J. Polym. Sci.* **37**, 323-336.
Oosawa, F., Fujime, S., Ishiwata, S., and Mihashi, K. (1973). *Cold Spring Harbor Symp. Quant. Biol.* **37**, 277-286.
Oosawa, F. and Kasai, M. (1962). *J. Mol. Biol.* **4**, 10-21.
Takebayashi, T., Morita, T., and Oosawa, F., (1977). *Biochim. Biophys. Acta* **492**, 357-363.
Umazume, T. and Fujime, S. (1975). *Biophys. J.* **15**, 163-180.
Yanagida, T. and Oosawa, F. (1975). *Eur. J. Biochem.* **56**, 547-556.
Yanagida, T. and Oosawa, F. (1978). *J. Mol. Biol.* **126**, 507-524.
Yanagida, T. and Oosawa, F. *J. Mol. Biol.* (in press).
Yanagida, T., Taniguchi, M., and Oosawa, F. (1974). *J. Mol. Biol.* **90**, 509-522.
Yoshino, S., Umazume, Y., Natori, R., Fujime, S., and Chiba, S. (1978). *Biophys. Chem.* **8**, 317-326.

Muscle Contraction: Its Regulatory Mechanisms (Ebashi, S. et al., eds.), pp. 173–180,
Japan Sci. Soc. Press, Tokyo / Springer-Verlag, Berlin (1980)

13

The Angles of Cross-bridges Bound to Nucleotide at Various Concentrations of Mg-(ε)-ATP

TOSHIO YANAGIDA

Summary: Measurements of polarized fluorescence from myosin heads bound to nucleotide in glycerinated muscle fiber were performed in the presence of various concentrations of fluorescent ADP and ATP analogs, ε-ADP and ε-ATP, to obtain the angles of bound nucleotides to the fiber axis and the angular distribution of them.

In the presence of ε-ADP (rigor), the maximum bound amounts (B_m) of nucleotides to the fiber were 270 μM and 220 μM, and K_d of bound ε-ADP were 360 μM and 80 μM at sarcomere lengths of 2.1 μm and 3.7 μm, respectively. The angles of absorption (ϕ_A) and emission dipoles (ϕ_E) of bound ε-ADP relative to the fiber axis were 68° and 66°, respectively, and the orientation distribution of the dipoles had a sharp maximum at a sarcomere length of 2.1 μm, when the ε-ADP concentration was less than 300 μM. As the ε-ADP concentration was increased, the orientation distribution of the dipoles became broad. At a sarcomere length of 3.7 μm, the orientation distribution was very broad.

In the presence of ε-ATP (+Mg), B_m was 210 μM and K_m of bound nucleotide was 200 μM at sarcomere length of 2.1 (pCa 4). Isometric tension was induced by ε-ATP with a maximum around 200 μM ε-ATP at pCa 8; the fibers were completely relaxed at 2 mM.

In a relaxed state, the orientation distribution for bound nucleotides were almost randomized, independently of sarcomere length.

During isometric contraction at 2 mM ε-ATP and 1 mM Ca, the orientation distribution for bound nucleotides was broad. However, at 50 μM ε-ATP, where only 1/3 of myosin heads were bound to nucleotides but 60% of maximum tension was induced, it had a sharp muximum when the sarcomere length was 2.1 μm, suggesting a rate-limiting reaction of the cross-bridge cycle; and then the values of ϕ_A and ϕ_E were 68° and 64°, respectively.

It was shown by Marston and Tregear (1972, 1973) that 2 mol of ADP were bound to one mol of myosin at saturation in a muscle fiber in the presence of ATP and ADP. It was claimed by Arata and Tonomura (1976) that the binding of ATP or ADP to myosin was heterogeneous because the ATPase activity reached a maximum where one mol of ADP was bound to one mol of myosin. Whether the binding of nucleotides to myosin is homogeneous or heterogeneous, the sliding theory predicts the movement of myosin cross-bridges during the actomyosin ATPase reaction in the muscle fiber. The signal from bound nucleotides should gave useful information about the movement of cross-bridges, if it occurs.

In this work, we have undertaken to measure the polarized fluorescence from a fluorescent analogue of ATP, ε-ATP or ε-ADP, bound to myosin in a glycerinated muscle fiber at sarcomere lengths of 2.1 and 3.7 μm. The method of polarized fluorescence is useful in studying the conformations of muscle proteins in the fiber; it was first employed by Morales and his co-workers to investigate the states of cross-bridges in the fiber (Morales and Borejdo, 1977). Recently we applied the same method to the study of the conformation of F-actin or thin filaments in the fiber after replacing ADP in F-actin with ε-ADP (Yanagida and Oosawa, 1978, 1980). In the present work, various concentrations of ε-ATP or ε-ADP are simply added to glycerinated muscle fibers in the absence and presence of Ca ions. The amount of bound ε-ADP is estimated from the fluorescence intensity, and the distribution of the angle of ε-ADP on myosin is calculated from four components of polarized fluorescence. On the basis of the experimental results, the conformation of cross-bridges in different physiological states is discussed.

MATERIALS AND METHODS

Glycerinated muscle fibers were prepared from rabbit psoas. A single fiber or a small bundle composed of 2–3 single fibers was used for measurement. When ε-ATP and/or ATP were used for relaxation or activation of the fiber, phosphocreatine kinase (PCK) of 2–5 mg/ml and phosphocreatine (PC) of 10 mM were coupled to them as a regenerating system of ε-ATP and/or ATP.

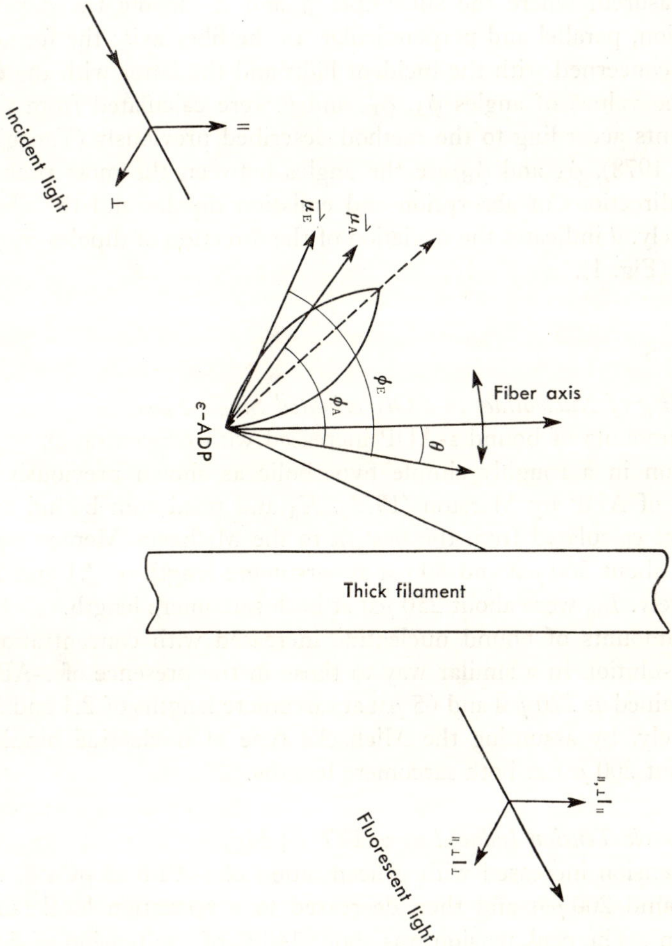

Fig. 1. The calculation of the polarized components of fluorescence.

The polarized fluorescence from muscle fibers was measured with a microspectrophotometer as described by Yanagida and Oosawa (1978). ε-ATP or ε-ADP were excited at 325 nm, and the emitted light was collected above 360 nm. The fluorescence due to ε-ATP or ε-ADP bound to myosin was obtained from the difference between the fluorescence from the fiber in a solution containing ε-ATP or ε-ADP and that in the other solution containing a 10–20-fold concentration of ATP or ADP in addition to ε-ATP or ε-ADP.

Four components of polarized fluorescence, ${}_{\parallel}I_{\parallel}$, ${}_{\parallel}I_{\perp}$, ${}_{\perp}I_{\parallel}$, and ${}_{\perp}I_{\perp}$, were measured, where the subscripts $\parallel$ and $\perp$ denote the direction of polarization, parallel and perpendicular, to the fiber axis; the former subscript is concerned with the incident light and the latter with the emitted light. The values of angles ϕ_A, ϕ_E, and θ were calculated from the four components according to the method described previously (Yanagida and Oosawa, 1978). ϕ_A and ϕ_E are the angles between the most probable or average directions of absorption and emission dipoles and the fiber axis, respectively. θ indicates the deviation of the direction of dipoles from these averages (Fig. 1).

RESULTS

1. *Binding of Nucleotides to a Glycerinated Single Fiber*

The amounts of bound ε-ADP increased with concentration of ε-ADP in solution in a roughly simple hyperbolic as shown previously in the presence of ADP by Marston (1973). K_d and maximum bound amounts (B_m) were calculated from the best fit to the Michaelis-Menten equation. K_d were about 360 μM and 80 μM at sarcomere length of 2.1 and 3.7 μm, respectively. B_m were about 250 μM at both sarcomere length.

The amounts of bound nucleotide increased with concentration of ε-ATP in solution in a similar way to those in the presence of ε-ADP. K_m were obtained as 220 μM and 65 μM at sarcomere lengths of 2.1 and 3.7 μm, respectively, by assuming the Michaelis type of nucleotide binding. B_m were about 200 μM at both sarcomere lengths.

2. *Isometric Tension Induced by ε-ATP (+Mg)*

The tension increased with concentration of ε-ATP at pCa 8, reached peak around 200 μM and then decreased to a relaxation level (substrate inhibition). The peak tension was about 140% of the tension of fibers activated fully in the presence of 2 mM ε-ATP (+Mg) and 1 mM Ca. The tension induced by ε-ATP under full activation was about 70% of that induced by ATP.

3. *Polarized Fluorescence in the Presence of ε-ADP*

Four components of polarized fluorescence from ε-ADP bound to myosin were measured in rigor at sarcomere lengths of 2.1 μm where thick filaments were fully overlapped with thin filaments and 3.7 μm where thick filaments were not overlapped. Table I gives the ratios of four components to each other; the angles, ϕ_A and ϕ_E, and the average of $\bar{\theta}(\theta)$ calculated from the values of the three ratios at less than 300 μM of ε-ADP. At a sarcomere length of 2.1 μm, the deviation of $\bar{\theta}$ from zero was very small (7°). This result suggests that the cross-bridges attach to thin filaments with a nearly single helical array, and the motions of both cross-bridges and ε-ADP on them are slight. The fluorescence polarization was almost independent of the concentrations of ε-ADP in solution from 20 μM to 300 μM, where the amount of bound ε-ADP varied from about 6 to 45% of total number of myosin heads. The values of ϕ_A and ϕ_E were obtained as about 68° and 66°, respectively. When the ε-ADP concentration was over 300 μM, the fluorescence polarization deviated from the values characteristic of a single helical array. At a sarcomere length of 3.7 μm, the calculated value of $\bar{\theta}$ was very large (~35°). This means that the orientation of bound ε-ADP is nearly completely disordered.

TABLE I

Ratios of Four Components of Polarized Fluorescence from Glycerinated Fibers in the Presence of ε-ADP and ε-ATP, and Calculated Values of ϕ_A, ϕ_E, and $\bar{\theta}$

The average values and standard deviations are for 5–18 measurements.

Condition	S.L. (μm)	$_{\parallel}I_{\perp}/_{\parallel}I_{\parallel}$	$_{\perp}I_{\parallel}/_{\parallel}I_{\parallel}$	$_{\perp}I_{\perp}/_{\parallel}I_{\parallel}$	ϕ_A (°)	ϕ_E (°)	$\bar{\theta}$ (°)
Rigor	2.1±0.1	1.7±0.1	2.3±0.3	6.8±0.8	68±2	66±1	7±1
ε-ADP 20, 100, 200, 300 μM	3.7±0.1	0.70±0.07	0.75±0.09	1.8±0.2	66±2	64±2	35±3
Relax	2.1±0.1	0.69±0.03	0.69±0.04	1.6±0.1	63±2	63±2	32±1
ε-ATP 2 mM, -Ca,	3.7±0.1	0.75±0.05	0.74±0.07	1.6±0.1	64±2	65±2	35±2
Active	2.1±0.1	0.60±0.02	0.70±0.40	1.5±0.1	63±2	66±2	35±1
ε-ATP 2 mM	3.7±0.1	0.66±0.04	0.68±0.04	1.5±0.1	63±2	63±2	32±1
ε-ATP 180 μM	2.1±0.1	1.1±0.1	1.6±0.1	5.4±0.3	70±1	66±1	18±1
ε-ATP 50 μM	2.1±0.1	1.6±0.1	2.2±0.1	6.1±0.7	68±1	64±1	8±1
	3.7±0.1	0.66±0.02	0.80±0.01	1.8±0.1	70±1	66±1	37±1

S. L., sarcomere length.

4. Polarized Fluorescence in the Presence of ε-ATP (Relaxed State)

Polarized fluorescence from glycerinated fibers in the presence of 2 mM ε-ATP (+Mg) and 2 mM EGTA was measured. The value of θ was large independently of the sarcomere length (Table I).

5. During Tension Development

1) Polarized fluorescence in the presence of a high concentration of ε-ATP: Polarized fluorescence from glycerinated fibers activated fully by 2 mM ε-ATP in the presence of 1 mM Ca was measured. The values of $\bar{\theta}$ were large and almost independent of the sarcomere length (Table I).

2) Polarized fluorescence in the presence of a low concentration of ε-ATP: The isometric tension was induced by the low concentration of ε-ATP independently of the concentration of Ca. To examine the states of nucleotides and myosin heads bound to them when the concentration of ε-ATP was low, the polarized fluorescence was measured in the presence of 50 μM ε-ATP in the solution containing a sufficient amount of the ε-ATP regeneration system (5 mg/ml PCK and 10 mM PC). Under this condition, about 1/3 of the total amount of myosin heads were expected to be bound to the nucleotides. The isometric tension induced by this concentration of ε-ATP was 80% of the peak tension at 200 μM ε-ATP and 90% of full tension. At a sarcomere length of 2.1 μm, the orientation distribution of bound nucleotides had a sharp maximum with about 8° of the $\bar{\theta}$, and the values of ϕ_A and ϕ_E were 68° and 64°, respectively (Table I). These results suggest that most of the cross-bridges which react with ε-ATP remain on the thin filament on the time-average with nearly identical angles to the thin filament axis under isometric conditions.

The value of $\bar{\theta}$ in the presence of 180 μM ε-ATP was intermediate between that at a low and that at a high concentration of ε-ATP in an activated state (Table I).

At a sarcomere length of 3.7 μm, the values of $\bar{\theta}$ were large.

DISCUSSION

The maximum bound amounts of nucleotides were 200–270 μM in the presence of ε-ADP and ε-ATP. Since the concentration of myosin molecules in rabbit psoas fiber was estimate to be about 120 μM (Tregear and Squire, 1973), these B_m correspond to almost two bound nucleotide molecules per myosin molecule.

Rodger and Tregear (1974) showed that the X-ray diffraction pattern was identical to the rigor pattern when ADP was bound to myosin. Therefore, the cross-bridge attachment to thin filament seems to remain identical to that in rigor when ε-ADP binds to myosin. Thus present results suggest the heterogeneous binding of two heads of myosin to thin filaments or two ε-ADP to two heads in rigor.

The polarized fluorescence from activated fibers in the presence of 2 mM ε-ATP and 1 mM Ca showed that the orientation distributions for nucleotides bound to myosin heads were broad. However, the orientation distribution for bound nucleotides had a sharp maximum at 50 μM ε-ATP at a sarcomere length of 2.1 μm, although 60% of maximum isometric tension was developed. This result showed that the cross-bridges reacting with ε-ATP remained on the thin filament with nearly an identical angle to the filament axis (there are a considerable number of cross-bridges forming rigor complex without nucleotides at low concentrations of ε-ATP, but they do not contribute to fluorescence), suggesting a rate-limiting re-

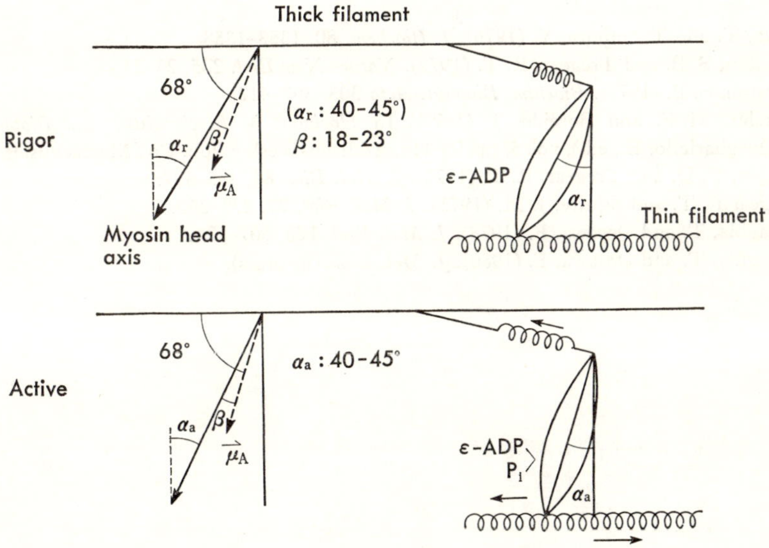

Fig. 2. The angles of the absorption dipole ($\vec{\mu}_A$) and the cross-bridge bound to it (AM-ε-ADP, AM-ε-ADP-P_i). The upper diagram shows those in rigor and the lower those during isometric contraction. The angle of the cross-bridge in rigor (α_r) is obtained as 40°–45° by X-ray diffraction and EM studies, the angle β is given as 18°–23° and the angle of the cross-bridges during isometric contraction (α_a) as 40°–45°. The arrows represent the directions of force of the rotational motion of the cross-bridge and the elasticities of myosin S-2 and thin filament.

action of the cross-bridge cycle. At 50 μM ε-ATP, about 1/3 of the myosin heads were bound to nucleotide. As the amount of bound nucleotide increased, the orientation distribution for bound nucleotides broadened. One may understand these results to mean that the cross-bridges reacting with ε-ATP are helically arrayed on the thin filament during isometric contraction on the time-average (because of the rate limiting reaction of its cycle) until all of one head of myosin is reacted with ε-ATP, but when other heads are bound to nucleotide at high concentration of ε-ATP, the distribution becomes broad because the orientations of two heads of myosin are not homogeneous. If so, from the values of ϕ_A and ϕ_E the cross-bridge angles in rigor and activation are obtained as shown in Fig. 2.

Acknowledgment

I should like to thank Professor F. Oosawa for his encouragement.

REFERENCES

Arata, T. and Tonomura, Y. (1976). *J. Biochem.* **80**, 1353–1358.

Marston, S. B. and Tregear, R. T. (1972). *Nature New Biol.* **235**, 23–24.

Marston, S. B. (1973). *Biochim. Biophys. Acta* **305**, 397–412.

Morales, M. F. and Borejdo, J. (1977). In *Horizons in Biochemistry and Biophysics* (Quaghariello, E., ed.), vol. 5, pp. 99–118, Addison-Wesley Pub. Co., Massachusetts.

Rodger, C. D. and Tregear, R. T. (1973). *J. Mol. Biol.* **84**, 495–497.

Tregear R. T. and Squire, J. M. (1975). *J. Mol. Biol.* **77**, 279–290.

Yanagida, T. and Oosawa, F. (1978). *J. Mol. Biol.* **126**, 507–524.

Yanagida, T. and Oosawa, F. (1980). *J. Mol. Biol.* (in press).

Muscle Contraction: Its Regulatory Mechanisms (Ebashi, S. et al., eds.), pp. 181–187, Japan Sci. Soc. Press, Tokyo / Springer-Verlag, Berlin (1980)

14

Dynamic Aspects of Muscle Thin Filaments

SHIGEO YOSHINO AND SATORU FUJIME

Summary: The muscle thin filament is not rigid but semi-flexible. Its flexibility, measured as the flexural rigidity of a rod, changes very much depending on concentrations of free divalent cations (Ca, Sr, and Mg ions) in the myofibrillar space. The midpoint of the change occurs at about 0.6 μM free Ca ions (depending strongly on concentrations of free Mg ions), or at about 60 μM free Sr ions in the absence of free Ca ions. The S-shaped dependence of the flexibility on concentrations of free Ca/Sr ions is very steep, suggesting that the cooperative structural change occurs at the level of the thin filament when Ca/Sr ions bind to troponin-C. The Ca-sensitivity of the flexural rigidity decreases by about 0.5 pCa per 1 mM free Mg ions, and saturation of this effect of Mg ions occurs above 0.8 mM. The temperature dependence of the flexural rigidity suggests that not only the absolute values of the flexibility in the presence and absence of free Ca ions but also their difference have some positive physiological meaning. All these results strongly suggest that the thin filament flexibility is surely involved in the elementary process of muscle contraction and its regulation.

The cooperative structural rearrangement at the quaternary level as well as within constituent subunits offers a very general molecular mechanism for biological function. In muscle contraction, the thin filament is regulated by Ca ions through its subunit (the tropomyosin/troponin system). On the other hand, from experimental and theoretical studies of the polymerization of actin, a certain kind of structural flexibility of F-actin has long been expected (*e.g.* Oosawa and Asakura, 1975). Since the strength of inter- and/or intra-molecular, noncovalent bonds of actin, tropomyosin and/or troponin fluctuates due to thermal agitation, the thin filament must be not rigid but semiflexible. Thus it is very important to investigate a correlation between the tension development and the flexibility of the thin filament. The F-actin flexibility was first detected by quasi-elastic light scattering (Fujime, 1970), and then the change in the F-actin flexibility in various conditions was extensively studied by the same technique (Fujime and Ishiwata, 1971; Ishiwata and Fujime, 1972; Fujime *et al.*, 1972). One of the important findings in these studies is that the flexibility of the *in vitro* reconstituted thin filament changes depending on micromolar concentrations of free Ca ions in the solution. (The interpretation in these studies was once questioned (Carlson and Fraser, 1974). But the apparent discrepancy was clarified to some extent (Maeda and Fujime, 1977), and a recent study essentially supports the earlier results (unpublished).)

Later, the flexibility of the *in vivo* thin filament was successfully studied by analyzing the *electro-optical effect* of Natori's fibers (Umazume and Fujime, 1975), and detailed studies were made on this effect from the theoretical (Fujime and Yoshino, 1978) and quantitative experimental (Yoshino *et al.*, 1978) points of view. The flexibility of the *in vivo* thin filament also changes depending on micromolar concentrations of free Ca ions in the myofibrillar space. During these studies, some other groups also measured the flexibility of F-actin and the thin filament by different techniques (Takebayashi *et al.*, 1977; Yanagida and Oosawa, 1978; Nagashima and Asakura, 1980). All the results in these studies were in agreement although the flexibility measured by one technique was slightly different from that measured by the other. Our technique is considered to be more quantitative than the others. In this paper, we briefly describe some of our new results.

EXPERIMENT

When an electric field is applied to an overstretched skinned fiber in liquid paraffin oil along its long axis with platinum electrodes, the intensities of all observable laser diffraction lines increase (Umazume and Fujime, 1975). For a sinusoidal field in time, *i.e.*, $E(t)=E_0 \sin(\omega t)$, the time-dependent part of the excess intensity is given by (Fujime and Yoshino, 1978)

$$\Delta I(t) = B_\omega \sin(\omega t - \phi_\omega) + B_{2\omega} \cos(2\omega t - 2\phi_{2\omega}) \qquad (1)$$

From the dispersion frequency ω_1 [rad/sec]$=2\pi\nu_1$ [Hz] where ϕ_ω or $\phi_{2\omega}=\pi/4$, the flexural rigidity of the thin filament ε is given by

$$\varepsilon = 3.1\times10^{-19}(\eta_T/\eta_{293})\omega_1 \qquad [\mathrm{dyn\cdot cm^2}] \qquad (2)$$

where η_T is the viscosity of water at T K.

The experimental set-up and the preparation method of muscle fibers were essentially the same as the previous ones (Yoshino *et al.*, 1978). The sarcomere length of the fibers was between 6 and 7 μm. The Ca ion concentration was regulated using a Ca^{2+}-EGTA buffer. The apparent binding constant between Ca ions and EGTA was assumed to be 1.0×10^6 M^{-1} at pH 6.8 and 20°C (Saida and Nonomura, 1978). This constant differs from the value of Schwarzenbach *et al.* (1957) by a factor of two. Thus, to keep consistency, the apparent binding constants of Ca-HEGTA, Mg-EGTA, Mg-HEGTA, and Sr-EGTA were assumed to be half of the corresponding values of Schwarzenbach *et al.* For example, the constant of Sr-EGTA was 3.3×10^2 M^{-1}. For ATP, the constants of Mg-ATP, Mg-HATP and K-ATP were from Kahn and Martell (1962). Solutions were prepared at 20±0.2°C.

RESULTS AND DISCUSSION

Figure 1 shows the S-shaped change in the flexural rigidity of the thin filament induced by free Ca/Sr ions. This change is much steeper than that of Ca/Sr-binding to (Ebashi and Endo, 1968), or the change in conformation of (van Eerd and Kawasaki, 1972), free troponin or troponin-C. This means that the cooperative structural change occurs at the level of the thin filament when Ca/Sr ions bind to troponin-C. The flexural rigidity in the presence of free Ca ions is nearly the same as that measured by other techniques, whereas that in the absence of free Ca ions is very much larger

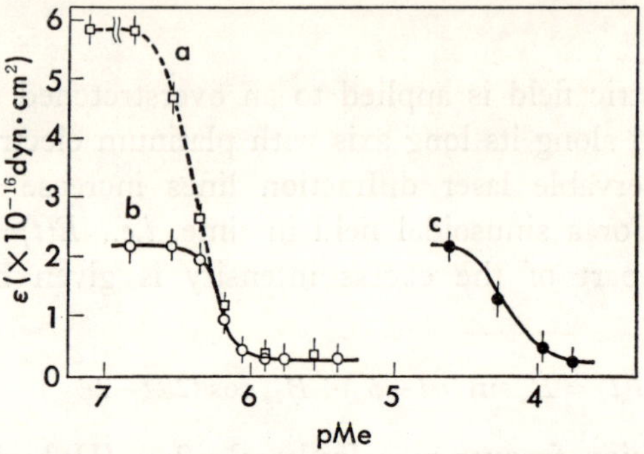

Fig. 1. The change in the flexural rigidity of the thin filament induced by Ca and Sr ions. The measurements were made at 20±0.2°C using mechanically skinned, glycerinated fibers equilibrated in a solution containing 4 mM $MgCl_2$, 4 mM ATP, 10 mM EGTA, appropriate amounts of $CaCl_2$ or $SrCl_2$, and 20 mM Tris maleate buffer, pH 6.8 at 20±0.2°C, ionic strength being adjusted to 0.15 by adding KCl. a, rabbit psoas muscle against pCa. b, frog semitendinosus muscle against pCa. c, frog semitendinosus muscle against pSr.

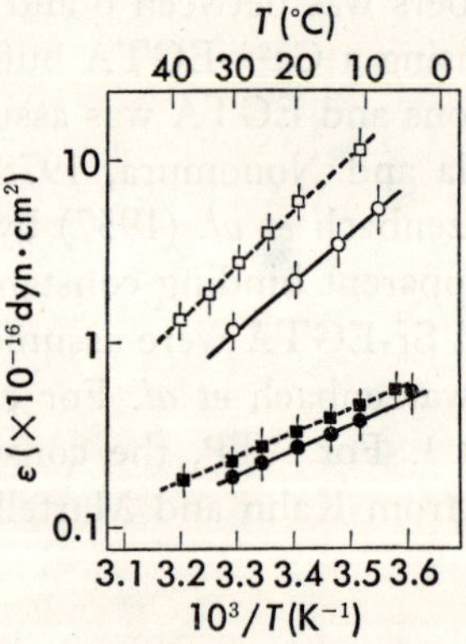

Fig. 2. The temperature dependence of the flexural rigidity of the thin filament in the presence and absence of free Ca ions. The experimental conditions were the same as in Fig. 1, except that the pCa's and pH indicated were those at 20±0.2°C. ○, ● frog semitendinosus muscle; □, ■ rabbit psoas muscle; ●, ■ pCa=5.5_6; ○, □ pCa=7.3_6.

than that in others: in our measurements at 20°C, the flexural rigidity in the absence of free Ca ions is 6 times (in the case of frog semitendinosus muscle) and 20 times (in the case of rabbit psoas muscle) larger than that in the presence of free Ca ions. The temperature dependence of the flexural rigidity (Fig. 2) is small in the presence of, and very large in the absence of, free Ca ions. Although the flexural rigidity of the thin filament of rabbit psoas muscle in the absence of free Ca ions is very large, its absolute values and the difference at 38°C are nearly the same as those of frog semitendinosus muscle at 20°C. If we consider the body temperature of 38°C for rabbit and the usual experience that frogs are active at the ambient temperature of 20°C, the above coincidence might not be accidental but many have some positive physiological meaning in muscle contraction and its regulation.

Tension development, superprecipitation and actomyosin ATPase are the result of the actin-myosin interaction, and in these processes the presence of Mg-ATP is essential. We tried to quantitatively examine the effect of free Mg ions on the thin filament flexibility. Since, however, the literature values of the binding constant between Mg ions and ATP do not agree, we measured in this experiment the flexural rigidity of the thin filament in the absence of ATP (Fig. 3). The midpoint of the change occurs at 0.2 mM in Fig. 3A (or 0.45 mM in Fig. 3C) free Mg ions, the inverse of which, $5\times10^3\ \text{M}^{-1}$ (or $2.2\times10^3\ \text{M}^{-1}$), is very close to the apparent binding constant between troponin-C and Mg ions estimated from the conformational change of troponin-C (Kawasaki and van Eerd, 1972). The effect of

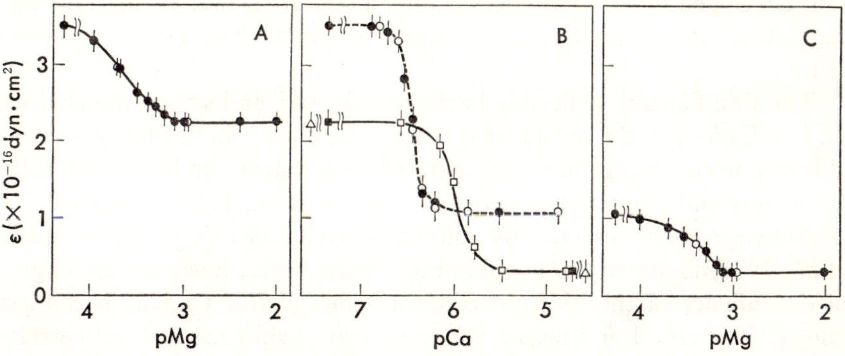

Fig. 3. Effect of free Mg ions on the flexural rigidity of the thin filament of frog semitendinosus muscle. Measurements were made at 20±0.2°C using mechanically skinned, glycerinated and Natori's fibers equilibrated in a solution containing 4 mM EGTA, appropriate amounts of $MgCl_2$ and $CaCl_2$, and 20 mM BES (N, N-bis (2-hydroxyethyl)-2-aminoethansulfonic acid) buffer, pH 7.2 at 20±0.2°C, ionic strength being adjusted to 0.15 by adding KCl. Vertical bars show standard errors. A: effect of free Mg ions at pCa=8.0_8–7.8_9. B: effect of free Ca ions at zero (○, ●) and 1 mM (□, ■) added $MgCl_2$. C: effect of free Mg ions at pCa=4.9_8–4.8_2. ●, ■ glycerinated; ○, □ Natori's. Note that (△) in B is from Natori's fibers *without* equilibration in the present solution.

Mg ions probably on troponin-C appears as a change in the flexural rigidity of the thin filament. This is the first observation so far as we know. Moreover, the S-shaped curve of the flexural rigidity moves to the right by about 0.5 pCa per 1 mM free Mg ions (Fig. 3B). This dependence is similar to those of the rate of superprecipitation and tension development (Ebashi and Endo, 1968). The saturation of the effect of Mg ions on the flexural rigidity occurs above 0.8 mM free Mg ions in both the presence and absence of free Ca ions, at least for frog semitendinosus muscle in the absence of ATP. Natori's fibers without equilibration in the artificial solution give a flexural rigidity equal to that in the presence of free Mg ions at a high/low pCa value when the applied electric field is very low/high (Δ in Fig. 3B). (It has been shown that the release of Ca ions from sarcoplasmic reticulum does/does not occur when the field strength is high/low (Yoshino *et al.*, 1978).) It is noted that the concentrations of free Mg ions in the solution of, say, 4 mM ATP and 1 mM $MgCl_2$ are much lower than the above "saturating" concentration of free Mg ions, whereas those in the solution of, say, 4 mM ATP and 4 mM $MgCl_2$ are nearly the same.

When the S-shaped curves of the flexural rigidity on concentrations of free Ca/Sr ions in Fig. 1 are replotted in terms of the *flexibility* defined by $\lambda = 3k_BT/4\varepsilon$ [cm^{-1}], their behavior becomes very similar to that of the tension developed by skinned fibers (Endo, 1972; Kitazawa, 1976). This fact suggests an important correlation between tension development and flexibility.

The thin filament is flexible in the presence of, and stiff in the absence of, free Ca ions in the myofibrillar space. The change in flexibility is probably due to the change in the strength of noncovalent bonds between actin monomers and tropomyosin/troponin (Fujime *et al.*, 1979). The so-called tropomyosin-shift observed by diffraction techniques (*e.g.*, Wakabayashi *et al.*, 1975) is one of the possible candidates. Since, however, the degree of dependence of this shift on concentrations of free Ca ions is not yet known, it is not clear whether the tropomyosin-shift has a *direct* correlation with the change in the flexibility of the thin filament.

Our method surely makes quantitative treatment possible in investigating the flexibility of the thin filament in the process of muscle contraction and its regulation.

Acknowledgment

We would like to thank Professor Fumio Oosawa for his encouragement and Professor Reiji Natori for his continuous support for, and interest in, this series of experiments.

REFERENCES

Carlson, F. D. and Fraser, A. B. (1974). *J. Mol. Biol.* **89**, 273–281.

Ebashi, S. and Endo, M. (1968). *Prog. Biophys. Mol. Biol.* **18**, 123.

Endo, M. (1972). *Nature* **237**, 211–213.

Fujime, S. (1970). *J. Phys. Soc. Japan* **29**, 751–759.

Fujime, S. and Ishiwata, S. (1971). *J. Mol. Biol.* **62**, 251–265.

Fujime, S., Ishiwata, S., and Maeda, T. (1972). *Biochim. Biophys. Acta* **283**, 351–363.

Fujime, S. and Yoshino, S. (1978). *Biophys. Chem.* **8**, 305–315.

Fujime, S., Yoshino, S., and Umazume, Y. (1979). In *Cross-bridge Mechanism in Muscle Contraction* (Sugi, H. and Pollack, G. H., eds.), pp. 51–67, Univ. Tokyo Press, Tokyo.

Ishiwata, S. and Fujime, S. (1972). *J. Mol. Biol.* **68**, 511–522.

Kahn, T.M.M. and Martell, A. E. (1962). *J. Phys. Chem.* **66**, 10–15.

Kawasaki, Y. and van Eerd, J. P. (1972). *Biochem. Biophys. Res. Commun.* **49**, 898–905.

Kitazawa, T. (1976). *J. Biochem.* **80**, 1129–1147.

Maeda, T. and Fujime, S. (1977). *J. Phys. Soc. Japan* **42**, 1983–1991.

Nagashima, H. and Asakura, S. (1980). *J. Mol. Biol.* **136**, 169–182.

Oosawa, F. and Asakura, S. (1975). In *Thermodynamics of the Polymerization of Proteins*, Academic Press, London and New York.

Saida, K. and Nonomura, Y. (1978). *J. Gen. Physiol.* **72**, 1–14.

Schwarzenbach, G., Senn, H., and Anderegg, G. (1957). *Helv. Chim. Acta* **40**, 1886–1900.

Takebayashi, T., Morita, Y., and Oosawa, F. (1977). *Biochim. Biophys. Acta* **492**, 357–362.

Umazume, Y. and Fujime, S. (1975). *Biophys. J.* **15**, 163–180.

van Eerd, J. P. and Kawasaki, Y. (1972). *Biochem. Biophys. Res. Commun.* **47**, 859–865.

Wakabayashi, T., Huxley, H. E., Amos, L. A., and Klug, A. (1975). *J. Mol. Biol.* **93**, 477–497.

Yanagida, T. and Oosawa, F. (1978). *J. Mol. Biol.* **126**, 507–524.

Yoshino, S., Umazume, Y., Natori, R., Fujime, S., and Chiba, S. (1978). *Biophys. Chem.* **8**, 317–326.

TROPONIN-LINKED REGULATION

Muscle Contraction: Its Regulatory Mechanisms (Ebashi, S. et al., eds.), pp. 191–206,
Japan Sci. Soc. Press, Tokyo / Springer-Verlag, Berlin (1980)

15

The Structure of Troponin-C and Thin Filament Regulation in Striated Muscle

J. GERGELY AND P. C. LEAVIS

Summary: This review deals with the structural basis of the interaction of Ca^{2+} with two classes of binding sites of troponin-C (TN-C). Evidence bearing on the assignment of high affinity sites to regions III and IV, and of low affinity sites to regions I and II is considered, particularly in the light of recent work on proteolytic fragments. Conformational changes taking place in TN-C on Ca^{2+}-binding to each class of binding sites and their relation to subunit-subunit interactions leading to the regulation of the actomyosin system are discussed.

It is an honor and a pleasure to dedicate this paper to Professors Kumagai and Natori, not only because of the long-standing ties of friendship and affection, but because of the straight line that leads from their discoveries to the topic of this contribution. Professors Kumagai and Natori deserve credit beyond their groundbreaking scientific contributions for creating an atmosphere, extending far beyond their own laboratories, that was so beneficial for the fluorishing of muscle science not only in Japan but also in the wider world community.

Natori's introduction of the skinned muscle fiber as an experimental object proved invaluable in the study of Ca^{2+} control. The classical paper by Kumagai, Ebashi, and Takeda (1955)—the latter is better known as Mrs. Ebashi—paved the way to the subsequent discoveries of Ebashi and his colleagues concerning the crucial role played by Ca^{2+} in the regulation of muscle contraction at the molecular level of the interaction of myosin and actin and the recognition of the existence of a regulatory complex consisting of tropomyosin and troponin. Ebashi and his colleagues went on to localize the two proteins on the thin filament and, on the basis of electron microscopic studies utilizing antibodies against troponin, estimated the molar ratios of troponin: tropomyosin: actin to be 1: 1: 7 and postulated that tropomyosin lies in the two-long-pitched grooves of the actin helix (Ebashi *et al.*, 1969). This picture was subsequently refined by electron microscopic image reconstructions and by X-ray diffraction (Moore *et al.*, 1970; Haselgrove, 1972; Parry and Squire, 1973), both indicating that in the relaxed muscle tropomyosin actually occupies a position somewhat to one side of the groove, closer to the edge of the filament, but moves toward the groove during contraction, presumably uncovering the site of actin-myosin interaction (Fig. 1). The suggested role of troponin in this scheme is to anchor the tropomyosin in the relaxed position in the absence of Ca^{2+} (Potter and Gergely, 1974; Hitchcock, 1973).

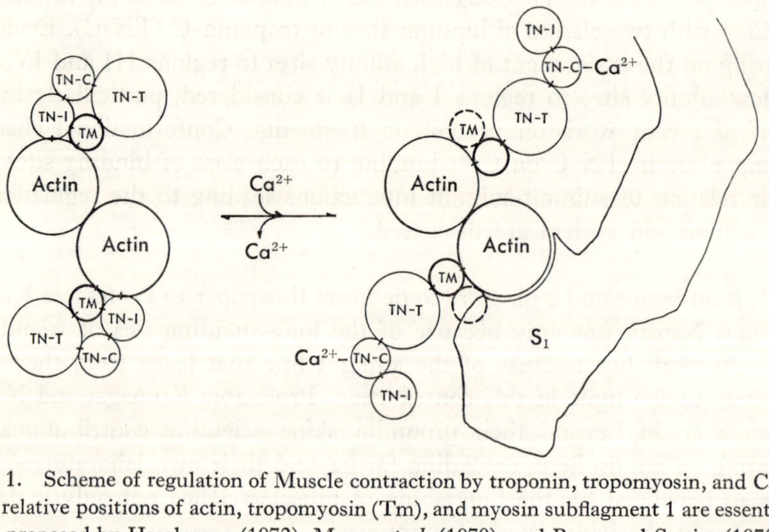

Fig. 1. Scheme of regulation of Muscle contraction by troponin, tropomyosin, and Ca^{2+}. The relative positions of actin, tropomyosin (Tm), and myosin subflagment 1 are essentially as proposed by Haselgrove (1972), Moore *et al.* (1970), and Parry and Squire (1973).

Troponin was at first considered a homogeneous protein (see Hartshorne and Dreizen, 1972) until Hartshorne and his colleagues (Hartshorne and Mueller, 1969; Hartshorne and Pyun, 1971), and Schaub and Perry (1969) showed that troponin could be separated into at least two components, other workers seeing as many as four (Ebashi *et al.*, 1971; Drabikowski *et al.*, 1972; Wilkinson *et al.*, 1972). The currently accepted subunit composition resulted from the work of Greaser and Gergely (1971, 1973), who isolated and characterized three troponin subunits and reconstituted them into a complex conferring Ca^{2+} sensitivity on the ATPase of the actomyosin-tropomyosin system. The last letter in the name of the subunit identifies its principal property; *viz.* troponin-C (TN-C) is the Ca^{2+}-binding subunit, troponin-T (TN-T), the tropomyosin binding subunit, and troponin-I (TN-I), the inhibitory subunit. Although the inhibition of actomyosin ATPase by TN-I alone is reversed by the addition of TN-C, there is no Ca^{2+} sensitivity (Greaser and Gergely, 1971). In the case of the complex containing all three subunits and tropomyosin, the inhibition, which is maximal at a troponin: tropomyosin: actin ratio is reversed by Ca^{2+} (Greaser and Gergely, 1973).

The details of intersubunit interactions in troponin still remain to be elucidated. It has been established that the dissociation of TN-C, TN-T, and TN-I in 6 M urea is reversible (Greaser and Gergely, 1971, 1973). Subsequent work indicated that it was possible to reconstitute the binary complexes TN-C-TN-T and TN-C-TN-I (Greaser and Gergely, 1971; Head and Perry, 1974) from purified subunits indicating that TN-C contained binding sites for the other two subunits. Direct interactions between TN-T and TN-I were implied by the existence of troponin B in early preparations (Hartshorne and Mueller, 1969; Hartshorne and Pyun, 1971; Schaub and Perry, 1969) and confirmed by the cross-linking experiments of Hitchcock (1975) but were not verified until the recent CD study of Horwitz *et al.* (1979). All of these studies, then, indicate that every troponin subunit interacts with each of the other two.

TROPONIN-C

Let us now turn to the details of the triggering moiety of the troponin complex, TN-C. Skeletal muscle TN-C has four binding sites for Ca^{2+} falling into two classes: the two sites of higher affinity for Ca^{2+} also competitively bind Mg^{2+} and have been referred to as the Ca^{2+}-Mg^{2+} sites ($K_{app}^{Ca}=2\times10^{7}$ M^{-1}, $K_{app}^{Mg}=5\times10^{3}$ M^{-1}); the two with a lower affinity

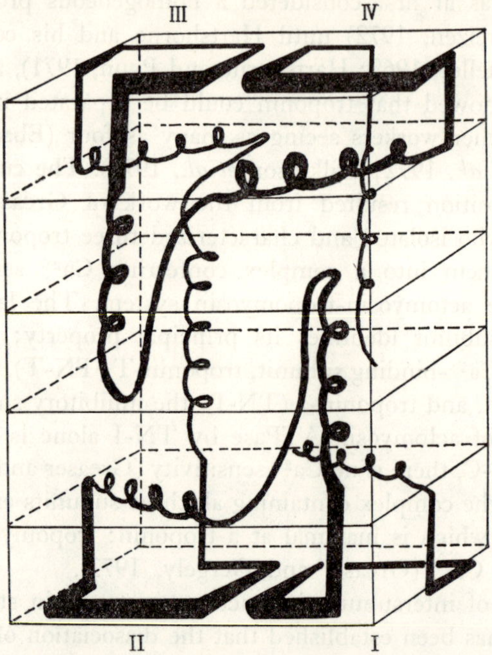

Fig. 2. Schematic diagram of TN-C structure showing the four Ca^{2+}-binding loops (I-IV) and their flanking helices. Based loosely on Kretsinger and Barry (1975).

are specific for Ca^{2+} ($K_{app}^{Ca} = 5 \times 10^5$ M^{-1}) (Potter and Gergely, 1975). The amino acid sequence of TN-C, containing 159 residues, shows extensive homologies with parvalbumin, a soluble Ca^{2+}-binding protein (Collins *et al.*, 1973). Parvalbumin has two Ca^{2+}-binding sites, whose structure has been identified by comparison of X-ray diffraction and sequence data (Coffee and Bradshaw, 1973; Kretsinger and Nockolds, 1973). TN-C reveals four regions, usually numbered I to IV, starting from the NH_2-terminus, each containing approximately 30 residues corresponding to the Ca^{2+}-binding sites in parvalbumin (Collins, 1974). Within each region there is, as in parvalbumin (Kretsinger and Nockolds, 1973), a Ca^{2+}-binding loops, rich in aspartic acid and glutamic acid residues which act as coordinating ligands for Ca^{2+}, flanked on either side by α-helical segments. Based on the positions of the various residues in parvalbumin deduced from X-ray diffraction data, Kretsinger and Barry (1975) have proposed a tentative TN-C structure, a schematic interpretation of which appears in Fig. 2.

The two classes of Ca^{2+}-binding sites appear to play different roles in the regulatory process. Briefly stated, it seems that Ca^{2+} or Mg^{2+} binding to the high affinity sites produces a conformation in TN-C that can then respond to binding of Ca^{2+} to the Ca^{2+}-specific sites of TN-C. The latter can be regarded as the event actually triggering actin-myosin interactions. This view is supported by several lines of evidence. The structure produced by binding of Ca^{2+} or Mg^{2+} to the Ca^{2+}-Mg^{2+} sites is likely to be present *in vivo* both in the relaxed and active state owing to the fact that these sites are always occupied by one or the other ion at the concentrations prevailing in muscle. Kinetic studies clearly implicate the low affinity Ca^{2+}-specific sites in the activation process. Thus the recently reported value of 8 sec^{-1} for the Mg^{2+} off-rate (Johnson *et al.*, 1979) would preclude activation by Ca^{2+} at the high affinity site which would be governed by the rate of displacement of Mg^{2+}. These studies also indicate that while Ca^{2+} binding to, and release from, the Ca^{2+}-specific sites can take place on a time scale consistent with activation and relaxation *in vivo*, the Ca^{2+} release rate from the Ca^{2+}-Mg^{2+} sites is far too slow to play a physiological role.

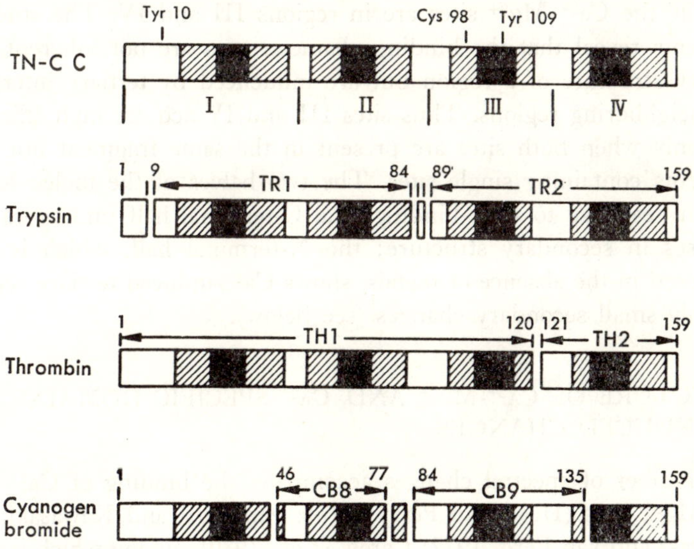

Fig. 3. Schematic diagram of TN-C (top) and the fragments isolated for study after digestion with trypsin, thrombin, and cyanogen bromide. Isolated fragments are indicated by arrows, the numbers indicating the included residues. Cross-hatched area, calcium-binding loops proper; shaded areas, putative flanking α-helices. The four regions are indicated by Roman numerals.

In view of the differing roles of the two classes of binding sites, it has been of interest to determine which sites are to be considered as the Ca^{2+}-Mg^{2+} and which as the Ca^{2+}-specific sites. Studies employing optical and spin probes attached to Cys 98, which is located within region III, have suggested that site III belongs to the Ca^{2+}-Mg^{2+} class (Potter *et al.*, 1976; Nagy and Gergely, 1979). Also the identification of site I as a low affinity Ca^{2+}-specific site is suggested by Ca^{2+}-binding studies on cardiac TN-C which reveal two high affinity Ca^{2+}-Mg^{2+} sites, but only a single low affinity rate (Leavis and Kraft, 1978). A comparison of the amino acid sequences of cardiac and skeletal TN-C shows conservative differences in all regions except in the Ca^{2+}-binding loop of region I (van Eerd and Takahashi, 1975). Amino acid substitutions involving key coordinating ligands in cardiac TN-C suggest that site I must be non-functional.

More direct assignments of the regions have come from Ca^{2+}-binding studies using proteolytic fragments of TN-C produced by digestion with CNBr, trypsin and thrombin (Leavis *et al.*, 1978; see Fig. 3), and by chemical modification of carboxylate containing sidechains (Sin *et al.*, 1978). These studies established that the Ca^{2+}-specific sites are in regions I and II while the Ca^{2+}-Mg^{2+} sites are in regions III and IV. The studies on fragments reveal that the binding characteristics are not inherent in the primary structure of a region but are influenced by tertiary interactions with neighboring regions. Thus sites III and IV acquire high affinity for Ca^{2+} only when both sites are present in the same fragment but not in fragments containing single sites. The two halves of the molecule differ in their response to Ca^{2+}-binding; the C-terminal half undergoes larger increases in secondary structure; the N-terminal half, which is highly structured in the absence of metals, shows Ca^{2+}-induced tertiary refolding but only small secondary changes (see below).

STRUCTURE OF Ca^{2+}-Mg^{2+} AND Ca^{2+}-SPECIFIC DOMAINS: Ca^{2+}-INDUCED CHANGES

A number of spectral changes accompany the binding of Ca^{2+} to the Ca^{2+}-Mg^{2+} sites (Head and Perry, 1974; van Eerd and Kawasaki, 1972; Kawasaki and van Eerd, 1972; Carew *et al.*, 1980), among which is a large enhancement of circular dichroism in the 222 nm region (van Eerd and Kawasaki, 1972; Kawasaki and van Eerd, 1972; Nagy and Gergely, 1979). This indicates an increase in the α-helical content from approximately 30% in the divalent cation-free protein to 50% on saturation of the

Ca^{2+}-Mg^{2+} sites, corresponding to the formation of two 8–10 residue α-helical regions (Nagy and Gergely, 1979). In addition, changes in circular dichroism and absorbance also occur in the near UV (Head and Perry, 1974; Nagy and Gergely, 1979), indicating the involvement of phenylalanine residues. On the basis of the distribution of phenylalanines, one of the helical segments formed has been located on the N-terminal side of site III and the other on the C-terminal side of site IV (helices E and H of Kretsinger and Barry, 1975).

Recently ^{1}H-NMR spectroscopy has been used to study the solution conformation of TN-C. In the spectrum of a folded protein, resonance lines arising from amino acids are shifted from their primary positions in the spectrum (*i.e.*, the positions they would occupy in a configuration in which they are accessible to the solvent). These so-called secondary shifts depend on H-bonding, shielding caused by nearby carboxyl groups, the degree of ionization of ionizable groups and ring current shifts produced by aromatic groups (see Levine *et al.*, 1979). In the case of TN-C, secondary shifts in a number of resonances indicate a Ca^{2+}-induced increase in secondary structure in the high affinity regions of the protein. Other changes detected by ^{1}H-NMR include Ca^{2+}-induced broadening of resonances assigned to tyrosine, phenylalanine, glutamate, and aspartate residues, all indicative of increased constraints on their movement (Levine *et al.*, 1977; Seamon *et al.*, 1977; Levine *et al.*, 1978). Many of these changes have also been observed in tryptic fragments of TN-C containing only the two Ca^{2+}-Mg^{2+} sites (Leavis *et al.*, 1978; Levine *et al.*, in preparation). Ca^{2+}-induced conformational changes in the C-terminal half of TN-C are also reflected in the behavior of various optical and spin probes attached to Cys 98, located within the N-terminal helix of region III (Potter *et al.*, 1976; Nagy and Gergely, 1979). Furthermore, the reactivity of Cys 98 to various thiol reagents is reduced by Ca^{2+} (Potter *et al.*, 1976; Nagy *et al.*, 1976, 1978). Other residues showing Ca^{2+}-dependent environmental changes are Lys 136 and 140 whose reactivity to acetylation is decreased by Ca^{2+} (Hitchcock, personal communication), His 125, whose pK is decreased (Levine *et al.*, 1976), and Tyr 109, which shows a change in interactions involving its phenolic hydroxyl group (Carew *et al.*, 1980).

In contrast to the large changes brought about by Ca^{2+} binding to the sites in the high affinity-class, binding to the Ca^{2+}-specific sites is generally believed to produce more subtle structural perturbations (Nagy and Gergely, 1979; Levine *et al.*, 1977, 1978; Seamon *et al.*, 1977). Our recent CD analysis (Nagy and Gergely, 1979) concluded that the observed far

UV changes may be interpreted as an addition of a peptide unit or two to one or more of the pre-existing helical segments in the N-terminal half of the molecule. According to two other reports (Johnson and Potter, 1978; Hinke *et al.*, 1978) up to 40% of the secondary change in the whole molecule results from Ca^{2+} binding to the low affinity sites in the pCa range of 6–5. The latter observations are difficult to reconcile with experiments at higher protein concentration in which spectral changes are essentially complete on the addition of 2 mol of Ca^{2+} per mol. There is, however, no doubt that significant changes in the tertiary structure are brought about by the occupation of the Ca^{2+}-specific sites in addition to whatever small changes may be induced in secondary structure. While ^{1}H-NMR studies provide little clear-cut evidence of backbone changes, broadening of several resonances corresponding to hydrophobic sidechains of leucine, isoleucine, and phenylalanine residues suggest local changes in their environment attributable to changes in the tertiary fold. In particular, the disappearance of ring-current shifted resonances arising from one or more phenylalanine residues upon the addition of the third and fourth Ca^{2+} to TN-C suggests the loss of hydrophobic contacts between pairs of phenylalanines (Levine *et al.*, 1977; Seamon *et al.*, 1977). Recently fresh light has been shed on this change by studies on TN-C fragments containing only the Ca^{2+}-specific regions and including only 5 of the 10 phenylalanine residues of the parent protein (TR2C, residues 9–84 and TH1, residues 1–120; Evans *et al.*, 1980). The hydrophobic sidechain signals in the spectra of the fragments are the same as in intact TN-C suggesting that the local environments of the relevant sidechains are retained in the fragments. A more detailed analysis of the upfield ring current shifted Phe resonances (δ=6.64 and 6.42 ppm, Fig. 4A) suggest that in the apo-protein, two pairs of phenylalanine residues, separated by some 50 residues in the primary structure, interact to stabilize the tertiary folds of the peptide, and on the basis of the tentative coordinates proposed by Kretsinger and Barry (1975) the most likely mutually stabilizing pairs are 19–72 and 23–75 (Fig. 4B). Binding of Ca^{2+} to sites I and II results in a loss of these contacts as well as changes of the environments of leucine and isoleucine residues whose interactions probably also contribute to the stability of the apoprotein. All the changes are consistent with the bending or sliding of helices A and D relative to each other.

The following picture of the high and low affinity domains of TN-C emerges from the studies described above. In the apo-protein, the domain of regions I and II, constituting the N-terminal half of the molecule, is

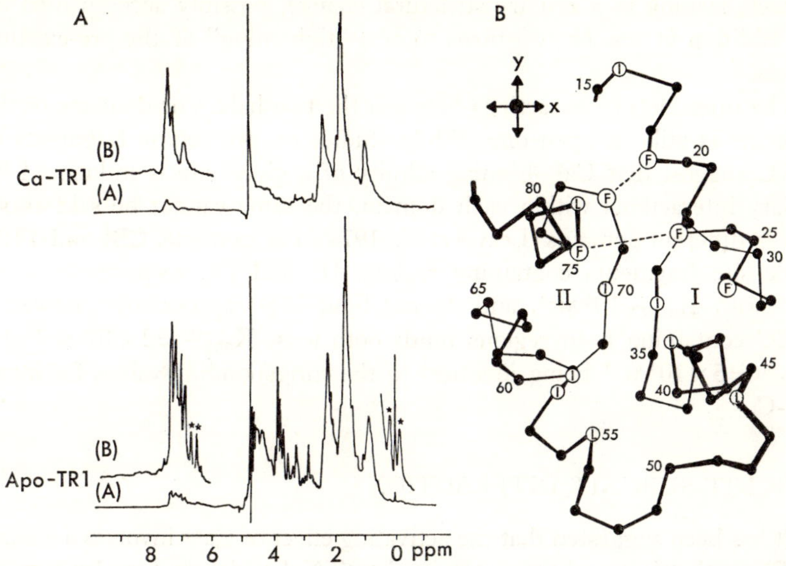

Fig. 4. A: (A) ^{1}H-NMR spectra of TR1 fragment of TN-C (residues 9–84) in the absence (apo-TR1) and presence of Ca^{2+} (Ca-TR1). (B) Resolution enhanced spectra of aromatic resonances and upfield methyl signals. In the spectrum of apo-TN-C, the stars indicate the positions of the two ring-current shifted phenylalanine signals ($\delta=6.64$ and 6.42 ppm) and the two upfield methyl signals corresponding to isoleucine ($\delta=0.19$ ppm) and leucine ($\delta=-0.16$ ppm). These shifted resonances all disappear upon addition of Ca^{2+}. For experimental details, see Evans *et al.* (1980). B: schematic representation of the conformation of regions I and II of TN-C involving helix-helix contacts between the N-terminal helix of region I (helix A) and the C-terminal helix (helix D) of region II in the absence of calcium. The structure is represented as two regions comprising a stretch of approximately 30 residues each in a helix-loop-helix configuration by analogy with parvalbumin. (Kretsinger and Barry, 1975).

highly organized with pre-formed α-helical segments, each containing 8–10 residues flanking each of the two sites. This structure is stabilized by the hydrophobic contacts involving phenylalanine, leucine, and isoleucine residues. In contrast, in regions III and IV constituting the C-terminal half, only two of the four potential helices (F and G) are actually coiled, the others existing in a disordered configuration. Binding of the first two Ca^{2+} ions to sites III and IV induces the coiling of helices E and H (Nagy and Gergely, 1979) which appears to account for ~90% of the increase in secondary structure in the molecule induced by Ca^{2+}. The subsequent binding of Ca^{2+} to regions I and II results in the loss of hydrophobic contacts leading to a tertiary structural change, possibly accompanied by the addition of one or two turns to or a "tightening" of the pre-existing helices.

The question of interactions between the two halves or domains of the molecule is still an open one. While studies on proteolytic fragments of TN-C suggest that Ca^{2+}-binding affinity of a given site is influenced by tertiary interactions within each domain, the same cannot be said about sites in different domains (Leavis *et al.*, 1978). For example, CB9 and TH2, single site fragments containing regions III and IV, respectively, bind Ca^{2+} with $K_{app} \sim 10^5$ M^{-1}, and do not bind Mg^{2+} appreciably; however, TR2C containing both regions binds both ions, $K_{app}{}^{Ca} = 2 \times 10^7$ M^{-1} and $K_{app}{}^{Mg} = 5 \times 10^3$ M^{-1} being identical to the corresponding values for intact TN-C.

SUBUNIT-SUBUNIT INTERACTIONS

It has been suggested that the activating effect of Ca^{2+} involves a change in TN-C that is somehow transmitted to TN-I which in turn leads to its release from a binding site on F-actin and tropomyosin with a subsequent movement of tropomyosin. A key problem that awaits solution is the precise nature of the TN-C-TN-I interaction and the intersubunit transmission of information upon Ca^{2+} binding. It is known from earlier work that complex formation between TN-C and TN-I is Ca^{2+}-dependent. However, once formed the complex does not dissociate upon the removal of the cation. Of particular interest is the reciprocity of Ca^{2+} and TN-I effects on TN-C; the increased Ca^{2+} affinity constant in the TN-I-TN-C complex compared with TN-C alone (Potter and Gergely, 1975) is mirrored by an apparent increase in the strength of the TN-C-TN-I interaction as deduced from the behavior of the complex in 6 M urea (Head and Perry, 1974). These observations suggest some sort of cooperativity between Ca^{2+} and TN-I binding to TN-C.

Perhaps the earliest work specifically directed towards clarification of the details of the TN-C-TN-I interaction was that of Syska *et al.* (1976), identifying two proteolytic fragments of TN-I that were shown to bind to TN-C-sepharose affinity columns. The two peptides, both highly basic, were identified as containing residues 1–47 and residues 97–116. The latter, interestingly, also was capable of binding to actin, partially inhibiting actomyosin ATPase activity.

Complementary experiments carried out using TN-C fragments both

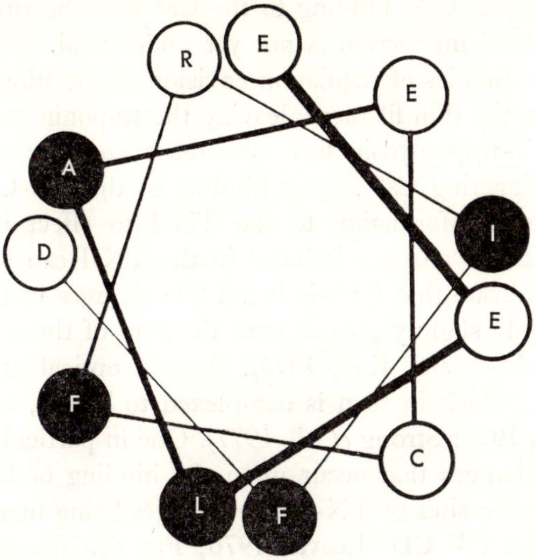

Fig. 5. End-on view of α-helix E (residues 93–103) of TN-C seen from the amino end of the sequence. Hydrophic amino acids are indicated by white letters on dark background. A, Ala; C, Cys; D, Asp; E, Glu; F, Phe; I, Ile; L, Leu; R, Arg (*cf.* Nagy and Gergely, 1979).

in our laboratory (Leavis *et al.*, 1978) and that of S. V. Perry (Weeks and Perry, 1978) showed that a region close to binding site III formed complexes with TN-I in the presence of Ca^{2+}. We were able to identify as the site of interaction residues 89–100, containing a number of negatively charged sidechains (Glu 92, 93, 94, and 97). Even more insight has been gained from the recent work of Weeks and Perry (1978) showing that the complexation of the small CB9 peptide of TN-C (residues 84–135) with TN-I blocked the 3′,5′-cyclic AMP-dependent phosphorylation of TN-I. This provides strong evidence for the binding of residues 89–100 of TN-C to the stretch of residues 97–116 of TN-I, since the phosphorylation occurs at Ser 117. It is tempting to speculate that the Ca^{2+} requirement for the TN-C-TN-I interaction is tied to the cation-induced folding of residues 92–102 of TN-C into an α-helix which results in the alignment of several glutamate sidechains along one side of the helix (see Nagy and Gergely, 1979; Fig. 5).

There is also a second region on TN-C, involving residues 46–55 (Leavis *et al.*, 1978) that appears to interact with TN-I. This interaction, as suggested by ^{1}H-NMR studies on TN-C (unpublished) and its fragments depends upon Ca^{2+} binding to the Ca^{2+}-specific sites. The significance of the second interaction is not yet understood.

Since current theories of regulation envision dissociation of TN-I from a binding site on the thin filament, leaving the troponin complex attached through a TN-T-tropomyosin link, one must consider the mechanism by which the triggering Ca^{2+}, upon binding to the TN-C moiety of the complex, transmits information to the TN-I to effect its dissociation. Large conformational changes induced in the TN-I can be ruled out by studies which indicate that Ca^{2+}-induced CD changes in the TN-C-TN-I complex are only slightly greater than the sum of those in the separate subunits (McCubbin and Kay, 1973). Several optical and spin probes bound to TN-I, which in turn is complexed to TN-C, respond to Ca^{2+} binding (Leavis, 1976; Strong *et al.*, 1977). One in particular, IAEDANS, shows spectral changes that occur upon the binding of Ca^{2+} to the low affinity Ca^{2+}-specific sites of TN-C, the changes being unaccompanied by any change in far UV CD (Leavis, 1976; Fig. 6). The extrinsic probes indicate Ca^{2+}-induced changes in the immediate environment of the probes; whether this reflects structural changes in TN-I *per se* or whether

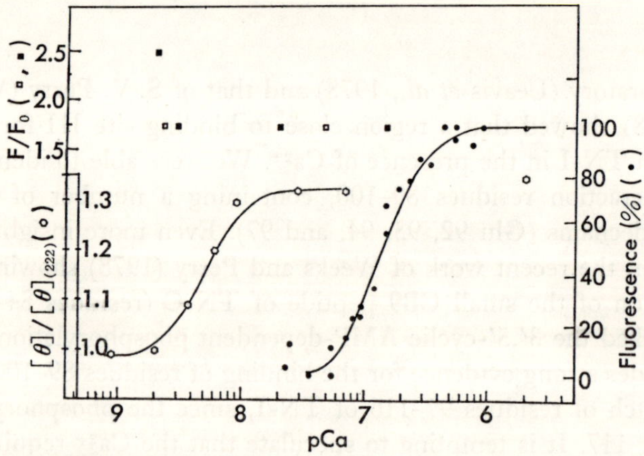

Fig. 6. Ca^{2+} binding to the binary complex TN-C-TN-I. For these experiments, TN-I was labelled with the fluorescent probe IAEDANS prior to its complexation with TN-C. Increments of Ca^{2+} were added directly to a cuvet containing the complex (0.1 mg/ml) in a solution consisting of 25 mM HEPES buffer, pH 7.0 or 7.5, 0.1 M KCl and 2 mM EGTA. Spectral parameters monitored were: ellipticity at 222 nm (○), IAEDANS fluorescence, $\lambda_{ex}=330$, $\lambda_{em}=550$ nm (●) and the tyrosine fluorescence intensity relative to that of apo-TN-C, $\lambda_{ex}=280$, $\lambda_{em}=300$ nm (■). Open squares indicate tyrosine fluorescence *vs.* the presence of 2 mM Mg^{2+}.

changes occur on the surface owing to altered interactions with TN-C, is as yet unclear. Recent acrylamide quenching studies on the complex do indicate that Ca^{2+} induces a "tightening," moving the two subunits closer together (Leavis and Tao, 1980).

TN-T, as previously mentioned, binds to both TN-C and TN-I. Preliminary studies using proteolytic fragments of TN-C indicate that the N-terminal half of the molecule binds to TN-T, the most likely binding site being located in region I (Grabarek *et al.*, manuscript in preparation). Complementary studies using TN-T fragments show that the TN-C-binding site involves residues 159–220, a basic region in the TN-T sequence (Pearlstone and Smillie, 1978). The exact site of TN-T binding to TN-I has not yet been determined. As far as the role of TN-T is concerned, in the simplest model, it would merely act as a connecting link between tropomyosin and the other troponin subunits.

Apart from the question whether the steric blocking mechanism is able to account for the inhibition of the actin-myosin interaction in the absence of Ca^{2+}, a mechanistic understanding of the role of the troponin complex in determining the position of tropomyosin is still incomplete. If for tropomyosin in the absence of troponin the active position on the thin filament is thermodynamically favored, then—on the simplest view—the three troponin components can be regarded as a lock holding tropomyosin in the relaxed position in resting muscle, which could be unlocked by Ca^{2+}. The TN-I-actin bond would contribute the actual anchoring connection and TN-T would function simply as a connector to tropomyosin. The fact that actin-myosin interactions do take place in the presence of the TN-C-TN-I complex but in the absence of Ca^{2+} is rather puzzling if one assigns an anchoring role to the binding of the TN-I to a site comprising actin and tropomyosin; a possible difficulty in this regard was noted by Potter and Gergely (1974) at the time this scheme was proposed. If, on the other hand, the thermodynamically favored position of tropomyosin is characteristic of the relaxed state, then one must assume that in addition to the dissociation of TN-I induced by Ca^{2+} binding to the TN-C subunit, there must be a change induced in either tropomyosin, actin, or both, which would result in reversal of the energy levels with TN-T acting as a channel of communication. Although the question of the favored tropomyosin position on the thin filament is not completely settled, evidence is accumulating from studies indicating cooperative binding of myosin heads to actin depending on the presence of tropomyosin (Katz, 1964; Eaton *et al.*, 1975; Shigekawa and Tonomura,

1973; Bremel *et al.*, 1972) and the complementary cooperative binding of tropomyosin to actin induced by myosin heads (Eaton, 1976), that suggests a dynamic role played by all of the thin filament proteins in the regulation of contraction.

Although great strides have been made in our understanding of the function of calcium in the regulation of muscle contraction since the pioneering work of Professors Kumagai and Natori some 25 years ago, challenging questions regarding details of the roles of all of the thin filament proteins remain to be solved.

Acknowledgment

The preparation of this manuscript and work in the authors' laboratories are supported by grants from NIH (HL-5949, HL-20464), the National Science Foundation and the Muscular Dystrophy Associations of America, Inc. P.C.L. is an Established Investigator of the American Heart Association.

REFERENCES

Bremel, R. D., Murray, J. M., and Weber, A. (1972). *Cold Spring Harbor Symp. Quant. Biol.* **37**, 267–275.

Carew, E. B., Leavis, P. C., Stanley, H. E., and Gergely, J. (1980). *Biophys. J.* (in press).

Coffee, C. J. and Bradshaw, R. A. (1973). *J. Biol. Chem.* **248**, 3305–3312.

Collins, J. H. (1974). *Biochem. Biophys. Res. Commun.* **58**, 301–308.

Collins, J. H., Potter, J. D., Horn, M. J., Wilshire, G., and Jackman, N. (1973). *FEBS Lett.* **36**, 268–272.

Drabikowski, W., Rafalowska, U., Dabrowska, R., Szpacenko, A., and Barylko, B. (1972). *FEBS Lett.* **19**, 259–263.

Eaton, B. L. (1976). *Science* **192**, 1337–1339.

Eaton, B. L., Kominz, D. R., and Eisenberg, E. (1975). *Biochemistry* **14**, 2718–2725.

Ebashi, S., Endo, M., and Ohtsuki, I. (1969). *Q. Rev. Biophys.* **2**, 351–384.

Ebashi, S., Wakabayashi, T., and Ebashi, F. (1971). *J. Biochem.* **69**, 441–445.

Evans, J. S., Levine, B. A., Leavis, P. C., Gergely, J., Grabarek, Z., and Drabikowski, W. (1980). *Biochim. Biophys. Acta* (in press).

Greaser, M. L. and Gergely, J. (1971). *J. Biol. Chem.* **246**, 4226–4233.

Greaser, M. L. and Gergely, J. (1973). *J. Biol. Chem.* **248**, 2125–2133.

Grabarek, Z., Drabikowski, W., Leavis, P. C., Rosenfeld, S. S., and Gergely, J. (in preparation).

Hartshorne, D. J. and Dreizen, P. (1972). *Cold Spring Harbor Symp. Quant. Biol.* **37**, 225–234.

Hartshorne, D. J. and Mueller, H. (1969). *Biochim. Bophys. Acta* **175**, 301–319.

Hartshorne, D. J. and Pyun, H. Y. (1971). *Biochim. Biophys. Acta* **229**, 698–711.
Haselgrove, J. C. (1972). *Cold Spring Harbor Symp. Quant. Biol.* **37**, 341–352.
Head, J. F. and Perry, S. V. (1974). *Biochem. J.* **137**, 145–154.
Hincke, M. T., McCubbin, W. D., and Kay, C. M. (1978). *Can. J. Biochem.* **56**, 384–395.
Hitchcock, S. E. (1975). *Biochemistry* **14**, 5162–5167.
Hitchcock, S. E., Huxley, H. E., and Szent-Györgyi, A. G. (1973). *J. Mol. Biol.* **80**, 825–836.
Horwitz, J., Bullard, B., and Mercola, D. (1979). *J. Biol. Chem.* **254**, 350–355.
Johnson, J. D., Charlton, S. C., and Potter, J. D. (1979). *J. Biol. Chem.* **254**, 3497–3502.
Johnson, J. D. and Potter, J. D. (1978). *J. Biol. Chem.* **253**, 3775–3777.
Katz, A. M. (1964). *J. Biol. Chem.* **239**, 3304–3311.
Kawasaki, Y. and van Eerd, J.-P. (1972). *Biochem. Biophys. Res. Commun.* **49**, 898–805.
Kretsinger, R. H. and Barry, C. D. (1975). *Biochim. Biophys. Acta* **405**, 40–52.
Kretsinger, R. H. and Nockolds, C. E. (1973). *J. Biol. Chem.* **248**, 3313–3326.
Kumagai, H., Ebashi, S., and Takeda, F. (1955). *Nature* **176**, 166–168.
Leavis, P. C. (1976). *Fed. Proc.* **35**, 1746.
Leavis, P. C. and Kraft, E. L. (1978). *Arch. Biochem. Biophys.* **186**, 411–415.
Leavis, P. C., Rosenfeld, S. S., Gergely, J., Grabarek, Z., and Drabikowski, W. (1978). *J. Biol. Chem.* **253**, 5452–5459.
Leavis, P. C. and Tao, T. (1980). *Fed. Proc.* **39**, 724.
Levine, B. A., Mercola, D., Coffman, D., and Thornton, J. M. (1977). *J. Mol. Biol.* **115**, 743–760.
Levine, B. A., Mercola, D., and Thornton, J. M. (1976). *FEBS Lett.* **61**, 218–222.
Levine, B. A., Moore, G. R., Ratcliffe, R. G., and Williams, R.J.P. (1979). In *Int. Rev. Biochem.* (Offoro, R. E., ed.), vol. 24, pp. 77–141, Univ. Park Press, Baltimore.
Levine, B. A., Thornton, J. M., Fernandes, R., Kelly, C. M., and Mercola, D. (1978). *Biochim. Biophys. Acta* **535**, 11–24.
McCubbin, W. D. and Kay, C. M. (1973). *Biochemistry* **12**, 4228–4232.
Moore, P. B., Huxley, H. E., and DeRosier, D. J. (1970). *J. Mol. Biol.* **50**, 279–295.
Nagy, B. and Gergely, J. (1979). *J. Biol. Chem.* **254**, 12732–12737.
Nagy, B., Potter, J. D., and Gergely, J. (1976). *Biophys. J.* **16**, 71a.
Nagy, B., Potter, J. D., and Gergely, J. (1978). *J. Biol. Chem.* **253**, 5971–5974.
Parry, D.A.D. and Squire, M. M. (1973). *J. Mol. Biol.* **75**, 33–55.
Pearlstone, J. R. and Smillie, L. B. (1978). *Can. J. Biochem.* **56**, 521–527.
Potter, J. D. and Gergely, J. (1974). *Biochemistry* **13**, 2697–2703.
Potter, J. D. and Gergely, J. (1975). *J. Biol. Chem.* **250**, 4628–4633.
Potter, J. D., Seidel, J. C., Leavis, P. C., Lehrer, S. S., and Gergely, J. (1976). *J. Biol. Chem.* **251**, 7551–7556.
Schaub, M. C. and Perry, S. V. (1969). *Biochem. J.* **115**, 993–1004.
Seamon, K. B., Hartshorne, D. J., and Bothner-By, A. A. (1977). *Biochemistry* **16**, 4039–4046.
Shigekawa, M. and Tonomura, Y. (1973). *J. Biochem.* **73**, 1135–1148.
Sin, I. L., Fernandes, R., and Mercola, D. (1978). *Biochem. Biophys. Res. Commun.* **82**, 1132–1139.

Strong, L. H., Fossel, E. T., and Gergely, J. (1977). *Biophys. J.* **17**, 118a.
Syska, H., Wilkinson, J. M., Grand, R.J.A., and Perry, S. V. (1976). *Biochem. J.* **153**, 375–387.
van Eerd, J.-P. and Kawasaki, Y. (1972). *Biochem. Biophys. Res. Commun.* **47**, 859–865.
van Eerd, J.-P. and Takahashi, K. (1975). *Biochem. Biophys. Res. Commun.* **64**, 122–127.
Weeks, R. A. and Perry, S. V. (1978). *Biochem. J.* **173**, 449–457.
Wilkinson, J. M., Perry, S. V., Cole, H. A., and Trayer, I. P. (1972). *Biochem. J.* **127**, 215–228.

Muscle Contraction: Its Regulatory Mechanisms (Ebashi, S. et al., eds.), pp. 207–220,
Japan Sci. Soc. Press, Tokyo / Springer-Verlag, Berlin (1980)

16

Structure and Function Relationships in Troponin and Calmodulin Regulatory Systems

S. V. PERRY

Summary: Detailed study of the structure and function relationships of calmodulin and troponin-C indicate that these proteins possess many properties in common. The differences that exist relate to the special role of troponin-C in I-filament regulation. It follows, therefore, that the troponin system of striated muscle is a specialized form of the system that has evolved for the regulation of processes through changes in the Ca^{2+} concentrations. Certain general features of such Ca^{2+} regulatory systems may be noted. The systems all appear to involve a protein with three to four binding sites with high affinity for calcium. Associated with the Ca^{2+}-binding proteins are several other proteins which fall into two classes as judged by the nature of their interaction with calmodulin. In addition in more complex systems the Ca^{2+} regulatory mechanism can be modulated by phosphorylation of the proteins that interact with the calcium-binding component and the phosphorylation can be under hormonal control.

Of the many biological phenomena regulated by a change in the intracellular Ca^{2+} concentration the contractile process in muscle is probably the best defined and understood. The regulatory system of muscle is highly specialized to respond rapidly to stimulation so that interactions between proteins can occur at a large number of sites distributed throughout the myofibril and thus lead to the high rate of ATPase activity that is associated with contraction. It is unlikely that the basic features of the mechanism are unique to muscle, indeed it is more probable that they have evolved from a simpler mechanism selected by evolution to take advantage of the unique properties of the calcium ion. Thus the basic features are probably characteristic of many, if not all, of the biological processes regulated by calcium. The purpose of this paper is to examine this premise in more detail and discuss some common aspects of troponin and calmodulin-regulated systems.

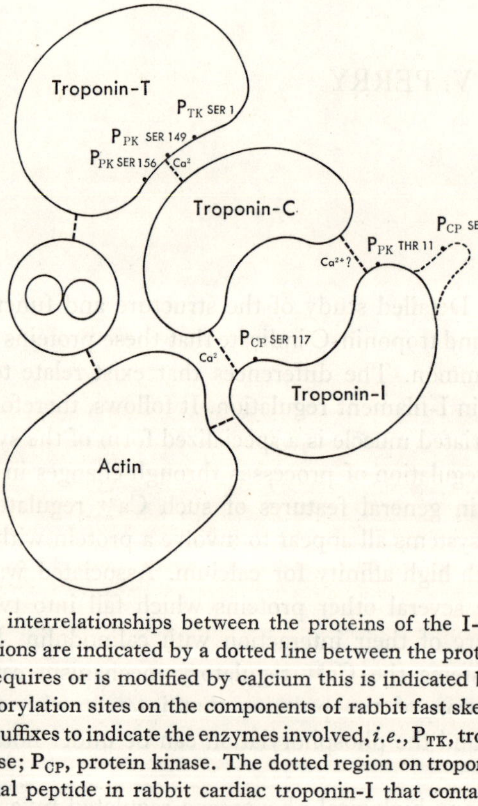

Fig. 1. Scheme of the interrelationships between the proteins of the I-filament. The well-established interactions are indicated by a dotted line between the proteins involved. Where the interaction requires or is modified by calcium this is indicated by the symbol for the ion. The phosphorylation sites on the components of rabbit fast skeletal troponin are indicated by P with suffixes to indicate the enzymes involved, *i.e.*, P_{TK}, troponin kinase; P_{PK}, phosphorylase kinase; P_{CP}, protein kinase. The dotted region on troponin I indicates the additional N-terminal peptide in rabbit cardiac troponin-I that contains the serine residue at position 20, the phosphorylation of which site changes the calcium-binding properties of cardiac troponin-C.

The interactions that occur between the I-filament proteins in striated muscle are schematically illustrated in Fig. 1. The system possesses a number of features that are of significance for its function. Interaction between the components occurs at discrete sites on the proteins. The functional sites retain to a large extent their properties when they are isolated as relatively small peptides by controlled cleavage of the proteins. Amino acid residues that can be phosphorylated are located close to the sites of interaction between the components. The response of the system to changes in Ca^{2+} concentration can be altered by covalent modification of components of the troponin complex other than the troponin-C itself (see Perry, 1979; Sperling *et al.*, 1979).

COMPARISON OF INTERACTION OF TROPONIN-C AND CALMODULIN WITH TROPONIN-I

The interaction involving the target protein for Ca^{2+} in muscle, troponin-C, and troponin-I, the protein that has the property of inhibiting the ATPase activity of actomyosin, would appear to be especially significant for the Ca^{2+} regulation of the ATPase of the system. Although interaction between these two proteins occurs in the absence of Ca^{2+} it is much strengthened in the presence of this divalent cation and probably involves two sites on each protein (Weeks and Perry, 1977, 1978; Leavis *et al.*, 1978; Perry, 1979). Complex formation between these two proteins is calcium-dependent even in the presence of 8 M urea, implying that the appropriate calcium-binding sites on troponin-C preserve their unique properties in strong dissociating conditions. This property is very unusual amongst proteins and it is not unreasonable to assume that it reflects certain unique aspects of the system that are features of this type of Ca^{2+}-regulated process. The existence of proteins forming similar Ca^{2+}-dependent complexes in high urea concentrations in other tissues might therefore be expected to indicate the presence of systems containing components that interact in a similar manner to troponin-I and troponin-C of muscle.

This conclusion appeared to have some basis when it was observed that complexes with these properties could be demonstrated in a variety of tissues other than striated muscle (Head *et al.*, 1977; Grand *et al.*, 1979) and that calmodulin, the widely distributed protein that is an essential component of a rapidly increasing number of Ca^{2+} activated enzyme systems, formed a similar complex with troponin-I (Grand *et al.*, 1979).

Calmodulin would not normally be expected to be complexed with troponin-I in tissues, but although some minor differences exist (Grand *et al.*, 1979; Vanaman and Perry, 1978) the interaction between these two proteins is very similar to that which occurs between troponin-C and troponin-I. On addition of calmodulins from various tissues the inhibitory action of troponin-I on the Mg^{2+}-stimulated ATPase of desensitized actomyosin can be neutralized (Amphlett *et al.*, 1976; C. M. Wall and S. V. Perry, unpublished results). Further, if Ca^{2+} was removed from these systems by the addition of EGTA the inhibitory action of the troponin-I was restored, *i.e.* complete Ca^{2+} sensitivity of the actomyosin ATPase was obtained with calmodulin and skeletal muscle troponin-I (Amphlett *et al.*, 1976). With troponin-C and troponin-I this is not the case and only partial Ca^{2+} sensitivity could be obtained with the two proteins alone (Amphlett *et al.*, 1976). For full restoration of Ca^{2+} sensitivity in the striated muscle system, troponin-T is required as well as troponin-I and troponin-C. Thus calmodulin possesses the property of being able to regulate the actomyosin ATPase in response to changing Ca^{2+} concentration.

Another aspect of the interaction between troponin-C and troponin-I from rabbit fast skeletal muscle is reflected in its effect on phosphorylation of the latter protein. In the presence of troponin-C in an equimolar ratio, phosphorylation of threonine 11, the major site of phosphorylation by

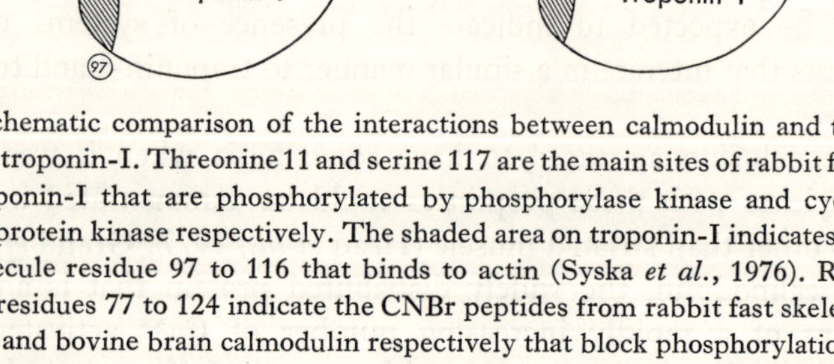

Fig. 2. Schematic comparison of the interactions between calmodulin and troponin-C with rabbit troponin-I. Threonine 11 and serine 117 are the main sites of rabbit fast skeletal muscle troponin-I that are phosphorylated by phosphorylase kinase and cyclic AMP-dependent protein kinase respectively. The shaded area on troponin-I indicates the region of the molecule residue 97 to 116 that binds to actin (Syska *et al.*, 1976). Residues 84 to 135 and residues 77 to 124 indicate the CNBr peptides from rabbit fast skeletal muscle troponin-C and bovine brain calmodulin respectively that block phosphorylation at serine 117 of troponin-I.

cyclic AMP-dependent protein kinase (Moir *et al.*, 1974; Huang *et al.*, 1974) is completely blocked (Perry and Cole, 1974; Cole and Perry, 1975). If the troponin-C is replaced by calmodulin, the effect on the phosphorylation of serine 117 by the cyclic AMP-dependent protein kinase is unchanged, inhibition being complete when calmodulin and troponin-I are present in a 1: 1 ratio. On the other hand, phosphorylation by phosphorylase kinase, presumably mainly at threonine 11 is much less effected (Vanaman and Perry, 1978). At a 1: 1 molar ratio inhibition was only about 10–20% and did not exceed 50% even in the presence of 7% molar excess of calmodulin. These studies suggest certain differences exist in the interaction of troponin-C and calmodulin with troponin-I. These are represented diagrammatically in Fig. 2.

SITES ON CALMODULIN INVOLVED IN INTERACTION WITH TROPONIN-I

The common features of the interaction of the two Ca^{2+}-binding proteins with troponin-I, particularly the inhibition of phosphorylation at serine 117, suggested that at least one homologous interaction site was present on the two Ca^{2+}-binding proteins. Previously we have shown that a region of the amino acid sequence of fast skeletal muscle troponin-C, located in the peptide obtained by CNBr digestion in peptide CB9, consisting of residues 84–135 was involved in interaction with troponin-I in the region close to serine 117. This conclusion was based on the finding that peptide CB9 neutralised the inhibitory activity of troponin-I, inhibited the phosphorylation of troponin-I catalysed by cyclic AMP-dependent protein kinase and formed a Ca^{2+}-dependent complex with a troponin-I Sepharose affinity column (Weeks and Perry, 1977, 1978). Taken in conjunction with the earlier investigations on troponin-C (Weeks and Perry, 1978; Leavis *et al.*, 1978), more recent studies with fragments obtained by controlled proteolysis with thrombin and trypsin suggest that the region represented by residues 84 to 100 may be of special importance in this interaction (Wall *et al.*, 1980).

As a consequence of the marked sequence homology that exists between troponin-C and calmodulin most of the methionine residues occur at similar positions in the amino acid sequence and CNBr digestion of the two proteins produces a similar series of peptides. The peptide produced by cleavage of the methionine at residues 76 and 124 of calmodulin would be equivalent to that consisting of residues 84 to 135, peptide CB9 (Collins *et al.*, 1973; Weeks and Perry, 1978) from fast skeletal muscle troponin-C. On complete digestion of bovine brain calmodulin with cyanogen bromide, however, the peptide consisting of residues 77 to 124 is not normally obtained because this protein contains an additional methionine in position 109 that is also cleaved under these conditions. Under conditions of partial digestion, however, the peptide band at position 109 is not cleaved and a peptide is obtained in low yield that is identified by amino acid and N-terminal analysis to consist of residue 77 to 124 (Vanaman and Perry, 1978; T. C. Vanaman and S. V. Perry, 1979 unpublished observations). When such digests were applied to a troponin-I-Sepharose affinity column in the presence of $CaCl_2$ low ionic strength, pH 7.8, the majority of the material passed through the column unretarded. On subsequent addition of EGTA to the eluting buffer a single peptide that was identified as identical to that previously shown to consist of residues 77–124 was eluted. On addition of high salt concentrations to the affinity column no further material was eluted. The results strongly suggest that the region represented by the peptide consisting of residues 77–124 is involved in Ca^{2+}-dependent binding to troponin-I. The general similarity of this sequence with that represented by peptide 84–135 of fast skeletal muscle troponin-C indicates that the interaction with troponin-I at the site close to serine 117 takes place with regions on troponin-C and calmodulin that are homologous.

The implications from the studies of phosphorylation catalysed by phosphorylase kinase already described are that calmodulin appears not to bind very strongly to N-terminal region of troponin-I in which the site specific for this enzyme is located (Figs. 1 and 2). On the other hand, the evidence is good that two sites are involved in the troponin-C-troponin-I reaction (Weeks and Perry, 1978; Leavis *et al.*, 1978). It is possible that this difference reflects a special feature of the interaction of troponin-C with troponin-I that is of significance for the mechanism of Ca^{2+} control in muscle in the troponin complex. Because of the relatively stronger binding of troponin-C, compared with calmodulin, to troponin-I an additional protein that interacts with troponin-C is required to modulate the interaction and at the same time attach the system to the I-filament *via* tropomyosin. Thus in muscle troponin-T, a protein for which so far no counterpart has yet been reported in non-muscle systems, is an essential component for complete Ca^{2+}-sensitivity regulated through the I-filament. The relatively weaker binding of skeletal muscle troponin-I with calmodulin is reflected in several of the properties discussed and is probably the reason why complete Ca^{2+} sensitivity of actomyosin ATPase activity is obtained in the presence of fast skeletal muscle troponin-I, calmodulin and tropomyosin only. Complete dissociation of the troponin-I-calmodulin complex is obtained in the absence of Ca^{2+} and hence the troponin-I is able to exert its full inhibitory effect. It may be speculated that when the calmodulin is replaced by troponin-C in the system, troponin-T must also be present to ensure that the troponin system is fully effective in inhibiting the actomyosin ATPase in the absence of Ca^{2+}.

CALMODULIN-BINDING PROTEINS

The existence of such apparently specific and complementary binding sites on calmodulin and troponin-I raises the question as to whether there exist proteins analogous to troponin-I in non-muscle tissues that contain relatively large amounts of calmodulin but no proteins with the properties of striated muscle troponin-I. So far the only clear demonstration of the presence of troponin-I has been in striated muscle, although reports have appeared of the presence of proteins with inhibitory activity in smooth muscle (Driska and Hartshorne, 1975; Grand *et al.*, 1977; Makioka and Hirabayashi, 1978) and in muscle capillaries with immunochemical properties similar to fast skeletal muscle troponin-I (Dhoot *et al.*, 1978).

As calmodulin is acidic in nature it would be expected to form complexes with a large number of basic proteins. From a number of investigations, including affinity chromatographic studies on whole tissue extracts using calmodulin linked to Sepharose, a number of proteins have been reported to bind calmodulin (see review by Cheung, 1980). Only a few of these proteins have the ability to form complexes that are stable in urea in the presence of Ca^{2+} and were, therefore, considered to be possibly analogous to troponin-C-troponin-I complex of muscle. A procedure for selectively isolating calmodulin-binding proteins with these properties was therefore devised (Grand and Perry, 1979).

Calmodulin made radioactive by labelling with 3-(4-hydroxy-5-[^{125}I]-iodophenyl)-N-succinimidyl propionate (Bolton and Hunter, 1973) retains the ability to form a complex with troponin-I (Grand and Perry, 1979). If a sample of calmodulin labelled in this manner was added to a homogenate of tissue in 6 M urea, radioactive complexes could be identified in the homogenate if calcium was present. Gel filtration on Sepharose 6B separated the homogenate into two peaks of radioactivity corresponding to the position of elution of the calmodulin-binding protein-calmodulin complexes were obtained. In all the non-muscle tissues studied the complexes present fell into two types that possessed apparent molecular weights of about 140,000 and 60,000. Skeletal and cardiac muscle extracts differed from those obtained from non-muscle tissues in that the peaks of radioactivity corresponded to complexes of apparent molecular weights of 35,000 and 37,000 respectively. These consisted of complexes of troponin-C and troponin-I and only traces of the complexes of higher apparent molecular weights characteristic of non-muscle tissues could be detected. On dissociation of the complexes obtained from non-muscle tissues a basic protein that migrated on sodium dodecyl sulphate electrophoresis with an apparent molecular weight of 140,000 was isolated from the higher molecular weight complex and two proteins of apparent molecular weight 77,000 and 61,000 were obtained from the lower molecular weight complex. These studies indicated that there were present in tissues at least three proteins that could form complexes with calmodulin with similar properties to the complexes that are formed by calmodulin and troponin-C with troponin-I.

Although the complexes of calmodulin with the calmodulin-binding protein clearly exist in appreciable amounts in the homogenates of all tissues studied, it cannot be concluded that they were present as such in the intact tissue. Nevertheless, by analogy with troponin-I and troponin-C in the troponin complex of muscle, it is probable that part of the calmodulin in non-muscle tissues is bound to the calmodulin-binding proteins.

INTERACTION SITES ON CALMODULIN

The calmodulin-binding proteins described above probably do not correspond to the enzymes that have been reported to require calmodulin as activators. The calmodulin-binding proteins occur in relatively larger amounts and so far fewer molecular species have been detected than is the case with calmodulin-activated enzymes. Also, some of these enzymes form complexes of different stabilities with calmodulin. For example, myosin light chain kinase of skeletal muscle, the best purified of the calmodulin-requiring enzymes, does not migrate on electrophoresis at pH 8.6 in non-dissociating conditions as a complex with calmodulin, let alone in urea (Nairn and Perry, 1979). On the other hand, in phosphorylase kinase, calmodulin is tightly bound to the enzyme complex (Cohen *et al.*, 1978).

The property of calmodulin forming a urea-stable Ca^{2+}-dependent

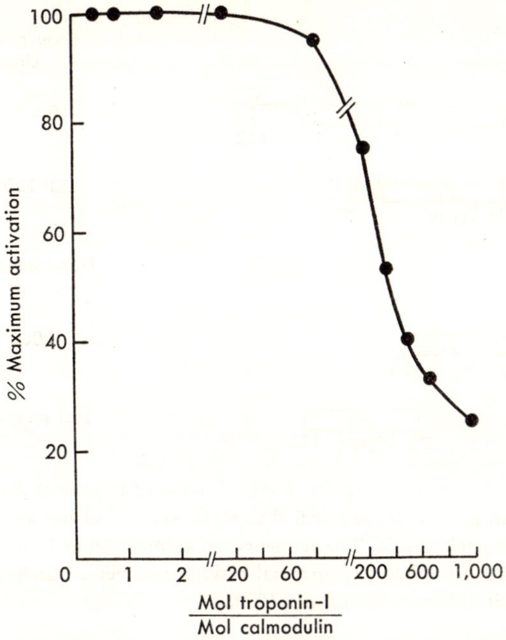

Fig. 3. Effect of troponin-I on the activation of myosin light chain kinase by calmodulin. Assays carried out as described by Nairn and Perry (1979) with pure rabbit skeletal muscle myosin light chain kinase (0.15 μg/ml) and bovine brain calmodulin (0.05 μg/ml). Activities expressed as a percentage of that obtained in the absence of rabbit fast skeletal muscle troponin-I (20–30 μmol/mg/min). (From Nairn, 1979).

complex with troponin-I and a weaker one with myosin light chain kinase (Nairn and Perry, 1979) suggests that different sites on calmodulin may be involved in these two interactions. This would appear to be the case for the calmodulin-troponin-I complex activates myosin light chain kinase and up to 100× molar excess of troponin-I can be present without significant inhibition of enzymic activity (Fig. 3).

To investigate this aspect further the ability of fragments obtained from calmodulin by selective cleavage procedures, to activate the myosin light chain kinase was investigated. The results are summarized in Fig. 4, from which it can be seen that the largest N-terminal fragment obtained by thrombin cleavage at the arginine residue at position 106 and which represents two-thirds of the molecule was two orders of magnitude less effective than the whole molecule. The complementary C-terminal fragment of calmodulin had no activity nor had the peptide consisting of

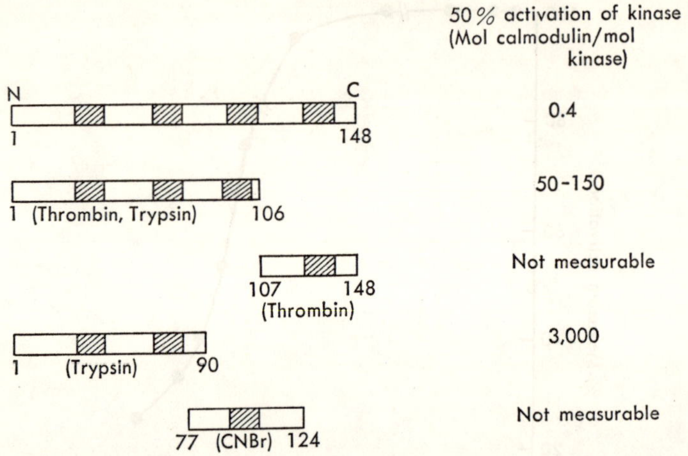

Fig. 4. Diagram illustrating the relative effectiveness of fragments derived from bovine brain calmodulin in activating the purified catalytic subunit of myosin light chain kinase from rabbit fast skeletal muscle. The sequences of calmodulin and the peptide fragments are represented with the N and C terminal residue numbers. Hatched regions indicate the Ca^{2+}-binding sites. (From Wall *et al.*, 1980).

residues 77 to 124 which was the troponin-I-binding region. These experiments together with the results of the interaction studies carried out with the whole proteins suggest that the myosin light chain kinase catalytic unit binds to calmodulin at a region other than that involved with interaction with troponin-I. The results taken with those obtained from studies on the CNBr peptides indicate that a site of interaction with troponin-I lies in the region of calmodulin represented by residues 77 to 106. Further the results also suggest that a large part of the calmodulin structure must remain intact for activation of the myosin light chain kinase.

In the light of these facts the interaction of calmodulin with the calmodulin-binding proteins and the enzyme systems it activates can be schematically represented as in Fig. 5. This figure also compares the interaction of troponin-C with the components of the troponin complex. Certain similarities can be noted in that both proteins possess sites involving urea-stable Ca^{2+}-dependent interactions for troponin-I or the calmodulin-binding proteins. As has been described above the latter site represents homologous regions in the two Ca^{2+}-binding proteins. The other site on calmodulin is involved in binding to myosin light chain kinase and probably other enzyme systems such as phosphodiesterase and

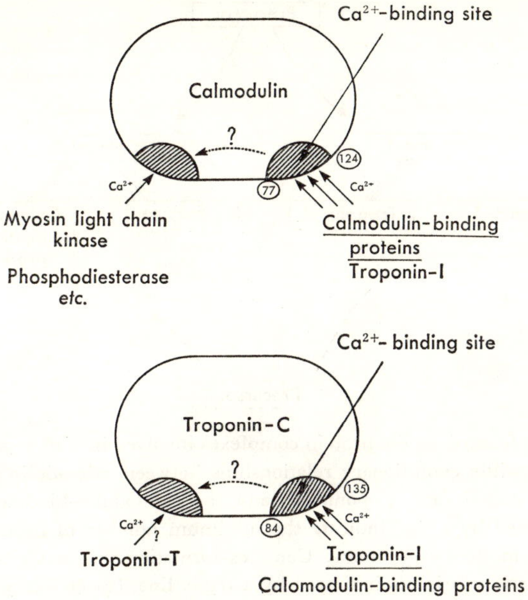

Fig. 5. Interaction sites on calmodulin and troponin-C. The triple arrows indicate urea-stable Ca^{2+}-dependent interactions.

thus bringing about their activation. The interaction also requires Ca^{2+} (Nairn and Perry, 1979) but is not urea-stable as is the interaction at the site that binds calmodulin-binding protein and troponin-I. The difference in stability may be a consequence of the need for the structure of the calmodulin molecule to be intact to activate enzymic catalytic units as was implied from the controlled cleavage experiments.

Although it maintains a homology to calmodulin, troponin-C has largely lost the ability to activate the enzyme systems (*cf.* Nairn and Perry, 1979), but it is conceivable that the region of the troponin-C molecule corresponding to the site in calmodulin that is concerned with enzyme interaction interacts with troponin-T. Certainly the requirement for Ca^{2+} and the urea instability of the complex between troponin-C and troponin-T is similar to that formed between calmodulin and myosin light chain kinase.

Consideration of these properties leads to the hypothesis that a basic similarity exists between the highly specialized troponin-C regulated system which is found only in striated muscle and the calmodulin-regulated

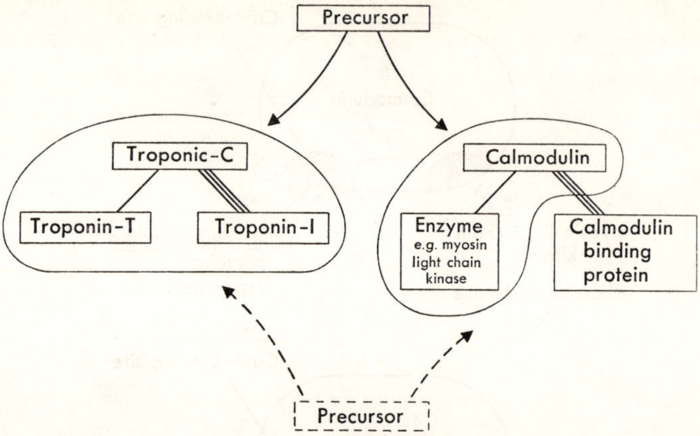

Fig. 6. Common features of the protein complexes involved in Ca^{2+} regulation. Scheme illustrating the possible evolutionary relationships between calmodulin and troponin-C and between troponin-I (and troponin-T) and the calmodulin-binding proteins. The components enclosed by a line indicate the minimum number of proteins required to give Ca^{2+} control in the two systems. Complex-formation that is Ca^{2+}-dependent and stable in strong urea solution is indicated by a triple line. (From Perry *et al.*, 1979).

systems in general. For the function of the troponin complex all three components are required for full Ca^{2+} sensitivity, unlike many of the simple calmodulin-regulated enzyme systems in which two components, calmodulin and the catalytic component alone, are required. Indeed it would appear that the specialization that has lead to the evolution of troponin-C has changed the nature of the interaction between the Ca^{2+}-binding protein and the component corresponding to the calmodulin-binding proteins in non-muscle tissues, troponin-I, so that a third protein, troponin-T, is required for Ca^{2+} sensitivity in the muscle system.

The role of the calmodulin-binding proteins in the calmodulin-containing enzyme systems is far from clear. Like troponin-T and troponin-I they are basic proteins and in view of the rather unique nature of the interaction of troponin-I and the calmodulin-binding proteins with calmodulin it is possible that they have evolved from a common precursor as is the case with troponin-C and calmodulin (Fig. 6).

Although many of the calmodulin-regulated enzyme systems appear, at the present state of our knowledge to be relatively simple, the phosphorylase kinase system is more complex and contains many of the elements represented in the scheme illustrated in Fig. 6. It is dangerous perhaps to draw too many comparisons with the troponin system, particularly as phosphorylase kinase consists of four rather than three subunits which, apart from calmodulin, have not been shown to exhibit any close functional similarities to the components of the troponin complex. Nevertheless, it is of interest to note that the calmodulin forms a strong urea stable, Ca^{2+}-dependent interaction with the 45,000-dalton γ-subunit, and a weaker Ca^{2+}-dependent interaction with the 130,000-dalton γ-subunit (P. Cohen, personal communication). To take the analogy further both α and β subunits of the phosphorylase kinase contain sites that are phosphorylated by protein kinase and thus modulate the activity of the system. Troponin-I in cardiac muscle and troponin-T also possess sites that when phosphorylated change the response of the system to Ca^{2+}. Further, phosphorylation of the cardiac troponin system and phosphorylase kinase are both under β-adrenergic control.

REFERENCES

Amphlett, G. W., Vanaman, T. C., and Perry, S. V. (1976). *FEBS Lett.* **72**, 163–168.

Bolton, A. E. and Hunter, W. M. (1973). *Biochem. J.* **133**, 52–539.

Cheung, W. Y. (1980). *Science* **207**, 19–27.

Cohen, P., Burchell, J., Foulkes, G., Cohen, P.T.W., Vanaman, T. C., and Nairn, A. C. (1978). *FEBS Lett.* **92**, 287–293.

Cole, H. A. and Perry, S. V. (1975). *Biochem. J.* **149**, 525–533.

Collins, J. H., Potter, J. D., Horn, M. J., Wiltshire, G., and Jackman, N. (1973). *FEBS Lett.* **36**, 268–272.

Dhoot, G. K., Gell, P.G.H., and Perry, S. V. (1978). *Exp. Cell Res.* **117**, 357–370.

Driska, S. and Hartshorne, D. J. (1975). *Arch. Biochem. Biophys.* **167**, 203–212.

Grand, R.J.A. and Perry, S. V. (1979). *Biochem. J.* **183**, 285–295.

Grand, R.J.A., Perry, S. V., and Weeks, R. A. (1977). In *Excitation Contraction Coupling in Smooth Muscle* (Casteels, R., Godfraindt, T., and Ruegg, J. C., eds.), Elsevier/North-Holland Biomedical Press, Amsterdam.

Grand, R.J.A., Perry, S. V., and Weeks, R. A. (1979). *Biochem. J.* **177**, 521–529.

Head, J. F., Weeks, R. A., and Perry, S. V. (1977). *Biochem. J.* **161**, 465–471.

Huang, T. S., Bylund, D. B., Stull, J. T., and Krebs, E. G. (1974). *FEBS Lett.* **42**, 249–252.

Leavis, P. C., Rosenfeld, S. S., Gergely, J., Graberek, Z., and Drabikowski, W. (1978). *J. Biol. Chem.* **253**, 5452–5459.

Makioka, A. and Hirabayashi, T. (1978). *J. Biochem.* **84**, 947–955.

Moir, A.J.G., Wilkinson, J. M., and Perry, S. V. (1974). *FEBS Lett.* **42**, 253–256.

Nairn, A. C. (1979). Ph.D. Thesis, University of Birmingham, Birmingham.

Nairn, A. C. and Perry, S. V. (1979). *Biochem. J.* **179**, 89–97.

Perry, S. V. (1979). *Biochem. Soc. Trans.* **7**, 593–617.

Perry, S. V. and Cole, H. A. (1974). *Biochem. J.* **141**, 733–743.

Perry, S. V., Grand, R.J.A., Nairn, A. C., Wall, C. M., and Vanaman, T. C. (1979). *Biochem. Soc. Trans.* **7**, 619–622.

Sperling, J. E., Feldmann, K., Meyer, H., Janke, U., and Heilmeyer, L.M.G. (1979). *Eur. J. Biochem.* **101**, 581–592.

Syska, H., Wilkinson, J. M., Grand, R.J.A., and Perry, S. V. (1976). *Biochem. J.* **153**, 375–387.

Vanaman, T. C. and Perry, S. V. (1978). Proceedings of the 7th European Conference on Muscle and Motility, Warsaw. Abstracts, p. 37.

Wall, C. M., Grand, R.J.A., Nairn, A. C., and Perry, S.V. (1980). *Biochem. J.* (submitted for publication).

Weeks, R. A. and Perry, S. V. (1977). *Biochem. Soc. Trans.* **5**, 1391–1392.

Weeks, R. A. and Perry, S. V. (1978). *Biochem. J.* **173**, 449–457.

Muscle Contraction: Its Regulatory Mechanisms (*Ebashi, S. et al., eds.*), *pp. 221–236, Japan Sci. Soc. Press, Tokyo / Springer-Verlag, Berlin* (*1980*)

17

Tropomyosin and the Various States of the Regulated Actin Filament

JOHN M. MURRAY, A. WEBER, AND
A. WEGNER

Summary: The position of tropomyosin on the periphery of the actin filament probably inhibits access of myosin to its actin binding site during relaxation, blocking completely the access of the subfragment 1-ATP (S-1-ATP) intermediate, reducing rigor affinity 10-fold and S-1-AMPPNP (β,γ-imido-ATP) binding 2 to 3-fold. We propose that the tropomyosin position modulates actin function also in the presence of calcium. Thus, the inhibition of actomyosin ATPase by tropomyosin-troponin in the presence of calcium may be due to steric hindrance by tropomyosin, which alternates with time its position between sites from the periphery and the groove. This competitive inhibition by tropomyosin disappears when the myosin saturation of the actin filament exceeds 5%. We propose that a sufficiently high degree of myosin saturation causes a complete shift of tropomyosin to the groove and that shift with the cooperation by the bound myosin tropomyosin that causes a conformational change of all 7 actin molecules increasing their affinity for myosin. Such a conformational change may be the basis for the increase in cofactor activity, relative to that of pure actin, that is associated with a high degree of saturation with rigor complexes and that we have called potentiation. The potentiated state also results from a high degree of saturation with S-1-AMPPNP, and it is characterized by a 5 to 7-fold reduction in the K_m for S-1 and a 5-fold increase in the rigor binding constant. A very low affinity of tropomyosin for single actin molecules permits rapid movements of tropomyosin over the actin surface although tropomyosin remains on the actin filament, because all tropomyosin molecules are bound to each other at the ends, connected into one single strand.

Pure actin filaments activate myosin ATPase and undergo syneresis with myosin. All actin molecules behave as individuals under these circumstances, simple enzyme cofactor kinetics prevail (Eisenberg and Moos, 1969), and there is no regulation. When the actin filaments contain tropomyosin, the actin monomers no longer act independently from each other, but their behavior is determined by the cooperative interactions between tropomyosin and actin (Bremel *et al.*, 1972; Weber and Murray, 1973; Murray *et al.*, 1975). When, in addition, troponin is also present, actin-myosin interaction is regulated by calcium (Ebashi and Ebashi, 1964; Ebashi and Endo, 1968). The nature of the cooperative interactions be-

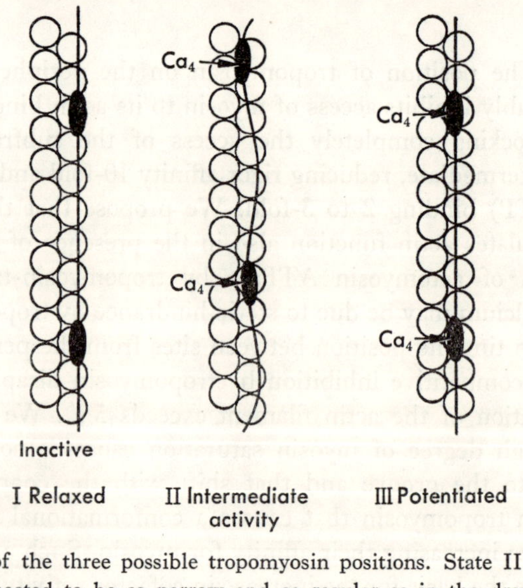

Fig. 1. Sketch of the three possible tropomyosin positions. State II bends arbitrary. They are not expected to be as narrow and as regular as in the drawing.

tween tropomyosin and actin varies with the position of tropomyosin on the actin filament. We propose in this paper that tropomyosin may occupy three major positions, thereby producing three different activity states of the actin filament (Fig. 1). When tropomyosin is firmly bound to the periphery of the filament, the filament is inactive, relaxed, in state 1. The state of activity is intermediate, state 2, when the tropomyosin molecule alternates its position between various binding sites on the actin surface (see below). Activity is very high, potentiated, in state 3, when tropomyosin is bound to the groove sites only. Spudich *et al.* (1972), Haselgrove (1972), and H. E. Huxley (1972), and Wakabayashi *et al.* (1975) produced evidence that tropomyosin can occupy either of two positions, on the periphery during relaxation or near the groove during contraction and rigor. On this basis, H. E. Huxley (1972) and Haselgrove (1972), and Parry and Squire (1973) explained the relaxed state by the elegant steric hindrance theory, which states that tropomyosin during relaxation blocks the access of myosin to the actin filament. Contraction is possible only after tropomyosin has vacated the peripheral sites as the result of calcium binding to troponin and has moved towards the groove away from the periphery. According to this theory tropomyosin should be capable of rapid motions over the actin surface so that it can leave the periphery within one msec or less, the time required for activation in very fast muscles. If the displaced tropomyosin retains its high diffusibility on the filament surface it can, after removal of the bound calcium, return troponin to its peripheral binding site on actin. Therefore, one may expect tropomyosin to be bound very loosely to individual actin molecules, in spite of the fact that tropomyosin is retained on the actin filament in the absence of significant concentrations of free tropomyosin in the medium or the cytoplasm. Tight binding to the actin filament combined with a low affinity for individual actin molecules, seems to be achieved by connecting all tropomyosin molecules into one single strand through binding at the overlapping ends (Cohen *et al.*, 1972; Stewart, 1975; Johnson and Smillie, 1977). As a result, the binding curve of tropomyosin to actin is highly cooperative (Wegner, 1979), the sigmoid foot of the curve indicating the low affinity of tropomyosin for the actin binding sites (Fig. 2). The binding constant of tropomyosin to the filament, including the overlap sites, is about 5×10^8 M^{-1} at 38.7°C (Table I). By contrast, the binding constant for one tropomyosin molecule to 7 actins is calculated to be 3.10^3 M^{-1} (3 M^{-1} for each segment to an individual monomer if binding is evenly distributed), assuming that the half saturation

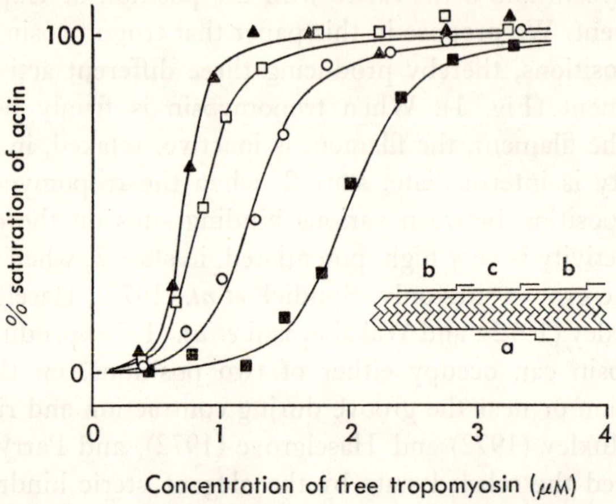

Fig. 2. Binding of $\alpha\alpha$-tropomyosin to actin filaments. Binding was measured by the increase in lightscattering (Wegner, 1979). 5.3 μM actin; 80 mM KCl; 1 mM $MgCl_2$; 5 mM triethanolamine, pH 7.5; 500 μM ATP; 2 mM DTT. Inset: scheme of the various modes of tropomyosin binding. (a) to actin only, binding constant K_1; (b) overlap at one end only, binding constant, K_2; (c) double overlap, binding constant K_3. ▲ 38.7°C; □ 39.4°C; ○ 40.6°C; ■ 41.9°C.

TABLE I

Binding to a Single Actin Molecule $\sqrt[7]{K_1} = 3\ \text{M}^{-1}$

Temperature (°C)	K_1 (M^{-1})	K_2 (M^{-1})	K_3 (M^{-1})
38.7	3.05×10^3	1.38×10^6	0.62×10^9
39.4	4.55×10^3	1.14×10^6	0.29×10^9
40.6	3.05×10^3	0.77×10^6	0.19×10^9
41.9	1.10×10^3	0.49×10^5	0.22×10^9

point is determined by single overlap binding, and the top of the curve by double overlap (Table I). Thus, it seems plausible that tropomyosin can move rapidly on the actin filament surface.

Haselgrove (1972) has shown that tropomyosin moves away from the periphery on the addition of calcium, independent of myosin attachment to the fiber, *i.e.*, in fibers at rest length and in fibers stretched beyond overlap. We propose that tropomyosin does not move into the groove but moves away from the periphery to many different sites on the actin surface to which it binds equally well and where it remains unless displaced by

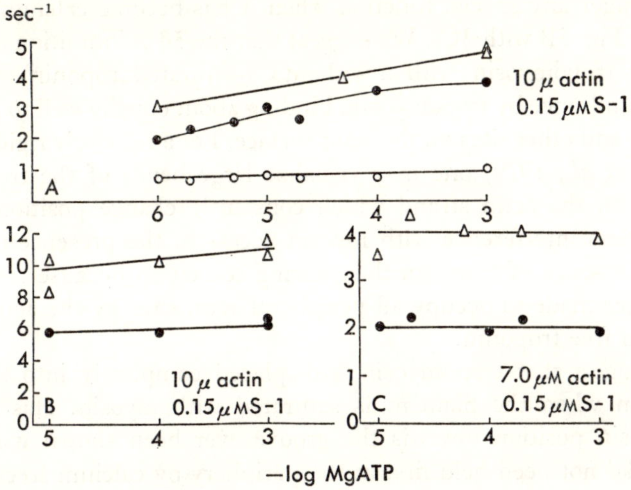

Fig. 3. ATPase activity of acto-S-1 with 50 to 60-fold actin excess over S-1. Comparison of pure actin with regulated actin in the absence, A, and in the presence, A and B, of calcium, and with tropomyosin actin C. 1 mM Mg excess, 10 mM imidazole, pH 7.0, 5.0 mM creatine phosphate, 1 mg/ml kinase, 25 mM ionic strength. The recent preparation of reconstituted regulated actin in A showed unusually poor relaxation. It was different from that used in Fig. 5. The difference in absolute values of ATPase activity between B and A and C are variations in S-1 activity. A and B: △ pure actin A; ● regulated A+Ca; ○ regulated A+EGTA. C: △ pure A; ● tropomyosin-A.

myosin. Such a view is compatible with the X-ray diffraction data (intensity changes of the second and third actin layer lines) which are qualitative, not quantitative; they only show a change in tropomyosin position but cannot distinguish between different degrees of shifting. Our proposal implies that tropomyosin still occupies a fraction of the peripheral sites and the proposal is based on the interpretation of ATPase measurements in the steric terms that H. E. Huxley and Haselgrove applied to the explanation of relaxation. Although it is not possible to measure the actin activation of myosin ATPase without any myosin attachment, it is possible to approach that condition by using a large excess of actin over myosin (less than 5% subfragment 1 (S-1) saturation of actin) (Fig. 3). Under those conditions, calcium saturated regulated actin filaments activate myosin ATPase only about half as much as does pure actin (Bremel *et al.*, 1972) (Fig. 3). The extent of activation of myosin ATPase is about the same for calcium activated regulated actin filaments and for tropomyosin-actin filaments without troponin (Murray *et al.*, 1975), suggesting that troponin has no longer any special function when it has become calcium saturated (compare Fig. 3B with 3C). We suggest that the 50% inhibition of ATPase activity by tropomyosin (with or without Ca-saturated troponin) may be due to steric hindrance by tropomyosin, binding about equally well to peripheral actin sites and other sites on the actin surface. For a relatively rigid molecule (Philipps *et al.*, 1979) one may visualize large bends of the tropomyosin molecule on the actin strand which constantly change position (Fig. 1). There is less interference with myosin access in the presence of calcium or in the absence of troponin than during relaxation because tropomyosin is no longer made to occupy all peripheral actin sites by the actin binding of calcium free troponin.

We suggest that tropomyosin is displaced completely into the groove by myosin when the filament is saturated with myosin. Has a shift of tropomyosin position towards the groove ever been shown when tropomyosin had not been held first at the periphery by calcium free troponin? Secondly, is there any evidence that myosin can displace tropomyosin from its position? The following experiment answers both questions positively. It was published by Vibert and his colleagues (1972) with the muscle anterior byssal retractor muscle (ABRM) of Mytilus a myosin regulated muscle. Nevertheless, the transition from the relaxed to the contracted state in this muscle is associated with similar changes in the actin layer lines as in vertebrate skeletal muscle. Since the mollusc muscle does not contain troponin one must look for another change in the actin filament that can be responsible for this tropomyosin translocation. During relaxation, there is no attached myosin on the actin filaments, because the calcium-free myosin is prevented from binding by its regulatory light chains. On addition of calcium, myosin attaches to the actin filament, suggesting that the tropomyosin shift may be caused by myosin attachment to the actin filament. One may state that during contraction the concentration of activated myosin is high enough for myosin to successfully compete with tropomyosin for the same actin sites.

In a troponin controlled muscle one may view relaxation as the result of tropomyosin's success in the competition for the actin sites because tropomyosin binding to the peripheral sites is enhanced by the binding of troponin to the periphery of the actin strand. In that case myosin may gain access to the peripheral sites only when its affinity is increased above the affinity of the myosin ATP intermediate for actin, for instance by the removal of the ATP or its intermediates. Displacement of tropomyosin by rigor complexes, in the absence of calcium, has been demonstrated by

Haselgrove (1972). He showed that tropomyosin is shifted towards the groove in calcium depleted frog muscle when the muscle goes into rigor. This experiment suggests that regulated actin filaments can be saturated with rigor complexes in the absence of calcium because nucleotide-free myosin successfully competes with tropomyosin for the peripheral actin sites.

In conclusion, it seems that X-ray diffraction data, our knowledge of rigor complex formation in the absence of calcium and some data on ATPase activity are compatible with the steric blockage hypothesis of H. E. Huxley and Haselgrove and Parry and Squire as extended by us and viewed in biochemical terms. In such terms the hypothesis states that tropomyosin and myosin compete for the same actin sites and that myosin must be capable of winning this competition when its concentration is sufficiently high under conditions where myosin binds strongly to actin. The data so far examined seem also to be compatible with the additional proposal that tropomyosin unaided by troponin has an affinity of its own for many sites on the actin surface, including the myosin binding sites.

In the second half of the paper we would like to examine in greater detail old and new ATPase activity measurements and S-1 actin binding data in order to see how far they all can be explained by a competition between tropomyosin and myosin for the same actin sites and by the view that tropomyosin determines the interaction between actin and myosin.

Reduced to its simplest form, the steric blockage hypothesis suggests that during relaxation there is a complete overlap between the tropomyosin and the myosin binding site. In that case, in the presence of calcium free troponin and tropomyosin the binding constant of myosin should be reduced by the same factor independent of the nucleotide content of the myosin. We tested this prediction by comparing directly the binding of nucleotide-free S-1 and S-1-AMPPNP (β,γ-imido-ATP) to relaxed actin filaments and indirectly of the S-1. ATP intermediate (S-1-ADP•P; Trentham *et al.*, 1976; Taylor, 1979).

To that purpose, we titrated regulated acto-S-1 with AMPPNP in the absence and presence of calcium and measured the dissociation of S-1 from actin (Fig. 1). Actin was in more than 30-fold excess over S-1, a condition where even at low ATP concentrations there are not enough rigor complexes to displace tropomyosin from the periphery. S-1 dissociation with increasing AMPPNP depends on the binding of AMPPNP to acto S-1 and the binding constant of S-1-AMPPNP to actin. Both constants, the AMPPNP and the S-1-AMPPNP binding constant, are obtained directly by extrapolation. The rigor complex binding constant can be calculated if the value for the binding constant of AMPPNP to S-1 is known. Alternatively, one can calculate the factor by which the rigor constant is reduced on removal of calcium based on the following relationships (AS, rigor complex; S, S-1; A, actin; N, AMPPNP; SN, S-1-AMPPNP).

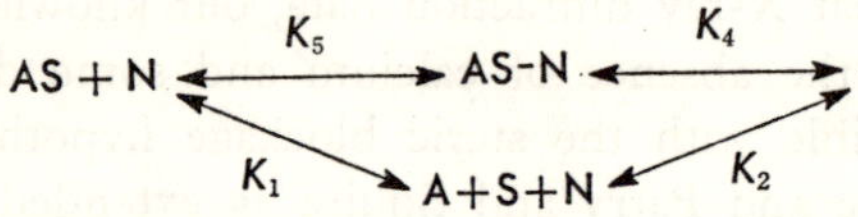

$$\frac{[AS]+[AS\text{-}N]}{[A][SN]} = \frac{K_4}{K_3[N]} + K_4\ ;\ K_4/K_3 = K_1/K_2 \quad ;\ \text{valid when the lowest concentration of } N = 100 \cdot \frac{1}{K_2}$$

$$\frac{K_4}{K_3}(\mathrm{EGTA}) \Big/ \frac{K_4}{K_3}(\mathrm{Ca}) = \frac{K_1(\mathrm{EGTA})}{K_1(\mathrm{Ca})}$$

The extent of dissociation of S-1 from regulated actin in the absence of calcium was greater and required lower concentrations of AMPPNP than in the presence of calcium (Fig. 5). The change in the apparent binding constants is given in Table II. Relaxed actin combined with AMPPNP-S-1 with an association constant that was only two to three times lower than for calcium activated actin, while the ATPase activity was reduced nearly 20-fold. We attribute the 20-fold reduction in ATPase activity to an equal decrease in the binding constant for the S-1-ATP-intermediate since during relaxation S-1 seems to be completely dissociated from actin. The calculated rigor complex constant was reduced 10-fold. This finding is not incompatible with the steric blockage model, it only rules out complete overlap of tropomyosin and myosin binding sites. In terms of the steric blockage model, these experiments suggest that the actin binding domains for the different myosin species are not identical and, therefore, are subjected to different degrees of steric hindrance by tropomyosin. In other words, one must assume that tropomyosin need not or not completely, vacate its site on the periphery of an actin molecule to allow the binding of either S-1-AMPPNP or the nucleotide-free S-1. In terms of a competition for the same actin sites on would have to say that tropomyosin and myosin compete only for a part of the myosin binding site and that it is a different part when myosin is nucleotide free or containing different nucleotides. If one considers O'Brien's suggestion (1978) that the peripheral tropomyosin and the myosin binding sites are at completely separate locations, one would have to suggest an allosteric interaction between the two binding sites. The surprising aspect is the great difference between the two nucleotide containing species the S-1 ATP intermediate and S-1-AMPPNP. If S-1-AMPPNP should be a valid analogue for S-1-ATP this would suggest a very considerable change in conformation of S-1 on hydrolysis of the bound ATP.

From a physiological point of view one may take notice that these data show that rigor complexes to relaxed actin are weaker than to active actin. The concept of competition between myosin and tropomyosin for the same site explains why in the absence of calcium myosin can attach to actin as rigor complexes and displace tropomyosin from the periphery. It does not explain why rigor complexes clear a path for the other myosin species, the ATP intermediate of myosin (myosin-ADP•P), *i.e.*, why in the absence of calcium there is true contraction at low ATP concentrations. (At ATP concentrations below ATP saturation rigor complexes are present by definition.) Contraction by cycling of bridges requires that there are free binding sites on actin where myosin-ADP•P can attach and cause shortening or tension development. However, before the concentration of ATP-free myosin is raised to a value where it competes successfully with tropomyosin for the peripheral actin sites tropomyosin occupies all peripheral sites (more or less) since ATPase activity is reduced to 3 to 5% and there is no shortening. Under such conditions, competition in simple biochemical terms implies that tropomyosin is absent only from those sites on the periphery that are occupied by a myosin molecule capable of successful competition, *i.e.*, with a high enough binding constant, such as a nucleotide free myosin. (If the myosin ATP intermediate by itself were capable of successful competition there would have been no relaxation in the first place.) How do we explain that nucleotide-free myosin can apparently clear some actin molecules of tropomyosin without substituting for tropomyosin on the same actin molecules? We assume that the tropomyosin molecule is relatively rigid. Tropomyosin displaced by rigor complexes from, let us arbitrarily say 3, out of 7 of its associated actin peripheral sites is too stiff to form sharp loops bending back to the peripheral sites between the rigor complexes. Biochemically we attribute the reversal of relaxation at low ATP to the cooperative effects of rigor complexes

(Bremel *et al.*, 1972; Weber and Murray, 1973). We now suggest that this cooperativity is based on the relative rigidity of the tropomyosin molecule. When we think of the tropomyosin movements discussed above we imagine very wide bends of the tropomyosin molecule—when fortified by calcium free troponin during relaxation—so that a single bend may include 14 actin molecules *i.e.*, be restricted to domains where the troponin molecule has temporarily dissociated from actin.

When we enumerated the evidence that leads us to believe that in the presence of calcium or in the absence of troponin some competition persists between tropomyosin and myosin for the same or part of the same actin binding site we included some ATPase measurements. The 50% inhibition of acto-S-1 ATPase activity by calcium saturated troponin-tropomyosin (Fig. 3) was not of the V_{max} but of the $K_{M\text{-actin}}$ type, *i.e.*, the kind of inhibition expected if one half of the actin sites had not been accessible to the myosin-ATP intermediate (Murray and Weber, 1980). Much weaker binding of tropomyosin to the peripheral sites than in the absence of calcium was indicated by the much weaker inhibition of ATPase activity, 50% as compared to 95–97% with good preparations of regulated actin. In agreement with these data we were not able to measure an inhibition of rigor complex formation by regulatory proteins in the presence of calcium and an S-1 saturation of the actin filaments below 5%.

What happens to acto-S-1 ATPase activity under conditions where S-1 competes successfully with tropomyosin and displaces it completely into the groove? One might expect that competitive inhibition by tropomyosin is overcome and the ATPase activity rises to the same level as with pure actin. That expectation is not fulfilled, the ATPase activity rises to much higher rates than in the absence of tropomyosin, with pure actin. Under these conditions tropomyosin seems to act as an activator of cofactor activity and we have called this state of elevated cofactor activity potentiation (Bremel *et al.*, 1972). Potentiation depends on tropomyosin-actin interactions only, and does not require the presence of troponin. Under the conditions of our earlier experiments tropomyosin displacement seemed to have been achieved by rigor complexes (Bremel *et al.*, 1972), whereby a considerably lower concentration of nucleotide-free S-1 was needed for displacement in the presence of calcium than in its absence. That is indicated by the observation that in the absence of calcium ATPase activity was raised above that of pure actin only at very low ATP concentrations (Fig. 4A and B) and that ATPase activity decreased more on the reduction

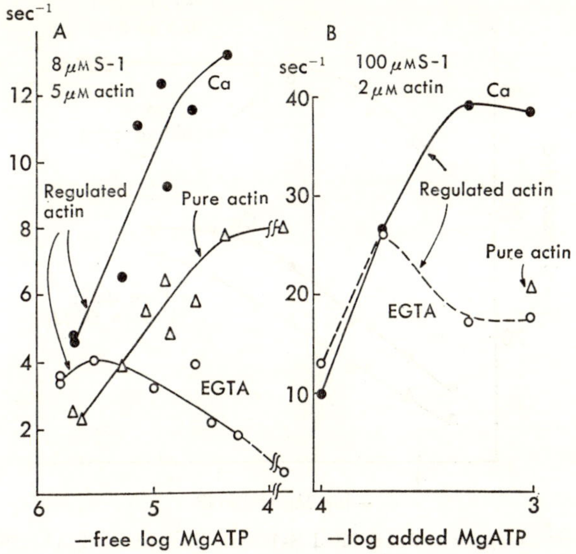

Fig. 4. Acto-S-1 ATPase with S-1 excess over actin. Conditions as in Fig. 3. In the presence of calcium, ATPase activity with regulated actin was higher than with pure actin below 1 mM ATP only, A and at all ATP concentrations with 100 μM S-1, B. In the absence of calcium, ATPase activity with regulated actin was higher than with pure actin only at very low ATP, A and B. B shows that with sufficiently large excess of S-1 over actin, and a high S-1 concentration, ATPase activities in the presence and absence of calcium become identical at low ATP concentrations as well as much higher (only 1 point shown here) than with pure actin.

TABLE II
Changes in Actin-binding Constants (Ratios) for S-1 and S-1-AMPPNP

	$K_{\text{AMP PNP}}$ (M^{-1})	$K_{\text{S-1-AMP PNP}}$ (M^{-1})
EGTA	1.1×10^4	2.2×10^4
Ca	3.3×10^3	6×10^4
Pure actin	5×10^3	8×10^4

	$K_{\text{S-1-AMP PNP}}$	ATPase	$K_{\text{assoc. rigor}}$
Ca/EGTA	2.7	19	9
Potentiated[a]/Pure	1.0		5

[a] The binding constants of S-1-AMPPNP and nucleotide-free S-1 to potentiated actin were obtained by the same procedures that were used for relaxed actin (Fig. 5) except that S-1 and actin were present in nearly equal concentrations.

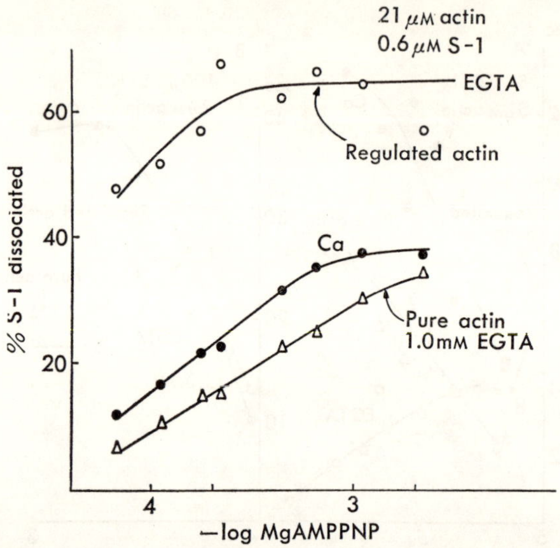

Fig. 5. Titration of pure and regulated S-1 with increasing MgAMPPNP. 5 mM Mg excess, pH 7.0, 60 mM ionic strength, 15°C. S-1 was labelled with H-3-NEM. The various binding constants, using pure actin, are similar to those determined by other investigators with native S-1. Dissociated S-1 was measured in the supernatant after 30 min centrifugation at 100,000 *g* (0.7 ml volume). S-1 and actin were corrected for their respective denatured fractions (5–7%). Regulated actin was reconstituted as previously described. (Bremel and Weber, 1972).

of the rigor complex concentration by increasing ATP levels in the absence than in the presence of calcium.

Our earlier data (Bremel *et al.*, 1972) show that potentiation results from a lowering of the K_M for S-1, below that for pure actin, *i.e.*, facilitation of the attachment of the myosin-ATP-intermediate. The affinity of nucleotide-free S-1 was also higher for potentiated than for pure actin (Table II), as indicated by a 5-fold lowering of the binding constant of AMPPNP to rigor complexes with potentiated actin (see above for equations) as compared with actin of intermediate activity or pure actin. In other words, myosin attaches more easily to actin filaments with tropomyosin near the groove than to filaments without tropomyosin at all. Only the affinity of S-1-AMPPNP was not increased when actin became potentiated. Nevertheless, complexes of S-1-AMPPNP with regulated actin also caused potentiation, as was shown by the following experiment. We compared potentiation as a function of ATP concentration in the absence and the presence of 5 mM Mg-AMPPNP, saturating 66% of the rigor complexes

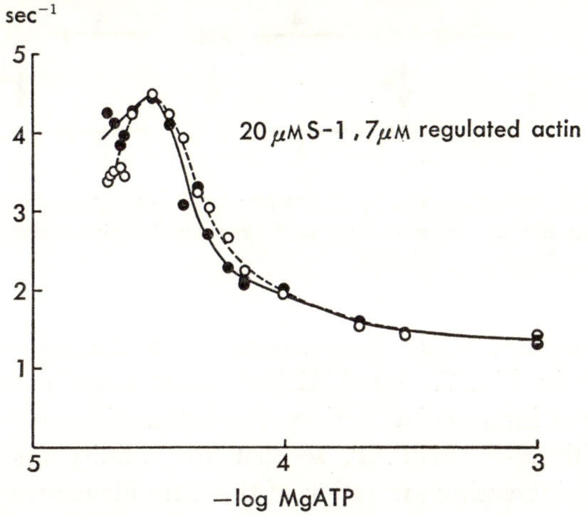

Fig. 6. Potentiation when 2/3 of the rigor complexes have been replaced with complexes containing S-1-AMPPNP. 25°C, 5 mM Mg excess, 10 mM imidazole pH 7.0, ionic strength maintained at 60 mM with KCl. Reconstituted regulated actin filaments.

with AMPPNP. Figure 6 shows that the presence of AMPPNP did not significantly alter the extent of potentiation, or the ATP concentration at which it began to disappear. That indicates that S-1-AMPPNP attached to actin was just as capable of causing potentiation as S-1 forming rigor complexes with actin. The reasoning is as follows. Potentiation diminished, and potentiated ATPase activity decreased at higher ATP concentrations, because the concentration of rigor complexes became more and more limiting. Since AMPPNP did not affect the extent of potentiation although it bound to 2/3 of the rigor complexes thereby substituting S-1-AMPPNP-actin for 2/3 of the rigor complexes, attached S-1-AMPPNP must have been just as effective as the rigor complexes. If only rigor complexes in this mixture were capable of causing potentiation, we would expect the potentiation to begin to diminish at a lower Mg-ATP concentrations with AMPPNP than without. This follows from the fact that in the presence of AMPPNP the concentration of rigor complexes was reduced to 1/3 of that in its absence. Therefore, a further reduction by dissociation by even very low ATP concentrations would lower their concentration below that necessary for potentiation.

The lack of change in the affinity of S-1-AMPPNP to potentiated actin compared to actin of intermediate activity (Table II), at first, seemed incompatible with their ability to cause potentiation. That it is not incom-

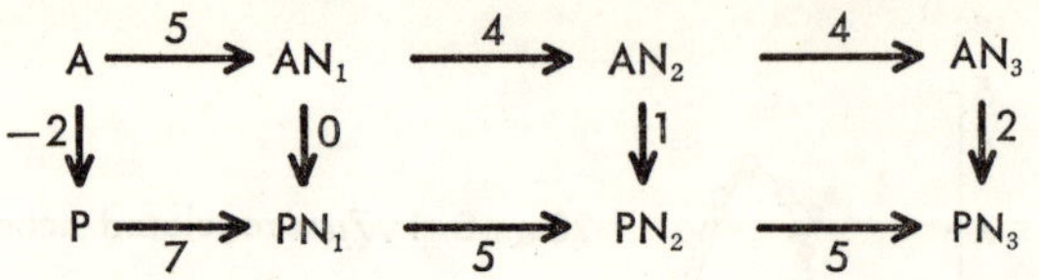

Fig. 7. Extended energy square. Numbers on arrows log K; positive numbers indicate energy release in the direction of arrow. A, regulated actin intermediate activity; P, potentiated actin; N, S-1-AMPPNP.

patible can be most easily demonstrated with the aid of the extended energy square in Fig. 7. S-1-AMPPNP must cause potentiation if the actin filament of intermediate activity (as distinct from potentiated) resists saturation with S-1-AMPPNP, so that the affinity for S-1-AMPPNP decreases with increasing saturation of the actin filament, while the potentiated filament maintains a higher binding constant. The decrease in the affinity of AMPPNP to the rigor complex of potentiated actin is due to the fact that more of its binding energy is needed to overcome the larger affinity difference between the nucleotide free S-1 and S-1-AMPPNP for potentiated actin. With respect to the physiological aspects of potentiation it is of interest that potentiation can becaused by nucleotide containing S-1; especially since S-1-AMPPNP has been reported to bind to actin at an angle different from that made by rigor complexes (Marston *et al.*, 1976). Whether the S-1 ATP intermediate can cause potentiation cannot be determined at this time. The reason is the high value of the K_M for S-1 which makes it necessary to use very high concentrations of S-1 excess. As a result, even a few percent of S-1 capable of forming ATP resistant rigor complexes (Pemrick and Weber, 1976) can saturate the actin filaments to a significant extent with rigor complexes (Fig. 4B). In our earlier paper (Bremel *et al.*, 1972) we were not aware of this artifact and assumed that potentiation at 1.0 mM ATP was caused by the S-1 ATP intermediate.

In conclusion, the cooperative interactions between tropomyosin and actin make the actin filament subject to regulation. When combined with troponin, actin interaction with myosin becomes calcium dependent. Without troponin, (as in myosin regulated muscles) or when troponin is calcium saturated, actin filaments show positive cooperativity in their response to myosin binding: attachment of myosin facilitates further myosin binding. This response is very different from the allosteric effects observed with pure actin, using fluorescence or nitroxide labels (Thomas *et al.*, 1979; Miki *et al.*, 1976). Tropomyosin seems to easily shift its position on the actin filament and may do so as a result of competition with myosin for the same actin sites. The position on the periphery is considered responsible for the dissociation of myosin from actin during relaxation. We suggest that the state of intermediate activity and potentiation may also be determined by the position of tropomyosin on the actin filament, causing a low degree of competitive inhibition in one position, and a conformational change leading to activation in the other.

Acknowledgment

We are very grateful to Dr. A. Gutfreund for a critical reading of this manuscript. This research was supported by grant HL 15692 and HL 15835.

REFERENCES

Bremel, R. D. and Weber, A. (1972). *Nature New Biol.* **238**, 97–101.

Bremel, R. D., Murray, J. M., and Weber, A. (1972). *Cold Spring Harbor Symp. Quant. Biol.* **37**, 267–275.

Cohen, D., Caspar, D.L.D., Parry, D.A.D., and Lucas, R. M. (1972). *Cold Spring Harbor Symp. Quant. Biol.* **36**, 205–216.

Ebashi, S. and Ebashi, F. (1964). *J. Biochem.* **55**, 604–613.

Ebashi, S. and Endo, M. (1968). *Prog. Biophys. Mol. Biol.* **18**, 123–183.

Eisenberg, E. and Moos, C. (1969). *J. Biol. Chem.* **24**, 2451–2456.

Greene, L. E. and Eisenberg, E. (1978). *Proc. Natl. Acad. Sci. U.S.* **75**, 54–58.

Haselgrove, J. C. (1972). *Cold Spring Harbor Symp. Quant. Biol.* **37**, 341–352.

Huxley, H. E. (1972). *Cold Spring Harbor Symp. Quant. Biol.* **37**, 361–376.

Johnson, P. and Smillie, L. B. (1977). *Biochemistry* **16**, 2264–2273.

Kendrick Jones, J., Lehman, W., and Szent-Györgyi, A. G. (1970). *J. Mol. Biol.* **54**, 313–326.

Marston, S. B., Rodger, C. D., and Tregear, R. T. (1976). *J. Mol. Biol.* **104**, 263–276.

Miki, M., Kouyama, T., and Mihashi, K. (1976). *FEBS Lett.* **66**, 98–101.

Murray, J. M., Weber, A., and Bremel, R. D. (1975). In *Calcium Transport in Contraction and Secretion* (Carafoli, E., Clementi, F., Drabikowski, W., and Margreth, A., eds.), pp. 489–496, North-Holland, Amsterdam.

Murray, J. M. and Weber, A. (1980). In *Excitation Contraction Coupling*, Jerry Lewis Symposium of the Muscle Dystrophy Association.

O'Brien, E. J., Morris, E. P., Seymour, J. V., and Couch, J. (1978). Abstr. 6th Int. Biophys. Congr., Kyoto, p. 311.

Parry, D.A.D. and Squire, J. M. (1973). *J. Mol. Biol.* **75**, 33–55.

Pemrick, S. and Weber, A. (1976). *Biochemistry* **15**, 5193–5198.

Phillips, G. N., Lattman, E. E., Cummins, P., Lee, K. Y., and Cohen, C. (1979). *Nature* **278**, 413–417.

Spudich, J. A., Huxley, H. E., and Finch, J. T. (1972). *J. Mol. Biol.* **72**, 619–632.

Stewart, M. (1975). *FEBS Lett.* **53**, 5–7.
Taylor, E. W. (1979). *Crit. Rev. Biochem.* **6**, 103–164.
Thomas, D., Seidel, J., and Gergely, J. (1979). *J. Mol. Biol.* **132**, 257–273.
Trentham, D. R., Eccleston, J. F., and Bagshaw, C. R. (1976). *Q. Rev. Biophys.* **9**, 217–281.
Vibert, P. J., Haselgrove, J. C., Lowy, J., and Poulson, F. R. (1972). *J. Mol. Biol.* **71**, 757–767.
Wakabayashi, T., Huxley, H. E., Amos, L. A., and Klug, A. (1975). *J. Mol. Biol.* **93**, 477–497.
Weber, A. and Murray, J. M. (1973). *Physiol. Rev.* **53**, 612.
Wegner, A. (1979). *J. Mol. Biol.* **131**, 839–853.

Muscle Contraction: Its Regulatory Mechanisms (Ebashi, S. et al., eds.), pp. 237–249, Japan Sci. Soc. Press, Tokyo / Springer-Verlag, Berlin (1980)

18

Functional Organization of the Troponin-Tropomyosin System

IWAO OHTSUKI

Summary: Immunoelectron microscopic study demonstrated that each of three troponin components distributed along the thin filament with a 38-nm periodicity. In one period, antitroponin-T formed a pair of striations, whereas antitroponin-C or -I formed one narrow transverse striation. Recently, it was clarified that two of these antitroponin-T striations are due to the particular portions of the molecule; the one on the Z-line side is derived from the T_1 region and the other on the free end side from the T_2. The position of antitroponin-T_2 was the same as those of antitroponin-C and -I. Antitroponin-T_1 was situated about 13 nm from the antitroponin-T_2 in the direction of Z line.

Some properties of the subfragments of troponin-T prepared by chymotrypsin treatment were investigated. Troponin-C and -I bound to troponin -T_2, but not to troponin-T_1. Tropomyosin, however, bound to both troponin -T_1 and -T_2. Action of troponin-T in maintaining the Ca-sensitivity of superprecipitation was replaced partially by troponin-T_2. Paracrystal analysis suggests that troponin components are arranged in the axial length of 7–9 nm.

Based on these findings, the molecular arrangement of troponin components and tropomyosin in the thin filament is discussed.

It has been established that a thin filament of striated muscle is an ordered aggregate of troponin, tropomyosin, and actin molecules (Ohtsuki *et al.*, 1967; Ebashi *et al.*, 1969; Ebashi, 1974; Ohtsuki, 1974). As shown in Fig. 1, end-to-end filaments of tropomyosin run in the grooves of double-stranded actin molecules which form the core of thin filaments. Each tropomyosin covers the length of seven globular actin molecules and attaches one troponin at the region about one-third of the molecular length from one end.

It has also been found that troponin is a complex of three components (*cf.* Ebashi, 1974; Ebashi *et al.*, 1972). Each component has its particular

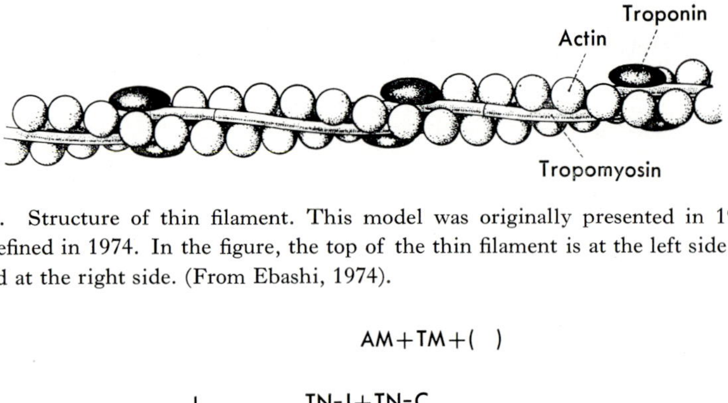

Fig. 1. Structure of thin filament. This model was originally presented in 1969 and then refined in 1974. In the figure, the top of the thin filament is at the left side and the Z band at the right side. (From Ebashi, 1974).

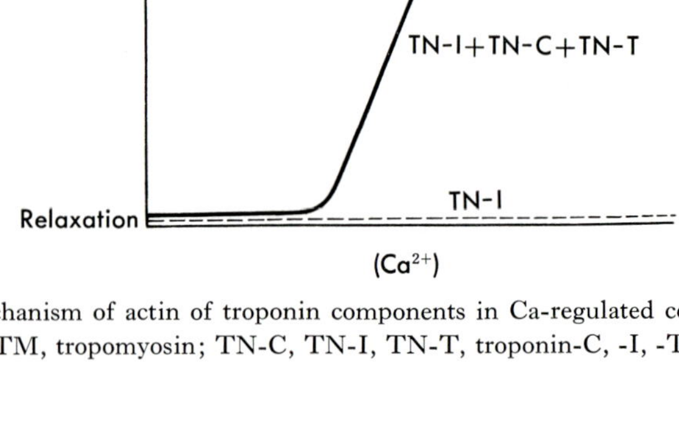

Fig. 2. Mechanism of actin of troponin components in Ca-regulated contraction. AM, actomyosin; TM, tropomyosin; TN-C, TN-I, TN-T, troponin-C, -I, -T.

function in troponin complex; troponin-C binds Ca ion, troponin-I inhibits actin-myosin interaction, and troponin-T binds to tropomyosin. Depression of the contractile response by troponin-I is released by troponin-C irrespective of Ca ion concentration. Further addition of troponin-T first restores the original Ca sensitivity of the contraction (Fig. 2). In this respect, regulatory activity in troponin complex can be said to be localized in troponin-T.

The present article reviews the studies on the molecular arrangement of troponin components and tropomyosin in the thin filament, with special reference to the role of troponin-T.

LOCALIZATION OF TROPONIN COMPONENTS IN THE THIN FILAMENT

Periodic distribution of troponin was first demonstrated by the use of antibody against whole troponin complex (Ohtsuki *et al.*, 1967; Ohtsuki, 1974). Subsequently, each troponin component was also shown to distribute with a 38-nm period (Ohtsuki, 1975). Antibodies against troponin-C and -I formed a narrow striation in each period, whereas antitroponin-T formed a pair of striations. Recently it was found that chymotrypsin was able to split troponin-T into two subfragments: troponin-T_1 and -T_2 (Ohtsuki, 1979, Fig. 3). Two striations of antitroponin-T were, then, revealed to be due to these particular portions of the molecule: the stria-

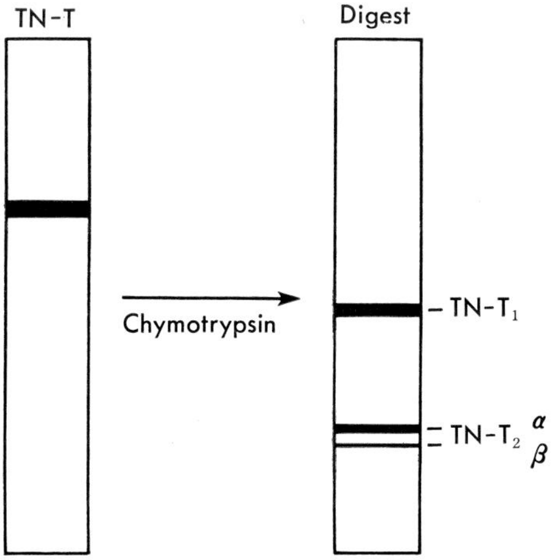

Fig. 3. SDS-gel electrophoretic pattern of chymotryptic troponin-T.

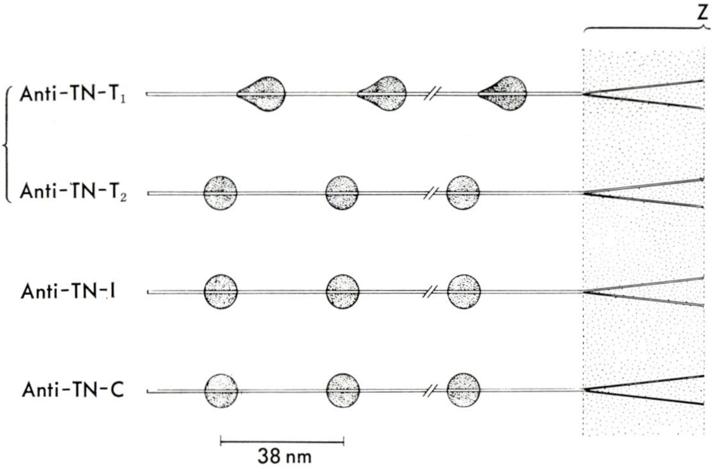

Fig. 4. Diagram illustrating the antibody distribution against troponin components. (From Ohtsuki, 1979).

tion on the Z-line side was derived from the T_1 region and the other from the T_2 region (Fig. 4). The position of antitroponin-T_2, the first band of which was about two-thirds of the period length, 27 nm, from the top of filaments, was the same as those of antibodies against troponin-C or -I. On the other hand, antitroponin-T_1 was situated about 13 nm from the antitroponin-T_2 in the direction of the Z line. This finding coincided with affinity chromatographic findings that troponin-C or -I interacted only with troponin-T_2 but not with troponin-T_1 (Katayama, 1979, Table I; Ohtsuki, 1979).

It has been established that troponin, and consequently troponin-T, binds to the region around the 190th cysteinyl residue of tropomyosin (total 284 residues), about two-thirds of the molecular length from the N-terminal (Ohtsuki, 1974; Stewart, 1975). The position of antitroponin at the free end portion of thin filament indicates that the tropomyosin molecule should associate along the thin filament as its N-terminal end is on the free side and its C-terminal end is on the Z-line side. In troponin-T, T_1, T_2 regions should be located at the N-terminal and C-terminal sides, respectively, of a filamentous tropomyosin molecule.

Localization of antitroponin-T_2, -C, and -I would then represent approximately the position of the troponin molecule. On the other hand, the position of antitroponin-T_1 was about 13 nm shifted to the Z-line direction from the antitroponin-T_2, and became wider on the specimens stained heavily with antibody, up to the width of 10 nm. This is explained by the fact that the antitroponin-T_1 molecule, whose size is approximately 10 nm, binds to troponin complex directionally only from the Z-line side. Thus, the striations usually observed by antitroponin-T_1 would be formed by the free ends of the oriented antibody molecule. Intense staining results in the visualization of the whole antibody molecule as the striation of about 10 nm. In other words, the largest portion of the troponin-T_1 region would be covered by other troponin components; only the portion facing the Z line is exosed. Considering that the size of the antibody molecule is comparable with, or a little larger than, that of troponin complex, the location of antibodies against troponin-C, -I, and T_2 in the same position would mean that these subunits are exposed on the surface of thin filament to the same extent rather than that they are situated at transversely the same position. A schematic illustration based on these con-

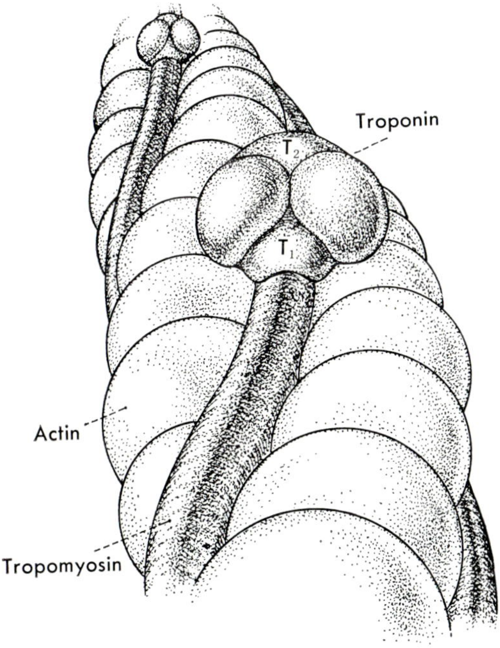

Fig. 5. Possible organization of troponin components in the thin filament. The view is from the Z-line side of the filament. Most of the T_1 region of troponin-T is covered by other components; it is only exposed in the direction of the Z line.

siderations is shown in Fig. 5. In this figure, thin filament is viewed obliquely from the Z-line side to the direction of the filament top. Most of the troponin-T_1 is covered by the other three subunits, the spatial arrangement of which has not yet been determined.

PROPERTIES OF CHYMOTRYPTIC TROPONIN-T SUBFRAGMENTS

Chymotrypsin was found to cleave troponin-T into two groups of subfragments: troponin-T_1 and -T_2 (Ohtsuki, 1979, Fig. 3). Troponin-T_1 is a single subfragment with a molecular weight of 22,000 daltons for rabbit skeletal muscle (26,000 daltons for chicken) on SDS gel pattern. The other subfragments, troponin-T_2, usually consisted of two fragments with molecular weight around 13,000 daltons. Since longer chymotryptic digestion always resulted in an increase of the smaller fragment (TN-T_2 β) density on SDS-gel and in a concomitant decrease of the larger fragment TN-T_2 α), the smaller subfragment should be a part of the larger subfragment.

Troponin-T_1 was an acidic peptide and could pass through a DEAE-Sephadex column in the presence of 6 M urea and 20 mM Tris-HCl (pH 8.0). On the other hand, troponin-T_2 was adsorbed to the DEAE-Sephadex in the above condition and eluted out on increasing NaCl concentration to 0.1 M. Troponin-T_2 fraction thus separated partially retained the Ca-sensitizing actin of the original troponin-T on superprecipitation. This action was not potentiated by the addition of troponin-T_1, which by itself slightly depressed the superprecipitation irrespective of Ca concentrations (Nakamura and Ohtsuki, 1979).

TABLE I
Interaction of Troponin-T Subfragments with Other Troponin Components and Tropomyosin

		Tropomyosin		Troponin-C				Troponin-I	
		NaCl 0.2 M	NaCl 0.4 M	NaCl 0.2 M Ca(+)	NaCl 0.2 M Ca(−)	NaCl 0.4 M Ca(+)	NaCl 0.4 M Ca(−)	NaCl 0.2 M	NaCl 0.4 M
Troponin-T_1		+	−	−	−	−	−	−	−
Troponin-T_2	α	+	−	+	+	+	−	+	+
	β	−	−	+	−	+	−	+	+

(Ohtsuki, 1979; Katayama, 1979).

Recently two troponin-T_2 subfragments have been further separated by the use of SE-Sephadex. The superprecipitation became sensitive to Ca ion only by the larger subfragment, not by the smaller one (Ohtsuki *et al.*, unpublished result).

The interaction of these subfragments with other components was investigated by affinity chromatography (Ohtsuki, 1979; Katayama, 1979, Table I). Both troponin-T_1 and -T_2 bound to tropomyosin at low NaCl condition. But the smaller subfragment of troponin-T_2, did not bind to tropomyosin. This may reflect the fact that binding to tropomyosin is essential for the Ca-sensitizing action to troponin-T.

Troponin-T_2 was also found to interact with troponin-C and troponin-I. Interaction between troponin-T_2 and -C was sensitive to Ca ion concentration. Essentially the same result was obtained with CNBr fragment of troponin-T (Pearlstone and Smillie, 1978). Interaction between troponin-T_2 and troponin-I was insensitive to the change in NaCl (or KCl) as well and Ca concentrations.

The results of amino acid analysis as well as of N-terminal determination indicate that troponin-T_1 and -T_2 are the N-terminal and C-terminal side fragments of troponin-T, respectively. The cleavage site was estimated to be amino terminal side of 159th leucyl residue (Katayama, 1979; Nakamura and Ohtsuki, unpublished result).

STRUCTURAL RELATION BETWEEN TROPONIN-T AND TROPOMYOSIN

Based on paracrystal analysis, troponin (or troponin-T) was concluded to be located at the region of tropomyosin about two-thirds of the molecular length, 26.5 nm, from one end (Ohtsuki, 1974). The molecular end of

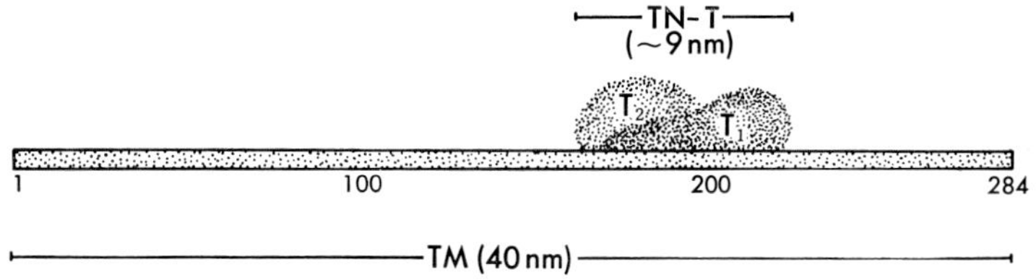

Fig. 6. Structual relation between troponin-T (TN-T) and tropomyosin (TM). T_1, N-terminal side region of troponin-T; T_2, C-terminal side region of troponin-T. The Numbers below tropomyosin are those of the amino acid residues, counting from the N-terminal.

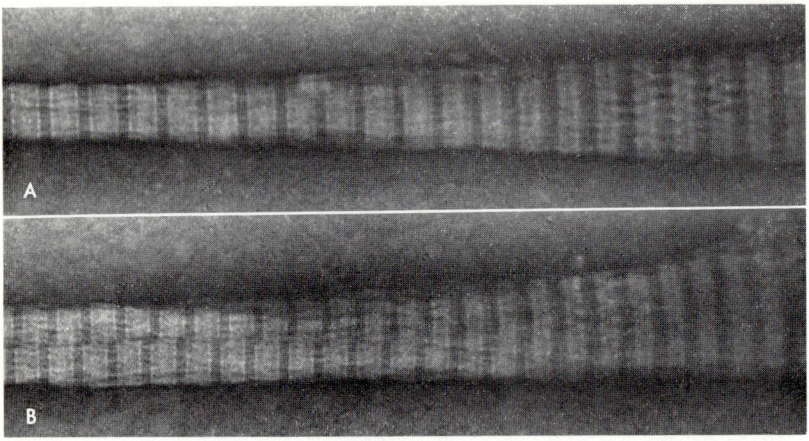

Fig. 7. Troponin-T-tropomyosin paracrystal. A, B: electron micrograph. Magnification ×125,000.

this side was also thought to be cationic. Determination of amino acid sequence and visualization of cysteinyl residue of tropomyosin on the paracrystalline structure led to the conclusion that troponin (or troponin-T) is located around the 190th cysteinyl residue of tropomyosin, the N-terminal region of which was shown to be rich in basic residues (Sodek *et al.*, 1972; Hodges *et al.*, 1972; Stewart, 1975).

Immunoelectron-microscopic results indicated that T_1 and T_2 regions of troponin-T are situated side by side axially along filamentous tropomyosin molecules; T_1 and T_2 regions are situated at the C- and N- terminal sides of tropomyosin (Fig. 6). This molecule thus should be rod-shaped rather than globular. It was suggested that the length of troponin-T is about 10 nm (Ohtsuki, 1975).

Figure 7 shows a troponin-T-tropomyosin paracrystal prepared in 10 mM $MgCl_2$ solution. The left half of the paracrystal is a usual Cohen-Longley type, called Type 1 in the present article. As this paracrystal moves to the right, it transforms into another type, called Type 2.

In Type 1, it has been shown that two boundary lines surrounding a 26.5-nm wide bright band (or a 13.5-nm narrow dark band) represent end-to-end bonding portions of two sets of tropomyosin filaments trending in opposite directions (Ohtsuki, 1974). A Type 2 paracrystal consists of alternative arrangement of 18 nm bright and 22-nm dark bands, between which rather wide boundary lines are present. These boundary lines should also be the bonding portions of tropomyosin filaments.

If tropomyosin molecules form continuous end-to-end filament all through the Type 1 and Type 2 paracrystals of Fig. 7 A, corresponding boundary lines in the two types of paracrystal should be regularly spaced at 40 nm internals. In Fig. 7 A, the left side lines of the wide bands in Type-1 and the right side lines of the wide bands in Type 2 are actually arranged at regular 40-nm intervals. On the other hand, the right side lines of Type 1 and the left side lines of Type 2 fade away in the transforming portion. This finding indicates that a set of tropomyosin filaments in the right direction continue through the Type 1 and Type 2 regions of the paracrystal and that tropomyosin molecules in the left direction terminate in the transforming portion (the direction of the tropomyosin molecule is defined by that of the arrow heads). Figure 7 B shows another type of the transforming portion of the paracrystal. In this case, one Type 2 paracrystal transforms into two Type 1 paracrystals which are spaced about 8 nm apart axially. The left side lines of narrow bands in Type 2 paracrystals and those on the right side in the upper Type 1 are spaced at 40 nm intervals. On the other hand, the right side lines of wide bands in the Type 2 and those on the left side in the lower Type 1 are regularly spaced. Thus, the two sets of oppositely directed filaments in Type 2 paracrystal continue in the two paracrystals of Type 1.

Troponin-T is known to increase to a density of about 9 nm in the middle of the wide band of the Type 1 paracrystal. This value determines the maximum length of the troponin-T molecule, for the two sets of oppositely oriented molecules are projected as about 9 nm wide. In the Type 2 paracrystal, troponin-T should locate in the narrow bright band, whose density seems relatively even over the whole band width of 18 nm.

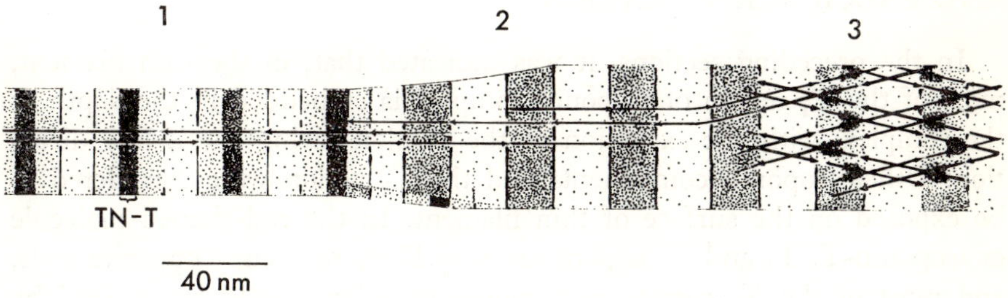

Fig. 8. Schematic illustration of troponin-T-tropomyosin paracrystal. Three types of paracrystals—1, 2, and 3 (indicated above the paracrystal structure)—and their mutual transformations are illustrated. Tropomyosin molecule is shown as a bar with an arrowhead whose length is 40 nm. Dark portions in both Type 1 and Type 2 are considered troponin-T locations.

This suggests that troponin-T is distributed almost all over the narrow band. The length of troponin-T would, therefore, be very near to 9 nm, at least more than 7 nm. This conclusion is further supported by other studies. Higashi and Ooi (1969) first showed that the Type 2 paracrystal transforms into a rhombic, or double diamond, network structure (Type 3), each filament of which would be a single tropomyosin molecule. Although Cohen *et al.* (1972) suggested that the cross-over points of this network were troponin-T, the present discussion indicates that the position of troponin-T corresponds to the part of the filament portion in which tropomyosin filament thickens over a length of about 10 nm (*cf.* Cohen *et al.*, 1972). Sometimes the thickened portion seemed dumbbell-shaped; this would reflect the molecular shape of troponin-T associated with tropomyosin. The relation between troponin-T and tropomyosin discussed above is schematically illustrated in Fig. 8.

Recently it was pointed out by Nagano *et al.* (1980) that the region around the 90–150th residues of the T_1 region can form a continuous triple stranded helical structure with the 155–220th residues of coiled-coil tropomyosin molecule. This suggest that the T_1 region interacts over the relatively wide range of tropomyosin side by side and that the definite portion of the T_1 region locates near tropomyosin.

Although the T_2 region also binds to tropomyosin, it loses affinity for tropomyosin when it is degraded into the slightly smaller subfragment (β). The T_2 region should bind to tropomyosin only through a small portion; most of it is situated relatively far from the tropomyosin.

MOLECULAR ARRANGEMENT OF TROPONIN COMPONENTS AND CALCIUM REGULATION

In the preceding sections, it was indicated that, in the thin filament, most of the T_1 region of troponin complex is embedded in the depth of the complex, with the exception of the portion facing the Z-line direction. Other troponin components and the T_2 region were considered to be exposed on the surface of thin filament. In the rod-shaped molecule of troponin-T, T_1 and T_2 regions are very likely to occupy opposite ends, and most of the T_1 region, in troponin complex, would be covered by troponin-C and troponin-I, not by troponin-T_2 (Figs. 4 and 6). There remains another possibility, however—that the T_1 region is in the space between tropomyosin and actin strands and only its tail portion is exposed to the surface.

In the absence of Ca ion, in muscle, troponin acts as an inhibitor for contraction. This inhibitory action is due to troponin-I. Although the stoichiometric amount of this component to tropomyosin shows full inhibitory action, it has been shown that troponin-I interacts strongly with actin and very weakly with tropomyosin. It is conceivable that troponin-I is able to form a stable inhibotory complex with tropomyosin only when it associates along the actin filament. At this inhibited state in the thin filament, tropomyosin filament was shown to deviate from the center of the long pitch groove of actin (Wakabayashi *et al.*, 1975).

According to the reconstructed image of the thin filament from electron micrograph, tropomyosin filament is on the clockwise-direction side of the groove if one look at the thin filament from the top (O'Brien *et al.*, 1978). Troponin-I most plausibly binds to actin at the position farthest from the groove center and would attract the tropomyosin molecule to the outside; troponin-I would be situated at the left side in the troponin complex in Fig. 4, and troponin-C is consequently on the right side.

When troponin-I forms complex with troponin-C (without troponin-T), it does not interact with tropomyosin-actin filament, and the contractile system always becomes activated irrespective of Ca concentration (Potter and Gergely, 1974; Ebashi *et al.*, 1972, Figs. 2 and 9). Further addition of troponin-T results in the restoration of original Ca sensitivity; the contraction is depressed at lower Ca concentration. Structually speaking, troponin-T arranges troponin-I to the fully inhibitory position in the absence of Ca ion (Fig. 9), by anchoring this component to tropomysoin. Although troponin-T_2 (without T_1) can anchor the C-I complex to tropomyosin, the complex would be arranged rather loosely. Then the troponin-

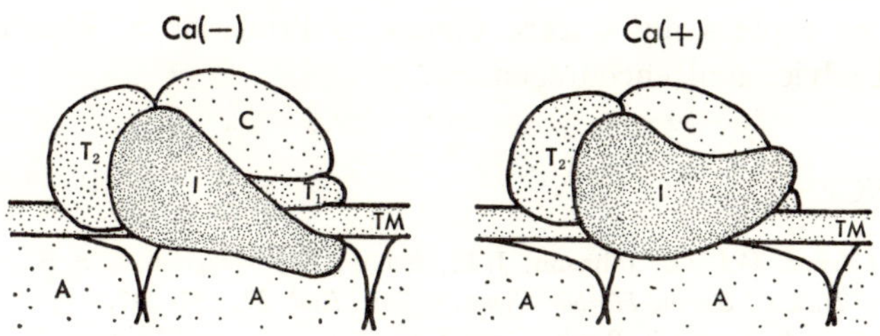

Fig. 9. Possible effect of Ca on the molecular arrangement of thin filament. In the absence of Ca, troponin-I interacts with tropomyosin and actin (inhibitory position). In the presence of Ca, troponin-C interacts strongly with troponin-I, and consequently troponin-I is detached from tropomyosin-actin (activated position). C, troponin-C; I, troponin-I; T_1, T_2, T_1 and T_2 regions of troponin-T; TM, tropomyosin; A, actin.

I could go into the inhibitory position only incompletely, and mostly would stay in position between inhibition and activation (Fig. 9). Wide range association of T_1 region with tropomyosin would help the T_2 region arrange the orientation of troponin-I for the inhibitory position in the absence of Ca ion (Fig. 9). The finding that a mere mixture of T_1 and T_2 subfragments did not restore the original Ca sensitivity, also indicates that the T_1 region exerts its effect through troponin-T_2.

The change in structure caused by Ca ion around troponin complex in the thin filament may include the lateral movement of tropomyosin filament, which is triggered by the detachment of troponin-I from tropomyosin-actin; interaction between troponin-I and troponin-C is strengthened instead. This movement seems to be concerned with the conduction of Ca-effect from tropomyosin to actin molecules apart from troponin complex (Fig. 1). The possibility that this lateral movement sterically regulates the actin-myosin interaction has been proposed (H. E. Huxley, 1972).

It is also conceivable that the rotation of tropomyosin along its filament axis is involved in the conduction of the Ca-effect. Since each tropomyosin molecule is supposed to interact with seven actins on each of two long pitch strands by its homologous regions distributed periodically along the coiled-coil structure (Ohtsuki and Wakabayashi, 1972; McLachlan and Stewart, 1975), its rotational movement would result in a cooperative change in interaction with seven actin molecules on each strands. This interaction change might also include, at least locally, the conformational change of each actin molecule.

Acknowledgment

I wish to express my sincere thanks to Professor S. Ebashi for his thoughtful advice and encouragement.

RERERENCES

Cohen, C., Caspar, D.L.D., Johnson, J. P., Nauss, K., Margossian, S. S., and Parry, D.A.P. (1972). *Cold Spring Harbor Symp. Quant. Biol.* **37**, 287–297.

Ebashi, S. (1974). In *Essays in Biochemistry* (Campbell, P. N. and Dickens, F., eds.), vol. 10, pp. 1–35, Academic Press, New York.

Ebashi, S., Endo, M., and Ohtsuki, I. (1969). *Q. Rev. Biophys.* **2**, 351–384.

Ebashi, S., Ohtsuki, I., and Mihashi, K. (1972). *Cold Spring Harbor Symp. Quant. Biol.* **37**, 215–223.

Higashi-Fujime, S. and Ooi, T. (1969). *J. Microsc.* **8**, 535.

Hodges, R. S., Sodek, J., Smillie, L. B., and Jurasek, L. (1972). *Cold Spring Harbor Symp. Quant. Biol.* **37**, 299–310.

Huxley, H. E. (1972). *Cold Spring Harbor Symp. Quant. Biol.* **37**, 361–376.

Katayama, E. (1979). *J. Biochem.* **85**, 1379–1381.

McLachlan, A. D. and Stewart, M. (1976). *J. Mol. Biol.* **103**, 271–298.

Nakamura, S. and Ohtsuki, I. (1979). Abstr. of Annual Meeting of Japan Biophysical Society (in Japanese), p. 80.

Nagano, K., Miyamoto, K., Matsumura, M., and Ohtsuki, I. (1980). *J. Mol. Biol.* in press.

O'Brien, E. J., Morris, E. P., Seymour, J. V., and Couch, J. (1978). Abstr. 6th Int. Biophys. Congr., Kyoto, p. 311.

Ohtsuki, I. (1974). *J. Biochem.* **75**, 755–765.

Ohtsuki, I. (1975). *J. Biochem.* **77**, 633–639.

Ohtsuki, I. (1979). *J. Biochem.* **86**, 491–497.

Ohtsuki, I. and Wakabayashi, T. (1972). *J. Biochem.* **73**, 369–377.

Ohtsuki, I., Masaki, T., Nonomura, Y., and Ebashi, S. (1967). *J. Biochem.* **61**, 817–819.

Pearlstone, J. C. and Smillie, L. B. (1978). *Can. J. Biochem.* **56**, 521–527.

Potter, J. D. and Gergely, J. (1974). *Biochemistry* **13**, 2697–2702.

Sodek, J., Hodges, R. S., Smillie, L. B., and Jurasek, L. (1972). *Proc. Natl. Acad. Sci. U.S.* **69**, 3800–3804.

Stewart, M. (1975). *Proc. Roy. Soc. Lond. B* **190**, 257–266.

Wakabayashi, T., Huxley, H. E., Amos, L. A., and Klug, A. (1975). *J. Mol. Biol.* **93**, 477–497.

Muscle Contraction: Its Regulatory Mechanisms (Ebashi, S. et al., eds.), pp. 251–258, Japan Sci. Soc. Press, Tokyo / Springer-Verlag, Berlin (1980)

19

Interactions among Thin Filament Proteins

EISAKU KATAYAMA

Summary: The mode of interaction among the protein constituents of the thin filament was studied in order to elucidate the detailed mechanism of Ca-regulation in the contraction of vertebrate skeletal muscle. Although several distinct interactions have been reported, some of them have not yet been fully investigated, and their physiological significance remains to be determined. The main reason for this difficulty might lie in the insolubility at low ionic strength of some components involved in the interaction. This obstacle was successfully overcome by immobilizing such proteins in a condition which allowed them to stay in soluble state. By the use of this technique, I could confirm and characterize some of the interactions which had not yet been studied sufficiently. Furthermore, the chymotryptic fragments of troponin-T were shown to be derived from the domains of different function. Based on these results and the consideration on the secondary structure of the assembly of involved components, a new model concerning the structure and function of regulatory proteins was proposed.

The contraction of vertebrate skeletal muscle is known to be regulated by Ca^{2+} through the conformational changes of actin containing thin filaments induced by the action of tropomyosin (TM) and three components of troponin: troponin-T (TN-T), troponin-I (TN-I), and troponin-C (TN-C). Since TN-C is the sole component that binds Ca^{2+} reversibly, the information derived from Ca^{2+} bound to TN-C must be transmitted to other constituents and then converted into the signal to produce the "on" or "off" state in thin filaments (see the review by Ebashi, 1974). To elucidate the detailed mechanism of Ca-regulation within thin filaments, it is necessary to examine the nature of the interactions among the protein constituents. Studies along this line have already been conducted in several laboratories (Potter and Gergely, 1974; Hitchcock, 1975). Their results more or less support the working hypothesis of the two site mechanism of Ca-regulation. According to this hypothesis, the TN complex binds to the thin filament through two distinct sites of attachment: one is the Ca-independent bond through TN-T and TM and the other the Ca-dependent one through TN-I and F-actin (FA).

Information from reported interactions is not enough for clear elucidation of the Ca-control mechanism. This seems to be due to the insolubility at low ionic strength of some of the components involved in the interactions: aggregation of a component may conceal its important interactions with other components. This obstacle can be overcome by immobilizing the protein when it is placed in a soluble state. The use of this method enabled me to show clearly some of the interactions which had been ignored or only equivocally shown.

On the other hand, TN-T molecules could be cleaved into two main fragments, T-1 and T-2, following the method of Ohtsuki (1979), by the limited chymotryptic digestion. The position of this cleavage was assigned to the bond between the 158th and 159th from the N-terminal: T-1 comprises the N-terminal whereas T-2 is the C-terminal (Katayama, 1976) in the primary structure of TN-T (Pearlstone *et al.*, 1976). Examination of the properties of these fragments in interaction with the other constituents indicated that they were derived from the two domains of different functions of TN-T.

The interactions included in building a model of a regulatory complex will be briefly described one by one. For general references, see the review by Ebashi (1974).

1. *FA-TM*

This interaction is well established and is not mentioned here.

2. TN-I–TN-C

This is one of the most significant interactions and is effective even in 6 M urea if Ca^{2+} is present. Sequestering Ca^{2+} with EGTA in this condition dissociates the complex (Perry *et al.*, 1972). Electrophoretically the TN-I–TN-C complex does not dissociate even in the absence of Ca^{2+}, unless urea is included in the solutions. The studies using peptides derived from these molecules suggest two different binding sites on both TN-I and TN-C molecules (Syska *et al.*, 1976; Weeks and Perry, 1978).

3. TM-TN-T

Troponin complex binds to TM through TN-T. The binding site of TN-T on TM was shown electron-microscopically as the region near Cys-190 (Stewart, 1975). On the basis of this observation and the analysis of the regularity of the primary sequence of TM (Sodek *et al.*, 1978; Stone and Smillie, 1978), McLachlan and Stewart (1976) predicted the region of the 197th through the 217th residue of TM as the binding site of TN-T. However, the electrophoretic examination of Ueno (1978) indicated the participation of the N-terminal side of Cys-190 for the binding of the TN complex.

Since we have made it possible to point out the location of given residue on the negatively stained image of polar paracrystal of TM (Katayama and Nonomura, 1979), we can indicate directly the binding region of the TN complex in the primary sequence of TM. A preliminary observation suggests the extension of the image of TN to the N-terminal side of Cys-

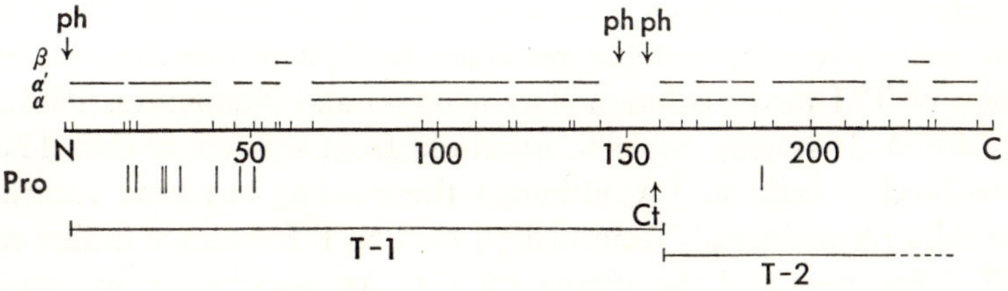

Fig. 1. Predicted secondary structure of TN-T by the algorithm of Chou and Fasman (1974). Numbers are the positions of each residue number. α, α', and β indicate the predicted regions of the α-helix, α-helix overestimated, and β-structure, respectively. Short vertical lines are the positions of prolyl residues. Ph's are the enzymatically phosphorylatable residues studied by Moir *et al.* (1977). Ct shows the position of the site of chymotryptic cleavage which has been tentatively determined (Katayama, 1979). Bars at the bottom indicate the fragments of TN-T.

190, in support of Ueno's (1978) findings (Katayama and Nonomura, 1979).

On the other hand, it is clear that the N-terminal fragments of TN-T, CNBl' (Jackson *et al.*, 1975), CB-2 (Pearlstone and Smillie, 1977) and T-1 (Ohtsuki, 1979; Katayama, 1979), have a specific affinity for TM. An estimation by CD measurement indicated that CB-2 is very rich in α-helix, as already predicted (Pearlstone *et al.*, 1976; see also Fig. 1). It should be noted that, when the primary structure of this part was projected on the α-helical conformation, an alternating distribution of acidic and basic zones could be detected, which fit the repeating pattern of charged residues on the surface of the coiled-coil structure of the TM molecule.

4. TN-T-TN-C

Ebashi *et al.* (1972) first reported the interaction of TN-T and TN-C as a Ca-dependent solubilization phenomenon at low ionic strength. As reported earlier (Katayama, 1979), TN-T can bind to the packed column of immobilized TN-C at an ionic strength around 0.4. When the crude digests of TN-T were applied to immobilized TN-I, T-2 fragments as well as the undigested TN-T were retained by the column, while T-1 was excluded (Katayama, 1979). The T-2 fragments bound were eluted by the application of 5 M urea (Katayama, 1979) or 2 mM EGTA (Ohtsuki, 1979) in the starting buffer. These results on the properties of C-terminal fragments of TN-T essentially agreed with those reported by Pearlstone and Smillie (1978).

5. FA-TN-T

F-actin precipitates from the solution by the addition of TN-T or TN at low ionic strength, and the resultant precipitate dissolves on further addition of TM to the solution (Drabikowski and Nonomura, 1968; Fujii *et al.*, 1973). It is reasonable to conclude from this effect that TN-T is able to bind directly to FA, although the binding has been regarded as equivocally physiological. Examination of TN-T fragments indicated that only T-1 fragment had the affinity for FA. As described in section 3, the structure of T-1 somewhat resembles that of TM with repeating pattern of charged residues. It is presumed that T-1 may bind to FA in a similar manner to that of TM.

6. FA-TN-I

The direct interaction of TN-I and FA without TM is reported in most of the binding studies (Potter and Gergely, 1974; Hitchcock, 1975). F-actin could bind well also to immobilized TN-I. While the stoichiometric amount of TN-I can inhibit the acto-myosin ATPase only in the presence of TM, it was observed that additional TN-I added in such a condition also bound to FA (Hitchcock, 1975). Judging from this fact, it cannot be said that all of the direct binding of TN-I and FA is site-specific.

7. *TM-TN-I*

Since the inhibition of acto-myosin ATPase or superprecipitation by TN-I is far more significant in the presence of TM than in its absence, it was naturally expected that TN-I might bind to TM directly. However, there had been no clear observation supporting this binding. As mentioned in the introductory section, the difficulty of this experiment seemes to be due to the insolubility of TN-I* at low ionic strength. Dissolution of TN-I adding salts would decrease the interaction of TM and TN-I. Affinity chromatography using immobilized TN-I indicated the fairly strong binding of TM, at least up to the ionic strength of 0.25 M.

8. *TN-T-TN-I*

No clear indication for the affinity between TN-I and TN-T had been noted in previous reports (Potter and Gergely, 1974; Hitchcock, 1975). As a result, the interaction of the two components was until recently presumed to be negligible. The reason the affinity of TN-T for TN-I was missed might be the aggregating tendency of both of the components in low-salt solutions. I could avoid this difficulty and show the strong interaction between them by the use of immobilization of TN-I and solubilization of TN-T at higher ionic strength (Katayama, 1979). The interaction of TN-T and TN-I was shown independently by Horwitz *et al.* (1979) by the use of CD spectroscopy and gel filtration at high ionic strength. Hincke *et al.* (1979) showed a similar interaction using proteins from cardiac muscle.

In addition, examination of the chymotryptic fragments of TN-T revealed that only T-2 had a specific affinity for immobilized TN-I. The binding of T-2 was strong and stable even at high ionic strength.

Table I is a summary of the results of interaction studies obtained mainly by affinity chromatography.

* Ebashi claimed that this insolubility was due to inappropriate preparation and that TN-I of good quality was fairly soluble even at low ionic strength.

TABLE I

Summary of Interaction Studies

+ and − indicate the presence and absence of affinities in each pair of components. Numbers indicate the sections in the text in which each interaction is mentioned.

		FA	TM	TN-T: T-1	TN-T: T-2	TN-I	TN-C
FA			++ 1)	+ 5)	− 5)	+ 6)	—
TM				++ 3)	+ 3)	+ 7)	—
TN-T	T-1					− 8)	− 4)
	T-2					++ 8)	++ 4)
TN-I							++ 2)
TN-C							

As shown in Fig. 1, prediction of the secondary structure from the primary sequence (Chou and Fasman, 1974) of TN-T gave many long α-helical segments especially in region from the N-terminal to the 143rd residue. A lot of proline located between the 16th and the 51st residues would arrest the helix elongation and might bend the peptide chain in this region. The predicted absence of an α-helix or β-structure in the segment from the 144th to the 155th residue, even if overestimated, probably indicates the existence of a loop region accessible to water. This region contains two enzymatically phosphorylatable residues, the reactivity of which was extensively investigated by Moir *et al.* (1977). Further, the peptide bond tentatively determined as the cleavage site of α-chymotrypsin is also included in this region, further supporting the prediction that this region is a loop.

Taking all the interactions newly shown as well as those already well established, and considering the secondary structure all together, a model was developed of the structure and function of regulatory complex.

The model is illustrated in Fig. 2. Troponin complex sits on the coiled-coil of the TM molecule. The N-terminal domain of TN-T, which is abnormally rich in proline, would be bent at a similar region to that found in the hinge region of the immuno-globulin molecule (Dayhoff, 1972). It links with the TM binding, or possibly the FA binding domain T-1, a part of which proved to be very helical by CD measurement. Connected

Fig. 2. Model for the structure of regulatory complex. See text.

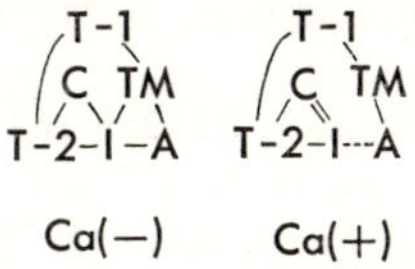

Fig. 3. Schematic presentation of TN action in regulatory complex.

with this helical region through the flexible segment or loop, the C-terminal domain of one third the length of the whole TN-T molecule holds both TN-I and TN-C, which were assumed to be globular and nearly of equal size.

Figure 3 is a scheme for TN from the functional point of view. Tropomyosin and TN-I always attach to T-1 and T-2, respectively. These bindings do not depend on the Ca^{2+} concentration. On the other hand, the binding of TN-I to other components strongly depends on Ca^{2+} concentration. When Ca^{2+} is absent, the affinity of TN-C for TN-I is weak and TN-C allows TN-I to bind to FA and TM simultaneously. In other words, TN-I cross-links FA and TM. Since the binding of TN complex to the FA-TM complex involves three sites of attachment in this "off" state, it is so tight that the interaction of FA and the active complex of myosin heads may well be blocked. When Ca^{2+} concentration is increased, Ca-bound TN-C pulls TN-I strongly away from FA and TM. Then the affinity of TN complex for thin filament decreases and the filament is switched "on"—*i.e.*, the myosin heads can react freely with FA.

Acknowledgment

I am indebted to Professor S. Ebashi, Drs. Y. Nonomura, I. Ohtsuki, and T. Wakabayashi for their interest and critical discussions.

REFERENCES

Chou, P. Y. and Fasman, G. D. (1974). *Biochemistry* **13**, 222–245.

Dayhoff, M. O. (ed.) (1972). *Atlas of Protein Sequence and Structure*, Nat. Biomed. Res. Found, Washington D.C.

Drabikowski, W. and Nonomura, Y. (1968). *Biochim. Biophys. Acta* **160**, 129–131.

Ebashi, S. (1974). In *Essays in Biochemistry* (Campbell, P.N. and Dickens, F., eds.), vol. 10, pp. 1–35, Academic Press, New York.

Ebashi, S., Ohtsuki, I., and Mihashi, K. (1972). *Cold Spring Harbor Symp. Quant. Biol.* **37**, 215–223.

Fujii, T., Kawamura, M., Yamamoto, K., and Maruyama K. (1973). *J. Biochem.* **73**, 269–280.

Hincke, M. T., McCubbin, W. D., and Kay, C. M. (1979). *Can. J. Biochem.* **57**, 768–775.

Hitchcock, S. E. (1975). *Eur. J. Biochem.* **52** 255–263.

Horwitz, J., Bullard, B., and Mercola, D. A. (1979). *J. Biol. Chem.* **254**, 350–355.

Jackson, P., Amphlett, G. W., and Perry, S. V. (1975). *Biochem. J.* **151**, 85–97.

Katayama, E. (1979). *J. Biochem.* **85**, 1379–1381.

Katayama, E. and Nonomura, Y. (1979). *J. Biochem.* **86**, 1495–1509.

McLachlan, A. D. and Stewart, M. (1976). *J. Mol. Biol.* **106**, 1017–1022.

Moir, A.J.G., Cole, A., and Perry, S. V. (1977). *Biochem. J.* **161**, 371–382.

Ohtsuki, I. (1979). *J. Biochem.* **86**, 491–497.

Pearlstone, J. R., Carpenter, M. R., Johnson, P., and Smillie, L. B. (1976). *Proc. Natl. Acad. Sci. U.S.* **73**, 1902–1906.

Pearlstone, J. R. and Smillie, L. B. (1978). *Can. J. Biochem.* **56**, 521–527.

Perry, S. V., Cole, H. A., Head, J. F., and Wilson, F. J. (1972). *Cold Spring Harbor Symp. Quant. Biol.* **37**, 251–262.

Potter, J. D. and Gergely, J. (1974). *Biochemistry* **13**, 2697–2703.

Sodek, J., Hodges, R. S., and Smillie, L. B. (1978). *J. Biol. Chem.* **253**, 1129–1139.

Stewart, M. (1975). *Proc. Roy. Soc. Lond. B* **190**, 257–266.

Stone, D. and Smillie, L. B. (1978). *J. Biol. Chem.* **253**, 1137–1148.

Syska, H., Wilkinson, J. M., Grand, J. A., and Perry, S. V. (1976). *Biochem. J.* **153**, 375–387.

Ueno, H. (1978). *J. Biochem.* **84**, 1559–1565.

Weeks, R. and Perry, S. V. (1978). *Biochem. J.* **173**, 449–457.

Muscle Contraction: Its Regulatory Mechanisms (Ebashi, S. et al., eds.), pp. 259–265, Japan Sci. Soc. Press, Tokyo / Springer-Verlag, Berlin (1980)

20

A Study on Troponin-T and Troponin-C Interaction by Cross-linking Reagents

OSAMU OHARA, SHO TAKAHASHI, AND TATSUO OOI

Summary: Cross-linking reactions using adipimidate were examined for solutions of troponin-T (TN-T) and troponin-C (TN-C) under various conditions. A cross-linked complex was obtained only at 0.4 M NaCl in the presence of Ca^{2+}, while no complex was formed at a high salt concentration and in the absence of Ca^{2+}, suggesting that a part of the interaction between TN-T and TN-C is electrostatic. After isolation of the cross-linked complex by sulfo-propyl (SP)-Sephadex C-25 column chromatography and by Sephacryl S-200 gel filtration, characterization of the complex and identification of the cross-linking sites were performed. The complex has a property similar to TN-C rather than TN-T, and circular dichroic spectrum of the complex does not change on the addition of Ca^{2+}. From SDS gel-electrophoresis of reaction mixtures after chemical cleavages of the complex by BNPB-skatole (2-(2-nitrophenylsulfenyl)-3-methyl-3-bromoindolenine) at two tryptophan residues in TN-T and by TNB-CN (2-nitro-5-thiocyano-benzoic acid) at a cystein residue in TN-C, the possible interaction site of TN-T is deduced as the region from 206 to 258, and the sites of TN-C as both N- and C-terminal sides.

Troponin isolated from rabbit skeletal muscle (Ebashi *et al.*, 1967) is composed of three subunits, troponin-T (TN-T), troponin-C (TN-C), and troponin-I (TN-I), the primary structures of which have been known (Pearlstone *et al.*, 1977; Collins, 1974; Wilkinson and Grand, 1975). Understanding of the molecular mechanism on regulation of muscle contraction may become possible through the identification of interaction sites for binding of tropomyosin, the determination of specific interaction sites on the troponin subunits in order to arrange them at proper positions in space, the conformational changes induced on binding of Ca^{2+} to TN-C, *etc.* Several studies along this line reported so far have been conducted using physico-chemical techniques to detect the changes in conformation (Ebashi *et al.*, 1972; van Eerd and Kawasaki, 1973; Potter and Gergely, 1974; Mani *et al.*, 1974), and using fragments obtained by limited proteolysis or chemical cleavage at specific chemical bonds to identify the interaction sites (Leavis *et al.*, 1977; Pearlstone and Smillie, 1977, 1978; Ohtsuki, 1979).

One promising technique for that purpose is the use of bifunctional reagents which would cross-link two subunits to give a protein complex fixed covalently. Since the reagents have definite molecular lengths of approximately 10 Å, it is feasible to identify interaction sites on the primary structures and the possible arrangement of subunits. This kind of information will be useful for the elucidation of the structure and function of a molecular assembly in space.

CROSS-LINKING OF TN-T, TN-C COMPLEX

For the systematic study of cross-linking reaction among three or more subunits, examinations are required of reactions occurring within and between the same subunits, between two different subunits, and so on. According to the preliminary experiments, cross-linking reactions of TN-T (or TN-I) give rise to multiple polymers of TN-T (or TN-I) molecules, indicating that TN-T (or TN-I) molecules in solution associate to form some aggregates. However, when TN-T and TN-C coexist in solution, the situation of cross-linking reaction is quite different, as described below.

In contrast to the low solubility of TN-T below a salt concentration of 0.3 M NaCl, a mixture of TN-T and TN-C is soluble over a wide range of salt concentrations. Figure 1 shows SDS gel-electrophoretic patterns of products obtained by the reaction of an equimolar mixture of TN-T and TN-C with adipimidate at 0.4 M and 1 M NaCl in the presence of 0.1 mM $CaCl_2$, and at 0.4 M NaCl in the absence of Ca^{2+}. The reactions at 1 M

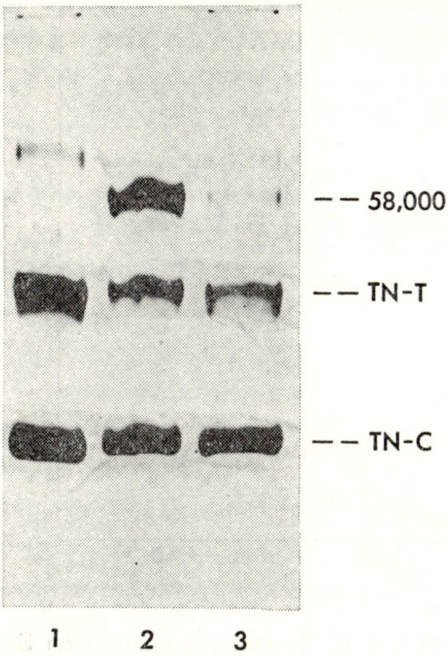

Fig. 1. SDS gel-electrophoretic patterns of the reaction mixtures of TN-T and TN-C treated with adipimidate at room temperature for 1 hr, in 50 mM triethanolamine-HCl buffer (pH 8.5). 1, 1 M NaCl, 0.1 mM $CaCl_2$; 2, 0.4 M NaCl, 0.1 mM $CaCl_2$; 3, 0.4 M NaCl, 1 mM EGTA.

NaCl in the presence and in the absence of Ca^{2+} do not produce a new band in the patterns, and only for the reaction at 0.4 M NaCl in the presence of Ca^{2+}, does a cross-linked complex of TN-T and TN-C appear as a new band which migrates at a rate corresponding to a molecular weight of 58,000. The result indicates that the interaction between TN-T and TN-C is weakened not only by the decrease in the Ca^{2+} concentration but also by the increase in the salt concentration. Therefore, the reaction condition for the later experiments was employed as 0.1 M NaCl in the presence of Ca^{2+}.

SEPARATION AND PURIFICATION OF CROSS-LINKED TN-T, TN-C COMPLEX

The reaction mixture of TN-T and TN-C with adipimidate contains unreacted or intramolecularly cross-linked materials in addition to the intermolecularly cross-linked complex (TXC). Therefore, the separation of TXC was performed in the following way: first, the reaction mixture was

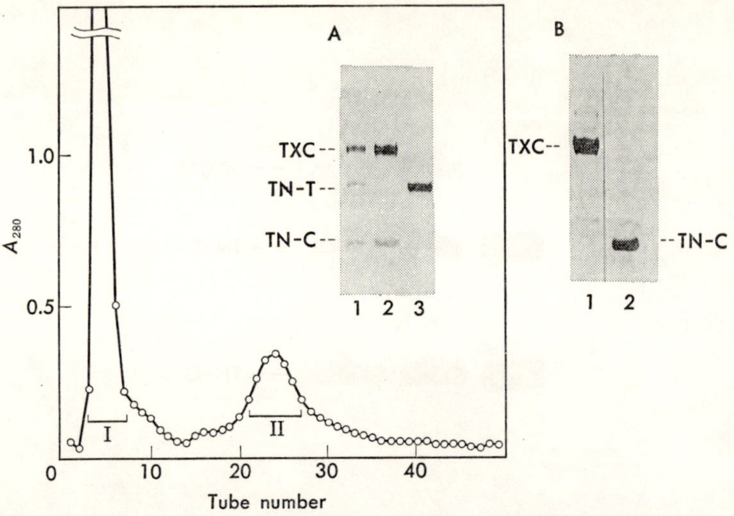

Fig. 2. A: the elution pattern of the reaction mixture on SP-Sephadex C-25 column in 6 M urea-40 mM sodium phosphate buffer (pH 6.0)-1 mM DTT, eluted with a gradient to 0.8 M NaCl. Inset represents the SDS gel-electrophoretic patterns of the fractions, I and II. B: the SDS gel-electrophoretic patterns of fractions obtained by Sephacryl S-200 gel filtration of fraction I to separate TXC and TN-C (in 1 M KCl-20 mM potassium phosphate buffer (pH 7.0)-1 mM DTT). The patterns of the fractions at peaks in the order of the elution are shown. A: 1, the cross-linked mixture of TN-T and TN-C; 2, fraction I; 3, fraction II (TN-T). B: 1, TXC; 2, recovered TN-C.

separated into TN-T and a mixture of TXC and TN-C by chromatography on SP-Sephadex C-25 in the presence of 6 M urea as shown in Fig. 2A. TXC and TN-C were eluted at the front of the elution, and TN-T appeared in retarded fractions, indicating that TXC has a property similar to TN-C rather than to TN-T. Then, Sephacryl S-200 gel filtration of the mixture of TXC and TN-C in 1 M KCl yielded the separation of the two components. SDS gel-electrophoretic patters of the purified fractions are shown in Fig. 2B.

Amino acid analyses on TXC, TN-T fraction, and TN-C fraction were performed to confirm the purity of TXC. The amino acid composition of TXC is coincident with the sum of those of TN-T and TN-C except for lysine residues. About ten lysine residues were converted to three N,N′-adipamidyl-bis-lysine residues (the reagent reacted with two lysines) and two N-adipamidyl-lysine residues (the reagent reacted with one lysine) by the reaction. For TN-T fraction, five to six lysine residues were reacted and 2.5 mol of N,N′-adipamidyl-bis-lysines were found, whereas TN-C fraction showed no loss in lysine residues. Therefore, three cross-linkages introduced in TXC may contain some intramolecular cross-linkages in TN-T portion.

CHARACTERIZATION OF THE COMPLEX

Now that the cross-linked complex of TN-T and TN-C is purified, several features of the complex may be examined under various conditions where an intact complex cannot be formed. TN-T does not interact with TN-C in the absence of Ca^{2+}, while TXC migrates in gel-electrophoresis in the absence of Ca^{2+} at the same rate as in the presence of Ca^{2+}, and the rate is identical with that of the intact complex formed in the presence of Ca^{2+}. Since TN-T is a basic protein and TN-C is an acidic protein, TXC has a property similar to TN-C.

Circular dichroic spectra of TXC with and without Ca^{2+} were the same. Thus, a conformational change deduced from the decrease in the magnitude of ellipticity on the addition of Ca^{2+} observed on the intact complex (Mani *et al.*, 1974) is not expected for TXC, presumably due either to some fixation of the conformation by the cross-linkage introduced or to a loss in Ca^{2+} binding capacity of TN-C molecule. In view of the cross-linking experiments on troponin by suberimidate (Hitchcock, 1975), which revealed the loss in regulational function in spite of the retension of Ca^{2+} binding capacity of TN-C, the cross-linkage would restrict the movement of the polypeptide chain.

We also examined the binding capacity of TXC to tropomyosin using an affinity column of tropomyosin. When TXC solution was applied to tropomyosin-attached Sepharose 4B column, the TXC was eluted gradually from the front, showing that the binding capacity of TXC to tropomyosin is not so high.

Therefore, the introduction of cross-linkages to TN-T and TN-C seems to fix the conformation of the complex.

IDENTIFICATION OF CROSS-LINKING SITES

For identification of cross-linking sites, we employed chemical cleavage by BNPS-skatole (2-(2-nitrophenylsulfenyl)-3-methyl-3-bromoindolenine) at two tryptophan residues (Pearlstone *et al.*, 1977) which are located at the 205th and 258th positions of TN-T, and by TNB-CN (2-nitro-5-

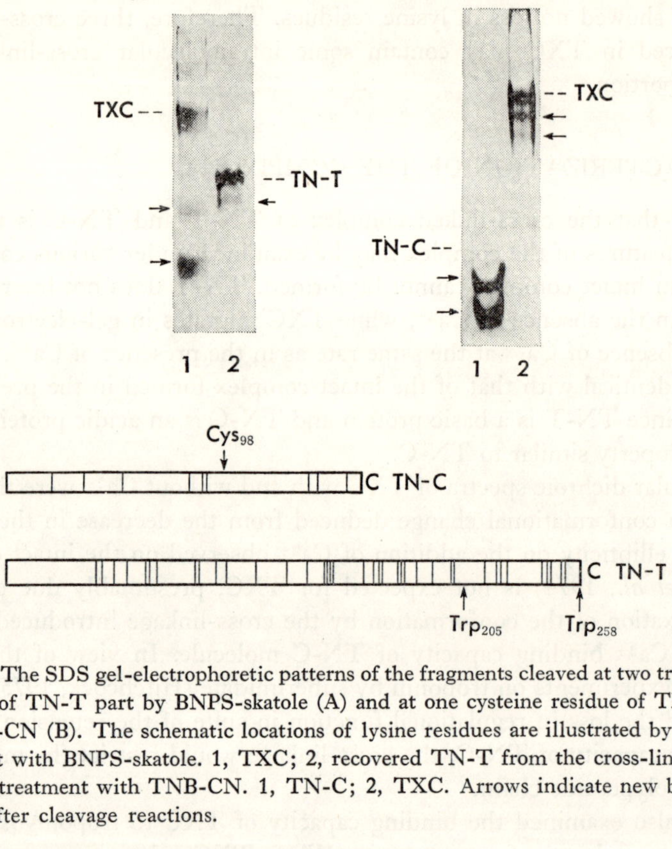

Fig. 3. The SDS gel-electrophoretic patterns of the fragments cleaved at two tryptophan residues of TN-T part by BNPS-skatole (A) and at one cysteine residue of TN-C part by TNB-CN (B). The schematic locations of lysine residues are illustrated by bars. A: treatment with BNPS-skatole. 1, TXC; 2, recovered TN-T from the cross-linked mixture. B: treatment with TNB-CN. 1, TN-C; 2, TXC. Arrows indicate new bands appeared after cleavage reactions.

thiocyano-benzoic acid) at a cysteine residue (Jacobson *et al.*, 1973) which is at the 98th position of TN-C. Figure 3 shows the results of SDS gel-electrophoresis of the reaction mixtures after the chemical cleavages of TXC. The appearance of new bands together with the comparison of bands corresponding to fragments of individual proteins by control experiments makes it possible to identify the location of the cross-linking sites.

The cleavage at tryptophan residues gave rise to two major fragments corresponding to molecular weights of 30,000 and 24,000. The fragment of 30,000 is the same fragment obtained by cleavage of TN-T, and the fragment of 24,000 is a new one corresponding to the TN-C+C-terminal part of TN-T. Therefore, the cross-linking site in TN-T is deduced as the C-terminal side from 206 to 258. As for the sites in TN-C, new bands of two fragments, one denser than the other, appeared after the cleavage at the cysteine residue. Since the apparent molecular weights of those fragments are 52,000 and 47,000, respectively, the sites of TN-C may be located at two positions, one on the N-terminal side (major) and the other on the C-terminal side (minor). These assignments are partly in coincidence with those obtained by the use of fragments (Pearlstone and Smillie, 1978; Leavis *et al.*, 1977).

In order to determine more detailed sites, isolation and identification of fragments containing cross-links are necessary, and work along this line is now in progress.

REFERENCES

Collins, J. H. (1974). *Biochem. Biophys. Res. Commun.* **58**, 301–308.

Ebashi, S., Ebashi, F., and Kodama, A. (1967). *J. Biochem.* **62**, 137–138.

Ebashi, S., Otsuki, I., and Mihashi, K. (1972). *Cold Spring Habor Symp. Quant. Biol.* **37**, 215–223.

Hitchcock, S. E. (1975). *Biochemistry* **14**, 5162–5267.

Jacobson G. R., Schaeffer, M. H., Stark, G. R., and Vanaman, T. C. (1973). *J. Biol. Chem.* **248**, 6583–6591.

Leavis, P. C., Drabikowski, W., Rosenfel, S., Grabarek, Z., and Gergely, J. (1977). In *Calcium Binding Protenis and Calcium Function* (Wasserman, R. H. *et al.*, eds.), pp. 281–283, Elsevier/North-Holland, Amsterdam.

Mani, R. S., McCubbin, W. E., and Kay, C. M. (1974). *Biochemistry* **13**, 5003–5007.

Ohtsuki, I. (1979). *J. Biochem.* **86**, 491–497.

Pearlstone, J. R., Johnson, P., Carpenter, M. R., and Smillie, L. B. (1977). *J. Biol. Chem.* **252**, 983–989.

Pearlstone, J. R. and Smillie, L. B. (1977). *Can. J. Biochem.* **55**, 1032–1038.

Pearlstone, J. R. and Smillie, L. B. (1978). *Can. J. Biochem.* **56**, 521–527.

Potter, J. D. and Gergely, J. (1974). *Biochemistry* **13**, 2697–2703.

Wilkinson, J. H. and Grand, R.J.A. (1975). *Biochem. J.* **149**, 493–496.

Muscle Contraction: Its Regulatory Mechanisms (Ebashi, S. et al., eds.), pp. 267–273,
Japan Sci. Soc. Press, Tokyo / Springer-Verlag, Berlin (1980)

21

Thermodynamics of the Interaction of Calcium with the Regulatory Protein System Relating to the Heat Production of Muscle

KAZUHIRO YAMADA

Summary: The heats produced in the interaction of troponin (TN) with Ca have been shown to coincide approximately with the activation heat production of muscle. When Ca is bound to TN at the neutral pH, no proton is released, but a dissociation of aggregated TN molecules occurs. The heat of the interaction of the TN-C-TN-I complex, the reconstituted thin filament, and of the interaction of TN-C with Ca are all reduced as compared with that of TN. This might indicate a positive enthalpy change for the aggregation of TN molecules. The binding of Ca to TN-C can be characterized as having a large positive entropy change except for the binding of the first 1 mol of Ca per mol of TN-C. The binding of this first 1 mol of Ca has a slightly positive entropy change and, therefore, causes TN-C to take a strongly ordered structure distinct from that associated with the further binding of Ca.

Owing to recent advances in the application of microcalorimetry to the reactions involving muscle proteins, the energetic aspects of these reactions underlying contraction can now be expected to contribute more directly to elucidating the mechanism of contraction rather than merely to formulating an equation for the total energy production.

The present discussion is on these energetic aspects of the "activation" processes, which may include Ca release and Ca-binding reactions, the interaction between proteins, and Ca transport with the ATP usage. In particular, however, the present study is confined to the interaction of Ca with troponin (TN), with the complex of TN-C and TN-I, with TN-C, and with the reconstituted thin filament.

HEAT OF INTERACTION OF TROPONIN WITH CA

The enthalpy titration method has been applied to study the heat of the binding of Ca to TN (Yamada *et al.*, 1976). The results are plotted in a different way in Fig. 1. Although there is uncertainty caused by side reactions, as will be described below, it has been shown that heat changes associated with Ca binding to TN approximately coincide with the activation heat production of muscle (Yamada *et al.*, 1976). The same study also showed that at neutral pH, proton exchanges of TN on Ca binding are

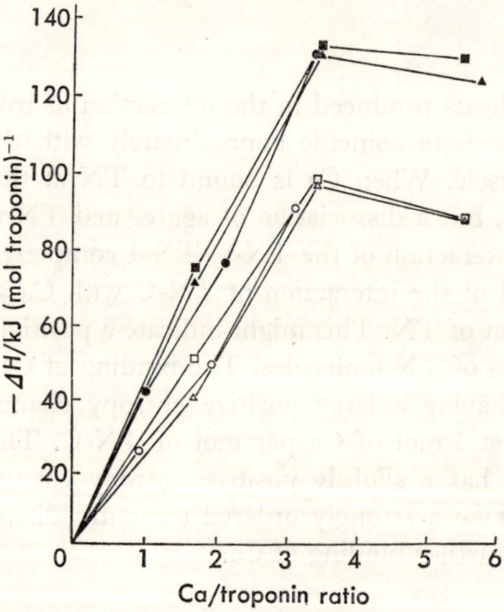

Fig. 1. Heat produced per mol of TN by Ca-TN interaction in 40 mM carnosine (●, ○), in 20 mM hepes (■, □) and in 20 mM pipes (▲, △), in 0.1 M KCl, at pH 7, 10°C. Closed symbols show measurements in 1–1.25 mM Mg and open ones in Mg-free solutions.

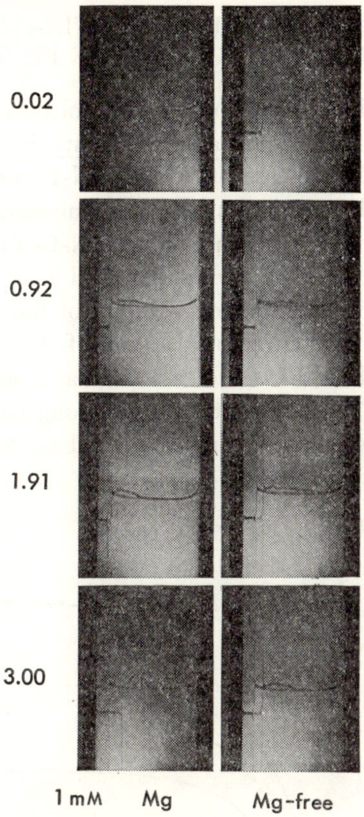

Fig. 2. Ultracentrifugation of TN after 26 min at a rotor speed of 60,000 rpm, in 1 mM Mg and Mg-free solution at pH 7 and 10°C. Figures on the left show Ca/TN ratio.

negligibly small. It should be noted, however, that in the presence of 1 mM Mg, a small number of protons are taken up by TN on Ca binding. This can be detected as an alkalinization of the solution, and might reflect structural changes of the molecule (Yamada *et al.*, 1976). Another point which causes difficulty in interpreting the results is that TN molecules aggregate when Ca is deprived (Wakabayashi and Ebashi, 1968; Yamada *et al.*, 1976). This is shown by the sedimentation diagram studied in parallel with calorimetry in Fig. 2. The dissociation of TN molecules on Ca binding may produce or absorb heat. This is an example of an "*in vitro* effect," which cannot occur in muscle, and there is no way we can measure the ΔH of this effect separately.

HEAT OF INTERACTION OF TROPONIN-C-TROPONIN-I COMPLEXA ND THE RECONSTITUTED THIN FILAMENT WITH Ca

As noted above, the aggregation of TN molecules, the extent of which is controlled by Ca, introduces an uncertainty in evaluating the ΔH of TN-Ca interaction. In the case of TN-C-TN-I complex, the aggregation of molecules on Ca deprivation is reduced (Yamada, unpublished results). Figure 3 shows the heat of interaction of TN-C-TN-I complex with Ca, which is reduced as compared with that of TN in Fig. 1. These results might indicate that the dissociation reaction of the aggregated TN molecules produces heat. The enthalpy titration of the reconstituted thin filament also indicates that this is the case (Yamada, unpublished results). It is probable, however, that the interaction among components of TN and among TN, tropomyosin and actin also contribute to the heat observed.

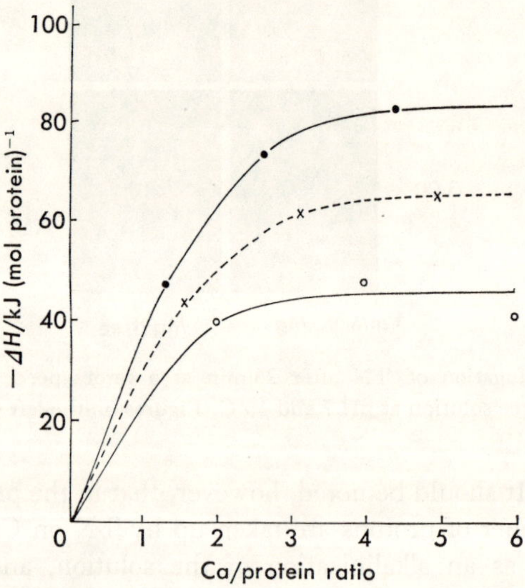

Fig. 3. Heat produced per mol of TN-C-TN-I complex and TN-C by Ca binding at pH 8.5 (16.6 mM glycylglycine) at 10°C. ● TN-C-TN-I complex in 2 mM Mg; ○ TN-C-TN-I complex in Mg-free solution; × TN-C in Mg-free solution.

HEAT OF INTERACTION OF TROPONIN-C WITH Ca

As is also seen in Fig. 3, the heat of interaction of TN-C, the Ca-receptive component of TN, with Ca is smaller than that of TN. This is particularly of interest because TN-C remains as single molecules irrespective of the level of Ca bound (Murray and Kay, 1972; Yamada, 1978). The enthalpy titration of TN-C with Ca has already been reported (Yamada, 1978). Potter *et al.* (1977) have also studied this but with different results. More recently, a calorimetric titration method was used to study ADP binding to myosin (Kodama and Woledge, 1976). We have applied this method to the enthalpy titration of TN-C with Ca, and in particular have studied the effect of temperature and the effect of Mg concentration on it (K. Yamada and K. Kometani, in preparation). Together with the high sensitivity of the LKB batch-mirocalorimeter used, the calorimetric titration method has enabled us to obtain more reliable results of higher resolution than have been reported earlier (Yamada, 1978). Figure 4 shows typical titration curves obtained by this method. The shape of the titration curves is unique in that heat is produced rapidly by adding Ca up to 1 mol per mol of TN-C; then the rate of heat production declines to a very low level until 2 mol Ca are reached, and rises again forming a plateau just beyond 4 mol Ca. As Ca association constants of TN-C are large, we can approximately ascribe these enthalpy titration curves to enthalpy changes associated with each of 4 Ca-binding sites of TN-C. The results are shown in Table I.

It should be noted in the table that the values of ΔS are positive and large (see Yamada, 1978), except for that for the binding of the first 1 mol

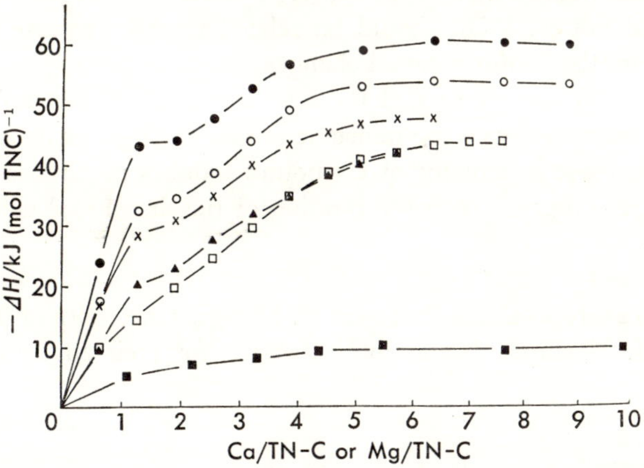

Fig. 4. Heat produced per mol of TN-C at pH 7 (13.3 mM pipes-NaOH) at 15°C. ● Ca binding in Mg-free solution; ○ Ca binding in 0.1 mM Mg; × Ca binding in 0.2 mM Mg; ▲ Ca binding in 1 mM Mg; □ Ca binding in 5 mM Mg; ■ Mg binding.

TABLE I
Thermodynamic Results of the Enthalpy Titration
The values for log K are from Potter and Gergely, 1975.

	Site	log K	$-\Delta H$/kJ mol^{-1}	$-\Delta G$/kJ mol^{-1}	ΔS/J mol^{-1} K^{-1}
Mg-free	1	7.32	38.0	40.37	8.23
	2	<7.32	16.0	<40.37	83.9
	3, 4	5.5	9.6	30.35	72.0
1 mM Mg	1	6.45	16.2	35.55	67.2
	2	6.45	10.7	35.55	86.3
	3, 4	5.04	11.1	27.8	57.9

of Ca per mol TN-C in Mg-free solution. In the case of divalent cation binding to biopolymers, the dominant driving force for the interaction is a positive entropy change, and this entropy increase arises from the release of water molecules from the hydrated cation (Krakauer, 1971). The binding of Ca to TN-C mostly shares this feature except for the binding to the first site. It is reasonable to assume, therefore, that the binding of Ca to the first site where no Mg is present also results in an increase in entropy. Thus only a slight increase in entropy associated with the binding of the first 1 mol of Ca per mol TN-C in Mg-free solution would be the sum of the increase of entropy associated with the binding and the decrease resulting from conformational changes. The Ca-induced changes in the circular dichroism and fluorescence of TN-C (*e.g.*, van Eerd and Kawasaki, 1972; Johnson and Potter, 1978) should be related to this decrease in entropy resulting from the conformational changes.

In short, only the binding of the first 1 mol of Ca per mol of TN-C in Mg-free solution causes the molecule to take strongly ordered structure. A further increase in amount of Ca bound induces much less conformational change compared with the binding of the first 1 mol of Ca.

Acknowledgment

It is a pleasure to thank Professor S. Ebashi, University of Tokyo, and Professor H. Mashima, Juntendo University, for their advice throughout this work.

REFERENCES

Johnson J. D. and Potter, J. D. (1978). *J. Biol. Chem.* **253**, 3775–3777.
Kodama, T. and Woledge, R. C. (1976). *J. Biol. Chem.* **251**, 7499–7503.

Krakauer, H. (1972). *Biopolymers* **11**, 811–828.
Murray, A. C. and Kay, C. M. (1972). *Biochemistry* **11**, 2622–2627.
Potter, J. D. and Gergely, J. (1975). *J. Biol. Chem.* **250**, 4628–4633.
Potter, J. D., Hsu, E. J., and Pownall, H.-J. (1977). *J. Biol. Chem.* **250**, 4628–4633.
van Eerd J.-P. and Kawasaki, Y. (1972). *Biochem. Biophys. Res. Commun.* **47**, 859–865.
Wakabayashi, T. and Ebashi, S. (1968). *J. Biochem.* **64**, 731–732.
Yamada, K., Mashima, H., and Ebashi, S. (1976). *Proc. Japan Acad.* **52**, 252–255.
Yamada, K. (1978). *Biochim. Biophys. Acta* **535**, 342–347.

Muscle Contraction: Its Regulatory Mechanisms (*Ebashi, S. et al., eds.*), *pp. 275–284, Japan Sci. Soc. Press, Tokyo / Springer-Verlag, Berlin (1980)*

22

Differentiation of Troponin in Developing Chicken Skeletal Muscle

TAKASHI OBINATA AND RYOICHI MATSUDA

Summary: The onset of troponin synthesis and the change in the type of troponin components during development of chicken skeletal muscles were investigated by immunological and electrophoretic methods. The three components of troponin are detectable in mononucleated myoblasts as well as in multinucleated myotubes in culture, but myogenic cells in mitotic state do not contain troponin. The troponin components are assembled into the thin filaments before or at the early phase of myofibrillogenesis. The electrophoretic patterns of troponin components change markedly during the development of chicken breast muscle; troponin-I and troponin-C do not change during development; leg-type troponin-T (TN-T) is present in the breast muscle of embryonic and young post-hatched stage, but it is drastically replaced with breast-type TN-T a week after hatching. The change in TN-T is correlated with that in troponin subunits during the development of breast muscle.

It is well known that skeletal muscle cells have a well-organized myofibrillar structure, and that contractile and regulatory proteins are assembled in this fine structure. Therefore, from the aspect of muscle cell differentiation, it is interesting to know when and how myofibrillar proteins are synthesized and polymerized into filamentous forms, and then packed into hexagonal, myofibrillar lattice during the course of muscle cell differentiation. There have been a number of studies on the differentiation of myofibrillar proteins during myogenesis (*cf.* Fischman, 1972). However, most of the previous works were focussed on the differentiation of contractile proteins, myosin and actin, while not many studies have been done on the differentiation of muscle regulatory proteins (Hitchcock, 1970; Llewellyn-Smith, 1978; Roy *et al.*, 1979; Obinata *et al.*, 1979). Many recent works have demonstrated that actin and myosin are widely distributed in various types of cells including myogenic and non-muscle cells: (Pollard and Weihing, 1972). In contrast, troponin has been discovered only in cross-striated muscle cells in the case of vertebrate animals (Ebashi and Endo, 1968), suggesting that troponin is a real muscle-specific protein. Therefore, troponin is considered to be a better marker for the investigation of myogenesis. We have investigated the differentiation of troponin during myogenesis. Since troponin is composed of three different components, troponin-T (TN-T), troponin-I (TN-I), and troponin-C (TN-C), and it has been shown that there are several varieties of troponin components in different muscle tissues (Wilkinson, 1978; Dhoot and Perry, 1979), we were interested in the following points:

1) At what stage of differentiation does troponin appear in myogenic cells?
2) Do the three components of troponin appear simultaneously?
3) When are troponin localized on thin filaments at the constant periodicity during myofibrillogenesis?

APPEARANCE OF TROPONIN COMPONENTS IN MYOGENIC CELLS *IN VITRO*

First of all, we investigated the synthesis of troponin components in myogenic cells in culture by using an immunological method. Antibodies against troponin components, TN-T, TN-I, and TN-C, were produced by injecting each purified component into rabbits with Freund's complete adjuvant. Immunoglobulin, IgG, was fractionated from the antisera by the standard method. The specificities of the antibodies in all cases were established by Ouchterlony's immunodiffusion test, by disc-immunoelectrophoresis, and by the demonstration of specific staining of isolated myofibrils (Obinata *et al.*, 1979). Suspension of embryonic skeletal muscle cells were obtained from breast muscle of 12-day old chick embryos by the standard procedure of dissociation with trypsin, and were cultured on plastic dishes as described by Shimada *et al.* (1967). After the cells were cultured for certain periods, they were fixed by 3.5% formaldehyde in

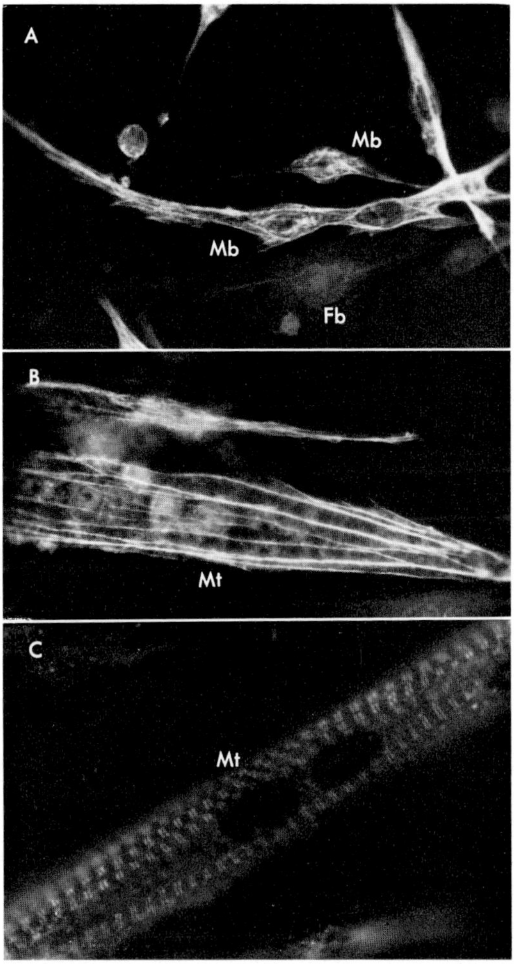

Fig. 1. Fluorescence micrographs of cultured muscle cells labeled with anti-TN-T and fluorescein isothiocyanate (FITC)-goat anti-rabbit IgG. Muscle was cultured from breast muscle of 12-day chick embryo. After 31 hr (A), 90 hr (B), or 6 days (C), cultures were fixed and stained with the antibodies. Mb, mononucleated myoblast; Mt, myotube; Fb, fibroblast.

phosphate-buffered saline (PBS) and were treated in an ascending ethanol series. Then, the cells were treated with antibodies against troponin components and subsequently with fluorescent-labeled goat antirabbit IgG (Lazarides, 1975). The cells, thus treated, were observed under a fluorescent microscope.

Figure 1 shows the immunofluorescent microscopic observations of cells treated with anti-TN-T antibody. At 31 hr culture *in vitro*, spindle-shaped mononucleated cells which are identified as presumptive myoblasts or myoblasts (Holtzer, 1970) are predominant in the dishes. Some of the myoblasts are attached to each other; these are designated as "attached myoblasts" in the following. Myotubes are already formed at this stage although they are relatively few in number. In addition, there are a small number of flat cells having trianglar or polygonal shape, which are regarded as fibroblasts. As seen in the figure, some of the spindle-shaped mononucleated cells were clearly stained by the treatment with anti-TN-T antibody, but others were not. Most of the "attached myoblasts" reacted with the antibody to give strong immunofluorescence. At 31 hr in culture, 50 to 60% of spidle-shaped mononucleated single cells and 80 to 90% of "attached myoblasts" were stained by the antibody. One hundred percent of the myotubes were stained with the antibody. The treatments with the antibodies against TN-I and TN-C gave essentially similar results as that with the anti-TN-T antibody. None of the fibroblasts which were coexistent in the culture reacted with the antibodies against any troponin components. In contrast, the anti-tropomyosin antibody stained fibroblast as well as myogenic cells, including myoblasts and myotubes, without exception.

The synthesis of troponin in myoblasts was definitely confirmed by observing the myoblasts which were arrested in a mononucleated state by adding EGTA to the culture medium. At about 20 hr of the EGTA-culture, only about 25% of spindle-shaped single cells, myoblast-like cells, contained troponin components, but the percent of the myoblasts stained by anti-troponin antibody increased with age. At about 100 hr in culture, almost all of them were stained by the antibodies against TN-T, TN-I, and TN-C as multinucleated myotubes. These results strongly indicate that the spindle-shaped mononucleated cells, myoblasts, have an ability to synthesize troponin before the onset of cell fusion, and the three components of troponin seem to be synthesized simultaneously during myogenesis.

Next, so as to answer the question of whether the cells in the proliferative stage, "presumptive myoblasts," synthesize troponin components, the following experiments were carried out. Cells in the S phase of the division cycle were marked by incorporation of ^{3}H-thymidine (0.5 μCi/ml) for 30 min at 31 hr of culture; then the cells were fixed and treated with the antibodies against troponin components for immunofluorescent microscopy. The cells with immunofluorescence were observed under a fluorescent microscope. Then the culture was treated with a autoradiographic emulsion. After processing for autoradiography, the cells which were observed previously under fluorescent microscopy were recovered under a microscope, and the incorporation of ^{3}H-thymidine into nuclei was examined. The mononucleated spindle-shaped cells which had nuclei labeled with ^{3}H-thymidine did not contain troponin components, while many of the cells having unlabeled nuclei were stained by the antibodies against troponin components. These results suggests that troponin-containing cells are post-mitotic. The cells in the proliferative stage, "presumptive myoblasts," might not synthesize troponin. Okazaki and Holtzer (1965) demonstrated that the synthesis of muscle myosin occurs in the post-mitotic myoblasts. The synthesis of troponin and muscle myosin seems to occur coordinately. Masaki and Yoshizaki (1972) previously demonstrated that various myofibrillar proteins appear simultaneously in the muscle tissue of developing chick embryo. The present result is consistent with their findings.

The intracellular structures stained with the antibodies against troponin components change with the developmental stages of myogenic cells, as seen in Fig. 1. In the mononucleated cells, myoblasts, or "attached myoblasts", the staining by the fluorescent antibody is associated with filamentous network structures (Fig. 1B); but in the case of well-developed myotubes, fluorescence appears to be associated with cross-striated myofibrillar structures (Fig. 1C). These observations suggest that troponin is localized in some filamentous structures, probably actin filaments, before or at an early phase of myofibrillogenesis; then the troponin-containing filaments are organized into myofibrillar structures. In order to demonstrate directly the location of troponin on the actin filaments in developing muscle cells, an immuno-electronmicroscopic technique was applied (Obinata *et al.*, 1979). In this investigation, the cultured cells were treated with 0.1% saponin in standard buffer for 20 min at room temperature (Ohtsuki *et al.*, 1978), and they were then incubated overnight at 4°C with anti-troponin antibody. The cells were fixed, embedded and sectioned for electronmicroscopy.

It was definitely demonstrated that troponin is located on actin filaments at a constant periodicity (about 38 nm) in the early phase of myofibril assembly as well as in the case of mature myofibril (Ohtsuki, 1967).

Troponin-containing thin filaments were isolated directly from young embryonic skeletal muscle. Troponin and tropomyosin on these filaments function to regulate the calcium-dependent interaction of the filaments with purified myosin (Obinata *et al.*, 1974).

CHANGES OF TROPONIN COMPONENTS DURING DEVELOPMENT

It has been shown that myofibrillar proteins exist in several different forms in vertebrate striated muscle. Polymorphic forms of troponin components have been demonstrated (see Dhoot and Perry, 1979). Perry and Cole (1974) and Wilkinson (1978) showed that chicken breast and leg muscles, mainly composed of fast skeletal muscle, contain different types

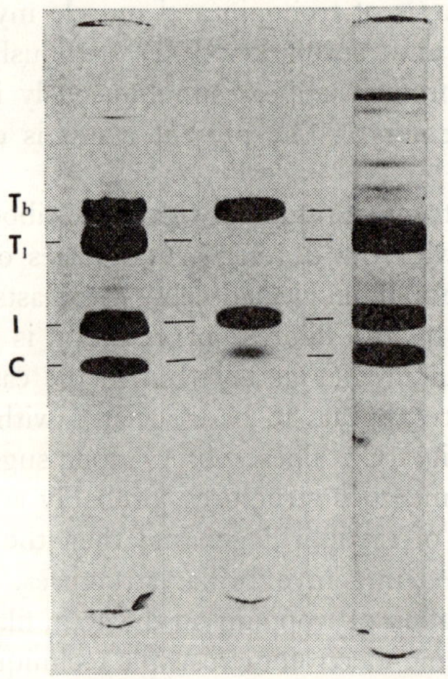

Fig. 2. SDS-acrylamide gel electrophoresis of troponin components from breast or leg muscle of young or adult chicken. 1d, 1-day postnatal chicken; ad, adult chicken; T_b, breast-type TN-T; T_l, leg-type TN-T; I, TN-I; C, TN-C.

of TN-T which are distinguishable by molecular weight. Developmental changes in myosin (Dow and Stracher, 1971; Masaki and Yoshizaki, 1974; Rubinstein *et al.*, 1977; Obinata *et al.*, 1980), actin (Whalen *et al.*, 1976), and tropomyosin (Roy *et al.*, 1979) in skeletal muscle have been pointed out. We, then, examined the changes in troponin components during muscle development. Troponin was extracted from the breast and leg muscle of chick embryo and postnatal chicken at various developmental stages, and was analyzed by electrophoretic and immunochemical methods. Troponin components of leg muscle do not change during development. In contrast, there is a marked difference in the patterns of troponin components in young and adult breast muscle. As seen in Fig. 2, during the development of breast muscle, the sizes of TN-I and TN-C stay constant, while two components (T_b and T_l in the figure) were present in the molecular weight range of TN-T. As judged by their electrophoretic mobilities, it seems that the faster migrating component (T_l) is identical with leg TN-T of the adult, and that T_b is identical with breast TN-T of the

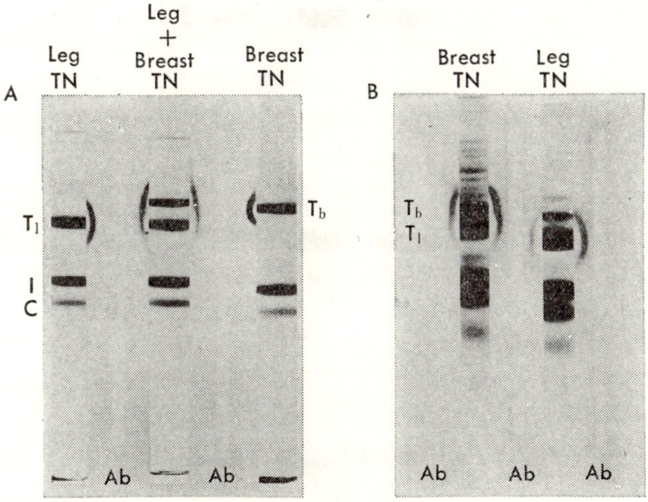

Fig. 3. Immunodiffusion combined with SDS-acrylamide gel electrophoresis of troponin from adult (A) or 2-day postnatal chicken (B) skeletal muscle. The immunoreactions of TN-T varieties with anti-breast TN-T antibody were tested. Electrophoretic acrylamide gels were embedded in agarose gel plate immediately after electrophoresis, and the antibody was applied in troughs parallel to the electrophoretic gels to cause immunoreaction. The electrophoretic patterns of each troponin were shown by inserting the photographs of electrophoretic gels stained by coomassie brilliant blue. Ab, anti-TN-T antibody. Other symbols are as in Fig. 2.

adult. Breast-type TN-T is immunologically partially identical with leg-type TN-T (Matsuda and Obinata, unpublished). Therefore, in the immunodiffusion test combined with SDS-acrylamide gel electrophoresis (Obinata *et al.*, 1980), the two types of TN-T formed immunoprecipitin lines against anti-breast TN-T antibody at the different positions corresponding to their electrophoretic mobilities (Fig. 3A). When troponin from breast muscle of young posthatched chickens was tested by immunodiffusion, two immunoprecipitin lines were formed at the positions where two types of TN-T-like components (T_b and T_l) were present. The pattern of dual immunoprecipitin lines with troponin from the breast muscle of young chicken was quite similar to that with the mixture of breast and leg TN-T of the adult chicken (compare Fig. 3A, middle, with Fig. 3B, left). There was a possibility that the smaller TN-T component (T_l) in the extract of the young breast muscle might be derived as a result of partial

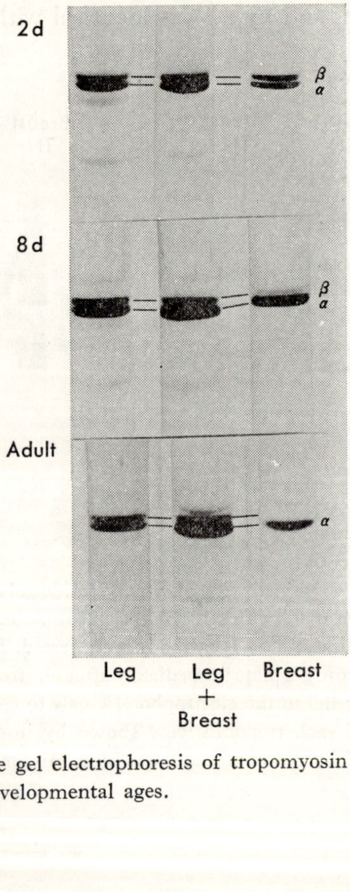

Fig. 4. SDS-acrylamide gel electrophoresis of tropomyosin from breast or leg muscle of chicken at various developmental ages.

proteolysis of the larger component (T_b) during the extraction procedure. However, this possibility was neglected because when young breast muscle was frozen by liquid nitrogen immediately after dissection out and then extracted with a hot SDS solution to minimize the degradation of troponin by proteolytic enzymes, T_l was detected as a major TN-T component in the extract by immunoelectrophoresis.

From these observations, it is concluded that leg-type TN-T is present as a major TN-T component in chicken breast muscle at the embryonic and young postnatal stage, and that then the leg-type TN-T is drastically replaced by the breast-type TN-T a week after hatching. Roy *et al.* (1979) have demonstrated that young chicken breast muscle contains α- and β-tropomyosin as adult leg muscle, while as development preceeds, the proportion of α-tropomyosin gradually increases so that in adult breast muscle α-tropomyosin is the only species present. In our observation, the change in the type of TN-T occurs at about same time as that in tropomyosin during development of breast muscle (Fig. 4). Since TN-T is a binding component of tropomyosin, the coordinate change in TN-T and tropomyosin seems to be reasonable. During the period when the changes in types of TN-T and tropomyosin occur, chicken breast muscle grows rapidly. What factor in the rapidly growing muscle controls the changes in TN-T and tropomyosin? This problem will be subjected to further investigation.

Acknowledgment

This study was supported by a grant from the Japanese Ministry of Education and by Grant No. 4, 1978, from the National Center for Nervous, Mental and Muscular Disorders (NCNMMD) of the Ministry of Health and Welfare, Japan.

REFERENCES

Dhoot, G. K. and Perry, S. V. (1979). *Nature* **278**, 714–718.

Dow, J. and Stracher, A. (1971). *Proc. Natl. Acad. Sci. U.S.* **68**, 1107–1110.

Ebashi, S. and Endo, M. (1968). *Prog. Biophys. Mol. Biol.* **18**, 123–183.

Fischman, D. A. (1972). In *Structure and Function of Muscle* (Bourne, G. H. ed.), pp. 75–148, Academic Press, New York.

Hitchcock, S. E. (1970). *Dev. Biol.* **23**, 399–423.

Holtzer, H. (1970). In *Cell Differentiation* (Schjeide, O. and Vellis, J. de, eds.), pp. 476–503, Van Nostrand-Reinhold, Princeton, New Jersey.

Lazarides, E. (1975). *J. Cell Biol.* **65**, 549–561.

Llewellyn-Smith, I. J. (1978). *Dev. Biol.* **67**, 40–53.
Masaki, T. and Yoshizaki, C. (1972). *J. Biochem.* **71**, 755–757.
Masaki, T. and Yoshizaki, C. (1974). *J. Biochem.* **76**, 123–131.
Matsuda, R. and Obinata, T. (in preparation).
Obinata, T., Hayashi, I., and Fischman, D. A. (1974). *Dev. Growth Diff.* **16**, 105–121.
Obinata, T., Shimada, Y., and Matsuda, R. (1979). *J. Cell Biol.* **81**, 59–66.
Obinata, T., Masaki, T., and Takano, H. (1980). *J. Biochem.* **87**, 81–88.
Ohtsuki, I. (1967). *J. Biochem.* **61**, 817–819.
Ohtsuki, I., Manzi, R. M., Palade, G. E., and Jamieson, J. D. (1978). *J. Microsc. Biol. Cell.* **31**, 119–126.
Okazaki, K. and Holtzer, H. (1965). *J. Histochem. Cytochem.* **13**, 726–739.
Perry, S. V. and Cole, H. A. (1974). *Biochem. J.* **141**, 733–743.
Pollard, T. D. and Weihing, R. R. (1974). *CRC Crit. Rev. Biochem.* **2**, 1–65.
Roy, R. K., Sreter, F. A., and Sarkar, S. (1979). *Dev. Biol.* **69**, 15–30.
Rubinstein, N. A., Pepe, F. A., and Holtzer, H. (1977). *Proc. Natl. Acad. Sci. U.S.* **74**, 4524–4527.
Shimada, Y., Fischman, D. A., and Moscona, A. A. (1967) *J. Cell Biol.* **35**, 445–453.
Whalen, R. G., Butler-Browne, G. S., and Gros, F. (1976). *Proc. Natl. Acad. Sci. U.S.* **73**, 2018–2022.

REGULATORY MECHANISMS IN SMOOTH MUSCLE

Muscle Contraction: Its Regulatory Mechanisms (*Ebashi, S. et al., eds.*),*pp. 287–301*,
Japan Sci. Soc. Press, Tokyo / Springer-Verlag, Berlin (1980)

23

Ca Regulation in Smooth Muscle and Phosphorylation: Some Properties of the Myosin Light Chain Kinase

D. J. HARTSHORNE, R. F. SIEMANKOWSKI,
AND M. O. AKSOY

Summary: The myosin light chain kinase from chicken gizzard is known to consist of two subunits. One subunit has a molecular weight of about 17,000 and has been identified as calmodulin. This has a regulatory function and it is responsible for the Ca^{2+} dependence of the kinase. The other component has a subunit molecular weight of about 105,000, and contains the active site for the transferase reaction. Neither component alone has any kinase activity. Since the myosin light chain kinase is thought to be an integral part of the regulatory mechanism in smooth muscle it is important that the enzyme is adequately characterized. In this paper some of its properties have been investigated and the following results obtained: 1) The amount of calmodulin required to fully activate the 105,000 subunit was estimated by two methods. The first was based on the measurement of kinase activity at different ratios of the two subunits, and the second method utilized the enhancement of fluorescence that occurred when dansyl aziridine labelled-calmodulin interacted with the larger subunit. Both methods gave a stoichiometry for the two components of 1: 1. 2) The dependence of kinase activity on the ATP concentration was studied and K_m and V_{max} values of between 60–70 μM and 5–15 μmol P transferred min^{-1} mg^{-1} kinase, respectively, were obtained. 3) It was shown also that free Mg^{2+} is not a requirement for kinase activity as Mg^{2+}: ATP ratios of less than unity were fully effective as a substrate for the kinase. An interesting observation was that Mn^{2+}-ATP appeared to function as a better phosphate donor than Mg^{2+}-ATP. 4) The effect of sulfhydryl modification by N-ethylmaleimide on the activity of the kinase was measured. It was found that as the extent of modification increased the kinase activity decreased. However, the kinase was not particularly susceptible to sulfhydryl modification and the loss of activity occurred only when approximately half of the molecule's 16 sulfhydryl groups were reacted. The presence of calmodulin during the reaction with N-ethylmaleimide slightly accelerated the loss of activity.

It is accepted that the level of Ca^{2+} within the smooth muscle cell determines the contractile activity of that cell. The mechanism whereby this control is exerted over the contractile apparatus, however, is not firmly established. The fundamental requirement for regulation in smooth muscle is for an activation of a dormant state in the presence of Ca^{2+}. This is illustrated by the finding that if myosin is purified from smooth muscle and assayed in the presence of actin and tropomyosin its Mg^{2+}-ATPase activity is low. Obviously it follows that in order to achieve the initiation of contraction (corresponding *in vitro* to the stimulation of Mg^{2+}-ATPase activity of the actomyosin) an additional factor(s) is required. In order for this factor to fulfill a regulatory role a further requirement is that the activation be achieved only in the presence of Ca^{2+}, *i.e.*, at the concentration necessary to promote contraction of the smooth muscle. It should be pointed out that the regulatory pattern of Ca^{2+} dependent activation of the dormant state is quite distinct from the regulatory mode in skeletal muscle which involves an inhibition of an active state in the absence of Ca^{2+}. This important distinction has been pointed out several times by Ebashi and his co-workers (Ebashi *et al.*, 1975a, b; 1976). That an activating factor is required is generally accepted, the controversy that exists is centered on the identification of the activator. Basically there are two theories: one that activation results from the phosphorylation of the 20,000-dalton light chains of myosin, and the alternate view, favored by Ebashi and colleagues, is that phosphorylation is not required for activation (Mikawa *et al.*, 1977a) and a separate regulatory system is involved (Mikawa *et al.*, 1977b) which has been termed "leiotonin" (Ebashi *et al.*, 1977; Mikawa *et al.*, 1978). Our preference favors the phosphorylation theory (Aksoy *et al.*, 1976; Chacko *et al.*, 1977; Górecka *et al.*, 1976; Ikebe *et al.*, 1977, 1978; Sobieszek, 1977a, b), and this article will be restricted to this viewpoint.

In the phosphorylation scheme there are two major components. One, the myosin light chain kinase (MLCK) which activates the system and another component, the myosin light chain phosphatase which is involved during the relaxation of the muscle and returns myosin to its dormant state. Although the existence of the phosphatase has been demonstrated several times (Chacko *et al.*, 1977; Frearson *et al.*, 1976; Ikebe *et al.*, 1977) the enzyme from smooth muscle has not been characterized. The phosphatase has been isolated from skeletal muscle (Morgan *et al.*, 1976) although it is not known if the skeletal and smooth muscle forms are identical. On the other hand the MLCK from several muscle sources has been isolated (Adelstein *et al.*, 1978; Dabrowska *et al.*, 1977, 1978; Nairn and Perry, 1979; Pires and Perry, 1977; Yazawa and Yagi, 1977, 1978; Waisman *et al.*, 1978) and many of its properties established. In our laboratory it was found (Dabrowska *et al.*, 1977) that the MLCK from chicken gizzard was composed of two components of subunit molecular weights, 105,000 and 17,000 daltons. Subsequently, it was shown that the smaller component was calmodulin (previous terms for this protein include modulator and calcium dependent regulator) and further it was demonstrated that calmodulins isolated from several sources were equally effective in functioning with the larger kinase subunit (Dabrowska *et al.*, 1978). The calmodulin requirement was also shown for the MLCK from turkey gizzard (Adelstein *et al.*, 1978), blood platelets (Hathaway and Adelstein, 1979) and from skeletal muscle (Yagi *et al.*, 1978; Nairn and Perry, 1979). Some difference in the size of the larger subunit of these kinases has been observed. The sequence of events that is thought to occur on activation of the MLCK (and this is based largely by analogy to the role of calmodulin in the activation of phosphodiesterase) is that Ca^{2+} is bound to calmodulin, and the Ca^{2+}-calmodulin complex in turn forms the active ternary complex with the larger kinase subunit. This reaction sequence for the MLCK is hypothetical at the moment, and must be validified by experimental findings. One aspect of this scheme, namely the stoichiometry of the protein components, has been examined in our laboratory and the results are presented below.

It is accepted that Ca^{2+} activates the MLCK by its interaction with calmodulin, but originally it was not known whether calmodulin was the sole Ca^{2+} target, or whether additional Ca^{2+} sites were involved. The most likely candidate for the additional Ca^{2+} sites is the myosin light chains, and this led to the possibility that Ca^{2+} binding to myosin might play a regulatory function. If this is not the case then the Mg^{2+}-ATPase activity of phosphorylated myosin would be activated by actin both in the presence and absence of Ca^{2+}; if, on the other hand, the Ca^{2+}-myosin interaction is involved in the control process then in the absence of Ca^{2+} phosphorylated myosin would be dormant. The former view was favored by Small and Sobieszek (1977a, b) and Ikebe *et al.* (1978), and the latter by Chacko *et al.* (1977). In order to decide between the two possibilities phosphorylated myosin should be prepared and its actin activated ATPase activity measured in the absence and presence of Ca^{2+}. Although the experimental design is simple, in practice the experiment was not as simple to perform; the major complication being due to contamination of the myosin by kinase and/or phosphatase. Thus the approach that we took (Sherry *et al.*, 1978) was to design experiments in which the influence of contaminant kinase or phosphatase was reduced. An example being the use of adenosine 5′-O-(3-thiotriphosphate) (ATPγS). This ATP analog serves as a substrate for the MLCK but the thiophosphorylated light chain is a poor substrate for the phosphatase (Gergely *et al.*, 1976; Gratecos and Fischer, 1974; Morgan *et al.*, 1976). The net result is that myosin becomes frozen in the thiophosphorylated state and may be assayed without the complications associated with contamination by the phosphatase. The experimental findings were that as the extent of thiophosphorylation increased the extent of Ca^{2+} sensitivity decreased with the ATPase activity being fixed in the active state. This suggested that phosphorylated myosin is active regardless or whether or not Ca^{2+} is present and does not support the idea that an additional regulatory function is associated with Ca^{2+} binding to myosin. Other experiments on the rate of dephosphorylation compared to ATPase rates were also consistent with this idea (Sherry *et al.*, 1978). Recently the effect of ATPγS on mechanically disrupted fibers of chicken gizzard (Hoar *et al.*, 1979) and rabbit ileum (Cassidy *et al.*, in press) was investigated and it was found that thiophosphorylation resulted in the loss of relaxation in the absence of Ca^{2+}. Thus it would appear that the myosin light chains do not serve a regulatory function by virtue of their cation binding ability, as they do in many invertebrate systems (Kendrick-Jones *et al.*, 1976; Lehman and Szent-Györgyi, 1975; Szent-Györgyi *et al.*, 1973). However, the work of Chacko *et al.* (1977) was done with vas deferens and it is possible that this smooth muscle is different from the others that have been examined.

Most of the earlier evidence in support of the phosphorylation theory came from biochemical studies on actomyosin or myosin. However, recently a number of experiments have been reported in which intact tissue or muscle fibers have been used. For example, Barron *et al.* (1979) using intact pig carotid arteries showed that norepinephrine or KCl induced contractures were associated with increases in light chain phosphorylation. Kerrick *et al.* (1980) using skinned fibers of rabbit ileum and pulmonary artery showed that the calmodulin antagonists trifluoroperazine and chlorpromazine induced relaxation of the muscle and that this was associated with the dephosphorylation of the myosin. A similar approach was also taken by Hidaka *et al.* (1979). Thus evidence in favor of the phosphorylation mechanism is being accumulated both at the biochemical level and using systems that more closely represent the living muscle. There is still, however, much to be learnt about the properties of the proteins involved and in this paper we will examine some aspects of the MLCK system.

CALMODULIN DEPENDENCE

The stoichiometry of the calmodulin-kinase complex was determined in two ways. The first was by measuring the amount of calmodulin necessary to yield maximum phosphotransferase activity. In order to do this a calmodulin free system was used. The heavier subunit of the MLCK was purified as described previously (Mrwa and Hartshorne, 1980) and the myosin light chains were separated from calmodulin by chromatography on DEAE-Sephacel under the conditions given by Nairn and Perry (1979). In these and for the next set of kinase assays the rates of phosphate transfer to the gizzard myosin light chains were determined under conditions where the enzyme concentration was rate limiting. Kinase activity was determined at varying concentrations of calmodulin for two levels of catalytic subunit, as shown in Fig. 1. Maximum activity was attained at approximately a 1: 1 molar stoichiometry.

In the second method which was used to determine the stoichiometry of the calmodulin-kinase complex the interaction between the two proteins was measured directly. This was achieved by the use of calmodulin labelled with dansyl aziridine. When the labelled calmodulin interacted with the kinase subunit the fluorescence of the dansyl group was enhanced.

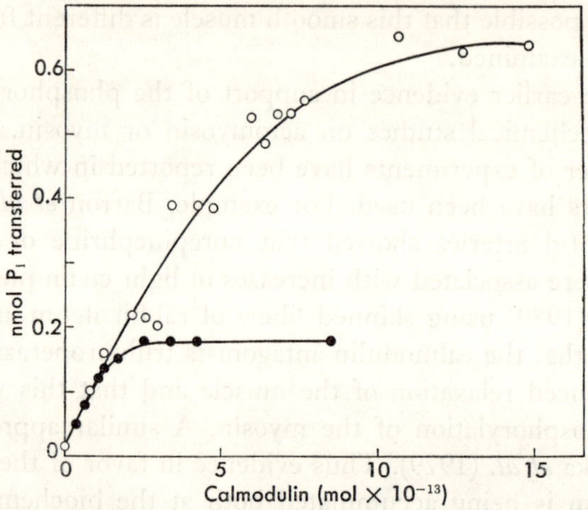

Fig. 1. Effect of calmodulin on the activity of the 105,000 kinase subunit. The assay conditions were: 4 mM $MgCl_2$, 1 mM γ-labelled [^{32}P] ATP (approx. 6,000 cpm per nmol) 25 mM tris-HCl (pH 7.6), 5×10^{-5} M $CaCl_2$, approx. 2 μM 20,000 light chain. Assay volume 0.5 ml. Temperature 25°C. In one experiment 105,000 subunit was constant at 12×10^{-13} mol (○), and in the other at 3×10^{-13} mol (●). In the former the assay time was 1.25 min and in the latter 1.5 min. The preparation of the 105,000 kinase subunit and calmodulin and the determination of ^{32}P incorporation was as described by Mrwa and Hartshorne (1980).

This was monitored at 495 nm. Using fluorescence enhancement to monitor the stoichiometry of the calmodulin-kinase complex a value of approximately 1:1 molar ratio was obtained (Fig. 2). Thus from both types of measurements it is clear that one molecule of calmodulin is required to activate one molecule of the catalytic (or larger) kinase subunit. A similar value was reported for the calmodulin requirement of the skeletal muscle kinase by Nairn and Perry (1979) and for turkey gizzard kinase (Adelstein, personal communication).

ATP AND Mg^{2+} REQUIREMENTS

Kinase activity was measured as a function of ATP concentration and the results are given in Fig. 3 in the form of a double-reciprocal plot. Two enzyme preparations were assayed. The K_m values obtained are 63 and 71 μM, and the V_{max} values are 5.2 and 13.4 μmol min^{-1} mg^{-1} kinase. In Table I these parameters are compared to those obtained for other MLCKs. In general the V_{max} values are similar for those kinases obtained from

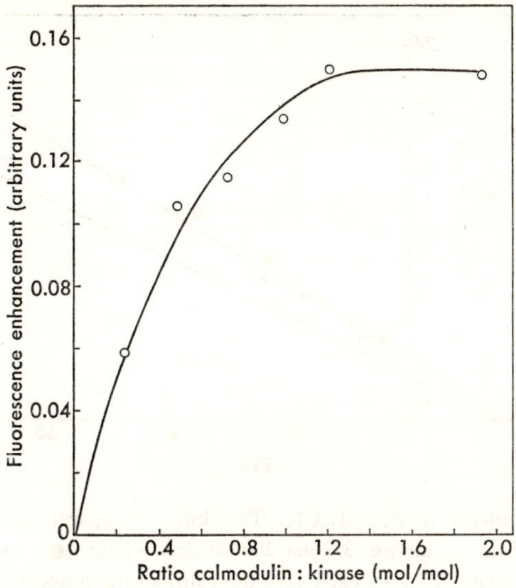

Fig. 2. Fluorescence enhancement of dansyl aziridine labelled-calmodulin on interaction with the 105,000 kinase subunit. The conditions were: 10 mM $MgCl_2$, 50 mM KCl, 25 mM tris-HCl (pH 7.6), volume 2 ml, temperature 25°C. The dansyl-calmodulin was constant at 7.5 μg/ml and the 105,000 subunit was varied from 23 to 152 μg/ml. Excitation and emission were fixed at 345 nm and 495 nm, respectively.

muscle sources, although our K_m value is slightly lower than the others. It is interesting that the K_m values indicate a relatively low affinity binding and it would be pertinent to compare this affinity with the K_m values obtained for the hydrolysis of Mg^{2+}-ATP by smooth muscle myosin and actomyosin. Relatively little data is available for the smooth muscle system using Mg^{2+}-ATP, although for gizzard myosin Onishi *et al.* (1978) reported a K_m of 1.3 μM and Malik (1978) calculated a K_m of approximately 300 μM. Thus there is obviously some discrepancy and the actual values must be established by future investigations. An observation that was made in our laboratory some years ago (Hartshorne *et al.*, 1977) was that at low concentrations of ATP (in the order of 10^{-5} M) the Ca^{2+} sensitivity of gizzard actomyosin was lost although measurable ATPase activity remained. One explanation for this behaviour is that the Ca^{2+} sensitivity was lost because the kinase was not effective (*i.e.*, the concentration of ATP was below the K_m value) and that the actomyosin could still bind and hydrolyse the ATP. This would imply that ATP is not inhibitory for actomyosin ATPase activity at low concentrations but be-

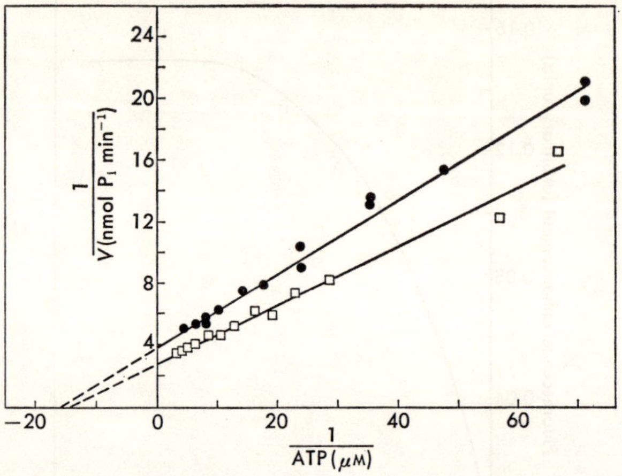

Fig. 3. ATP dependence of the MLCK. Two kinase preparations were assayed. The conditions were: 63 ng/ml kinase, 3.2 μM 20,000 light chain (●), and 69 ng/ml kinase, 6.5 μM 20,000 light chain (□). In each case the calmodulin concentration was at approx. 10-fold molar excess over the kinase. Each point represents a rate determined over a period of approx. 1.5 min. Other conditions and procedures as given in Fig. 1.

TABLE I
Comparison of the Properties of Different Myosin Light Chain Kinases

Source of MLCK	K_m (μM ATP)	$V_{max.}$[a] (μmol/min/mg kinase)	References
Chicken gizzard	60–70	5–13	This study
Turkey gizzard	—	4.8[b]	Adelstein *et al.* (1978)
Skeletal muscle	200–400	15–30	Pires and Perry (1977) and Nairn and Perry (1979)
Skeletal muscle	280	4.3	Yazawa and Yagi (1978)
Blood platelets	121	3.1	Hathaway and Adelstein (1979)
Rat myoblasts	—	0.077	Scordilis and Adelstein (1978)

[a] All values given for 25°C.
[b] Adjusted from the value of 4.3 obtained at 24°C.

comes inhibitory at higher concentrations; and it would also imply that the inhibition by ATP of actomyosin ATPase activity is removed by phosphorylation. A similar theme has been suggested by Onishi and Watanabe (1979) using ITP as a substrate for actomyosin.

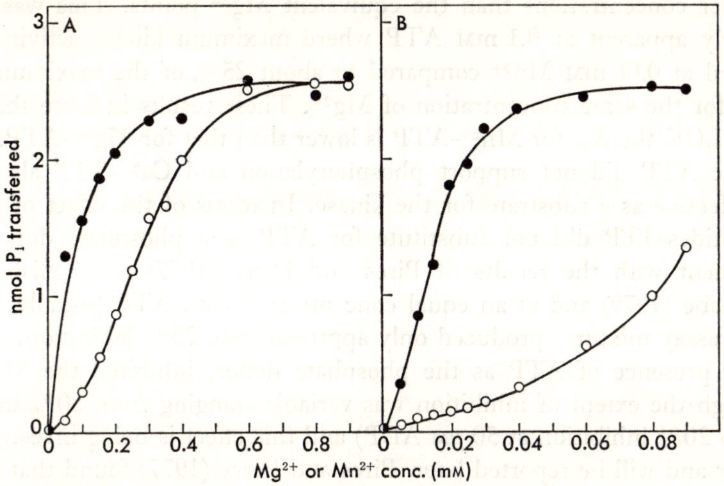

Fig. 4. The Mg^{2+} and Mn^{2+} dependence of the MLCK. A: 1 mM ATP. B: 0.1 mM ATP. The assay conditions were: kinase, 1.3 μg/ml, 20,000 light chain 5 μM, calmodulin at least 10-fold molar excess. Assays were carried out for 5 min at 25°C. Other conditions and procedures as in Fig. 1. ○ Mg^{2+} assay; ● Mn^{2+} assay.

Another variable that was investigated was the Mg^{2+} to ATP ratio required for optimal kinase activity. The motivation for these experiments came from the well known observations that maximum actomyosin-ATPase activity or tension development required Mg^{2+} concentrations in excess of the ATP concentration (Filo *et al.*, 1965; Russell, 1973). The obvious point to test was whether or not this Mg^{2+}-dependence was reflected in the activity of the kinase. Experiments were done at two levels of ATP, 1 mM and 0.1 mM, and the MLCK activity was measured at different Mg^{2+} concentrations using gizzard myosin light chains. The results given in Fig. 4 are surprising in that the kinase did not exhibit a marked Mg^{2+} dependency. Although there is a requirement for Mg^{2+} to form the substrate, Mg^{2+}-ATP, the kinase will tolerate relatively high levels of free ATP before an inhibition is shown. For example, in the assays done at 1 mM ATP full activity was given at 0.6 mM Mg^{2+}, that is at a free ATP concentration of not less than 0.4 mM. In the assays done at 0.1 mM ATP the Mg^{2+} dependency was more marked, but this was probably because the Mg^{2+}-ATP concentration was approaching K_m for the MLCK.

In addition to Mg^{2+} we investigated the equivalent experiments done with Mn^{2+}. At both levels of ATP the use of Mn^{2+} gave a higher activity at lower concentrations than the equivalent Mg^{2+} points. This was particularly apparent at 0.1 mM ATP where maximum kinase activity was attained at 0.06 mM Mn^{2+} compared to about 25% of the maximum activity for the same concentration of Mg^{2+}. These results indicate that for the MLCK the K_m for Mn^{2+}-ATP is lower than that for Mg^{2+}-ATP.

Free ATP did not support phosphorylation and Ca^{2+}-ATP also was not effective as a substrate for the kinase. In terms of the effect of other nucleotides ITP did not substitute for ATP as a phosphate donor, in agreement with the results of Pires and Perry (1977) and Onishi and Watanabe (1979) and at an equal concentration with ATP (usually 1 mM in the assay mixture) produced only approximately 20% inhibition. ADP, in the presence of ATP as the phosphate donor, inhibited the MLCK although the extent of inhibition was variable (ranging from 80% inhibition to 20% inhibition at 50 μM ADP) and this effect is being investigated further and will be reported later. Pires and Perry (1977) found that ADP (in the mM range) inhibited the MLCK from skeletal muscle. IDP, AMP, cyclic AMP, dibutyryl cyclic AMP, cyclic GMP, and adenosine, at concentrations not less than 100 μM, did not affect the kinase activity. Similarly, potassium phosphate (10 mM) was without effect. When pyrophosphate was tested it was found that inhibition resulted because of the formation of the insoluble Mg^{2+} pyrophosphate complex and full activity was restored when additional Mg^{2+} was added.

SULFHYDRYL SENSITIVITY

The reason that these studies were undertaken was because it has been known for many years that the Ca^{2+} dependence of smooth muscle actomyosin is a labile property and is rapidly lost under non-reducing conditions (Sparrow and van Bockxmeer, 1972). It was therefore of interest to determine whether or not the loss of Ca^{2+} sensitivity reflected the modification of the sulfhydryl groups of the MLCK. Initially iodoacetamide was chosen as the sulfhydryl reagent; it was found however that this reagent was not particularly reactive with the kinase (*e.g.*, approximately 80% of the original kinase activity was retained following reaction at 25°C, 10 mM iodoacetamide, pH 8.4). N-ethylmaleimide (NEM) was then tested as an alternative reagent, and was found to be more effective than iodoacetamide. Figure 5 shows the effect of NEM, at concentrations from 2 to 10 mM, on the activity of the MLCK. (The conditions for the reaction are given in the figure legend.) Clearly there is a progressive loss

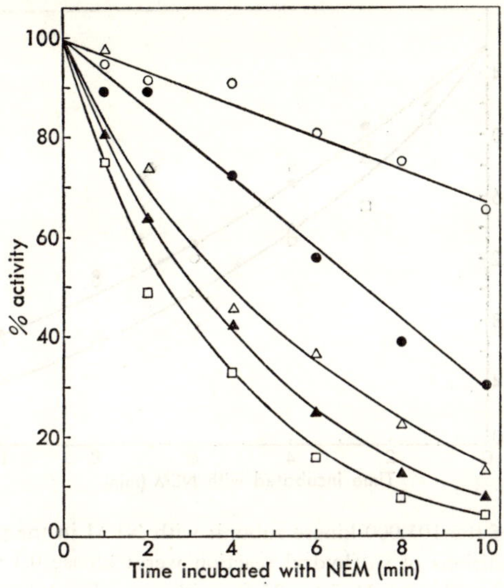

Fig. 5. Reaction of the 105,000 subunit of the MLCK at different levels of NEM. The sulfhydryl reaction was carried out at a kinase concentration of 0.1 mg/ml at 25°C in 50 mM potassium phosphate (pH 7.6) at the following concentrations of NEM: 2 mM (○), 4 mM (●), 6 mM (△), 8 mM (▲), and 10 mM (□). At the times indicated the reaction was stopped by adding 0.075 ml aliquots to 1 ml 20 mM dithiothreitol, 5% sucrose. ^{32}P incorporation was then determined after 6 min at 25°C using kinase at 0.14 μg/ml, and 20,000 light chain at 5 μM. Calmodulin was present at approx. a 10-fold molar excess. Other conditions and procedures as in Fig. 1.

of kinase activity as the extent of sulfhydryl modification increases. The reaction as shown in Fig. 5 was carried out on the purified 105,000 subunit of the kinase, and following the completion of sulfhydryl modification, myosin light chains and calmodulin were added for the assay of kinase activity, in this case in the presence of Ca^{2+}. In the absence of Ca^{2+} (*i.e.*, in the presence of EGTA) the kinase was not active and no increase in activity occurred on reaction with NEM. Thus any reduction in Ca^{2+} sensitivity as a result of sulfhydryl modification of the kinase would occur by reducing the plus Ca^{2+} kinase activity rather than by increasing the minus Ca^{2+} level.

The influence of the other components of the assay system on the reaction of the kinase with NEM was also tested. It was found that the presence of calmodulin during the NEM reaction slightly increased the extent of inhibition of kinase activity. This is illustrated in Fig. 6, for the

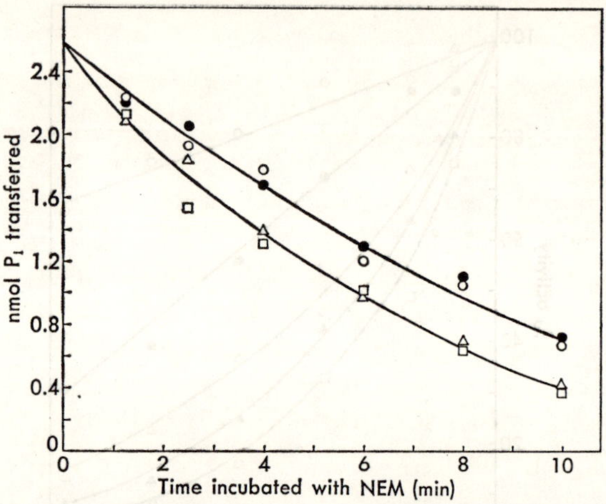

Fig. 6. Reaction of the 105,000 kinase subunit with NEM in the presence and absence of calmodulin. Conditions of sulfhydryl reaction were: kinase 0.1 mg/ml, 4 mM NEM, 50 mM potassium phosphate (pH 7.6), 25°C, with no calmodulin (○), and with calmodulin at a molar ratio to kinase of 1: 1 (△) and 2: 1 (□), and 2: 1 plus 2 mM EGTA (●). Reaction with NEM was stopped and kinase activity determined as in Fig. 5.

1: 1 and 2: 1 calmodulin to kinase molar ratios. In the presence of EGTA the effect of calmodulin was eliminated, as shown in Fig. 6. The extent of sulfhydryl reaction under the conditions as shown in Fig. 6 was estimated using ^{14}C-labelled NEM. The kinase contained 16 mol 1/2 cystine per mol kinase (using 105,000 as the molecular weight) as determined following reduction and reaction of the kinase with excess iodoacetic acid in 5 M guanidine hydrochloride. The kinase activity was lost when approximately nine sulfhydryl groups were reacted. The extent of sulfhydryl modification was not markedly affected by the presence or absence of calmodulin. The presence of light chains, ATP and Mg^{2+}-ATP (up to 1 mM) during the NEM reaction did not significantly alter the reaction pattern. Thus it appears that although the kinase is sensitive to the modification of its sulfhydryl groups it is unlikely that a specific sulfhydryl group is involved in the phosphotransferase reaction. The loss of enzyme activity could more logically be explained by a change occurring in the kinase molecule as a result of extensive sulfhydryl modification. The loss of Ca^{2+} sensitivity which occurs in actomyosin therefore is probably not due to an oxidation of the kinase molecule and as pointed out recently by Seidel (1979) would more likely be due to an oxidation of myosin.

CONCLUSION

One of the reasons why the above studies were undertaken was that it would be interesting to compare the properties exhibited by the isolated kinase with the ATPase characteristics of smooth muscle actomyosin. Particularly since phosphorylation of the myosin light chains is thought to be a prerequisite for actin-activated Mg^{2+}-ATPase activity of myosin. However many of the properties described above cannot be directly related to the steady state ATPase activity of actomyosin. The Mg^{2+}-dependence is probably the most striking example. Actomyosin ATPase activity shows a marked requirement for free Mg^{2+} whereas this is not the case for the kinase activity. It appears then that the rate of myosin phosphorylation is not reflected by the steady-state ATPase activity of actomyosin. This is not surprising, as one would expect the influence of phosphorylation to be detected during the initial phases of the ATPase sequence, *i.e.*, in that period where the myosin is "activated" by phosphorylation. Given this assumption a knowledge of some of the properties of the kinase, such as those listed above, could be useful in understanding the kinetics of actomyosin ATPase activity as they might allow the discrimination between different phases of the ATPase time sequence.

REFERENCES

Adelstein, R. S., Conti, M. A., Hathaway, D. R., and Klee, C. B. (1978). *J. Biol. Chem.* **253**, 8347–8350.

Aksoy, M. O., Williams, D., Sharkey, E. M., and Hartshorne, D. J. (1976). *Biochem. Biophys. Res. Commun.* **69**, 35–41.

Barron, J. T., Bárány, M., and Bárány, K. (1979). *J. Biol. Chem.* **254**, 4954–4956.

Cassidy, P. S., Hoar, P. E., and Kerrick, W.G.L. *J. Biol. Chem.* (in press).

Chacko, S., Conti, M. A., and Adelstein, R. S. (1977). *Proc. Natl. Acad. Sci. U.S.* **74**, 129–133.

Dabrowska, R., Aromatorio, D., Sherry, J.M.F., and Hartshorne, D. J. (1977). *Biochem. Biophys. Res. Commun.* **78**, 1263–1272.

Dabrowska, R., Sherry, J.M.F., Aromatorio, D. K., and Hartshorne, D. J. (1978). *Biochemistry* **17**, 253–258.

Ebashi, S., Mikawa, T., Hirata, M., Toyo-oka, T., and Nonomura, Y. (1977). In *Excitation-contraction Coupling in Smooth Muscle* (Casteels, R., Godfraind, T., and Rüegg, J. E., eds.), pp. 325–334, Elsevier/North-Holland, Amsterdam.

Ebashi, S., Nonomura, Y., Kitazawa, T., and Toyo-oka, T. (1975a). In *Calcium Transport in Contraction and Secretion* (Carofoli, E., Clementi, F., Drabikowski, W., and Margreth, A., eds.), pp. 405–414, North-Holland, Amsterdam.

Ebashi, S., Nonomura, Y., Toyo-oka, T., and Katayama, E. (1976). In *Symposia of the*

Society for Experimental Biology, Calcium in Biological Systems (Duncan, C. J., ed.), pp. 349–360, Cambridge Univ. Press, London.

Ebashi, S., Toyo-oka, T., and Nonomura, Y. (1975b). *J. Biochem.* **78**, 859–861.

Filo, R. S., Bohr, D. F., and Rüegg, J. C. (1965). *Science* **147**, 1581–1583.

Frearson, N., Focant, B.W.W., and Perry, S. V. (1976). *FEBS Lett.* **63**, 27–32.

Gergely, P., Vereb, G., and Bot, G. (1976). *Biochim. Biophys. Acta.* **429**, 809–816.

Górecka, A., Aksoy, M. O., and Hartshorne, D. J. (1976). *Biochem. Biophys. Res. Commun.* **71**, 325–331.

Gratecos, D. and Fischer, E. H. (1974). *Biochem. Biophys. Res. Commun.* **58**, 960–967.

Hartshorne, D. J., Abrams, L., Aksoy, M. O., Dabrowska, R., Driska, S., and Sharkey, E. M. (1977). In *The Biochemistry of Smooth Muscle* (Stephens, N. L., ed.), pp. 513–532, Univ. Park Press, Baltimore.

Hathaway, D. R. and Adelstein, R. S. (1979). *Proc. Natl. Acad. Sci. U.S.* **76**, 1653–1657.

Hidaka, H., Yamaki, T., Totsuka, T., and Asano, M. (1979). *Mol. Pharmacol.* **15**, 49–59.

Hoar, P. E., Kerrick, W.G.L., and Cassidy, P. S. (1979). *Science* **204**, 503–506.

Ikebe, M., Aiba, T., Onishi, H., and Watanabe, S. (1978). *J. Biochem.* **83**, 1643–1655.

Ikebe, M., Onishi, H., and Watanabe, S. (1977). *J. Biochem.* **82**, 299–302.

Kendrick-Jones, J., Szentkiralyi, E. M., and Szent-Györgyi, A. G. (1976). *J. Mol. Biol.* **104**, 747–775.

Kerrick, W.G.L., Hoar, P.E., and Cassidy, P. S. (1980). *Fed. Proc.* (in press).

Lehman, W. and Szent-Györgyi, A. G. (1975). *J. Gen. Physiol.* **66**, 1–30.

Malik, M. N. (1978). *Biochemistry* **17**, 27–32.

Mikawa, T., Nonomura, Y., and Ebashi, S. (1977a). *J. Biochem.* **82**, 1789–1791.

Mikawa, T., Nonomura, Y., Hirata, M., Ebashi, S., and Kakiuchi, S. (1978). *J. Biochem.* **84**, 1633–1636.

Mikawa, T., Toyo-oka, T., Nonomura, Y., and Ebashi, S. (1977b). *J. Biochem.* **81**, 273–275.

Morgan, M., Perry, S. V., and Ottaway, J. (1976). *Biochem. J.* **157**, 687–697.

Mrwa, U. and Hartshorne, D. J. (1980). *Fed. Proc.* (in press).

Nairn, A. C. and Perry, S. V. (1979). *Biochem. J.* **179**, 89–97.

Onishi, H. and Watanabe, S. (1979). *J. Biochem.* **86**, 569–573.

Onishi, H., Yamada, Y., Ikebe, M., Suzuki, H., and Watanabe, S. (1978). *J. Biochem.* **83**, 129–135.

Pires, E.M.V. and Perry, S. V. (1977). *Biochem. J.* **167**, 137–146.

Russell, W. E. (1973). *Eur. J. Biochem.* **33**, 459–466.

Scordilis, S. P. and Adelstein, R. S. (1978). *J. Biol. Chem.* **253**, 9041–9048.

Seidel, J. C. (1979). *Biochem. Biophys. Res. Commun.* **89**, 958–964.

Sherry, J.M.F., Górecka, A., Aksoy, M. O., Dabrowska, R., and Hartshorne, D. J. (1978). *Biochemistry* **17**, 4411–4418.

Small, J. V. and Sobieszek, A. (1977a). *Eur. J. Biochem.* **76**, 521–530.

Small, J. V. and Sobieszek, A. (1977b). In *Excitation-contraction Coupling in Smooth Muscle* (Casteels, R., Godfraind, T., and Rüegg, J. C., eds.), pp. 385–393, Elsevier/North-Holland, Amsterdam.

Sobieszek, A. (1977a). In *The Biochemistry of Smooth Muscle* (Stephens, N. L., ed.), pp. 413–443, Univ. Park Press, Baltimore.

Sobieszek, A. (1977b). *Eur. J. Biochem.* **73**, 477–483.

Sparrow, M. P. and van Bockxmeer, F. M. (1972). *J. Biochem.* **72**, 1075–1080.

Szent-Györgyi, A. G., Szentkiralyi, E. M., and Kendrick-Jones, J. (1973). *J. Mol. Biol.* **74**, 179–203.

Waisman, D. M., Singh, T. J., and Wang, J. H. (1978). *J. Biol. Chem.* **253**, 3387–3390.

Yagi, K., Yazawa, M., Kakiuchi, S., Ohshima, M., and Uenishi, K. (1978). *J. Biol. Chem.* **253**, 1338–1340.

Yazawa, M. and Yagi, K. (1977). *J. Biochem.* **83**, 287–289.

Yazawa, M. and Yagi, K. (1978). *J. Biochem.* **84**, 1259–1265.

Muscle Contraction: Its Regulatory Mechanisms (Ebashi, S. et al., eds.), pp. 303–313, Japan Sci. Soc. Press, Tokyo / Springer-Verlag Berlin (1980)

24

Regulation of Smooth Muscle Myosin Kinase Activity by Phosphorylation and Dephosphorylation

ROBERT S. ADELSTEIN, MARY D. PATO, AND MARY ANNE CONTI

Summary: Myosin light chain kinase from turkey gizzard smooth muscle can undergo reversible phosphorylation. Cyclic AMP-dependent protein kinase catalyzes the incorporation of 1 mol of phosphate per mol of myosin kinase. Phosphorylation of myosin kinase decreases the binding of Ca^{2+}-calmodulin to myosin kinase and thus favors an inactive form of the kinase.

Dephosphorylation of myosin kinase is catalyzed by a phosphatase purified from turkey gizzard smooth muscle. Dephosphorylation of myosin light chain kinase results in an increase in kinase activity, returning the activity to the same level as that found prior to phosphorylation of the kinase.

A rise in the concentration of cyclic AMP results in the relaxation of many different smooth muscles. Using the above results on the reversible phosphorylation of myosin kinase, we present a biochemical mechanism that might account for this *in vivo* observation.

Regulation of smooth muscle contraction-relaxation is not entirely understood at present. Two major regulatory systems have been described.

One involves the reversible phosphorylation of the 20,000-dalton light chain of myosin. The second involves a complex of novel proteins, leiotonin A and leiotonin C. Proponents of both regulatory systems (myosin phosphorylation and leiotonin) agree that smooth muscle myosin differs from skeletal and cardiac muscle myosin in that as isolated, purified smooth muscle myosin cannot be activated by actin in the presence of Mg^{2+} at low (0.03 M) concentrations of KCl. They also agree that a rise in Ca^{2+}, from about 10^{-7} M to 10^{-5} M initiates smooth muscle actin-myosin interaction. Disagreement centers about the nature of the proteins with which Ca^{2+} interacts and how this interaction results in actin-activation of the myosin ATPase activity.

Papers from the laboratories of Drs. Ebashi and Hartshorne, which are included in this volume, also address this subject, so we will summarize the two regulatory systems as follows:

1) Myosin phosphorylation: the covalent phosphorylation of the 20,000-dalton light chain of myosin is required for actin-activation of smooth muscle myosin ATPase activity. Ca^{2+} initiates actin-myosin interaction by binding to the calcium-binding protein calmodulin, which in turn activates the kinase catalyzing the phosphorylation of myosin. Myosin phosphorylation has been shown to regulate actin-activation of myosin ATPase activity in experiments utilizing myosin from the following smooth muscles: gizzard (Sobieszek and Small, 1977; Gorecka *et al.*, 1976; Onishi and Watanabe, 1979) vas deferens (Chacko *et al.*, 1977), stomach (Small and Sobieszek, 1977), uterus (Lebowitz and Cooke, 1979), arterial wall (Di Salvo *et al.*, 1978), and placenta (Huszar and Bailey, 1979).

2) Leiotonin: in place of myosin phosphorylation, regulation of gizzard, stomach, and aorta myosin requires three proteins in addition to actin, for activation of myosin ATPase activity (Ebashi *et al.*, 1978). These are tropomyosin, leiotonin A, and leiotonin C. Leiotonin C is a calcium-binding protein, which is distinct from calmodulin (Mikawa *et al.*, 1978). Ca^{2+} presumably initiates actin-myosin interaction by binding to leiotonin C, which interacts with leiotonin A and tropomyosin resulting in actin-activation of myosin ATPase activity. Phosphorylation of myosin is not required in this regulatory system. There is, of course, the distinct possibility that smooth muscle contains *both* regulatory systems, and that each might operate under a unique set of circumstances.

Our laboratory has been interested in how phosphorylation of myosin regulates actin-myosin interaction. Our interest was first directed to this area by finding that phosphorylation of platelet myosin was required for actin-activation of the platelet myosin ATPase activity (Adelstein and Conti, 1975). Subsequently, a similar regulatory system was shown to be present in proliferative myoblasts (Scordilis and Adelstein, 1977) and in macrophages (Trotter and Adelstein, 1979). Our initial interest in smooth muscle involved myosin isolated from guinea pig vas deferens where evidence for the presence of two regulatory systems was found (Chacko *et al.*, 1977). Although actin-activation of vas deferens myosin ATPase activity was found to require phosphorylation, the actin-activated ATPase of the phosphorylated myosin free of kinase and phosphatase was higher when Ca^{2+} was present in the assay mixture than in the absence of Ca^{2+} (*i.e.*, upon addition of EGTA). This type of "dual regulation" was shown not to be present in gizzard smooth muscle using purified gizzard myosin (Sherry *et al.*, 1978) and chemically skinned smooth muscle fibers (Hoar *et al.*, 1979).

Our present studies center on the enzyme catalyzing the phosphorylation of turkey gizzard myosin. Turkey gizzard myosin light chain kinase is a Ca^{2+}-dependent, cyclic AMP-independent kinase that specifically catalyzes the phosphorylation of the 20,000-dalton light chain of myosin (Conti and Adelstein, 1980). This myosin kinase, similar to skeletal muscle myosin kinase (Yagi *et al.*, 1978) has been shown to require the ubiquitous calcium-binding protein calmodulin (mol. wt. 16,500) for activity (Dabrowska *et al.*, 1977). In turkey gizzard, the active myosin kinase is composed of Ca^{2+}-calmodulin and a 130,000-dalton protein which confers specificity on the enzyme and contains at least a part, if not all of the active site.

This paper reviews our recent work on the regulation of myosin light chain kinase by the catalytic subunit of cyclic AMP-dependent protein kinase. Three points will be emphasized: a) phosphorylation of myosin light chain kinase by the catalytic subunit of cyclic AMP-dependent protein kinase results in decreased myosin kinase activity (Adelstein *et al.*, 1978). b) The decreased myosin kinase activity is the result of a decrease in the ability of the phosphorylated kinase to bind Ca^{2+}-calmodulin (Conti and Adelstein, 1980). c) A phosphatase has been isolated from turkey gizzard smooth muscle which dephosphorylates myosin light chain kinase. Dephosphorylation of myosin kinase restores the activity of the enzyme, presumably by strengthening the binding to Ca^{2+}-calmodulin. The above findings can be used to construct a scheme explaining how a rise in cyclic AMP might directly effect smooth muscle relaxation *via* the myosin phosphorylation-dephosphorylation regulatory system.

PHOSPHORYLATION OF MYOSIN LIGHT CHAIN KINASE

Figure 1 is a time course for phosphorylation of myosin kinase by the catalytic subunit of cyclic AMP-dependent protein kinase. (The protein kinase was a gift of Dr. E. H. Fischer). Approximately 1 mol of phosphate is incorporated per mol of myosin kinase. Note that addition of the specific inhibitor of protein kinase to the phosphorylating mixture, or omission of the protein kinase from the mixture results in only low levels of phosphorylation.

Evidence that the phosphorylation of myosin kinase might be confined to a single site on the molecule is shown in Fig. 2. This is a radioautogram of a tryptic digest of ^{32}P-labeled myosin light chain kinase. Note that a single radioactive peptide (mol. wt. 22,000) is released following brief tryptic digestion. The fact that all the radioactivity is confined to a single tryptic peptide is consistent with phosphorylation occurring at a single site. Since it is unlikely that there is just one serine or threonine residue

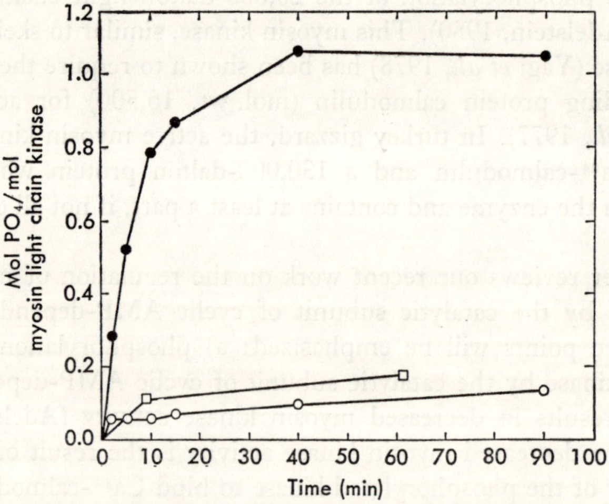

Fig. 1. Time dependence of incorporation of ^{32}P from γ-$AT^{32}P$ into myosin light chain kinase by the catalytic subunit of cyclic AMP-dependent protein kinase. ^{32}P incorporation into myosin kinase was assayed in the presence of added protein kinase (●) or of protein kinase plus the specific inhibitor of protein kinase (□), or in the absence of both protein kinase and inhibitor (○). Assays were performed by Millipore filtration and location of radioactive label in the 130,000-dalton band of the myosin kinase confirmed by radioautography of slab gels. (For details see Conti and Adelstein, 1980, from which this is taken.)

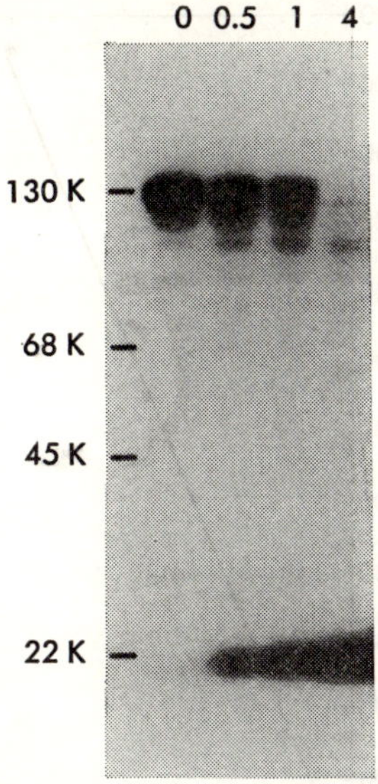

Fig. 2. Radioautography of ^{32}P-labeled myosin light chain kinase prior to and after digestion with trypsin. A 0.2% SDS, 7.5 to 20% gradient polyacrylamide slab gel of aliquots removed at 0 (before the addition of trypsin) 0.5, 1, and 4 min after the addition of trypsin at 0°C was used for radioautography following electrophoresis. The molecular weights shown (K=1,000) correspond to the location of standards which were stained with Coomassie brilliant blue after electrophoresis of the same gel. (Previously published: Adelstein *et al.*, 1978.)

in this peptide this does not constitute proof for phosphorylation at a single site. Sequence studies will be necessary to establish that a unique amino acid residue is being phosphorylated.

Phosphorylation of myosin light chain kinase by either the catalytic subunit of protein kinase, or by a crude preparation of the holoenzyme which is activated by the addition of cyclic AMP, results in a decrease in myosin light chain kinase activity (Adelstein *et al.*, 1978; Conti and Adelstein, 1980). The mechanism by which phosphorylation of myosin kinase decreases its enzymatic activity has been examined by titrating the unphosphorylated and phosphorylated forms of the enzyme with calmodulin in the presence of 0.2 mM $CaCl_2$ and measuring the rate at which the respective kinases can phosphorylate myosin light chains. Figure 3 is a double reciprocal plot of the data from the calmodulin titration curves of the phosphorylated and unphosphorylated kinase. As can be seen from the intercept of the x-axis, phosphorylation of the kinase results in a 5-fold

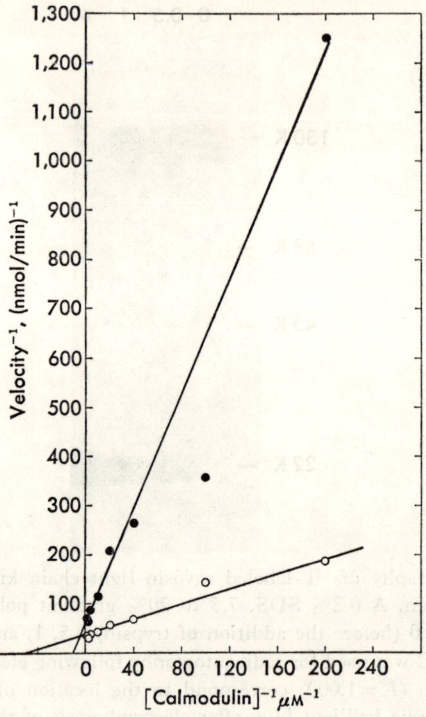

Fig. 3. Double reciprocal plot of calmodulin concentration *vs.* myosin kinase activity. The data for this plot was obtained from titration curves of the phosphorylated (●) and unphosphorylated (○) kinases with calmodulin. Myosin light chain kinase activity was assayed by the ability to transfer ^{32}P from γ-$AT^{32}P$ to smooth muscle myosin light chains at 24°C. Kinase concentration was 10^{-9} M and assay time was typically 3–5 min. Incorporation was linear with time under the assay conditions. Calmodulin concentration was determined by $\varepsilon^{1\%}_{230nm} = 25$ and appropriate dilutions made into 1 mg/ml bovine serum albumin monomer. (For details see Conti and Adelstein, 1980.)

decrease in the apparent binding constant of calmodulin for the myosin kinase. There is a smaller (1.5 fold) effect on the maximum velocity of the two enzymes. Recent experiments have indicated that the difference in the binding constant for calmodulin, between the phosphorylated and unphosphorylated kinase is closer to 20-fold while the V_{max} still differs by 1.5-fold (Conti and Adelstein, manuscript in preparation).

In summary, phosphorylation of the 130,000-dalton component of myosin kinase results in a decrease in the rate at which the kinase phosphorylates myosin light chains. This decreased activity results from a decrease in the apparent binding of calmodulin, which is required for activity, by the phosphorylated kinase.

DEPHOSPHORYLATION OF MYOSIN KINASE

Having found that cyclic AMP-dependent phosphorylation of myosin light chain kinase decreased myosin kinase activity, it was of interest to study the reverse reaction, dephosphorylation of myosin kinase:

$$\text{P-myosin kinase} + H_2O \xrightarrow{\text{Phosphatase}} \text{Myosin kinase} + P_i$$

and to confirm that upon dephosphorylation the specific activity of the myosin kinase would increase.

Recently Dr. Mary D. Pato of our laboratory isolated a phosphatase from turkey gizzards that dephosphorylates myosin light chain kinase at a rapid rate. The final step in the purification of this phosphatase was chromatography on an affinity column of myosin light chains which were first coupled to Sepharose 4B and then phosphorylated with adenosine 5′-O-(3-thiotriphosphate) (ATPγS).

A single peak of phosphatase activity was eluted from this affinity column. Gel electrophoresis of this peak in SDS-polyacrylamide revealed three polypeptide chains with molecular weights of 60,000, 55,000, and 38,000. A scan at 584 nm showed that the bands were present in equimolar ratios, assuming that the polypeptide chains bind Coomassie blue with the same affinity.

The possibility that all three polypeptide chains are part of the phosphatase is suggested by the following experiment. An aliquot of the phosphatase purified by affinity chromatography was electrophoresed in a polyacrylamide tube gel in the absence of SDS. The gel was sliced into 2 mm slices and, following elution, it was assayed for phosphatase activity. A single peak of phosphatase activity was eluted from the gel. When aliquots across this peak were electrophoresed on an SDS-polyacrylamide slab gel, all three polypeptide bands (60,000, 55,000, and 38,000 daltons) were found to be present in equimolar amounts and to correlate with phosphatase activity.

The structure of this enzyme is very similar to that proposed for phosphorylase phosphatase by Lee (Lee *et al.*, 1977) and experiments are presently underway to see if these two enzymes are related.

Figure 4 is a time course for dephosphorylation of a number of sub-

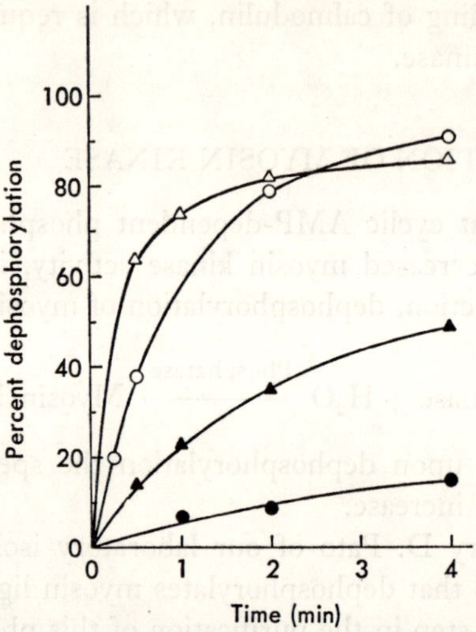

Fig. 4. Time course of dephosphorylation of various phosphorylated substrates by a purified phosphatase isolated from smooth muscle. The molar concentrations of all the substrates were the same (3 μM) with the exception of the myosin kinase, which was 10-fold lower. Assays were carried out in 50 mM Tris-HCl, 10 mM $MgCl_2$, 0.1 mM $CaCl_2$, pH 7.0 at 30°C. Samples were removed at the times indicated, precipitated with TCA and counted following Millipore filtration. ○ myosin light chain; △ myosin kinase; ▲ casein; ● histone.

TABLE I
The Effect of Phosphorylation and Dephosphorylation on Myosin Kinase Activity

	Specific activity[a] (μmol/mg/min)
Prior to phosphorylation	4.25
Phosphorylated	0.42
Following dephosphorylation	4.12

[a] 6.7×10^{-9} M calmodulin and 0.2 mM $CaCl_2$; activity was measured by determining the transfer of ^{32}P from γ-AT^{32}P to the 20,000-dalton light chain of smooth muscle myosin. (Conti and Adelstein, 1980).

strates by this phosphatase. Of particular interest was the rapid rate with which this enzyme catalyzed dephosphorylation of myosin kinase. Indeed, under the conditions employed in this particular assay, the initial rate of dephosphorylation was nonlinear, indicating it was greater than that shown.

Having prepared a phosphatase that could dephosphorylate myosin kinase, it was of interest to compare the specific activity of the phosphorylated and dephosphorylated kinase. The major question we sought to answer is whether the effect of phosphorylation of myosin kinase on calmodulin-binding could be reversed by dephosphorylation of the kinase. Table I presents data comparing the specific activity of myosin light chain kinase a) prior to phosphorylation, b) with 1 mol of phosphate incorporated/mol of kinase, and c) following dephosphorylation.

As shown in Table I, dephosphorylation of myosin light chain kinase restored the specific activity of the kinase to its original rate, presumably by restoring the ability of the kinase to bind calmodulin at the tighter level seen prior to phosphorylation of the kinase.

These experiments indicate that modulation of myosin kinase activity by cyclic AMP phosphorylation is a reversible phenomenon.

REGULATION OF SMOOTH MUSCLE CONTRACTION BY Ca^{2+}-CALMODULIN AND CYCLIC AMP

Figure 5 is a scheme summarizing the role of Ca^{2+}-calmodulin and cyclic AMP-dependent phosphorylation in regulating smooth muscle contrac-

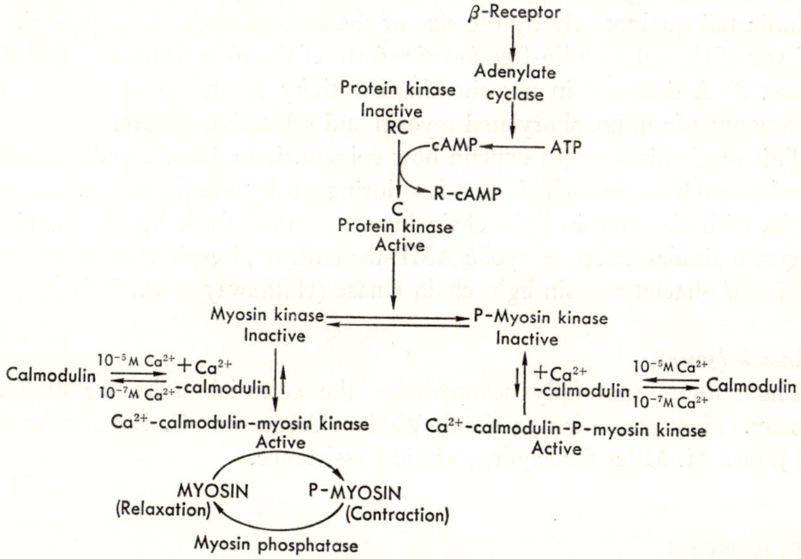

Fig. 5. Regulation of smooth muscle contraction by Ca^{2+}-calmodulin and cyclic AMP. The scheme is discussed in detail in the text (p. 312). It has been published previously, Conti and Adelstein, 1980 and Adelstein and Eisenberg, 1980.

tion and relaxation. The essential features of this scheme are: 1) only the phosphorylated form of myosin (P-myosin at lower left) can interact with actin and undergo contraction. Unphosphorylated myosin (myosin) cannot interact with actin. 2) A rise in Ca^{2+} from 10^{-7} M to 10^{-5} M initiates smooth muscle contraction by binding to calmodulin (lower left). The Ca^{2+}-calmodulin complex binds to myosin kinase converting the enzyme from the inactive to the active form. 3) The active myosin kinase phosphorylates myosin resulting in smooth muscle contraction. 4) Lowering the Ca^{2+} concentration below 10^{-7} M dissociates the Ca^{2+}-calmodulin-kinase complex and results in the inactive form of the kinase. Since myosin light chain phosphatase(s) do not require Ca^{2+} for activity, P-myosin is dephosphorylated, and smooth muscle relaxation ensues.

Modulation of smooth muscle relaxation by cyclic AMP is shown in the upper right portion of the diagram: 1) epinephrine interacts with the β-receptor and stimulates adenylate cyclase activity resulting in a rise in cyclic AMP. Increased levels of cyclic AMP result in activation of protein kinase by binding to the regulatory subunit (R) (for a recent review on protein kinases see Krebs and Beavo, 1979). The free catalytic subunit of protein kinase (C), phosphorylates myosin light chain kinase. 2) Phosphorylation of myosin kinase weakens the binding for calmodulin and, as indicated qualitatively by the size of the arrows, shifts the equilibrium in favor of the calmodulin-free *inactive* form of the phosphorylated myosin kinase. 3) A decrease in myosin kinase activity results in an increase in the amount of unphosphorylated myosin and relaxation ensues.

This mechanism might explain how epinephrine relaxes certain smooth muscles, such as bronchial muscles during an asthmatic attack. Recent results with the myosin light chain kinase purified from human platelets suggest a similar effect of cyclic AMP-dependent phosphorylation on the activity of platelet myosin light chain kinase (Hathaway *et al.*, 1979).

Acknowledgment

The authors gratefully acknowledge the editorial assistance of Exa Murray. They also wish to thank William Anderson, C. Robert Eaton, and James M. Miles for expert technical assistance.

REFERENCES

Adelstein, R. S. and Conti, M. A. (1975). *Nature* **256**, 597–598.

Adelstein, R. S. and Eisenberg, E. (1980). *Annu. Rev. Biochem.* **49**, 921–956.

Adelstein, R. S., Conti, M. A., Hathaway, D. R., and Klee, C. B. (1978). *J. Biol. Chem.* **253**, 8347–8350.

Chacko, S., Conti, M. A., and Adelstein, R. S. (1977). *Proc. Natl. Acad. Sci. U.S.* **74**, 129–133.

Conti, M. A., and Adelstein, R. S. (1980). *Fed. Proc.* **39**, 1569–1573.

Dabrowska, R., Aromatorio, D., Sherry, J.M.F., and Hartshorne, D. J. (1977). *Biochem. Biophys. Res. Commun.* **78**, 1263–1272.

Di Salvo, J., Gruenstein, E., and Silver, P. (1978). *Proc. Soc. Exp. Biol. Med.* **158**, 410–414.

Ebashi, S., Mikawa, T., Hirata, M., and Nonomura, Y. (1978). *Proc. N.Y. Acad. Sci.* **307**, 451–461.

Gorecka, A., Aksoy, M. O., and Hartshorne, D. J. (1976). *Biochem. Biophys. Res. Commun.* **71**, 325–331.

Hathaway, D., Eaton, C. R., and Adelstein, R. S. (1980). In *Proceedings of the International Workshop on Regulation of Coagulation*, pp. 271–276, Elsevier/North-Holland, New York.

Hoar, P. E., Kerrick, W.G.L., and Cassidy, P. (1979). *Science* **204**, 503–506.

Huszar, G. and Bailey, P. (1980). *Am. J. Obstet. Gynecol.* **135**, 718–726.

Krebs, E. G. and Beavo, J. A. (1979). *Annu. Rev. Biochem.* **48**, 923–959.

Lebowitz, E. and Cooke, R. (1979). *J. Biochem.* **85**, 1489–1494.

Lee, E.Y.C., Mellgren, R. L., Killilea, S. D., and Aylward, J. H. (1977). In *Regulatory Mechanisms of Carbohydrate Metabolism* (Esmann, V., ed.), pp. 327–346, Pergamon Press, Oxford.

Mikawa, T., Nonomura, Y., Hirata, M., Ebashi, S., and Kakiuchi, S. (1978). *J. Biochem.* **84**, 1633–1636.

Onishi, H. and Watanabe, S. (1979). *J. Biochem.* **85**, 457–472.

Scordilis, S. P. and Adelstein, R. S. (1977). *Nature* **268**, 558–560.

Sherry, J.M.F., Gorecka, A., Aksoy, M. O., Dabrowska, R., and Hartshorne, D. J. (1978). *Biochemistry* **17**, 4411–4418.

Small, J. V. and Sobieszek, A. (1977). *Eur. J. Biochem.* **76**, 521–530.

Sobieszek, A. and Small, J. V. (1977). *J. Mol. Biol.* **112**, 559–576.

Trotter, J. A. and Adelstein, R. S. (1979). *J. Biol. Chem.* **254**, 8781–8785.

Yagi, K., Yazawa, M., Kakiuchi, S., Ohshima, M., and Uenishi, K. (1978). *J. Biol. Chem.* **253**, 1338–1340.

Muscle Contraction: Its Regulatory Mechanisms (Ebashi, S. et al., eds.), pp. 315–328, Japan Sci. Soc. Press, Tokyo / Springer-Verlag, Berlin(1980)

25

Myosin-linked Regulation in Chicken Gizzard Muscle and in Clam Foot Muscle*

SHIZUO WATANABE

Summary: Some observations in support of the phosphorylation-dephosphorylation mechanism of calcium regulation in chicken gizzard muscle are presented: a) the ATPase and superprecipitation reactions of skeletal acto-gizzard "unphosphorylated" myosin (A-UM) and those of skeletal acto-gizzard "phosphorylated" myosin (A-PM) were studied as a function of the ATP concentration. The effect of gizzard native tropomyosin (NTM) on the two reactions was thus shown to be accounted for by a calcium-independent effect of tropomyosin (TM) and by a calcium-

* The study was conducted in collaboration with H. Onishi, K. Nakamura, M. Ikebe, Y. Yamada, H. Kajiyama, H. Suzuki, G. Ashiba, T. Aiba, T. Asada, I. Ohtsuka, T. Kato, S. Iijima, M. Kasai, H. Uchiwa, H. Anzai, K. Haraki, S. Suzuki, J. Umeda (Tokyo Institute of Technology), K. Arai, K. Konno, I. Kimura, K. Nishita, T. Ojima, O. Ohtani, and K. Takahashi (Hokkaido University). The content of this presentation was reported partly by H. Onishi at the International Congress of Biochemistry (July 10, 1979; Tronto, Canada), partly by G. Ashiba at the annual meeting of Japan. Soc. Fisheries (October 5, 1979; Hakodate, Japan), partly by M. Kasai at the annual meeting of Japan. Soc. Biochem. (October 8, 1979; Tokyo, Japan), and partly at the annual meeting of Japan. Soc. Biophys. (October 25, 1979; Nagoya, Japan).

dependent effect of myosin light chain kinase. b) Rabbit skeletal troponin (TN) (in the presence of TM) was shown to act effectively on superprecipitation of A-PM but not on that of A-UM. c) Using glycerinated fibers of gizzard muscle, it was shown that ATP-induced tension development was enhanced by calcium whereas ITP-induced tension development was not.

Two findings with clam foot muscle are also presented: a) Mg-ATPase activity of clam foot myosin alone with no actin was sensitive to calcium, and this calcium sensitivity was dependent on the "EDTA-removable" light chain subunit (regulatory light chain) of clam foot myosin. b) "Uncoupling" of the ATPase reaction from the superprecipitation reaction was observed along the processes of desensitization and resensitization.

In 1970, Kendrick-Jones *et al.* reported their great discovery that the regulatory protein in scallop adductor muscle is not troponin (TN) but an EDTA-removable light-chain subunit of myosin. Their discovery stimulated the idea that calcium regulation in gizzard muscle is also myosin-linked (Bremel, 1974). Moreover, evidence had been accumulated by 1977 for the proposal that enzymic phosphorylation and dephosphorylation of gizzard myosin are responsible for contraction and relaxation, respectively (Sobieszek, 1977; Hartshorne *et al.*, 1977; Adelstein *et al.*, 1977; Ikebe *et al.*, 1977).

CHICKEN GIZZARD MYOSIN

We began a study of gizzard myosin in the spring of 1975. The aim of the study was to find a molecular explanation for the known fact that very high concentrations of magnesium are required for contraction of glycerinated fibers of smooth muscle (Filo *et al.*, 1965). However, achieving little along the lines of the initial aim, we soon found ourselved engaged in a study of the mechanism of calcium regulation in gizzard muscle after one of our students, studying superprecipitation of rabbit skeletal acto-chicken gizzard myosin, happened to obtain the results favoring the phosphorylation-dephosphorylation mechanism of calcium regulation. The first report of his results appeared in 1977 (Ikebe *et al.*). In that first report, we made four suggestions:

1) Calcium is required primarily for activation of myosin light chain kinase (MLCK). 2) Tropomyosin (TM) makes no contribution to calcium sensitivity; it activates superprecipitation equally well in the absence of

calcium and in its presence. 3) Enzymic phosphorylation and dephosphorylation are required for contraction and relaxation, respectively. 4) Skeletal native tropomyosin (NTM, actin-linked regulation) may act effectively only after gizzard myosin is phosphorylated.

Since that report, we have obtained some supporting results (Ikebe *et al.*, 1978; Onishi *et al.*, 1979 a; Onishi *et al.*, 1979 b; Aiba *et al.*, 1979; Onishi *et al.*, 1979 c). A few of these results will be presented below.

1. *Superprecipitation of Skeletal Acto-gizzard Unphosphorylated Myosin (A-UM) as a Function of the ATP Concentration*

Figure 1 shows the following: a) A-UM alone did not show any superprecipitation ability at any concentrations of ATP we tested, from 0.3 μM to 1 mM ATP in the presence of an ATP-regenerating system of puruvate kinase. b) Tropomyosin was capable of activating superprecipitation of A-UM at low concentrations of ATP, 1 and 10 μM ATP. The TM-activated superprecipitation showed no calcium sensitivity. c) Gizzard NTM activated superprecipitation not only at low concentrations of ATP but also at high concentrations. The NTM activated superprecipitation at low concentrations of ATP was not sensitive to calcium, whereas that at high concentrations of ATP was sensitive to calcium.

2. *Superprecipitation of Skeletal Acto-gizzard Phosphorylated Myosin (A-PM) as a Function of the ATP Concentration*

Figure 2 shows the following: a) with A-PM alone, superprecipitation occurred at all the ATP concentrations tested except at 1 mM ATP, and no calcium sensitivity was detected in the superprecipitation of A-PM. b) Tropomyosin was now capable of enhancing superprecipitation of A-PM not only at low concentrations of ATP but also at high concentrations. Accordingly, the effect of TM was no longer different from that of NTM. c) Moreover, superprecipitation of APM in the presence of TM and that in the presence of NTM were both independent of calcium.

On the basis of the results shown in Figs. 1 A and 1 B, it may be concluded that the effect of gizzard NTM can be satisfactorily accounted for by two independent effects; calcium-independent activation by TM, and calcium-dependent phosphorylation by MLCK.

3. *ATPase Reaction*

Figure 2 A shows that in the early stage of the ATPase reaction catalyzed by skeletal acto-gizzard unphosphorylated myosin with gizzard NTM

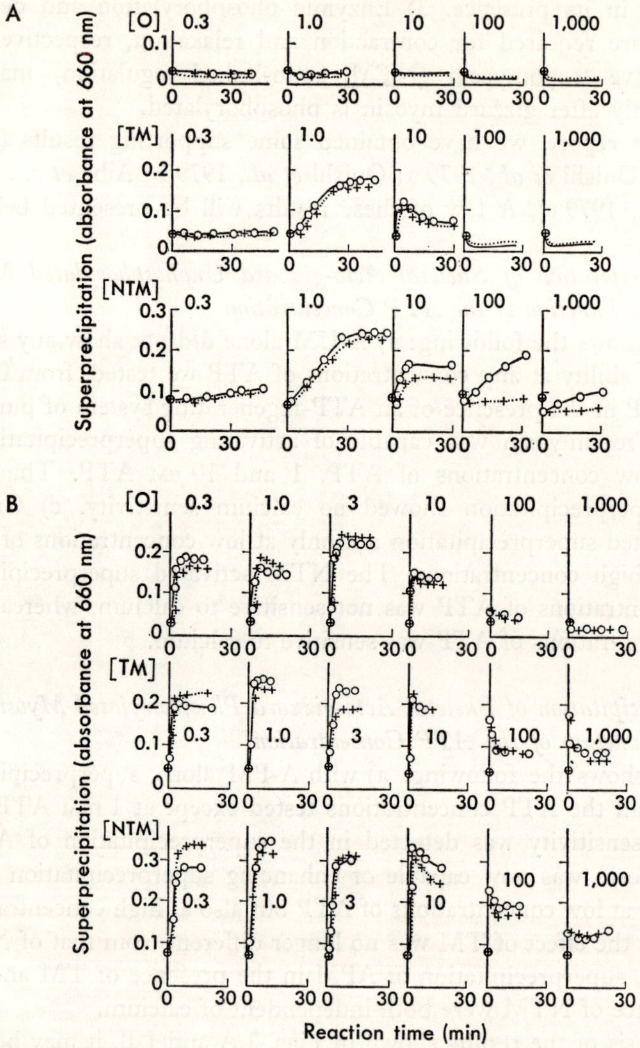

Fig. 1. Acto-unphosphorylated myosin (A) and A-PM (B) in the superprecipitation reaction. 0.1 mg/ml rabbit skeletal actin, 0.1 mg/ml gizzard myosin (UM or PM), 0.07 mg/ml gizzard (or skeletal) TM or gizzard NTM. 0.1 mg/ml pyruvate kinase, 1 mM phoshoenol pyruvate, 70 mM KCl, 20 mM Tris-maleate (pH 6.8), 10 mM $MgCl_2$, and 0.2 mM $CaCl_2$ plus 0.1 mM EGTA (○) or 1 mM EGTA (+). 28°C.

(A-UM-NTM), "burst" hydrolysis of ATP and "lag" in ATP hydrolysis were observed at low concentrations of ATP and at its high concentrations,

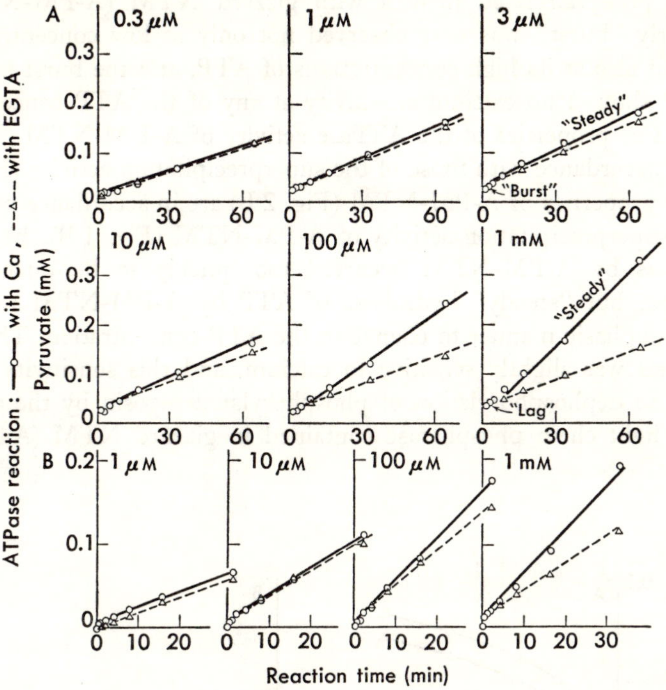

Fig. 2. Acto-unphosphorylated myosin (A) and A-PM (B) in the ATPase reaction. Conditions are the same as described for Fig. 1.

respectively. According to Watanabe (1970) and Yasui *et al.* (1965), burst hydrolysis of ATP is closely associated with superprecipitation, and according to Onishi and Watanabe (1979 a), lag in ATP hydrolysis occurs in the period during which the 20,000-dalton light chain of gizzard myosin is phosphorylated by MLCK. "Burst" and "lag" were both followed by "steady" hydrolysis of ATP. If one looks at the early stage of ATP hydrolysis (burst and lag), one may conclude that the ATPase activity of A-UM-NTM responds in a biphasic manner to change in the ATP concentration. On the other hand, if one looks at the subsequent (steady) stage of ATP hydrolysis, one may conclude that the ATPase activity of A-UM-NTM responds in a monophasic manner to change in the ATP concentration. Calcium sensitivity was detected only at high concentrations of ATP.

Figure 2 B shows the ATPase reaction catalyzed by skeletal acto-gizzard phosphorylated myosin with gizzard NTM (A-PM-NTM). a) The early "burst" was now observed not only at low concentrations of ATP but also at its high concentrations of ATP, and the burst hydrolysis of ATP showed no calcium sensitivity at any of the ATP concentrations tested. The properties of the ATPase activity of A-UM-NTM (Fig. 2 A) were in accordance with those of the superprecipitation activity (Fig. 1 A), and the properties of A-PM-NTM (Fig. 2 B) are in accordance with those of the superprecipitation activity of A-PM-NTM (Fig. 1 B). b) "Burst" hydrolysis by A-PM-NTM occurred too quickly to be quantitatively evaluated, but "steady" hydrolysis of ATP by A-PM-NTM responded in a monophasic manner to change in the ATP concentration. The steady hydrolysis was slightly sensitive to calcium, and this sensitivity is attributable to dephosphorylation of phosphorylated myosin by the action of myosin light chain phosphatase contained in gizzard NTM (Aiba *et al.*, 1979).

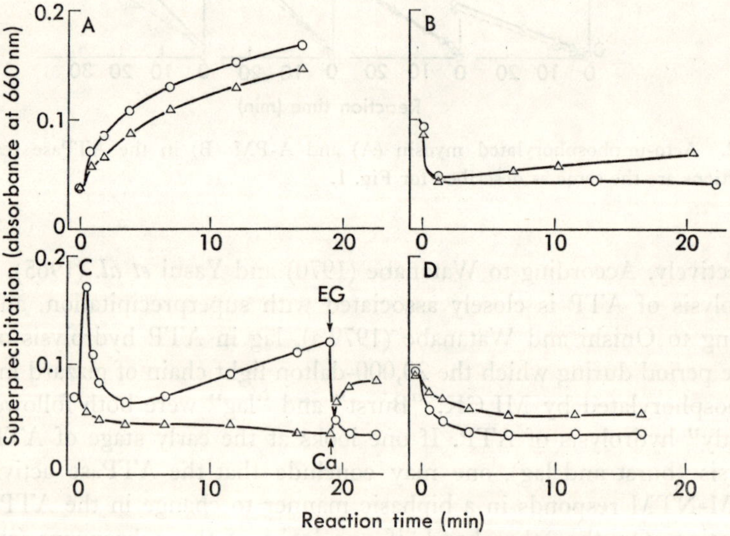

Fig. 3. Effect of skeletal TN on superprecipitation of A-UM-TM and on that of A-PM-TM. 0.1 mg/ml rabbit skeletal actin, 0.1 mg/ml gizzard myosin (UM or PM), 0.01 mg/ml skeletal TM, 0.01 mg/ml skeletal TN, 60 mM KCl, 20 mM Tris-maleate (pH 6.8), 10 mg $MgCl_2$, 0.5 mM ATP, and 0.2 mM $CaCl_2$ with 0.1 mM EGTA (○) or 0.2 mM EGTA (△). 28°C. A: A-PM+TM. B: A-UM+TM. C: A-PM+TM+TN. D: A-UM+TM+TN.

4. Troponin from Skeletal Muscle

As shown above in Fig. 1 A, acto-unphosphorylated myosin in the presence of TM (A-UM-TM) showed no superprecipitation activity in such a high concentration of ATP as 0.5 mM, either in the presence of calcium or in its absence (Fig. 3 B). We have now observed, as shown in Fig. 3 D, that addition of rabbit skeletal TN made no difference in the superprecipitation activity of A-UM-NTM. We also showed in a previous figure (Fig. 1 B) that A-PM-TM was capable of superprecipitating in a high concentration of ATP, and that the superprecipitation reaction was not sensitive to calcium (Fig. 3 A). However, we have now observed, as shown in Fig. 3 C, that addition of skeletal troponin made A-PM-TM sensitive to calcium; that is, superprecipitation of A-PM-TM-TN was strongly inhibited in the absence of calcium and enhanced in the presence of calcium. Moreover, the calcium-enhanced superprecipitation was reversed on subsequent addition of EGTA, and the reversal occurred without accompanying dephosphorylation of PM (data not shown). In the latter connection, we are now studying if the ATPase reaction is decelerated on the EGTA-induced reversal superprecipitation.

GLYCERINATED FIBERS OF GIZZARD MUSCLE

ATP- and ITP-induced tension developments in glycerinated fibers of chicken gizzard muscle were studied. As shown in Fig. 4, ATP and magnesium without calcium induced tension development in glycerinated fibers (Mg-tension), and a subsequent addition of calcium caused a further increase in tension (Ca-tension). Removal of calcium by adding EGTA caused decrease in Ca-tension to the level of Mg-tension. We then used ITP instead of ATP, and observed that magnesium without calcium also induced tension development, but that a subsequent addition of calcium failed to further enhance tension development. Since ITP did not serve as a substrate for MLCK (Onishi *et al.*, 1979 b), phosphorylation was not supposed to occur in ITP media even in the presence of calcium. In fact, in separate experiments in which suspensions of a large number of glycerinated fibers and urea gel electrophoresis were employed, we were able to show that in ITP media, no phosphorylation was detectable at any time during the tension measurement, and that in ATP media, no phosphorylation of the regulatory light chain occurred at time *b* (Mg-tension), nearly complete phosphorylation occurred at time *c* (Ca-tension), and complete dephosphorylation occurred at time *d* (EGTA-relaxation).

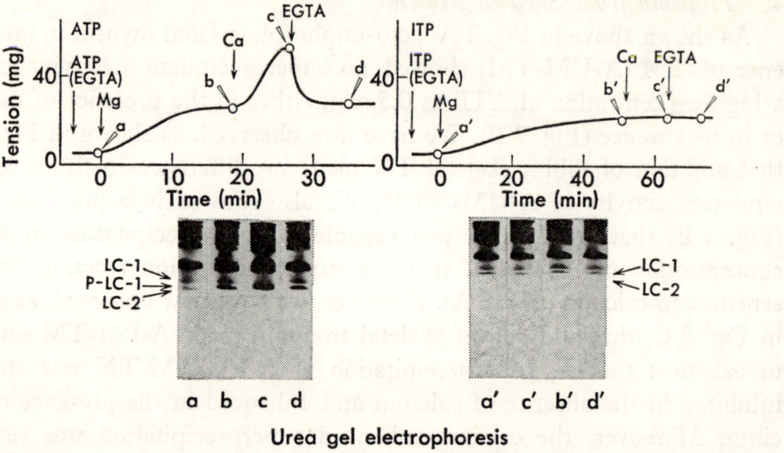

Fig. 4. ATP- and ITP-induced tension developments in glycerinated fibers of chicken gizzard muscle. The basic medium contained 150 mM KCl, 20 mM imidazole-HCl (pH 6.8), and 0.3 mM DTT. (↓ATP (EGTA)) 5 mM ATP and 2 mM EGTA. (↓Mg) 10 mM $MgCl_2$, 5 mM ATP or ITP, and 2 mM EGTA. (↓Ca) 0.2 mM $CaCl_2$, 0.1 mM EGTA, 10 mM $MgCl_2$, and 5 mM ATP or ITP. (↓EGTA) 2 mM EGTA, 10 mM $MgCl_2$, and 5 mM ATP or ITP. (↓ITP (EGTA)) 5 mM ITP and 2 mM EGTA. Urea gel electrophoresis (Perrie, Smillie, and Perry (1973) *Biochem. J.* **135**, 151–164): 8 M urea, 10% polyacrylamide, amido black staining, and 480 μg protein per gel rod of 0.5×5 cm. Samples for the electrophoretic analysis were picked at ▽ marks with a, b, c, and d and with a′, b′, c′, and d′.

It is therefore strongly suggested that calcium regulation of ATP-induced tension development in glycerinated fibers involves phosphorylation and dephosphorylation of myosin light chains. Moreover, the fact that the excess concentrations of magnesium with no calcium induced tension development may imply that the number of actin-myosin binding (cross-bridge) in the absence of calcium was sufficiently large to produce tension. It seems interesting, therefore, to investigate (for example, by measuring the stiffness of fibers as a measure of actin-myosin bindings) whether calcium-activated tension development involves primarily increase in the number of actin-myosin bindings, or whether it is associated with increase in ATP hydrolysis rather than increase in the number of actin-myosin binding.

In the latter connection, it is noteworthy that Mg-ATPase activity and Mg-ITPase activity of gizzard phosphorylated myosin (PM with no actin) were higher than those of gizzard unphosphorylated myosin (UM with no actin) (Ikebe *et al.*, 1978; Onishi *et al.*, 1979 b). Moreover, as will be shown shortly (see also Asada *et al.*, 1979), Mg-ATPase activity of clam myosin with no actin is strongly activated by very low concentrations (~10 μM) of calcium.

CLAM FOOT MYOSIN

Our study of clam foot myosin began in the fall of 1976. The aim of the study was to elucidate the mechanism of a possible cooperation between two heads of a myosin molecule. The cooperation was suggested, by Kendrick-Jones *et al.* (1976), to be at work in molluscan muscles. However, here again, we haven't had even a clue to begin studying the cooperation, and have found ourselves studying instead an apparent uncoupling of the ATPase reaction from the superprecipitation reaction.

1. *Calcium-activation of Myosin-ATPase*

The ATPase activity and the superprecipitation activity (Watanabe, 1970) of clam foot myosin B were measured by varying the ATP concentration at a fixed concentration (10 mM) of $MgCl_2$. The results obtained are shown in Fig. 5 A. a) The activities responded in a biphasic manner to change in the ATP concentration, giving the maximal activities at 10 μM to 1 mM. b) Calcium sensitivity of the ATPase activity was clearly detected at all the ATP concentrations we tested, from 1 μM to 1 mM. c) When the ATP concentration was higher than about 0.1 mM, practically no superprecipitation occurred, whereas the ATPase activity still showed a fairly strong response to calcium.

The two activities were also measured by varying the $MgCl_2$ concentration from 1 μM to 1 mM at a fixed concentration (1 mM) of ATP, and essentially the same results as described above were obtained (Fig. 5 B), thus indicating that Mg-ATP complex is the chemical species responsible for the biphasic (activation and inhibition) response.

Essentially the same results as described above (a, b, c) were also obtained by using rabbit skeletal acto-clam foot myosin (1: 1 by weight) in place of clam foot myosin B (Fig. 6). This time we also measured Mg-ATPase activity of clam myosin alone, and found, as shown at the left-hand side in Fig. 6, that it was activated by 0.1 mM calicum at all the ATP concentrations tested. Moreover, when the ATP concentration was higher than approximately 0.1 mM, calcium-activated ATPase activity of myosin alone was nearly equal to that of acto-myosin. In other words,

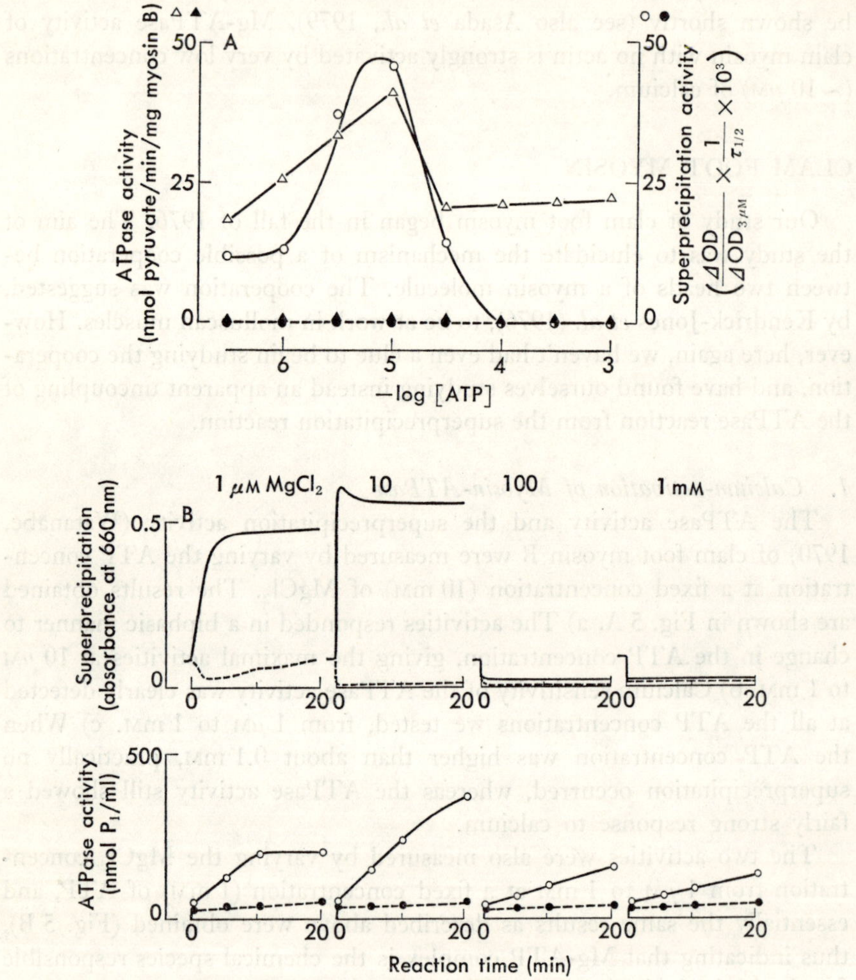

Fig. 5. The ATPase activity and the superprecipitation activity of clam foot myosin B as a function of the ATP concentration (A), and as a function of the magnesium concentration (B). (A) 0.2 mg/ml or (B) 0.3 mg/ml clam myosin B, 50 mM KCl, 20 mM Trismaleate (pH 6.5), (A) 10 mM $MgCl_2$ with various concentrations of ATP (0.1 mg/ml pyruvate kinase and 1 mM phosphoenol pyruvate) or (B) 1 mM ATP with various concentrations of $MgCl_2$. ○, △ 0.3 mM $CaCl_2$ with 0.2 mM EGTA; ●, ▲ 0.2 mM EGTA. 20°C.

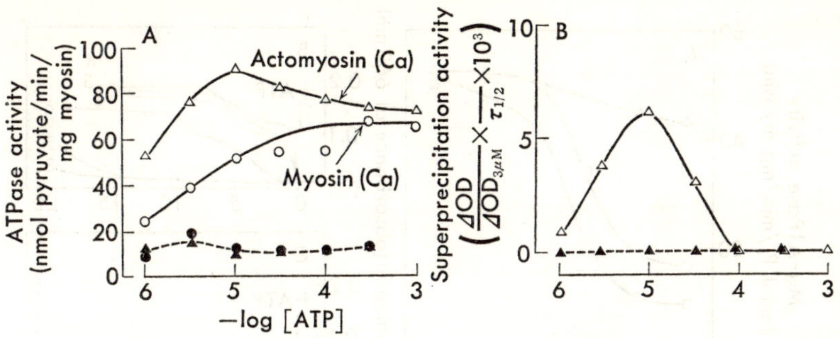

Fig. 6. The ATPase activity and the superprecipitation acitivity of actomyosin, and the ATPase activity of myosin alone as a function of the ATP concentration. 0.1 mg/ml rabbit skeletal actin with 0.1 mg/ml clam foot myosin (△▲) or 0.1 mg/ml clam foot myosin with no actin (○●). ○, △ 0.1 mM $CaCl_2$; ●, ▲ 1 mM EGTA. Other conditions are the same as described in Fig. 5.

practically no actin-activation of myosin-ATPase was observed in the ATP concentration where no superprecipitation occurred, and superprecipitation occurred only in the ATP concentration where actin-activation of myosin-ATPase was observed to occur.

2. *Desensitization and Resensitization*

EDTA-wash and recombination of the EDTA-removable light-chain subunit of clam foot myosin were conducted by using the procedures reported by Szent-Györgyi *et al.* (1973) for scallop myosin B and myofibrils. We then obtained evidence for that EDTA-removable light-chain subunit (LC-1) of clam foot myosin is responsible for the calcium sensitivity we found in the ATPase activity of clam myosin alone. Figure 7 A shows a) that calcim activation of "unwashed" myosin (UM) occurred at very low concentrations of calcium, b) that when clam myosin was washed ten times with 10 mM EDTA at 0°C, the washed myosin (desensitized myosin, DM) lost most of its calcium sensitivity, and c) that the calcium sensitivity returned when the light chain subunit once removed by EDTA-wash was added back to UM (obtaining resensitized myosin, RM).

Combining these myosins (UM, DM, and RM) with skeletal actin in a weight ratio of 1/1, superprecipitation of the combined system was studied, and an unexpected result was obtained. As shown in Figs. 7 B, C, and D, much of the superprecipitation ability was lost in acto-DM (A-DM), and the lost ability was regained by acto-RM (A-RM). Since

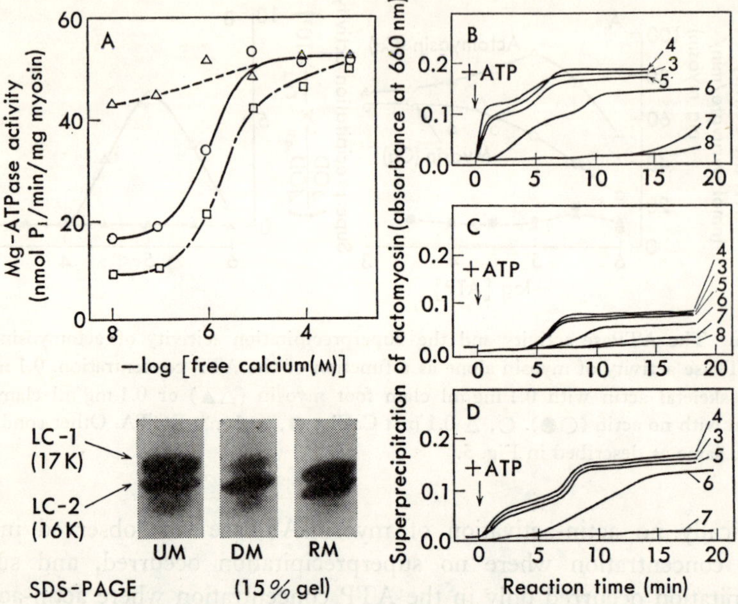

Fig. 7. Desensitization and resensitization of clam foot myosin. A: ATPase activity of myosin. △ DM, ○ UM; □ RM. B, C, and D: superprecipitation of actomyosin: rabbit skeletal actin with clam foot myosin (1:1 in a weight ratio), 10 mM $MgCl_2$, 0.1 mM ATP, 50 mM KCl, and 20 mM Tris-maleate (pH 6.5). Numerical figure attached at each of turbidity curves in the pCa value. B: A-UM. C: A-DM. D: A-RM.

in the presence of calcium, the ATPase activity of DM was as high as that of UM, and since in the absence of calcium, that of DM was much higher than that of UM, we had expected to observe that superprecipitation of A-DM occurs in the absence of calcium as well as in its presence. This unexpected result was also obtained when clam foot myosin B was subjected to the desensitization procedure. The molar ratio of LC-1 to LC-2 estimated by SDS-gel electrophoresis was approximately 1.0 in unwashed myosin B, 0.66 in 5-time washed myosin B, 0.5 in 10-time washed one, and 0.34 in 20-time washed one. Figure 8 shows that the superprecipitation activity suddenly dropped when the LC-1 / LC-2 ratio was reduced to less than about 0.5, whereas the ATPase activity stayed at a high level in the presence and absence of calcium. It appeared therefore that removal of more than half of the LC-1 content resulted in "uncoupling" of the ATPase reaction from the superprecipitation reaction. With this result, we thought that we had finally obtained a clue to a study of

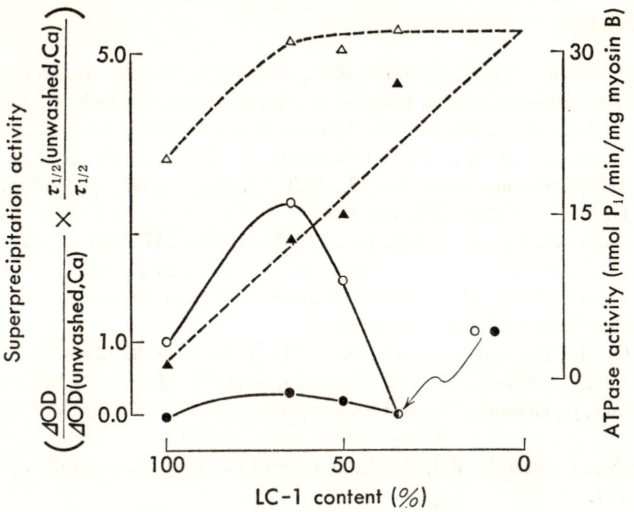

Fig. 8. Desensitization of myosin B from clam foot muscle. 0.1 mg/ml Clam foot myosin B, 50 mM KCl, 20 mM Tris-maleate (pH 6.5), 10 mM $MgCl_2$, and 0.1 mM ATP, 20°C. The light chain ratio of LC-1/LC/2 was estimated by SDS gel electrophoresis (see Fig. 7). Triangles are for ATPase, and circles for superprecipitation. △, ○ 0.1 mM $CaCl_2$; ▲, ● 1 mM EGTA.

the cooperation between two heads of a myosin molecule. However, Nishita *et al.* (1979) with Akazaca adductor myosin, and Konno *et al.* (1979; see Poster Session of this conference) with scallop adductor myosin, have recently observed that not one but 2 mol of regulatory light chains per mol of myosin are required in order for desensitized myosin (in combination with rabbit skeletal actin) to regain the full activity of superprecipitation and the full sensitivity to calcium that undesensitized myosin had. Since EDTA-wash of myosin B results in removal of not only LC-1 (EDTA-light chains) but also other components of myosin B such as TM and paramyosin, the effect of EDTA-wash is not entirely attributable to removal of LC-1. On the other hand, it is fairly certain that EDTA-wash of myosin causes only removal of EDTA-light chains. Presently we are investigating the effect of paramyosin and that of TM on the contractile activities of clam actomyosin.

Finally we would like to just mention that a common feature for myosin-linked regulations in chicken gizzard muscle and in clam foot muscle could possibly be in that the primary action of calcium is to activate myosin-ATPase rather than to induce actin-myosin binding.

REFERENCES

Adelstein, R. S., Conti, M. A., Scordilis, S. P., Chacko, S., Barylko, B., and Trotter, J. A. (1977). In *Excitation-contraction Coupling in Smooth Muscle* (Casteels, R., Godfraind, T., and Rüegg, J. C., eds.), pp. 359-366, Elsevier/North-Holland, Amsterdam.

Aiba, T., Ohtsuka, I., Onishi, H., and Watanabe, S. (1979). *J. Biochem.* **86**, 1275-1282.

Asada, T., Ashiba, G., and Watanabe, S. (1979). *J. Biochem.* **85**, 1543-1546.

Bremel, R. D. (1974). *Nature* **252**, 404-407.

Filo, R. S., Bohr, D. F., and Rüegg, J. C. (1965). *Science* **147**, 1581-1583.

Hartshorne, D. J., Gorecka, A., and Aksoy, M. O. (1977). In *Excitation-contraction Coupling in Smooth Muscle* (Casteels, R., Godfraind, T., and Rüegg, J. C., eds.), pp. 377-384, Elsevier/North-Holland, Amsterdam.

Ikebe, M., Onishi, H., and Watanabe, S. (1977). *J. Biochem.* **82**, 299-303.

Ikebe, M., Aiba, T., Onishi, H., and Watanabe, S. (1978). *J. Biochem.* **83**, 1643-1655.

Kendrick-Jones, J., Lehman, W., and Szent-Györgyi, A. G. (1970). *J. Mol. Biol.* **54**, 313-326.

Kendrick-Jones, J., Szentkiralyi, E. M., and Szent-Györgyi, A. G. (1976). *J. Mol. Biol.* **104**, 747-775.

Konno, K., Arai, K., and Watanabe, S. (1979). *J. Biochem.* **86**, 1639-1650.

Nishita, K., Ojima, T., and Watanabe, S. (1979). *J. Biochem.* **86**, 663-673.

Onishi, H. and Watanabe, S. (1979a). *J. Biochem.* **85**, 457-472.

Onishi, H. and Watanabe, S. (1979b). *J. Biochem.* **86**, 569-573.

Onishi, H., Iijima, S., Anzai, H., and Watanabe, S. (1979c). *J. Biochem.* **86**, 1283-1290.

Sobieszek, A. (1977). *Eur. J. Biochem.* **73**, 477-483.

Szent-Györgyi, A. G., Szentkiralyi, E. M., and Kendrick-Jones, J. (1973). *J. Mol. Biol.* **74**, 179-203.

Watanabe, S. (1970). *J. Biochem.* **68**, 913.

Yasui, T. and Watanabe, S. (1965). *J. Biol. Chem.* **240**, 98-104.

Muscle Contraction: Its Regulatory Mechanisms (*Ebashi, S. et al., eds.*), *pp. 329–345*,
Japan Sci. Soc. Press, Tokyo / Springer-Verlag, Berlin (*1980*)

26

Leiotonin, Regulatory Protein of Smooth Muscle

YOSHIAKI NONOMURA

Summary: 1. The Calcium regulatory proteins, leiotonin complexes, A and C, were prepared from the supernatant of chicken gizzard myosin B washed with 1 mM $NaHCO_3$. Addition of a leiotonin complex to the desensitized actomyosin with tropomyosin activated the superprecipitation of this system in the presence of calcium without phosphorylation of the P light chain. Leiotonin C was relatively easily distinguished from calmodulin by alkaline slab gel electrophoresis. It was confirmed using this technique that purified myosin B reserves leiotonin C, but not calmodulin.

2. Thin filament fraction was isolated from fresh chicken gizzard using differential centrifugation under various conditions. Purity and uniformity were checked by electron microscopy and SDS-gel electrophoresis. This fraction activated myosin ATPase in the presence of Ca^{2+}, while actin with tropomyosin alone could not activate it.

3. Phosphorylation of P light chain in living smooth muscle was examined with autoradiography. Although many protein bands were phosphorylated in the fiber fixed at contracted state, thc phosphorylation of

P light chain was not so clear. The chemically skinned fiber showed the phosphorylation of the P light chain in the fiber fixed at one time contraction, but its extent was very minor compared with unphosphorylated one. The phosphorylation of the P light chain remained in the fiber fixed in a relaxed state after repetitive contraction and relaxation.

At present the intracellular trigger for contraction in smooth muscle is well known to be Ca^{2+}, but direct evidence, as in the case of striated muscle, was not established until the introduction of the chemically skinned fiber with saponin. The smooth muscle skinned fiber shows clearly that the repetition of contraction and relaxation is regulated by calcium concentrations (Endo *et al.*, 1977; Saida and Nonomura, 1978). Since the first approach to the calcium regulation of smooth muscle in the protein level (Ebashi *et al.*, 1966), a number of trials for preparing troponin-like proteins have been conducted by several laboratories. It is, however, becoming clear that smooth muscle does not contain a troponin system for calcium regulation like that in striated muscle. Instead, two different views have been proposed for explanation of the regulatory mechanism in smooth muscle. These two concepts are schematized in Fig. 1.

Before the studies of the first group started, Perry and his collaborators

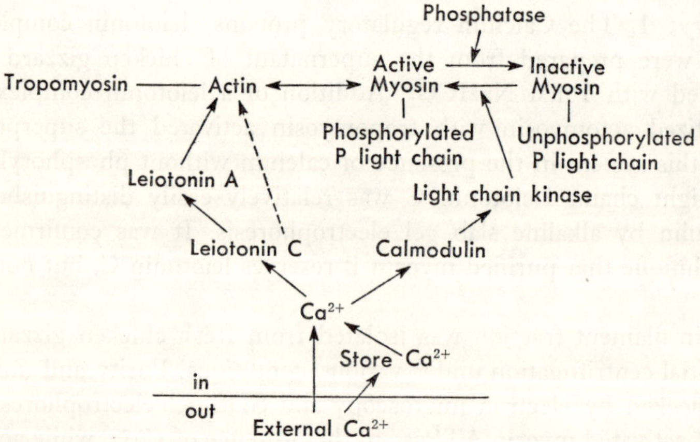

Fig. 1. Scheme illustrating the mechanisms proposed for Ca^{2+} regulation on the basis of two different concepts. The solid arrows indicate the direction of binding, interaction, and activation. The dashed arrow suggests the possibility of binding and interaction. Tropomyosin has a special position in both concepts because it activates an interaction of the actomyosin system, though its mechanism has not been explained yet.

discovered that myosin light chain kinase phosphorylates the P light chain present in all vertebrate myosin, including smooth muscle (Pires *et al.*, 1974; Frearson and Perry, 1975; Frearson *et al.*, 1976). On the basis of these works, several groups have proposed that the contraction-relaxation of smooth muscle is regulated by phosphorelation and dephosphorylation of the P light chain in a Ca^{2+}-dependent way (Górecka *et al.*, 1976; Sobieszek and Small, 1976; Chacko *et al.*, 1977; Bremel *et al.*, 1977; Ikebe *et al.*, 1978). This concept is demonstrated in the right-hand side of Fig. 1.

According to this concept, smooth muscle myosin and actin do not interact in the unphosphorylated state of the P light chain; once phosphorylated, the actomyosin system becomes activated irrespective of the presence or absence of Ca^{2+}. Hartshorne's group (Dabrowska *et al.*, 1978) added a further important contribution for Ca^{2+} dependency to this system. They separated calmodulin from the light chain kinase; the Ca^{2+}-dependent activation of the light chain kinase was found to be induced by the binding of calmodulin in a Ca^{2+}-concentration range of more than 10^{-6} M

On the other hand, we could isolate the Ca^{2+}-regulatory protein, leiotonin from the native tropomyosin fraction of smooth muscle, and this factor activates the actomyosin system with tropomyosin in the presence of Ca^{2+} (Mikawa *et al.*, 1977). Furthermore, an acidic protein with low molecular weight was isolated, and thus leiotonin is known to be composed of an 85,000-dalton actin-binding protein, leiotonin A and a 17,000-dalton acidic protein, leiotonin C (Mikawa *et al.*, 1978 b). Only one of them could not activate the actomyosin-tropomyosin system. Leiotonin C is very like calmodulin, but essentially different in various properties: no activation of phosphodiesterase (Mikawa *et al.*, 1978 b); slightly different mobility of SDS gel electrophoresis (Mikawa *et al.*, 1978 b); no tri-methyl-lysine but the presence of cysteine residue; different peptide map pattern (Mikawa *et al.*, 1978 a). Furthermore, leiotonin C in actin filament shows a strong Ca^{2+} binding affinity (Hirata *et al.*, 1980). Our view is presented in the left-hand side of Fig. 1.

Recently some reports in support of the phosphorylation concept have been published by two groups dealing with living and skinned fibers. Barany's group showed, using living arterial smooth muscle, that the P light chain was rapidly phosphorylated in the case of Ca^{2+} requirement (Barron *et al.*, 1979). Kerrick's group reported using the disrupted muscle (Hoar *et al.*, 1979) and the chemically skinned fiber (Cassidy *et al.*, 1979) that the phosphorylation of the P light chain occurred in accordance with the phosphorylation concept.

In the present study, leiotonin was prepared from the supernatant of myosin B washed with low ionic strength. The purpose was to start the preparation of leiotonin from materials with as low a content of light chain kinase and calmodulin as possible. Further, I tried to separate thin filament fraction in a natural form to confirm the presence of thin filament-linked regulation (Nonomura *et al.*, 1980). Finally, critical studies were examined with the intact and skinned fiber to check the phosphorylation of the P light chain in contraction and relaxation.

MATERIALS AND METHODS

Chicken gizzard was used for the preparation of contractile proteins. Myosin B was prepared essentially according to Nonomura and Ebashi (1974), with a slight modification: the use of ultracentrifugation for isolating microsome fraction. All the following procedures were carried out at 4°C. Minced tissues were homogenized in the solution containing 0.4 M KCl and 30 mM Tris-HCl (pH 7.4) by a Vertis homogenizer (Vertis Co., Ltd.) for a short time at high speed. After ATP was added to the extracted solution (final concentration, more than 3 mM), the dissociated solution was immediately centrifuged at 30,000 rpm for 45 min. The supernatant contained mainly myosin B, while the resultant precipitates were composed cell fragments, nuclei, mitchondria, and microsomes including ribsomes. Without such ultracentrifugation the myosin B fraction would be contaminated by microsomes during the succeeding purification procedure. Purification of myosin B was carried out with repetitions of solubilization of myosin B at high ionic strength (0.4 M KCl) and precipitation by dilution down to 60 mM KCl. Sometimes 1 to 3 mM $MgCl_2$ was added to the solution during the precipitation procedure.

Myosin and actin were prepared from chicken gizzard myosin B according to Ebashi (1976), but sometimes actin was purified by the method of Spudich and Watt (1971) from the acetone powder of the residue after extraction of myosin B. Tropomyosin was purified by the same method as skeletal muscle tropomyosin (Ebashi *et al.*, 1968) from native tropomyosin fraction of chicken gizzard (Ebashi *et al.*, 1975). SDS gel electrophoretic patterns of these proteins are shown in Fig. 9.

Superprecipitation and ATPase were measured by the procedure described in the RESULTS section.

The conditions for checking the phosphorylation in the intact fiber and the skinned fiber with autoradiography are also described in the RESULTS section. The chemically skinned fiber was prepared with appropriate concentrations of saponin according to Saida and Nonomura (1978).

SDS gel electrophoresis was carried out by a discontinuous Tris-glycine system according to Laemmli (1971), alkaline gel electrophoresis according to Davis (1964), and urea gel electrophoresis by the method of Perrie and Perry (1970). A mini-slab apparatus (Wakamori Co., Ltd.) was used for electrophoresis.

Samples were negatively stained with 4% uranyl acetate. Electron micrographs were taken by a JEM 100 B-4 type electron microscopy with accelerating voltagd of 80 KV.

Brain calmodulin was kindly supplied by Professor Kakiuchi, Osaka University.

RESULTS

1. *Isolation of Leiotonin Complex from Myosin B*

Since smooth muscle myosin B shows calcium sensitivity in superprecipitation and ATPase, the fraction providing the calcium sensitivity to actomyosin was considered to be isolated from myosin B. Before starting the extraction of calcium sensitizing factor, the supernatant at centrifugation in 60 mM KCl was collected during purification and was called S^0. The extraction was carried out by washing myosin B with low ionic strength solution. After myosin B was diluted with 1 mM $NaHCO_3$, the suspension was centrifuged at 7,000 rpm for 15 min. The resultant supernatant was named S^1, and the precipitate P^1. The same procedure was repeated, and thus second washing supernatants S^2, S^3 . . . and/or second washing precipitates P^2, P^3, . . . were obtained. S-fractions were concentrated by an appropriate method. It was known from SDS gel electrophoretic pattern of S and P that tropomyosin began to be extracted from the 1st step of washing, and large amounts were detached from myosin B in the second step. The concentrated S fraction was applied to DEAE-52 cellulose column chromatography and eluted in steps by 0.1 M, 0.25 M, and 0.4 M KCl buffered with Tris-HCl (pH 7.4). The fractions obtained at each of steps are named D^1, D^2, and D^3, respectively.

As shown in Fig. 2, the elution by 0.25 M KCl of S^3 (named S^3D^2 fraction) contained tropomyosin and other proteins, perhaps leiotonin complex. Arrows point to the positions of leiotonin A (about 85,000 daltons) and leiotonin C (17,000 daltons). S^3D^2 fraction was further applied to hydroxyapatite column chromatography for isolating tropomyosin from leiotonin

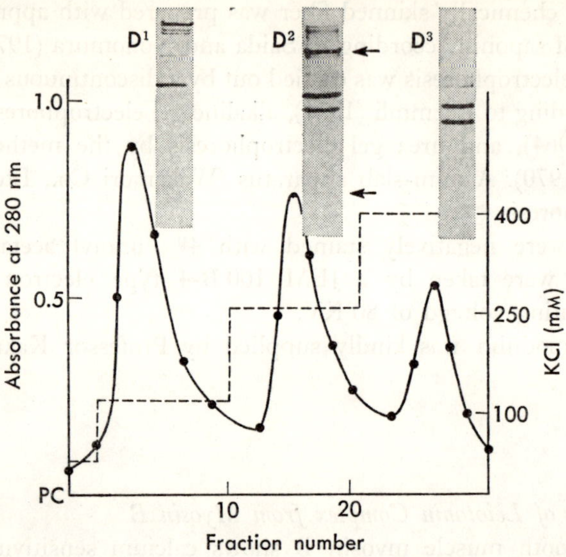

Fig. 2. DEAE-cellulose chromatography of the supernatant of myosin B washed with 1 mM $NaHCO_3$ three times (S^3 fraction). The sample was applied to a 1×10 cm column equilibrated with 15 mM Tris-HCl, pH 7.4 and eluted stepwisely by 0.1 M, 0.25 M, and 0.4 M KCl buffered with Tris-HCl. SDS gel electrophoretic patterns at each peak are shown as D^1, D^2, and D^3, respectively. Arrows point to the positions of leiotonin A and C.

complex. The fraction was eluted continuously by 50 mM to 300 mM KCl buffered with Tris-HCl (pH 7.4). Two main peaks were obtained at around 80 mM and 180 mM KCl, and the fraction at 180 mM KCl was determined to be tropomyosin from the SDS gel electrophoretic pattern. The eluted fraction at 80 mM KCl (see Fig. 9 d for SDS gel electrophoretic pattern) showed remarkable activation on superprecipitation in the presence of tropomyosin and calcium when the fraction was added to P^3 myosin B, while P^3 myosin B scarcely showed activation in the presence of tropomyosin and calcium (Fig. 3 A). The superprecipitated samples collected by centrifugation were applied to the urea gel electrophoresis to check phosphorylation of the P light chain. Figure 3 B clearly showed that the P light chains in the both cases in the presence of calcium of Fig. 3 A were not phosphorylated, but that the P light chain in control myosin B without washing at low ionic strength showed a phosphorylated form by the same procedure as in the previous two cases.

From this result the fraction by hydroxyapatite column chromatograph

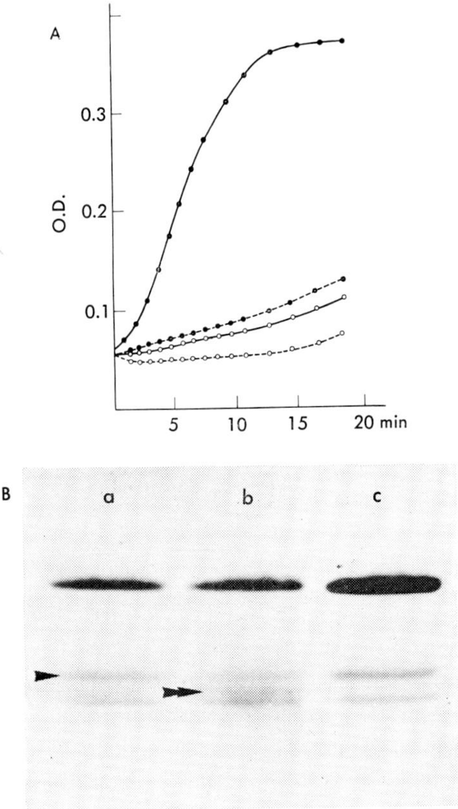

Fig. 3. Superprecipitation of desensitized myosin B and the additional effect of leiotonin complex (A) and their urea gel patterns (B). Desensitized myosin B was prepared by washing myosin B with 1 mM $NaHCO_3$ three times (P^3 fraction). Superprecipitation was measured by the method of Ebashi (1961) at 660 nm. 0.6 mg/ml protein concentration was used. The solution contained 50 mM KCl, 8 mM $MgCl_2$, and 15 mM Tris-maleate (pH 6.8). The reaction was started by addition of 0.5 mM ATP. A: --- desensitized myosin B with 0.01 mg/ml tropomyosin; —— desensitized myosin B with 0.01 mg/ml tropomyosin and 0.002 mg/ml hydroxyapatite fraction eluted with 80 mM KCl of S^3D^2 fraction (leiotonin complex, see Fig. 9d); ● +Ca (10^{-5} M); ○ −Ca (10^{-7} M). B: urea gel electrophoretic patterns of samples after superprecipitation. After superprecipitation the samples were centrifuged and the precipitates were solved in a solution containing 8 M urea, 0.15 M Tris-glycine (pH 8.4) and 0.2% 2-mercaptethanol. a, the sample with adding leiotonin in the presence of Ca^{2+}; b, the usual myosin B in the presence of Ca^{2+} (although the superprecipitation was not shown, it was fully activated); c, the sample of desensitized myosin B with tropomyosin in the presence of Ca^{2+}. Single arrow indicates the position of the unphosphorylated P light chain, and double arrow shows the position of the phosphorylated one.

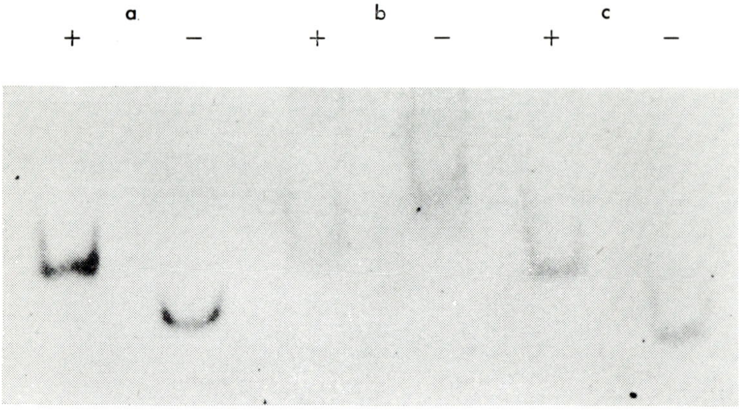

Fig. 4. Alkaline gel electrophoretic patterns of the fractions obtained by DEAE column chromatography. a, brain calmodulin; b, S^3D^2 fraction; c, S^0D^3 fraction; +, the sample was applied in the presence of calcium (0.1 mM Ca^{2+}); −, in the absence of calcium (1 mM EGTA).

of S^3D^2 is considered mainly to be leiotonin complex, and not to contain light chain kinase and calmodulin. To confirm this point, alkaline gel electrophoresis is useful. Mikawa *et al.* (1978 a) suggested that leiotonin C showed the same tendency toward mobility as troponin C in alkaline gel, moving faster in the presence of calcium than in its absence. Calmodulin is known to behave in the opposite fashion, moving solwer in the presence of calcium than in its absence. Figure 4 shows that the fraction obtained form S^3D^2 behaves like troponin-C, while the fraction of S^0D^3 moves like brain calmodulin.

Each fraction was tested for the presence of leiotonin C or calmodulin by the alkaline gel mentioned above. The presence of calmodulin was confirmed in the fractions of S^0D^3 and S^1D^3, while leiotonin C was recognized in the fractions of S^1D^2, S^2D^2, and S^3D^2. S^2D^3 and S^3D^3 should signs of the presence of both calcium-binding proteins, but not so clearly.

Although this study is a preliminary one, relatively many calmodulins are considered to loosely coexist in crude myosin B fraction, but in the purification procedure almost all the calmodulins are eliminated from myosin B. On the other hand, leiotonin C binds relatively tightly to myosin B, but gradually leaves myosin B at low ionic strength. We can conclude from the present study that leiotonin complex plays an important role in the calcium sensitivity of chicken gizzard myosin B.

2. *Separation of Thin Filament Fraction*

It is well known that separation of thick and thin filament is extremely difficult and practically impossible in skeletal muscle (Hardwicke and Hanson, 1971). However, smooth muscle myosin has some special characteristics: necessity of magnesium ions for filament formation, and ribbon formation in relatively high magnesium concentrations (Nonomura, 1979). On the other hand, there are a more thin filaments than with thick, as recognized fro the cross sectioned electron microscopical view of chicken gizzard (Fig. 5 A). The ratio of thin to thick filaments is about 8 to 1 in the cross-sectioned area. Utilizing these characteristics, we can partially separate thin filaments from thick filaments.

Fresh minced muscle was homogenized by a Vertis homogenizer in a solution containing 60 mM KCl, 20 mM Tris-HCl (pH 7.4), 1 mM EGTA,

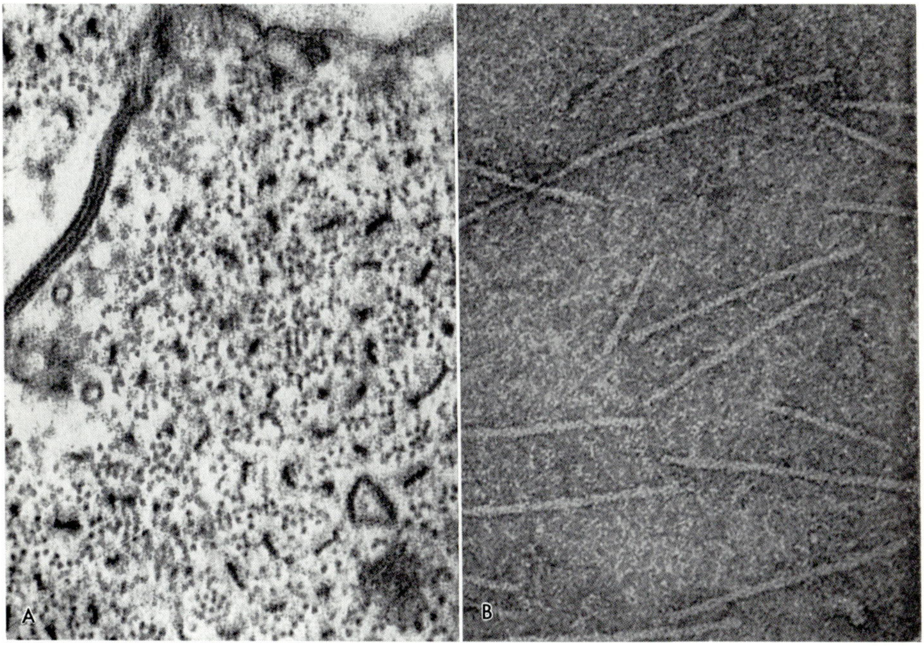

Fig. 5. Electron micrographs of section of chicken gizzard and negative staining of separated thin filaments. A: cross-sectioned image of chicken gizzard. Materials were fixed in the resting state according to the usual method. Notice abundant distribution of thin filaments compared with thick filaments. Structure of nexus and dense body can be seen. B: negatively stained image of thin filament fraction. Negatively stained with 4% uranyl acetate. Notice filaments with short length.

10 mM $MgCl_2$, and 5 mM ATP for a short time. The homogenized solution was immediately centrifuged at 10,000 rpm for 20 min. A large amount of myosin was precipitated in ribbon form together with cell debris, nuclei, and mitchondria. Microsomes were also precipitated as aggregates in such high concentrations of magensium.

Next, the supernatant including many thin filaments was centrifuged at 40,000 rpm for 90 min. The pellet was suspended in a solution including 60 mM KCl, 3 mM $MgCl_2$, and 10 mM Tris-maleate (pH 6.8). The remaining myosin binds to thin filaments under these conditions. Then it was eliminated by centrifugation at 8,000 rpm for 20 min. Further purification was attained by repeating the procedure after the ultracentrifugation mentioned above. Figure 6 shows the SDS gel electrophoretic pattern of the purified thin filament and other products in the purification process. The thin filament is composed of actin, tropomyosin, and other minor components. A negatively stained image of this fraction showed homogenous distribution of typical thin filaments (Fig. 5 B), but their length was relatively short, being less than 1 μm. This thin filament fraction retained calcium sensitivity because the addition of thin filament fraction to purified myosin gave calcium sensitivity to myosin ATPase, while purified actin

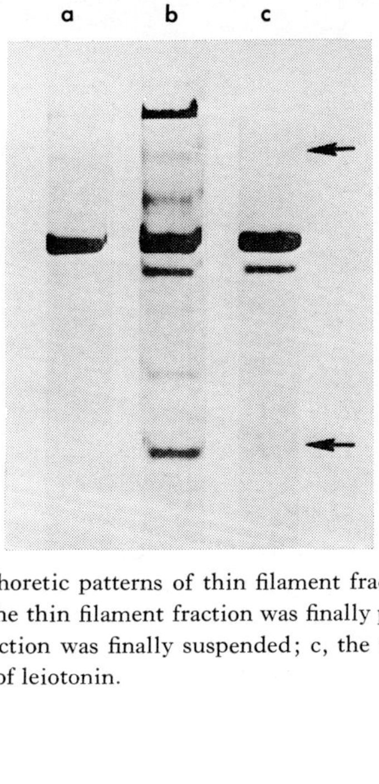

Fig. 6. SDS gel electrophoretic patterns of thin filament fraction and the related fraction. a, supernatant after the thin filament fraction was finally precipitated; b, precipitate when the thin filament fraction was finally suspended; c, the thin filament fraction. Arrows indicate the position of leiotonin.

TABLE I
ATPase of Reconstructed Actomyosin System
ATPase was measured in solutions containing 60 mM KCl, 20 mM Tris-HCl (pH 7.2), 8 mM $MgCl_2$, and 1 mM ATP according to Youngburg and Youngburg (1930). Calcium concentrations were calculated by Saida and Nonomura (1978) with calcium-EGTA buffer: +Ca, 4×10^{-5} M; −Ca, 10^{-7} M.

Components for ATPase assay	Specific activity of ATPase		Ratio of +Ca/−Ca
	+Ca	−Ca	
Myosin+actin+tropomyosin[a]	17.0	12.3	1.38
Myosin+thin filament[b]	33.8	13.7	2.47

[a] 0.4 mg/ml myosin, 0.1 mg/ml actin, and 0.01 mg/ml tropomyosin were used.
[b] 0.4 mg/ml myosin, 0.1 mg/ml thin filament fraction were used.

and myosin showed no calcium sensitivity in ATPase even with the addition of tropomyosin, as shown in Table I (Nonomura *et al.*, 1980).

This result supports the view that the calcium-sensitive site exists at least on the thin filament, as we have insisted up to the present.

3. *Does Phosphorylation of the P Light Chain Occur in Contracted State in the Intact and Skinned Fibers?*

Essentially according to the procedure of Barany's group (Barron *et al.*, 1979) the teased fibers of chicken gizzard were incubated overnight in Locke's solution including [^{32}P] orthophosphate. One sample was isometrically contracted by replacing Na^+ in Locke's solution with K^+, and the other was relaxed by adding 2 mM $MnCl_2$ (Nonomura *et al.*, 1966). Both samples were fixed by adding perchloric acid (PCA) and then homogenized by a small grass homogenizer in a solution containing Locke's solution with 10% PCA and centrifuged at 10,000 rpm for 30 min. The resultant supernatant was checked for incorporating on of [^{32}P] orthophosphate into ATP in the cell, while the pellet was solved in the solution containing 0.125 M Tris-HCl (pH 7.0), 2% SDS, 10% glycerol, and 0.2% 2-mercaptoethanol. The SDS slab gel electrophoresis of solved sample was carried out with the standard sample of chicken gizzard myosin and tropomyosin. After staining and drying, autoradiography of the specimen was performed. As shown in Fig. 7, [^{32}P] was incorporated into ATP and many protein bands were labeled, but the band corresponding to the P light chain was rarely phosphorylated. The band near the P light chain was recognized to be strongly labeled. Unless autoradiography is used

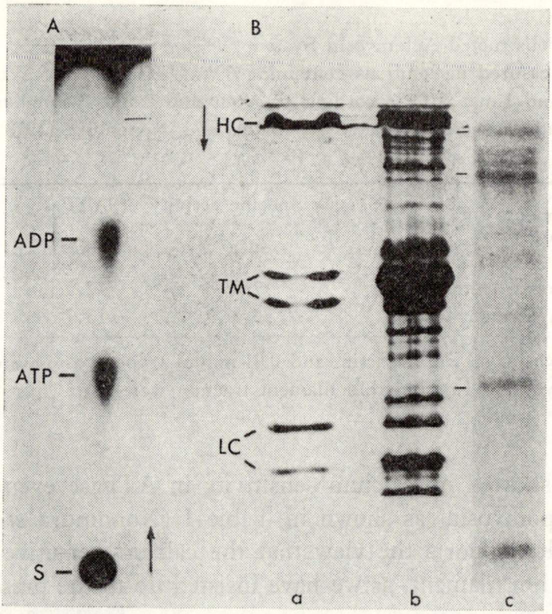

Fig. 7. Autoradiography of thin layer chromatography and SDS slab gel electrophoresis of intact chicken gizzard. The tissue incubated with [^{32}P] orthophosphate was fixed in a contracted state in high K^+, homogenized, and then separated into supernatant and precipitate. A: thin layer chromatography of the supernatant. [^{32}P] was transferred to ATP. B: SDS electrophoretic pattern of proteins solubilized from the precipitate. a, marker sample of myosin and tropomyosin; b, coomassie blue-stained proteins; c, autoradiograph of the dried sample of b.

with the standard marker, this labeled band might be misunderstood as the P light chain.

As the author has reported already (Saida and Nonomura, 1978), the chemically skinned fiber of guinea pig taenia coli could continue the cycle of contraction and relaxation (depending on calcium concentration) for a long time. Using γ [^{32}P] ATP in the relaxing or contracting solution, the specimen could easily check its phosphorylation of proteins. Contraction was induced by increasing calcium concentration up to 10^{-5} M, and relaxation by keeping calcium less than 10^{-7} M. The treatment of the specimen was essentially the same as that of the intact fiber; in the case of urea gel electrophoresis, the precipitated sample was solved in a solution containing 8 M urea, 0.125 M Tris-HCl (pH 7.0), 10% glycerol, and 0.2% 2-mercaptoethanol. The autoradiograph shows clearly the phosphorylation

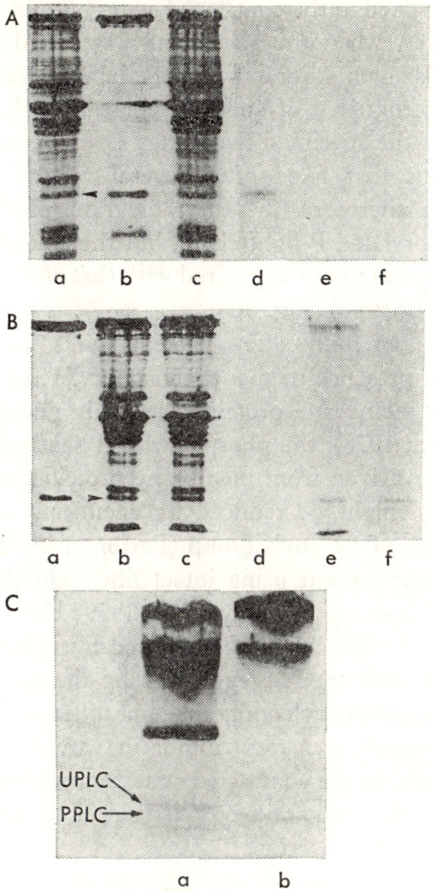

Fig. 8. Autoradiographs of SDS slab gel electrophoresis of chemically skinned fiber of guinea pig's taenia coli. A: SDS slab gel electrophoresis of the skinned fiber fixed in contracted (a and d) or relaxed (c and f) state. The skinned fiber was contracted and relaxed only one time in the presence of [^{32}P] ATP. b and d have myosin as a marker. a, b, and c are coomassie blue-stained samples and d, e, and f are autoradiograph of dried gels corresponding to a, b, and c. B: the skinned fiber fixed in contracted (b and d) or relaxed (c and f) state after repetition of contraction and relaxation. a and d have myosin as a marker. Other conditions are the same as in B. C: urea gel electrophoresis of the skinned fiber fixed in contracted state one time in the presence of [^{32}P] ATP. a, coomassie blue-stained sample; b, autoradiograph of dried gel of a. The arrow UPLC points at the position of the unphosphorylated P light chain, and PPLC points at that of the phosphorylated P light chain.

of the P light chain fixed in the contracted state, but not in the relaxed state (Fig. 8 A). However, when the specimens were fixed after repetition of contraction and relaxation several times, the labeled feature of the relaxed state was different from that of one-time relaxation. As shown in Fig. 8, the P light chain of the specimen fixed after several contractions remained phosphorylated, but that fixed after several relaxations also remained phosphorylated. Furthermore, the autoradiograph of the urea gel shows the labeled phosphorylated P light chain, but the amount of the phosphorylated P light chain was minor compared with that of the unphosphorylated one on the stained urea gel (Fig. 8C).

In the skinned fiber, only a few protein bands were phosphorylated, compared with the presence of many phosphorylated bands in the intact fibers. Since many membrane proteins are easily phosphorylated (Schulman and Greengard, 1978), the phosphorylated bands observed in the intact fibers might be drived from membrane proteins, and thus the phosphorylated proteins might be reduced by chemically skinned treatment. At least the evidence of Barany's group (Barton *et al.*, 1979) could not be confirmed by our experiment using intact fiber. On the other hand, the evidence of Kerrick's group (Cassidy *et al.*, 1979; Hoar *et al.*, 1979) could be partly reproduced with chemically skinned fiber, but the result of the repeated relaxation could not be explained by the relaxation mechanism on the basis of dephosphorylation. The phosphorylation of the P light chain, once it occurred, was very minor in extent compared with the unphosphorylated one in the contracted state from the result of the stained urea gel. Details of the study will be reported elsewhere (Nonomura and Mikawa, unpublished manuscript).

DISCUSSION

In this study I have examined our concept of the thin filament-linked calcium regulatory system in smooth muscle from some different approaches. We have previously prepared regulatory proteins from the native tropomyosin fraction which was extracted at low ionic strength and salted out with ammonium sulfate (Ebashi *et al.*, 1975). Although this fraction is a good source of a large number of regulatory proteins, it also contains light chain kinase and calmodulin (except for leiotonin complex); in fact, Sovieszek and Small (1976) named this fraction the kinase-tropomyosin fraction. Myosin B is a useful starting material because calmodulin is easily eliminated from myosin B during the purification procedure main-

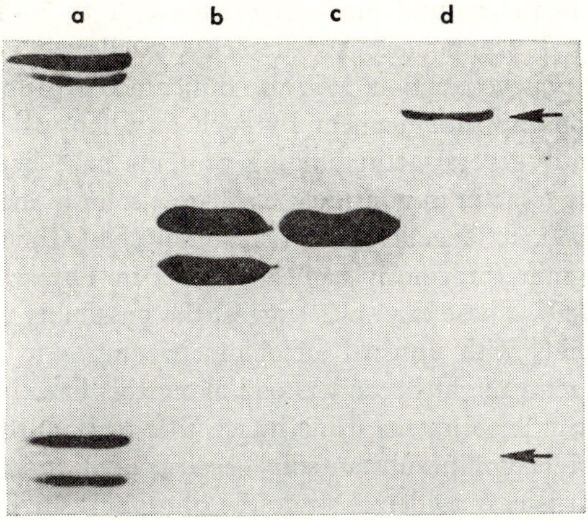

Fig. 9. SDS slab gel electrophoretic pattern of contractile proteins in chicken gizzard smooth muscle. a, myosin; b, tropomyosin; c, actin; d, leiotonin complex. Arrow points to the position of leiotonin A and leiotonin C.

taining a certain ionic strength, more than 60 mM KCl. Leiotonin is resistant, holding on fast to myosin B during the usual purification procedure, but it is released from myosin B at low ionic strength, and then myosin B becomes desensitized. Leiotonin complex could be obtained from the supernatant of myosin B washed with 1 mM $NaHCO_3$. (In actuality, I am describing obtaining leiotonin C, not calmodulin; both calcium binding proteins are easily identified by alkaline gel, but leiotonin A cannot be simply identified from light chain kinase.) Although our fraction can be confirmed functionally to not include kinase fraction, I would like to search for a better method to identify leiotonin A and kinase easily.

The SDS gel electrophoretic pattern of structural proteins in smooth muscle is shown in Fig. 9. Calcium regulation can be perfectly reconstructed from these four contractile proteins.

This strongly supports our concept that the mixture of the thin filament fraction and myosin induces calcium sensitivity to actomyosin ATPase. Generally speaking, preparation of the thin filament fraction is still difficult, and the yeild is small. Since this experiment is preferable for the thin filament-linked system, the preparation method should be improved so that we can obtain much larger amounts easily. Here is our weak point: not being able to show definitely the presence of leiotonin in actin and tropomyosin (see Fig. 6) because the ratio of leiotonin to actin is extremely low in the natural mode of existence, 1 to 50–100 (Ebashi *et al.*, 1978, 1979). It seems curious that such a small ratio could affect the function without enzyme action. It is also difficult to explain how leiotonin could localize on the thin filament by such an abnormal stoichiometry. However, recently several actin binding proteins have been reported to exist at low ratio to actin in non-muscle cells; gelsolin in macrophage (Yin and Stossel, 1979), actinogelin in Ehrlich tumor cell (Mimura and Asano, 1979), and fragmin (previously named actinin) in Physarum (Hasegawa and Hatano, 1980). These examples suggest the possibility of the presence of special proteins with unusual stoichiometric ratios to actin. Binding of leiotonin to actin might be understood along this line.

It is difficult in experiments using intact fiber or skinned fiber to show the existence of leiotonin substantially because of its minor contents in the cell, while it is easy to demonstrate the phosphorylation of the P light chain because of its labeling property. But there is on the contrary, some tendency to overestimate its amount. The result with skinned fiber has corresponded to the concept of phosphorylation-dephosphorylation in the case of simultaneous contraction-relaxation. On the urea gel, however, the P light chain was not fully phosphorylated, but its phosphorylation was very minor. Recently Kerrick's group seems to agree partly with this evidence (Cassidy *et al.*, 1979). If hereafter such minor phosphorylation of the P light chain is enough for smooth muscle contraction and if its role is interpreted as a kind of trigger for contraction, it should be outside the general discussion of the calcium regulatory mechanism.

REFERENCES

Barron, J. T., Barany, M., and Barany, K. (1979). *J. Biol. Chem.* **254**, 4954–4956.

Bremel, R. D. (1974). *Nature* **252**, 405–406.

Bremel, R. D., Sobieszek, A., and Small, J. V. (1977). In *The Biochemistry of Smooth Muscle* (Stephens, N. L., eds.), pp. 533–549, Univ. Park Press, Baltimore.

Cassidy, P., Phyllis, E., and Kerrick, W.G.L. (1979). *J. Biol. Chem.* **241**, 11148–11153.

Chacko, S., Conti, M. A., and Adelstin, R. S. (1977). *Proc. Natl. Acad. Sci. U.S.* **74**, 129–133.

Dabrowska, R., Sherry, J.M.F., Aromatorio, D. K., and Hartshorne, D. J. (1978). *Biochemistry* **17**, 235–257.

Ebashi, S. (1961). *J. Biochem.* **50**, 236–244.

Ebashi, S. (1976). *J. Biochem.* **79**, 229–231.

Ebashi, S., Iwakura, H., Nakajima, H., and Nakamura, R. (1966). *Biochem. Z.* **345**, 201–211.

Ebashi, S., Kodama, A., and Ebashi, F. (1968). *J. Biochem.* **65**, 465–477.

Ebashi, S., Mikawa, T., Hirata, M., and Nonomura, Y. (1978). *Ann. N.Y. Acad. Sci.* **307**, 451–461.

Ebashi, S., Nonomura, Y., Mikawa, T., Hirata, M., and Saida, K. (1979). In *Cell Motility: Molecules and Organization* (Hatano, S., Ishikawa, H., and Sato, H., eds.), pp. 225–237, Univ. Tokyo Press, Tokyo.

Ebashi, S., Toyo-oka, T., and Nonomura, Y. (1975). *J. Biochem.* **78**, 859–861.

Endo, M., Kitazawa, T., Yagi, S., Iino, M., and Kakuta, Y. (1977). *In Excitation-contraction Coupling in Smooth Muscle* (Casteels, R., Godfraind, T., and Rüegg, J. C., eds.), pp. 199–209, Elsevier/North-Holland, Amsterdam.

Frearson, N., Focant, B.W.W., and Pery S. V. (1976). *FEBS Lett.* **63**, 27–32.

Frearson, N. and Perry, S. V. (1975). *Biochem. J.* **151**, 99–107.

Górecka, A., Aksoy, M. O., and Hartshorne, D. J. (1976). *Biochem. Biophys. Res. Commun.* **71**, 325–331.

Hardwicke, P.M.D. and Hanson, I. (1971). *J. Mol. Biol.* **59**, 509–516.

Hasegawa, T. and Hatano, S. (1980). *Biochemistry* (in press).

Hirata, M., Mikawa, T., Nonomura, Y., and Ebashi, S. (1980). *J. Biochem.* **87**, 369–378.

Hoar, P. E., Kerrick, W.G.L., and Cassidy, P. S. (1979). *Science* **204**, 503–506.

Ikebe, M., Aiba, T., Onishi, H., and Watanabe, S. (1978). *J. Biochem.* **83**, 1643–1656.

Laemmli, U. K. (1970). *Nature* **227**, 680–685.

Mikawa, T., Hirata, M., Nonomura, Y., and Ebashi, S. (1978a). Abstr. 6th Int. Congr. Biophys., Kyoto, p. 233.

Mikawa, T., Nonomura, Y., Hirata, M., Ebashi, S., and Kakiuchi, S. (1978b). *J. Biochem.* **84**, 1633–1636.

Mikawa, T., Toyo-oka, T., Nonomura, Y., and Ebashi, S. (1977). *J. Biochem.* **81**, 273–275.

Mimura, N. and Asano, A. (1979). *Nature* **282**, 44–48.

Nonomura, Y. (1979). In *Cell Motility: Molecules and Organization* (Hatano, H., Ishikawa, H., and Sato, H., eds.), pp. 467–482, Univ. Tokyo Press, Tokyo.

Nonomura, T. and Ebashi, S. (1974). In *Methods in Pharmacology* (Daniel, E. E. and Paton, D. M., eds.), vol. 3, pp. 141–162, Plenum Press, New York.

Nonomura, Y., Mikawa, T., and Ebashi, S. (1980). *Proc. Japan Acad.* **56**, 178-183.

Nonomura, Y., Hotta, Y., and Ohashi, H. (1966). *Science* **152**, 97–99.

Perrie, W. T. and Perry, S. V. (1970). *Biochem. J.* **119**, 31–38.

Pires, E., Perry, S. V., and Thomas, M.A.V. (1974). *FEBS Lett.* **41**, 292-296.

Saida, K. and Nonomura, Y. (1978). *J. Gen. Physiol.* **72**, 1–14.

Schulman, H. and Greengard, A. P. (1978). *Proc. Natl. Acad. Sci. U.S.* **75**, 5432-5436.

Sobieszek, A. and Small, J. V. (1976). *J. Mol. Biol.* **101**, 75–92.

Spudich, J. A. and Watt, S. (1971). *J. Biol. Chem.* **246**, 4866–4871.

Yin, H. L. and Stossel, T. P. (1979). *Nature* **281**, 583-586.

Youngburg, G. E. and Youngburg, M. V. (1930). *J. Lab. Clin. Med.* **16**, 158–166.

Muscle Contraction: Its Regulatory Mechanisms (*Ebashi, S. et al., eds.*), *pp. 347–357*,
Japan Sci. Soc. Press, Tokyo / Springer-Verlag, Berlin (1980)

27

"Freezing" of Activated and Depressed Thin Filaments of Skeletal and Smooth Muscles

TAKASHI MIKAWA

Summary: The reconstituted thin filaments from actin, tropomyosin and troponin of rabbit skeletal muscle were chemically cross-linked with glutaraldehyde in the presence and absence of calcium ions. Although the presence and absence of calcium ions during cross-linking had no significant influence on the extent of intermolecular bridging, thin filaments cross-linked in the presence of calcium ions were always in the active, "on" state, whereas those cross-linked in the absence of calcium ions were always in the inactive the "off" state. Separation of thin filaments into these two types by cross-linking required the presence of troponin. Glutaraldehyde intramolecularly cross-linked with troponin-C and hindered its structural change by calcium ions. Glutaraldehyde modification thus appeares to be useful in preserving, or "freezing," the thin filament in an "on" or "off" state. Interaction of the cross-linked thin filament with scallop myosin was still controlled by calcium ions, but ATPase was activated more effectively by the thin filament frozen in an "on" state than in an "off" state. The thin filaments of smooth muscle were prepared directly from chicken gizzard. The cross-linking of the separated thin filaments with glutaraldehyde also froze them in "on" and "off" states in activating the myosin ATPase of smooth muscle.

Calcium-regulation of skeletal muscle contraction is mediated by tropomyosin and troponin which are located on the thin filament (Ebashi and Endo, 1968). These proteins inhibit the actin-myosin-ATP interaction in the absence of calcium ions. There should exist, at least, two states in the thin filament; one is the activated state in the presence of calcium ions, and the other the depressed state in the absence of calcium ions. Recently many investigators have shown differences in the structural arrangement of the thin filament between the two states (Haselgrove, 1973; H. E. Huxley, 1973; Hanson *et al.*, 1973; Oosawa *et al.*, 1973; Parry and Squire, 1973; Yanagida *et al.*, 1974; Wakabayashi *et al.*, 1975; Gillis and O'Brien, 1975). Although it has not yet been proven that these differences are the essential phenomena underlying the regulation of contraction, two types of model have been proposed to explain the regulatory mechanism. One is that the regulatory proteins, tropomyosin and troponin, control the actin-myosin-ATP interaction by changing the conformation of the F-actin (Hanson *et al.*, 1973; Oosawa *et al.*, 1973). The other is the steric blocking model (Haselgrove, 1973; H. E. Huxley, 1973; Parry and Squire, 1973) in which inhibition of the actin-myosin-ATP interaction results from the position of tropomyosin rather than from structural change in F-actin.

One approach to understanding the role of the structural difference in the thin filament proteins between the two states in the regulation would be to black the structural changes claimed to be essential for regulation, and then to test whether thereby the regulation is lost or not. It has previously been shown that glutaraldehyde treatment produces two types of thin filament; one type, treated by the reagent in the presence of calcium ions, is active in interaction with myosin, whereas the other, treated in the absence of calcium ions, is inactive; in both cases, the interaction with myosin is no longer controlled by calcium ions (Mikawa, 1979a). In addition, the ability of the smooth muscle thin filament reconstituted from actin and regulator proteins, tropomyosin and leiotonin (Ebashi *et al.*, 1977; Mikawa *et al.*, 1978), is also frozen in the active and inactive states as a result of glutaraldehyde treatment (Mikawa, 1979b). These results indicate that the thin filament treated with glutaraldehyde is devoid of regulatory nature, retaining the properties of the condition in which it is frozen.

In the following some properties of cross-linked thin filament of rabbit skeletal muscle will be described. I will also refer to my attempt to freeze the native thin filament of smooth muscle.

GENERAL FEATURES OF CROSS-LINKING WITH GLUTARALDEHYDE

Thin filaments of skeletal muscle were reconstituted from actin, tropomyosin and troponin with a weight ratio of 4: 1: 1 (Fig. 1); in this paper these reconstituted thin filaments will be called "thin filament." About 1 mg per ml of the thin filament was incubated with 0.1% glutaraldehyde in 0.1 M KCl, containing 3 mM $MgCl_2$ and 50 mM HEPES, pH 7.2, at 25°C. Glutaraldehyde intremolecularly bridged the thin filament proteins and thereby most components did not enter the gel matrix of 13.5% on

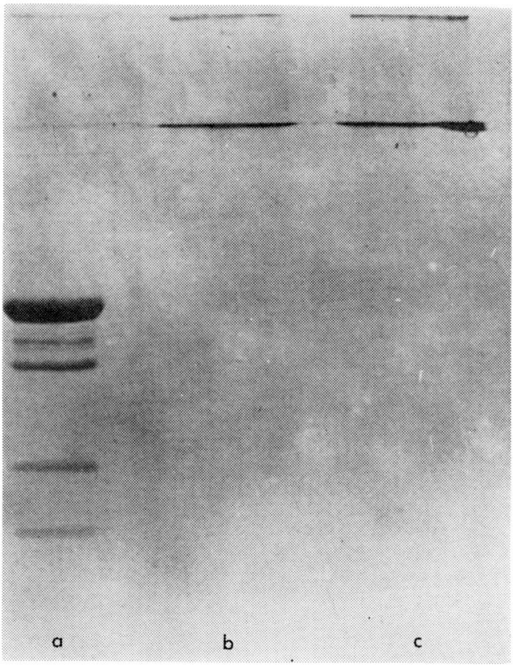

Fig. 1. SDS-gel electrophoretic patterns of the reconstituted thin filament cross-linked in either the presence or absence of calcium ions. a, thin filament of rabbit skeletal muscle reconstituted from actin, tropomyosin, and troponin with a weight ratio of 4: 1: 1; b, thin filament cross-linked in the absence of calcium ions; c, this filament cross-linked in the presence of calcium ions.

electrophoresis in the presence of SDS (Fig. 1). The extent of cross-linking was not influenced by the presence or absence of calcium ions under the cross-linking reaction so far examined by the electrophoretic

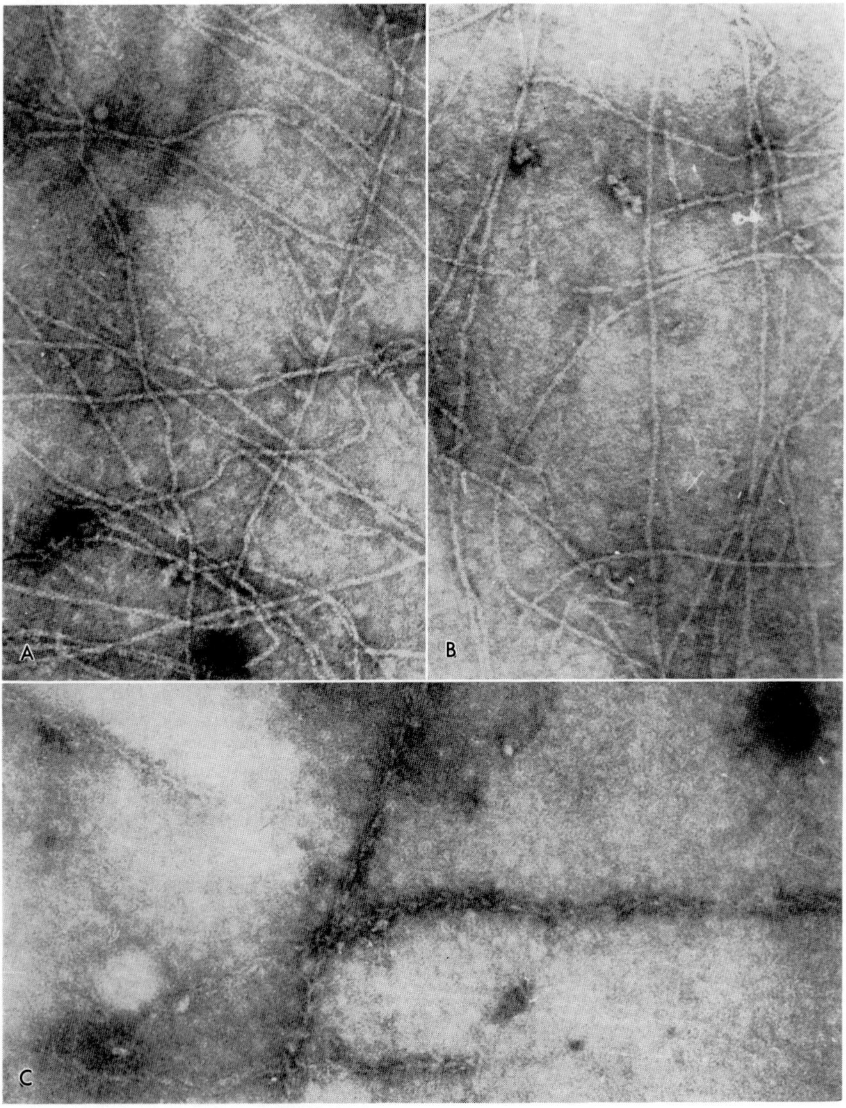

Fig. 2. Electron micrographs of the cross-linked thin filament. A: unmodified thin filament. B: cross-linked thin filament. C: cross-linked thin filament decorated with heavy meromyosin.

patterns (Fig. 1). Under the electron microscope, no distinct difference was observed between filamentous structure before and after modification with glutaraldehyde (Fig. 2). The cross-linked thin filaments were decorated with heavy meromyosin and showed "arrowhead" structure in 0.6 M KCl containing 3 mM $MgCl_2$ and 20 mM Tris-HCl, pH 7.6 (Fig. 2C).

TIME COURSE OF EFFECTS OF CROSS-LINKING ON THE INTERACTION OF THE THIN FILAMENT WITH MYOSIN

Figure 3 describes the time course of effects of cross-linking on the ability of the thin filament to activate myosin ATPase. Cross-linking in the presence of calcium ions resulted in a gradual loss of inhibition by the removal of calcium ions in the interaction with myosin. On the other hand, cross-linking in the absence of calcium ions resulted in a decrease of the extent in activation by calcium ions. This time course was in accord with the expectation derived from the mechanism proposed for the regulation by the troponin-tropomyosin system, *i.e.*, troponin induces the depression of the myosin-actin-ATP interaction, and calcium ions remove

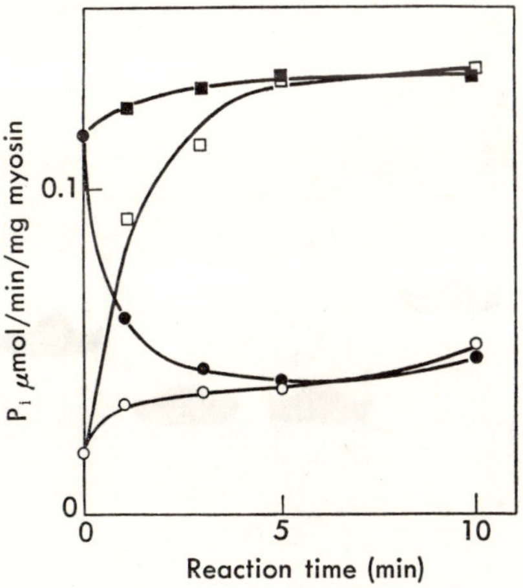

Fig. 3. Myosin ATPase activities activated by the thin filament which cross-linked for varying time. ATPase activities were assayed in a solution containing 0.22 mg/ml thin filament, 0.2 mg/ml myosin, 0.1 M KCl, 3 mM $MgCl_2$, 1 mM ATP, 20 mM Tris-maleate, pH 6.8, at 25°C. ○, ● thin filament cross-linked in the absence of calcium ions; □, ■ thin filament cross-linked in the presence of calcium ions. ○, □ 1×10^{-7} M Ca^{2+}; ●, ■ 1.3×10^{-5} M Ca^{2+}.

the depression (Ebashi and Endo, 1968). In fact, the omission of troponin during cross-linking produced only active filament.

CROSS-LINKING OF TROPONIN-C

As a preliminary study to clarify whether the above effects of cross-linking with glutaraldehyde involve hindrance of the calcium-regulated structural change of individual protein in the thin filament, a calcium-binding component of troponin, troponin-C, was first subjected to treatment with glutaraldehyde. Structural differences between the two states—*i.e.*, in the presence and absence of calcium ions, can be monitered by measuring the mobilities of troponin-C with and without calcium ions on electrophoresis in 6 M urea at pH 8.6 (Head and Perry, 1974). The binding of calcium ions to troponin-C produces an increase in mobility (Fig. 4). If troponin-C is cross-linked, its electrophoretic mobility is not influenced by calcium ions. The mobilities of troponin-C cross-linked in the presence and absence of calcium ions were similar to those of untreated troponin-C of with and without calcium ions, respectively (Fig. 4), though slightly faster. Thus the calcium-regulated structure of the thin filament proteins, at least of troponin-C, can be stabilized by chemical cross-linking with glutaraldehyde.

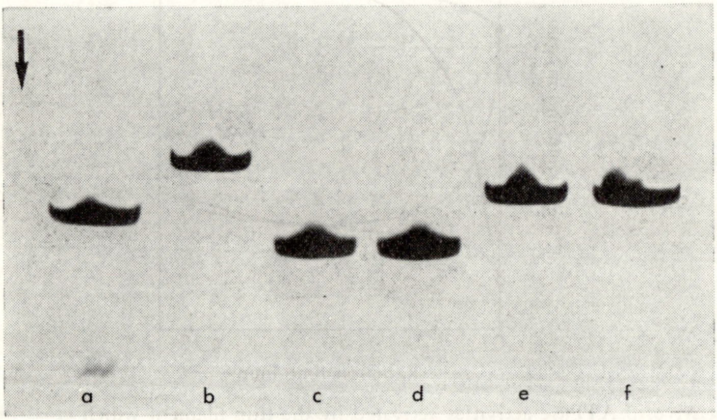

Fig. 4. Urea-alkaline gel electrophoresis of cross-linked troponin-C. a, b, unmodified troponin-C; c, d, troponin-C cross-linked in the presence of calcium ions; e, f, troponin-C cross-linked in the absence of calcium ions. a, c, e, electrophoresed in the presence of calcium ions; b, d, f, electrophoresed in the absence of calcium ions.

INTERACTION OF ACTIN-LINKED AND MYOSIN-LINKED REGULATION

Besides the regulation by tropomyosin and troponin associated with the actin in skeletal muscle (*i.e.*, actin-linked regulation), a myosin-linked type of regulation has been discovered in scallop striated muscle in which the resting state is maintained by inactivating the function of myosin (Kendrick-Jones *et al.*, 1970; Szent-Györgyi *et al.*, 1973). Recent studies have suggested that both actin-linked control and myosin-linked control operate simultaneously in many kinds of muscles of protostamia (Lehman *et al.*, 1973; Lehman and Szent-Györgyi, 1975). It is not yet certain, however, that interaction of scallop myosin with actin can be controlled by tropomyosin and troponin, although this is the case in the interaction of EDTA-light chain deficient myosin, or "desensitized" myosin, with actin (Lehman *et al.*, 1973).

As described in Table I, an active myosin of the scallop (*i.e.*, in the presence of calcium ions, reacted to the full with the thin filament frozen in the activated state) whereas its interaction with that frozen in the depressed state was distinctly reduced. Although inactive myosin of the scallop (*i.e.*, in the absence of calcium ions) reacted only slightly with both types of frozen thin filament, the resting state seemed to be maintained more effectively by the thin filament frozen in the depressed state than by that in the activated state. These results indicate that tropomyosin and troponin are able to control the interaction of actin with scallop myosin, and hence it is conceivable that double regulation would be advantageous in bringing about a complete resting state.

TABLE I
Interaction of Cross-linked Thin Filament with Scallop Myosin
ATPase activities were assayed in the solution containing 0.15 mg/ml scallop myosin, 0.13 mg/ml cross-linked thin filament, 75 mM KCl. Others were the same as described in the legend to Fig. 3.

Type of thin filament	ATPase activity (nmol/min/mg)	
	1×10^{-7} M Ca^{2+}	1.3×10^{-5} M Ca^{2+}
Cross-linked in the presence of calcium ions	32	185
Cross-linked in the absence of calcium ions	11	45

CROSS-LINKING OF SMOOTH MUSCLE THIN FILAMENT

The regulation in both skeletal and scallop muscles is operated by the inhibition of the actin-myosin-ATP interaction in the absence of calcium ions. Although the regulated site varies, these two types of regulation show overall a similarity in function. In contrast to the above, Ebashi and his colleagues (Ebashi *et al.*, 1975, 1977; Mikawa *et al.*, 1977a) have discovered that the interaction of actin and myosin in the presence of ATP corresponds to the resting state, and the contraction requires the presence of calcium-dependent activator in addition to actin and myosin. The third factors responsible for the activation in the presence of calcium ions are the actin-binding proteins, tropomyosin and leiotonin (Ebashi *et al.*, 1977, 1978, 1979; Mikawa *et al.*, 1977a, 1978; Hirata *et al.*, 1977). Although this activator type of regulation mediated by tropomyosin and leiotonin is the direct opposite of the derepressor type of regulation by tropomyosin and troponin, their regulatory sites reside in the thin filament. It has been shown that each of the two states in the reconstituted actin-tropomyosin-leiotonin

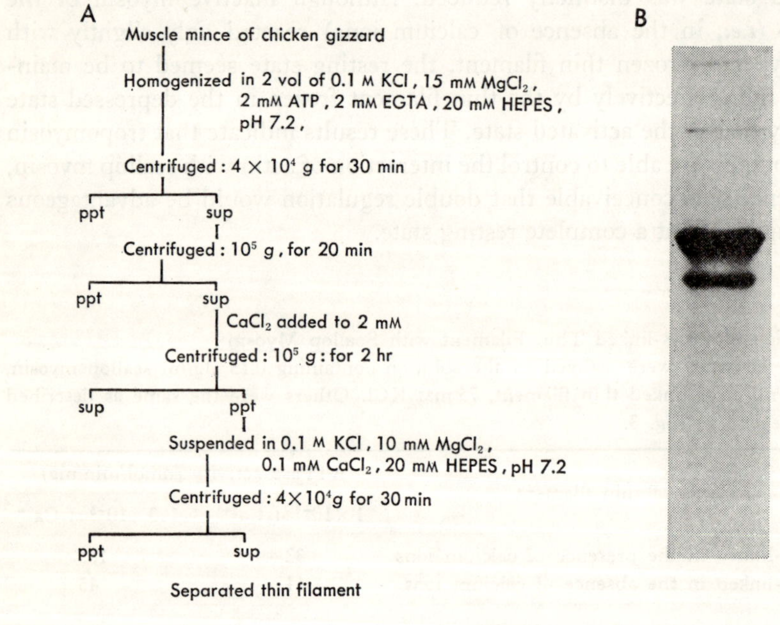

Fig. 5. A: procedure for preparing the native thin filament from smooth muscle. B: SDS-gel electrophoretic pattern of separated thin filament.

TABLE II
Effects of Leiotoin, Tropomyosin, and Native-tropomyosin on ATPase of Smooth Muscle Myosin Activated by Cross-linked Thin Filament
ATPase activities were assayed in the solution containing 0.2 mg/ml separated thin filament, 0.2 mg/ml myosin of gizzard, 50 mM KCl, 8 mM $MgCl_2$. Leiotonin, tropomyosin, and native-tropomyosin were added to 0.01, 0.05, and 0.05 mg/ml, respectively. Others were the same as described in the legend to Fig. 3.

	ATPase activity (nmol/min/mg)	
	1×10^{-7} M Ca^{2+}	1.3×10^{-7} M Ca^{2+}
a) Separated thin filament	7	53
b) Cross-linked in the presence of Ca^{2+}	50	49
c) Cross-linked in the absence of Ca^{2+}	11	10
d) b) plus leiotonin and tropomyosin	49	51
e) c) plus leiotonin and tropomyosin	9	11
f) b) plus native-tropomyosin	51	59
g) c) plus native-tropomyosin	9	15

complex can also be visualized as a result of glutaraldehyde treatment (Mikawa, 1979b).

As shown in Fig. 5, the thin filament was directly separated from chicken gizzard. The major components of the thin filament were actin and tropomyosin; other minor components were also visible by electrophoresis in the presence of SDS. Table II describes the interaction of the separated thin filament with myosin of smooth muscle. The unmodified thin filaments were able to activate myosin ATPase only in the presence of calcium ions. If the thin filaments were cross-linked in the presence of calcium ions, they always activated the myosin ATPase even in the absence of calcium ions, whereas those cross-linked in the basence of calcium ions lost the ability to activate myosin ATPase. Further additions of excess tropomyosin and leiotonin which were purified from chicken gizzard (Mikawa *et al.*, 1978) did not restore the calcium-dependent interaction with myosin (Table II), indicating that the glutaraldehyde treatment had frozen the regulation by leiotonin.

However, if native tropomyosin (Ebashi *et al.*, 1975) was used in place of leiotonin and tropomyosin, a somewhat different response in the reaction of the cross-linked thin filament with myosin could be seen that is, phosphate liberation from ATP was slightly increased in the presence of calcium ions (Table II). It is possible that this effect could be the manifestation of the activation of myosin ATPase by the phosphorylation of the myosin light chain (Chacko *et al.*, 1977; Sherry *et al.*, 1978; Ikebe *et al.*,

1978) because of the strong contamination of this fraction by myosin light chain kinase (Aksoy *et al.*, 1976), but we have not yet succeeded in demonstrating the activation of actomyosin ATPase mediated by the light chain phosphorylation (Ebashi *et al.*, 1977, 1978, 1979; Mikawa *et al.*, 1977; Hirata *et al.*, 1977).

Acknowledgment

The author is deeply indebted to Professor S. Ebashi for his helpful advice and discussion.

REFERENCES

Aksoy, M. O., Williams, D., Sharkey, E. M., and Hartshorne, D. J. (1976). *Biochem. Biophys. Res. Commun.* **69**, 35–41.

Chacko, S., Conti, M. A., and Adelstein, R. S. (1977). *Proc. Natl. Acad. Sci. U.S.* **74**, 129–133.

Ebashi, S. and Endo, M. (1968). *Prog. Biophys. Mol. Biol.* **18**, 123–183.

Ebashi, S., Toyo-oka, T., and Nonomura, Y. (1975). *J. Biochem.* **78**, 859–861.

Ebashi, S. Mikawa, T., Hirata, M., Toyo-oka, T., and Nonomura, Y. (1977). In *Excitation-contraction Coupling in Smooth Muscle* (Casteels, R., Godfraind, T., and Rüegg, J. C., eds.), pp. 325–334, Elsevier/North-Holland, Amsterdam.

Ebashi, S., Mikawa, T., Hirata, M., and Nonomura, Y. (1978). *Ann. N.Y. Acad. Sci.* **307**, 451–461.

Ebashi, S., Nonomura, Y., Mikawa, T., Hirata, M., and Saida, K. (1979). In *Cell Motility: Molecules and Organization* (Hatano, S., Ishikawa, H. I., and Sato, H., eds.), pp. 225–237, Univ. Tokyo Press, Tokyo.

Gillis, J. M. and O'Brien, E. J. (1975). *J. Mol. Biol.* **99**, 445–459.

Hanson, J., Sednev, V., O'Brien, E. J., and Bennett, P. M. (1973). *Cold Spring Harbor Symp. Quant. Biol.* **37**, 311–318.

Haselgrove, J. C. (1973). *Cold Spring Harbor Symp. Quant. Biol.* **37**, 341–352.

Head, J. F. and Perry, S. V. (1974). *Biochem. J.* **137**, 145–154.

Hirata, M., Mikawa, T., Nonomura, Y., and Ebashi, S. (1977). *J. Biochem.* **82**, 1793–1796.

Huxley, H. E. (1973). *Cold Spring Harbor Symp. Quant. Biol.* **37**, 361–376.

Ikebe, M., Aiba, T., Onishi, H., and Watanabe, S. (1978). *J. Biochem.* **83**, 1643–1655.

Kendrick-Jones, J., Lehman, W., and Szent-Györgyi, A. G. (1970). *J. Mol. Biol.* **54**, 313–326.

Lehman, W., Kendrick-Jones, J., and Szent-Györgyi, A. G. (1973). *Cold Spring Harbor Symp. Quant. Biol.* **37**, 319–330.

Lehman, W. and Szent-Györgyi, A. G. (1975). *J. Gen. Physiol.* **66**, 1–30.

Mikawa, T. (1979a). *Nature* **278**, 473–474.

Mikawa, T. (1979b). *J. Biochem.* **85**, 879–881.

Mikawa, T., Toyo-oka, T., Nonomura, Y., and Ebashi, S. (1977a). *J. Biochem.* **81**, 273–275.

Mikawa, T., Nonomura, Y., and Ebashi, S. (1977b). *J. Biochem.* **82**, 1789–1791.

Mikawa, T., Nonomura, Y., Hirata, M., Ebashi, S., and Kakiuchi, S. (1978). *J. Biochem.* **84**, 1633–1636.

Miki, M. (1979). *Biochim. Biophys. Acta.* **578**, 96–106.

Oosawa, F., Fujime, S., Ishiwata, S., and Mihashi, K. (1973). *Cold Spring Harbor Symp. Quant. Biol.* **37**, 277–285.

Parry, D.A.D. and Squire, J. M. (1973). *J. Mol. Biol.* **75**, 33–55.

Sherry, J.M.F., Gorecka, A., Aksoy, M. O., Dabrowska, R., and Hartshorne, D. J. (1978). *Biochemistry* **17**, 4411–4418.

Szent-Györgyi, A. G., Szentkiralyi, E. M., and Kendrick-Jones, J. (1973). *J. Mol. Biol.* **74**, 179–203.

Wakabayashi, T., Huxley, H. E., Amos, L. A., and Klug, A. (1975). *J. Mol. Biol.* **93**, 477–497.

Yanagida, T., Taniguchi, M., and Oosawa, F. (1974). *J. Mol. Biol.* **90**, 509–522.

Muscle Contraction: Its Regulatory Mechanisms (Ebashi, S. et al., eds.), pp. 359–371,
Japan Sci. Soc. Press, Tokyo / Springer-Verlag, Berlin (1980)

28

Calmodulin in Smooth Muscle

JAMES F. HEAD, ROBERT J. BIRNBAUM,
AND BENJAMIN KAMINER

Summary: Calmodulin in vertebrate smooth muscle (chicken gizzard) is considered in terms of its purification and characterization as well as its possible role in the regulation of smooth muscle, taking into account its potentially multifunctional role. In particular its interaction with smooth muscle phosphodiesterase was studied. At least three forms of phosphodiesterase were isolated, only one of which was calmodulin activatable. This form could be purified by means of a calmodulin-Sepharose affinity column. Preliminary evidence is presented on the identification of other calmodulin interacting proteins in smooth muscle by means of calmodulin affinity chromatography. Further studies on the interaction of these proteins with calmodulin will lead to a better understanding of the functional role of calmodulin in smooth muscle.

Two important milestones in the elucidation of the role of calcium in excitation-contraction coupling in muscle were the "invention" of the skinned muscle fiber by Natori (1954) and the discovery by Kumagai *et al.* (1955) that relaxation could be brought about by a microsomal fraction.

These important contributions of Drs. Kumagai and Natori, to whom this article is dedicated, helped lead to the development of our present day understanding of the role of calcium in excitation-contraction coupling, (see Ebashi, 1976 and Endo, 1977). The more precise role of calcium in triggering the interaction of myosin and actin became more apparent when Ebashi (1963) found that "native tropomyosin" conferred calcium sensitivity on striated muscle actomyosin. The components of this "native" protein, tropomyosin and troponin, make up a thin filament linked regulatory system which has been extensively studied in striated muscle (see reviews, Ebashi *et al.*, 1969 and Ebashi, 1977). Extraction of a similar "native tropomyosin" from gizzard muscle (Ebashi *et al.*, 1966) led to the assumption that vertebrate smooth muscle also had a thin filament linked regulatory system. However, a number of controversial opinions on the regulation of vertebrate smooth muscle actomyosin have since developed. Some evidence in support of thin filament linked regulation has been published (*e.g.*, Carstens, 1971; Sparrow and Van Bockxmeer, 1972; Ito and Hotta, 1976; Ebashi *et al.*, 1975; Ebashi, 1979; Nonomura *et al.*, 1980), but many studies have not found troponin or troponin-like components in vertebrate smooth muscle and have suggested that calcium regulation in smooth muscle is linked to the thick, myosin containing, filaments (*e.g.*, Bremel, 1974; Sobieszek and Bremel, 1975; Sobieszek and Small, 1976; Mrwa and Ruegg, 1975; Sobieszek, 1977; Chacko *et al.*, 1977; Gorecka *et al.*, 1976; Dabrowska *et al.*, 1978; Ikeba *et al.*, 1978; Adelstein *et al.*, 1978; Barron *et al.*, 1979; Hoar *et al.*, 1979).

A troponin-C-like protein was identified and isolated by Head *et al.* (1977a) from vertebrate smooth muscle, using alkaline urea gel electrophoresis and troponin-I Sepharose affinity chromatography. The identification and isolation was based on the finding that in skeletal muscle troponin-I and troponin-C form a stable complex in 8 M urea in the presence of calcium (Perry *et al.*, 1972; Head and Perry, 1974). We pursued the characterization of this TN-C-like protein from gizzard smooth muscle and found it to be extremely similar, if not identical, to brain modulator protein (now designated calmodulin) with respect to molecular weight, calcium binding, amino acid composition, peptide map, and the ability to activate phosphodiesterase (Head *et al.*, 1977b). A similar troponin-C-like protein from chicken gizzard has also been described by Dabrowska *et al.* (1978) and Makioka and Hirabayashi (1979).

It has now been shown that calmodulin is a ubiquitous protein. It has been isolated from vertebrate and invertebrate tissues and also from plants. Initially identified as an activator of phosphodiesterase (Cheung, 1967, 1970; Kakiuchi *et al.*, 1970), this protein is now believed to be a multifunctional molecule involved in the activation of many enzyme systems including adenylate cyclases in brain (Brostrom *et al.*, 1975), and pancreatic tissue (Valverde *et al.*, 1979), guanylate cyclase in tetrahymena (Nagao *et al.*, 1979), Ca-Mg ATPases in erythrocyte membranes (Jarrett and Penniston, 1977; Gopinath and Vincenzi, 1977) and brain (Sobue *et al.*, 1979a). It augments active calcium transport systems in erythrocyte membranes (Larsen and Vincenzi, 1979), synaptic membranes (Kuo *et al.*, 1979) and in cardiac sarcoplasmic reticulum (Katz and Remtulla, 1978), NAD kinases in plants (Anderson and Cormier, 1978) and sea urchin eggs (Epel, 1980) and phospholipase A_2 in platelets (Wong and Cheung, 1979). Calmodulin is a subunit of the phosphorylase kinase holoenzyme (Cohen *et al.*, 1978) and is an activator of several other kinases including myosin light chain kinases in striated muscle (Yagi *et al.*, 1978; Barylko *et al.*, 1978; Nairn and Perry, 1979) and smooth muscle (Dabrowska *et al.*, 1978) and kinases involved in the phosphorylation of synaptosomal membrane proteins and membrane proteins from various other tissues (Schulman and Greengard, 1978a and b). Calmodulin has been found to promote disaggregation of polymerized tubulin (Marcum *et al.*, 1978) and it may be associated with tubulin in the mitotic apparatus (Welsh *et al.*, 1978, 1979). It is also able to substitute, *in vitro* at least, for troponin-C in a reconstituted calcium dependent striated muscle actomyosin system (Amphlett *et al.*, 1976; Dedman *et al.*, 1977; Head *et al.*, 1977b). A comprehensive review on calmodulin has been published recently by Cheung (1980).

Our studies are aimed at determining which of these potential functions calmodulin serves in the smooth muscle cell and how it influences the function of this tissue.

SMOOTH MUSCLE CALMODULIN

Using a striated muscle troponin-I-Sepharose 4B affinity column, calmodulin can be purified from smooth muscle homogenates in a single step, Fig. 1 (Head *et al.*, 1977a). The protein purified in this way has all the characteristics of bovine brain calmodulin—a molecular weight of 17,000 daltons, a high acidic residue to basic residue ratio, high phenylalanine to

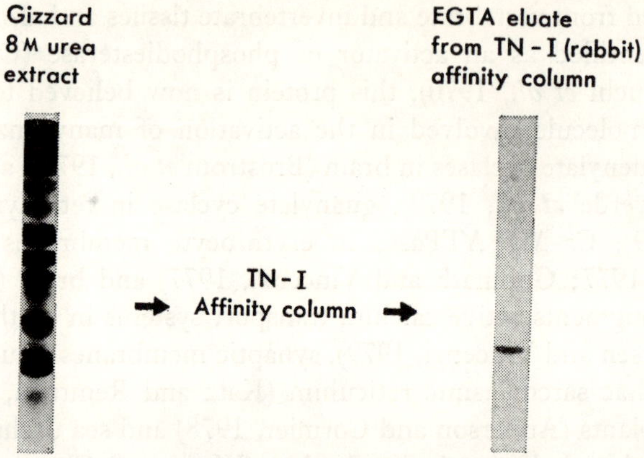

Fig. 1. SDS gel showing purification of gizzard calmodulin by means of a troponin-I (TN-I) (rabbit)-Sepharose 4B affinity column. Gel on left shows bands from gizzard extracted in 8 M urea. Gel on right shows a single band obtained from the same urea extract after passage through a TN-I-Sepharose 4B affinity column in the presence of calcium and elution with an EGTA buffer. Sample and wash buffer 8 M urea, 20 mM Tris-HCl, 1 mM $CaCl_2$, 15 mM 2-mercaptoethanol pH 7.8. Elution buffer: same but 5 mM EGTA in place of $CaCl_2$. Electrophoresis: 10% acrylamide, 0.1% SDS. 0.1 M sodium phosphate pH 7.0.

tyrosine ratio, no cysteine, binds four calcium ions per mol, activates bovine brain phosphodiesterase and has a similar peptide map to the brain protein (Head *et al.*, 1977b).

Densitometric scanning of alkaline urea gels of whole tissue homogenates indicates that calmodulin represents 0.4% of the total protein in the chicken gizzard (Head *et al.*, 1977b). This represents a calcium sensitive switching mechanism of considerable capacity compatible with the number of potential roles described for calmodulin. In Fig. 2 we have drawn a schematized representation of a number of enzymic activities occuring in the smooth muscle cell. Extrapolating from the results obtained from the many studies of calmodulin activity in other tissues we have indicated sites at which calmodulin might have a potential regulatory role in smooth muscle. This scheme is highly speculative and is almost certainly incomplete as far as the number of sites of calmodulin action are concerned. It may also indicate sites of calmodulin regulation in smooth muscle where none exists in reality. It has become evident that the ability of calmodulin to serve a function in one tissue does not necessarily mean that it will serve the same function in another. Calmodulin was first described as an activator of brain phosphodiesterase, however, it is apparent that there is not a calmodulin

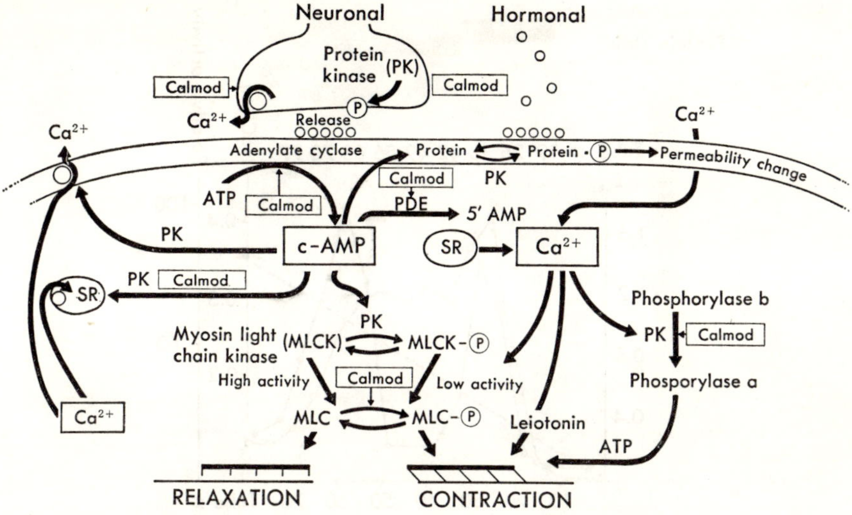

Fig. 2. Hypothetical scheme of possible sites of action of calmodulin in the regulation of smooth muscle.

dependent phosphodiesterase in all cells. For instance calmodulin from the unfertilized egg of the sea urchin, *Arbacia punctulata*, is able to activate bovine brain phosphodiesterase as effectively as bovine brain calmodulin, however, it has no effect on the activity of the phosphodiesterase of the *Arbacia* egg itself (Head *et al.*, 1979). Similar observations have also been made in studies of the electric organ of the eel (Childers *et al.*, 1977) and the earthworm (Waisman *et al.*, 1978). It becomes of interest therefore to consider the polymorphic forms of phosphodiesterase in relation to their interaction with calmodulin. As part of our studies on the role of calmodulin in smooth muscle we have investigated the effect of calmodulin on phosphodiesterase activity in this tissue.

SMOOTH MUSCLE CYCLIC NUCLEOTIDE PHOSPHODIESTERASE

In gizzard, as in brain and many other tissues, there are several polymorphic forms of phosphodiesterase. These different forms of the enzyme may be separated, to some extent, from one another, and also from calmodulin by ion exchange chromatography on DEAE cellulose at pH 7.8 in the absence of calcium (Fig. 3). The enzyme elutes on a salt gradient principally as two peaks of activity at 0.15 M and 0.25 M NaCl respectively,

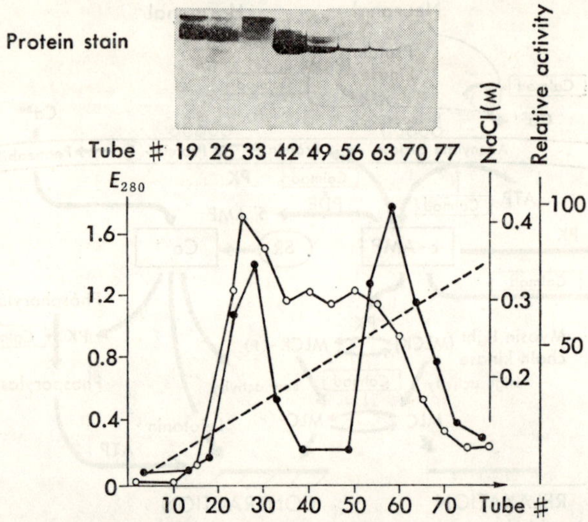

Fig. 3. Chromatography of gizzard cyclic nucleotide phosphodiesterase on DEAE cellulose. ○ E280; ● relative phosphodiesterase activity (determined by method of Boudreau and Drummond, 1975); - - - NaCl gradient (0–0.4 M). Column buffer: 20 mM Tris-HCl, 1 mM EDTA, 15 mM 2-mercaptoethanol, pH 7.8. Slab gel shows protein stain of eluted samples.

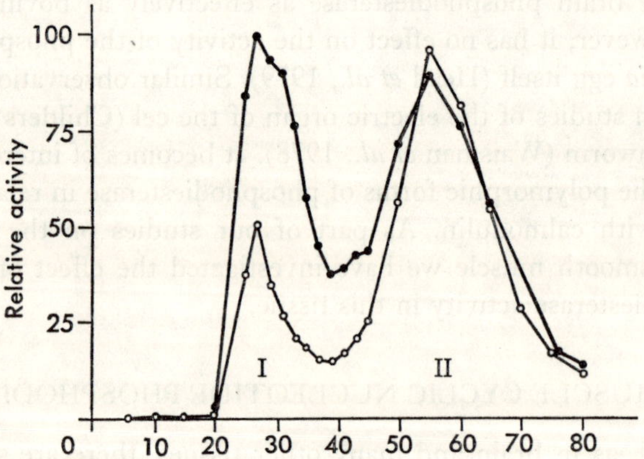

Fig. 4. Relative phosphodiesterase activity of samples eluted from DEAE cellulose column as in Fig. 3, showing activation of peak I by calmodulin. ● plus calmodulin in assay mixture; ○ without calmodulin.

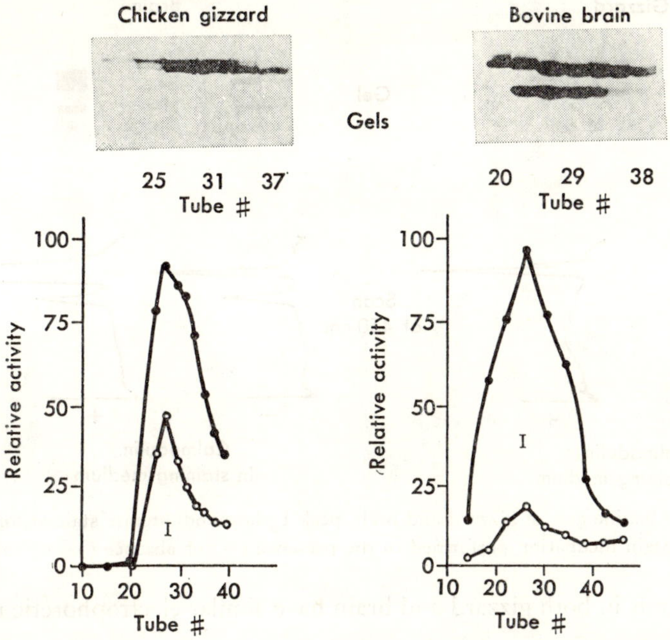

Fig. 5. Relative phosphodiesterase activity of peak I of chicken gizzard (left) and bovine brain (right) eluted from DEAE cellulose column as in Fig. 3. ● plus calmodulin in assay mixture; ○ without calmodulin. Alkaline slab gels (8% acrylamide) of eluted sample stained for phosphodiesterase activity by the method of Goren *et al.* (1971).

calmodulin elutes at about 0.3 M NaCl. Only the first peak of enzyme is activated on addition of calmodulin and calcium (Fig. 4). For comparative purposes bovine brain phosphodiesterase was similarly fractionated and also showed calmodulin sensitivity only in the peak I enzyme. Gels of samples from peak I (from gizzard and brain) stained for phosphodiesterase activity by the method of Goren *et al.* (1971), show double banded enzyme staining (Fig. 5) indicating at least two forms of the enzyme in this peak. To determine if only one of these enzymic forms or both are calmodulin sensitive we carried out the enzymic stain in the presence of calmodulin. This caused an increase in the stain intensity of only one of the bands in both gizzard and brain peak I enzyme (Fig. 6) suggesting that only this form of the enzyme is calmodulin activated. In the gizzard sample it is the faster of the two enzymic bands which is activated and in the brain it is the slower of the pair. From gel scans it appears that the calmodulin sen-

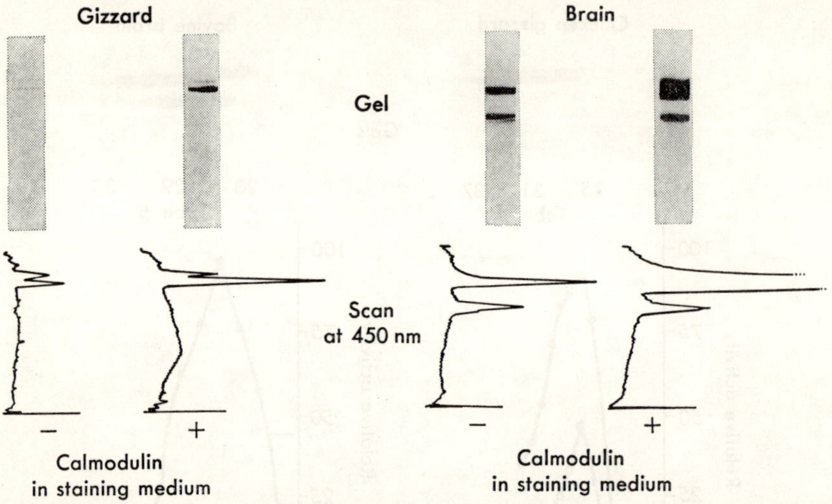

Fig. 6. Alkaline gels of gizzard and brain peak I phosphodiesterase stained for enzymic activity. Stain incubation performed in the presence (+) or absence (−) of calmodulin.

sitive bands in both gizzard and brain have similar electrophoretic mobility (Fig. 6).

To further purify the calmodulin activatable form of phosphodiesterase we applied pooled samples of peak I enzyme to a calmodulin-Sepharose 4B affinity column in the presence of calcium. After washing the column with calcium containing buffer the calcium dependent interaction of calmodulin with bound phosphodiesterase was disrupted by eluting with EGTA buffer. As shown in Fig. 7 the enzyme eluting in the EGTA buffer corresponds, both in brain and gizzard, to the calmodulin dependent form of the enzyme identified previously.

On SDS gel electrophoresis the EGTA eluate of the calmodulin affinity column shows the gizzard phosphodiesterase fraction to contain a major 80,000-dalton band and a few minor bands (Fig. 8). The 80,000-dalton protein migrates close to the phosphodiesterase band on alkaline gels (Head *et al.*, 1977b) but apparently is distinct from the phosphodiesterase since the 80,000-dalton material also binds in a calcium dependent manner to a bovine serum albumen—Sepharose 4B matrix (Fig. 8), and when eluted from this albumen affinity column has no phosphodiesterase activity. The relationship of the minor bands evident on SDS gels to phosphodiesterase activity is the subject of further study.

In the gizzard it is evident that we have at least three forms of phos-

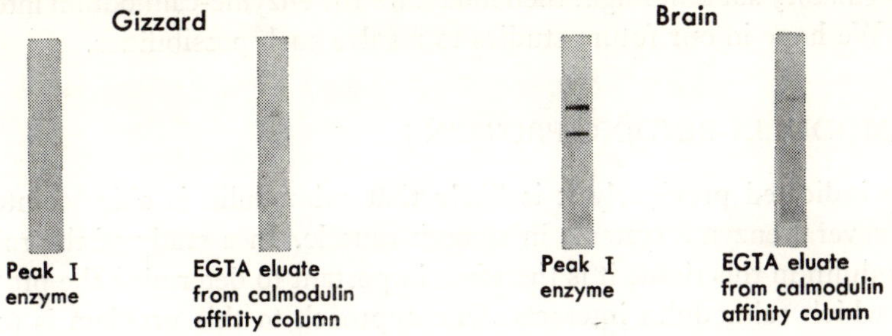

Fig. 7. Alkaline gels of peak I phosphodiesterase before and after purification on calmodulin affinity column. Gels are stained for phosphodiesterase activity. Affinity column —sample and wash buffer: 200 mM NaCl, 20 mM Tris, HCl, 1 mM $CaCl_2$, 15 mM 2-mercaptoethanol, pH 7.8. Elution buffer: same but 5 mM EGTA in place of $CaCl_2$.

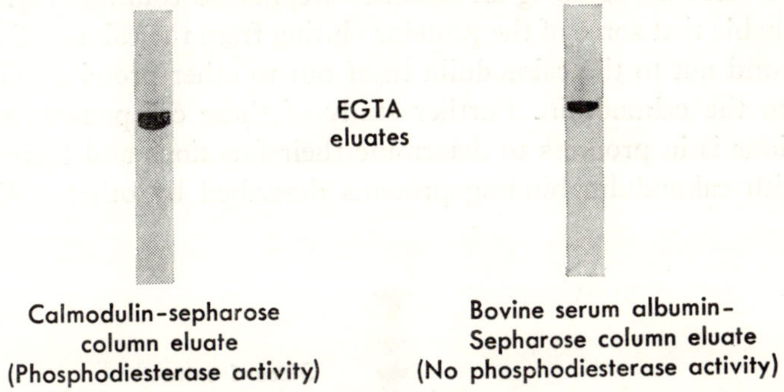

Fig. 8. SDS gels of EGTA eluates from calmodulin affinity column (left) and albumen affinity column (right). The sample applied to each column was peak I phosphodiesterase (see Fig. 3). Conditions for affinity chromatography as described in Fig. 7. SDS gels: 10% acrylamide, 0.1% SDS, 0.1 M sodium phosphate, pH 7.0.

phodiesterase, two in our peak I fraction and at least one in peak II; only one of these forms is calmodulin activatable. Murtagh and Bhalla (1979) have described four forms of phosphodiesterase in arterial smooth muscle; again only one form was calmodulin activatable. The existence of a subunit structure for brain calmodulin dependent phosphodiesterase has been described (Klee *et al.*, 1979). The two forms of enzymes in our peak I could therefore be structurally related and might represent different combinations of subunits, one band might represent a catalytic subunit with a regulatory subunit attached and the other band a catalytic subunit alone;

the regulatory subunit might then modulate the enzyme-calmodulin interaction. We hope in our future studies to resolve such possibilities.

CALMODULIN-BINDING PROTEINS

As indicated previously it is likely that calmodulin is able to interact with several enzyme systems in smooth muscle. In a study of the role of calmodulin in this tissue it is therefore important to determine the proteins with which calmodulin interacts. One approach to this problem is to see which proteins in gizzard are able to bind in a calcium dependent manner to a calmodulin—Sepharose 4B affinity column. As shown in Fig. 9 many proteins bind in such a way when a sample of whole gizzard extract is applied to the column. Some of these proteins are, however, not binding specifically to calmodulin but rather to the Sepharose matrix, as shown by a control carried out using an albumen-Sepharose column (Fig. 9). It is also probable that some of the proteins eluting from the calmodulin column were bound not to the calmodulin itself but to other proteins which were bound to the calmodulin. Further study of these components and their interactions is in progress to determine their functions and their relationships with calmodulin binding proteins described by others (Wang and

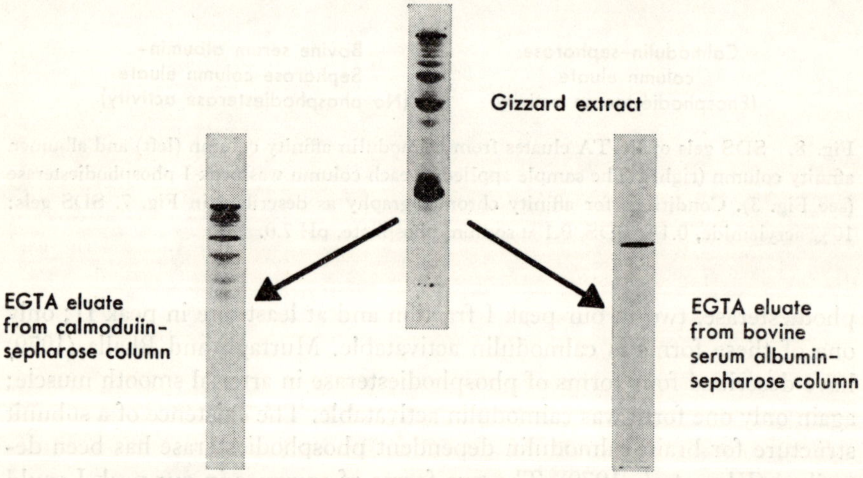

Fig. 9. SDS gels of gizzard samples before and after affinity chromatography on calmodulin (left) and albumen (right)—Sepharose affinity columns. Gizzard extract in 200 mM NaCl, 20 mM Tris-HCl, 15 mM 2-mercaptoethanol pH 7.8. Affinity chromatography conditions as in Fig. 7.

Desai, 1977; Klee and Krinks, 1978; Sharma *et al.*, 1979; Sobue *et al.*, 1979b; Grand and Perry, 1979; Wood *et al.*, 1980).

CONCLUSION

The potential roles of calmodulin in smooth muscle are apparently various. Which or how many of these roles it actually serves *in vivo* remains unclear at present. *In vitro* observations may not necessarily reflect events occuring in the cell. The function of calmodulin with respect to a given enzyme system will depend on a number of parameters many of which are still poorly defined. These include the amount of calmodulin available in the cell compared to the number of potential sites of interaction, the location of calmodulin with respect to these sites and the relative binding constants of calmodulin for one site of interaction in comparison to another. As these factors are defined and as functions are ascribed to the various "calmodulin binding proteins" the functional role of calmodulin in smooth muscle will be better understood.

REFERENCES

Adelstein, R. S., Conti, M. A., and Hathaway, D. R. (1978). *J. Biol. Chem.* **253**, 8347–8350.

Amphlett, G. W., Vanaman, T. C., and Perry, S. V. (1976). *FEBS Lett.* **72**, 163–168.

Anderson, J. M. and Cormier, M. J. (1978). *Biochem. Biophys. Res. Commun.* **84**, 595–602.

Barron, J. T., Barany, M., and Barany, K. (1979). *J. Biol. Chem.* **254**, 4954–4956.

Barylko, B., Kuznicki, J., and Drabikowski, W. (1978). *FEBS Lett.* **90**, 301–304.

Boudreau, R. J. and Drummond, G. I. (1975). *Anal. Biochem.* **63**, 388–399.

Bremel, R. D. (1974). *Nature* **252**, 404–407.

Brostrom, C. O., Huang, Y. C., Brekenridge, B. M., and Wolff, D. J. (1975). *Proc. Natl. Acad. Sci. U.S.* **72**, 64–68.

Carsten, M. E. (1971). *Arch. Biochem. Biophys.* **147**, 353–357.

Chacko, S., Conti, M. A., and Adelstein, R. S. (1977). *Proc. Natl. Acad. Sci. U.S.* **74**, 129–133.

Cheung, W. Y. (1967). *Biochem. Biophys. Res. Commun.* **29**, 478–482.

Cheung, W. Y. (1970). *Biochem. Biophys. Res. Commun.* **38**, 533–538.

Cheung, W. Y. (1980). *Science* **207**, 19–27.

Childers, S. R., Sitarayama, A., Egrie, J. C., Campbell, J. A., and Siegel, F. L. (1977). In *Calcium Binding Proteins and Calcium Function* (Wasserman, R. H. *et al.*, eds.), p. 127–134, Elsevier/North-Holland, New York.

Cohen, P., Buchell, A., Faulkes, J. G., Cohen, P.T.W., Vanaman, T.C., and Nairn, A. C. (1978). *FEBS Lett.* **92**, 287–293.

Dabrowska, R., Sherry, J.M.F., Aramatorio, D. K., and Hartshorne, D. J. (1978). *Biochemistry* **17**, 253–258.

Dedman, J. R., Potter, J. D., and Means, A. R. (1977). *J. Biol. Chem.* **252**, 2437–2440.
Ebashi, S. (1963). *Nature* **200**, 1010.
Ebashi, S. (1976). *Annu. Rev. Physiol.* **38**, 293–313.
Ebashi, S. (1977). In *Search and Discovery a Tribute to Albert Szent-Györgyi* (Kaminer, B., ed.), pp. 77–89, Academic Press, New York.
Ebashi, S. (1979). In *Advances in Pharmacology and Therapeutics* (Stoclet, J., ed.), vol. 3, pp. 81–98, Pergamon Press, New York.
Ebashi, S., Endo, M., and Ohtsuki, I. (1969). *Q. Rev. Biophys.* **2**, 351–384.
Ebashi, S., Iwakura, H., Nakajima, R., Nakamura, R., and Ooi, Y. (1966). *Biochem. Z.* **345**, 201–211.
Ebashi, S., Toyo-oka, T., and Nonomura Y. (1975). *J. Biochem.* **78**, 859–861.
Endo, M. (1977). *Physiol. Rev.* **57**, 71–108.
Epel, D. (1980). In *38th Symposium Society Development Biology* (Subtelny, S., ed.), Academic Press, New York (in press).
Gopinath, R. M. and Vincenzi, F. F. (1977). *Biochem. Biophys. Res. Commun.* **77**, 1203–1209.
Gorecka, A., Aksoy, M. O., and Hartshorne, D. J. (1976). *Biochem. Biophys. Res. Commun.* **71**, 325–331.
Goren, E. N., Hirsch, A. H., and Rosen, O. (1971). *Anal. Biochem.* **43**, 156–161.
Grand, R.J.A. and Perry, S. V. (1979). *Biochem. J.* **183**, 285–295.
Head, J. F., Mader, S., and Kaminer, B. (1977b). In *Calcium Binding Proteins and Calcium Function* (Wasserman, R. H. *et al.*, eds.), pp. 275–277, Elsevier/North-Holland, New York.
Head, J. F., Mader, S., and Kaminer, B. (1979). *J. Cell Biol.* **80**, 211–218.
Head, J. F. and Perry, S. V. (1974). *Biochem. J.* **137**, 145–154.
Head, J. F., Weeks, R. A., and Perry, S. V. (1977a). *Biochem. J.* **161**, 465–471.
Hoar, P. E., Kerrick, W.G.L., and Cassidy, P. S. (1979). *Science* **204**, 503–506.
Ikeba, M., Aiba, T., Onishi, H., and Watanabe, S. (1978). *J. Biochem.* **83**, 1643–1655.
Ito, N. and Hotta, K. (1976). *J. Biochem.* **80**, 401–403.
Jarrett, H. W. and Penniston, J. T. (1977). *Biochem. Biophys. Res. Commun.* **77**, 1210–1216.
Kakiuchi, S., Yamazaki, R., and Nakajima, H. (1970). *Proc. Japan Acad.* **46**, 587–592.
Katz, S. and Remtulla, M. A. (1978). *Biochem. Biophys. Res. Commun.* **83**, 1373–1379.
Klee, C. B., Crouch, T. H., and Krinks, M. H. (1979). *Biochemistry* **18**, 722–729.
Klee, C. B. and Krinks, M. H. (1978). *Biochemistry* **17**, 120–126.
Kumagai, H., Ebashi, S., and Takeda, F. (1955). *Nature* **176**, 166.
Kuo, C., Ichida, S., Matsuda, T., Kakiuchi, S., and Yoshida, H. (1979). *Life Sci.* **25**, 235–240.
Larsen, F. L. and Vincenzi, F. F. (1979). *Science* **204**, 306–309.
Makioka, A. and Hirabayashi, T. (1979). *J. Biochem.* **85**, 967–975.
Marcum, J. M., Dedman, J. R., Brinkley, B., and Means, A. R. (1978). *Proc. Natl. Acad. Sci. U.S.* **75**, 3771–3775.
Mrwa, U. and Ruegg, J. C. (1975). *FEBS Lett.* **60**, 81–97.
Murtagh, T. J. and Bhalla, R. C. (1979). *Arch. Biochem. Biophys.* **196**, 465–474.
Nagao, S., Suzuki, Y., Watanabe, Y., and Nozawa, Y. (1979). *Biochem. Biophys. Res. Commun.* **90**, 261–268.

Nairn, A. C. and Perry, S. V. (1979). *Biochem. J.* **179**, 89–97.
Natori, R. (1954). *Jikeikai Med. J.* **1**, 119–126.
Nonomura, Y. (1980). This volume, pp. 329–345.
Perry, S. V., Cole, H. A., Head, J. F., and Wilson, F. (1972). *Cold Spring Harbor Symp. Quant. Biol.* **37**, 251–262.
Schulman, H. and Greengard, P. (1978a). *Nature* **271**, 478–479.
Schulman, H. and Greengard, P. (1978b). *Proc. Natl. Acad. Sci. U.S.* **75**, 5432–5436.
Sharma, R. K., Desai, R., Waisman, D., and Wang, J. (1979). *J. Biol. Chem.* **254**, 4276–4282.
Sobieszek, A. (1977). *Eur. J. Biochem.* **73**, 477–483.
Sobieszek, A. and Bremel, R. D. (1975). *Eur. J. Biochem.* **55**, 49–60.
Sobieszek, A. and Small, J. V. (1976). *J. Mol. Biol.* **102**, 75–92.
Sobue, K., Ichida, S., Yoshida, H., Yamazaki, R., and Kakiuchi, S. (1979a). *FEBS Lett.* **99**, 199–202.
Sobue, K., Muramoto, Y., Yamazaki, R., and Kakiuchi, S. (1979b). *FEBS Lett.* **105**, 105–109.
Sparrow, M. D. and Van Bochxmeer, J. (1972). *J. Biochem.* **72**, 1075–1080.
Valverde, I., Vandermeers, A., Anjaneyulu, R., and Malaisse, W. J. (1979). *Science* **206**, 225–227.
Waisman, D. M., Stevens, F. C., and Wang, J. H. (1978). *J. Biol. Chem.* **253**, 1106–1113.
Wang, J. H. and Desai, R. (1977). *J. Biol. Chem.* **252**, 4175–4184.
Welsh, M. F., Dedman, J. R., Brinkley, B. R., and Means, A. R. (1978). *Proc. Natl. Acad. Sci. U.S.* **75**, 1867–1871.
Welsh, M. J., Dedman, J. R., Brinkley, B. R., and Means, A. R. (1979). *J. Cell Biol.* **81**, 624–634.
Wong, P.Y.K. and Cheung, W. Y. (1979). *Biochem. Biophys. Res. Commun.* **90**, 473–480.
Wood, J. G., Wallace, R. W., Whitaker, J. N., and Cheung, W. Y. (1980). *J. Cell Biol.* **84**, 66–76.
Yagi K., Yazawa, M., Kakiuchi, S., Oshima, M., and Uenishi, K. (1978). *J. Biol. Chem.* **253**, 1338–1340.

MYOSIN-LINKED REGULATION

Muscle Contraction: Its Regulatory Mechanisms (Ebashi, S. et al., eds.), pp. 375–389,
apan Sci. Soc. Press, Tokyo / Springer-Verlag, Berlin (1980)

29

Role of Regulatory Light Chains in Myosin-linked Regulation

ANDREW G. SZENT-GYÖRGYI

Summary: Calcium sensitivity of tension generation and of ATPase activity in scallop muscles require the presence of regulatory light chains. Both regulatory light chains must be bound by an intact myosin molecule for the inhibition of ATPase activity in the absence of calcium, although single headed myosin is also regulated. Myofibrils that lost all the regulatory light chains by treatment with EDTA at elevated temperatures, hybridize stoichiometrically with foreign regulatory light chains. Regulatory light chains of some myosin (molluscs, urechis, gizzard, limulus, and cricket) form functional hybrids and restore calcium sensitivity to denuded myofibrils. Regulatory light chains of vertebrate striated, cardiac and lobster myosins form hybrids that are not regulated. Antibodies against essential light chains abolish calcium sensitivity by selectively activating the ATPase in the absence of calcium. It is suggested that both regulatory and essential light chains are involved in myosin linked regulation and that only molecules that retained the subfragment 1-2 (S-1–S-2) hinge region are regulated. X-ray diffraction and electron microscopic studies indicate that the kind of rigor links formed depend on regulatory light chains.

It is a great privilege to participate in a symposium in honor of Professors Natori and Kumagai. Their work has greatly influenced my own science. Professor Kumagai's famous Nature paper with the Ebashis (1955) represented the first significant advance in the isolation and characterization of the "relaxing factor" discovered by Marsh (1951) in Bailey's laboratory. Their discovery started the train of events that led to the present structural and biochemical formulation of regulation of muscle contraction. Dr. Natori's skinned fiber preparation (1954) serves as a unique link between the approaches of biochemists and physiologists and continues to gain in importance in present and future studies of how muscle contracts and is regulated.

My interest during the past years has focused on myosin linked regulation. In this system the actin binding site of myosin is regulated by calcium. Regulated myosins are unable to combine with pure actin in the absence of calcium (Kendrick-Jones *et al.*, 1970) and the regulatory component is a low molecular weight subunit of myosin (Szent-Györgyi *et al.*, 1973). In the presence of the regulatory light chain the actin activated Mg-ATPase activity of myosin is depressed and the inhibition is reversed by calcium or by removal of the light chains (Szent-Györgyi *et al.*, 1973). Although myosin linked regulation is widely distributed in the animal kingdom (Lehman and Szent-Györgyi, 1975), scallop myosin is rather unique in that its regulatory light chains can be detached reversibly by EDTA (Kendrick-Jones *et al.*, 1976). One of the two regulatory light chains of myosin are readily removed by EDTA in the cold, accompanied by a loss of calcium sensitivity as judged by ATPase activity. Desensitized myosins rebind regulatory light chains in the presence of magnesium and regain calcium sensitivity (Szent-Györgyi *et al.*, 1973). Desensitized scallop myosins hybridize with regulatory light chains of other myosins and the myosin that contains a mol of scallop regulatory light chain and a mol of foreign light chain ("mixed hybrid") is fully regulated. All myosins so far tested contain a particular light chain that is regulatory in mixed hybrids even if the source of the light chain is a non-regulated myosin (Kendrick-Jones, 1974; Kendrick-Jones *et al.*, 1976); hence this class of light chains are designated as regulatory light chains. Specific, high affinity calcium binding by regulated myosin requires the presence of regulatory light chains, and the binding depends on the interaction of the heavy chains and the regulatory light chains. Calcium binding by myosin was originally thought to be a requirement for myosin linked regulation and although this is true in many instances where calcium directly reverses the inhibitory

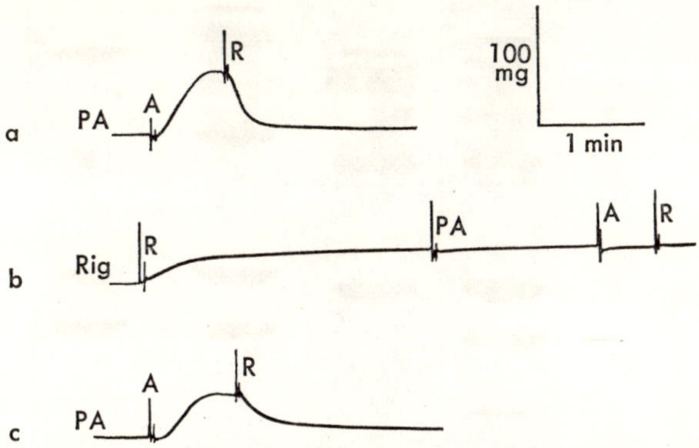

Fig. 1. Tension records from chemically skinned muscle fibre bundles from the striated adductor of *Placopecten magellanicus*. a, activation-relaxation sequences before desensitization. b, same fibre bundle after desensitization by EDTA treatment at 25°C. c, same fibre bundle after resensitization with scallop light chain. In parallel experiments regulatory light chains after EDTA treatments could not be detected (15 experiments), after resensitization over 80% of the original regulatory light chain content was regained (6 experiments). PA, pre-activating solution, $[Ca^{2+}] \sim 10^{-8}$ M; A, activating solution, $[Ca^{2+}] = 10^{-4.8}$ M; R, relaxing solution, $[Ca^{2+}] < 10^{-10}$ M; Rig, rigor solution, $[Ca^{2+}] < 10^{-10}$ M, no ATP. (Moisescu, 1976). (Data from Simmons and Szent-Györgyi, 1980).

action of the light chains, calcium binding may not be obligatory for those myosins that are switched on by the phosphorylation of light chains. Since removal of only one of the two regulatory light chains resulted in the loss of inhibition it was thought that cooperation between the two myosin heads was obligatory for maintaining the relaxed state (*cf.* Szent-Györgyi, 1975).

The concepts of myosin linked regulation and of the role of regulatory light chains were based on the interpretation of the calcium dependence of the actin activated ATPase activity. Since it is possible to remove regulatory light chains from skinned fibre bundles, myosin control can also be studied from the calcium dependence of tension generation. Scallop fibre bundles develop tension only in the presence of calcium (Simmons and Szent-Györgyi, 1978) (Fig. 1). A 30–60 min treatment at room temperature with EDTA removes a considerable portion of the regulatory light chains and the fibres now develop full tension in a calcium free relaxing solution. Fibres depleted in regulatory light chains are no longer able to relax. After an about 30–60 min incubation with scallop regulatory light chains the full complement of regulatory light chain content and full cal-

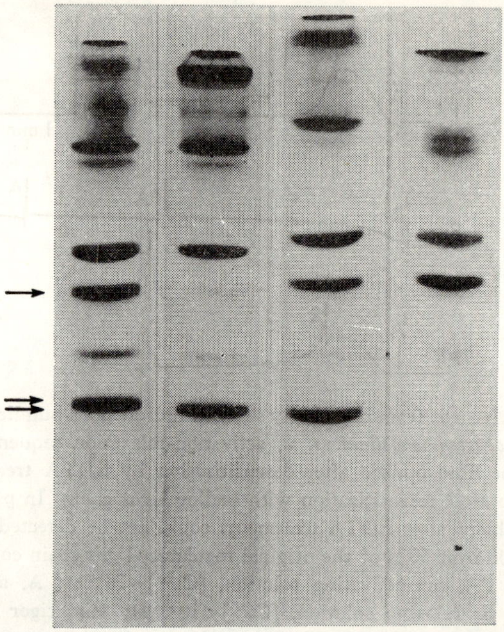

Fig. 2. Reversible dissociation of regulatory light chains from *Aequipecten* myofibrils at 35°C. a, intact myofibrils. b, Myofibrils treated for 5′ with 10 mm EDTA at 35°C in 40 mM NaCl, 5 mM phosphate, pH 7.0. c, EDTA treated myofibrils incubated with excess regulatory light chains in the presence of 1 mM $MgCl_2$ for 3 hr followed by repeated washings to remove unbound light chains. d, supernatant after EDTA treatment. 10% urea acrylamide gel electrophoresis. Arrow regulatory light chains; double arrow; essential light chains. (Data from Chantler and Szent-Györgyi, 1980).

cium sensitivity is regained (Simmons and Szent-Györgyi, 1978). The results fully support the ATPase studies and indicate in a more direct manner that in Placopecten at least, the major mode of regulation is by myosin and that regulatory light chains are obligatory for the calcium dependence of tension generation. This is important since the steady state ATPase depends not only on the association of actin and myosin, the step blocked by the regulatory light chain, but on other kinetic constants as well that may or may not influence tension.

At about 30°C, short EDTA treatment removes both regulatory light chains from myofibrils (Chantler and Szent-Györgyi, 1980) (Fig. 2). In the presence of magnesium, original light chain content can be restored. Since light chains are bound to myofibrils with high affinity, all the light

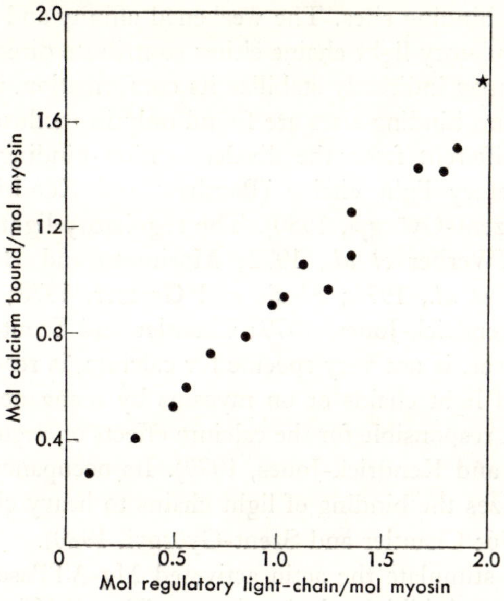

Fig. 3. Restoration of calcium binding upon readdition of regulatory light chains to 35°C-desensitized *Aequipecten* myofibrils. Calcium binding measured at 2.1×10^{-6} M free calcium in the presence of 1 mM $MgCl_2$, pH 7.0 in 40 mM NaCl. Light chain ratios from densitometry of urea gels. (Data from Chantler and Szent-Györgyi, 1980).

chain added is bound until the stoichiometry of 2 mol of regulatory light chain per myosin is reached. It is possible thus to obtain myofibrils at any desired regulatory light chain to heavy chain ratios and to study the role of regulatory light chains on ATPase activity and on calcium binding more in detail. Moreover, hybrids that contain only foreign regulatory light chains can be readily constructed from myofibrils that are fully denuded of regulatory light chains (pure hybrids) (Chantler *et al.*, 1979).

Specific, high affinity calcium binding (measured at 10^{-6} M free calcium, in the presence of 1 mM $MgCl_2$ at pH 7.0) is proportional to regulatory light chain content (Chantler and Szent-Györgyi, 1980), 2 mol of calcium is bound per mol of myosin in untreated or fully reconstituted myofibrils (Fig. 3). The site of this calcium binding is likely to be the heavy chains of myosin (Chantler and Szent-Györgyi, 1980; Bagshaw and Kendrick-Jones, 1979). In myofibrils, stripped of regulatory light chains, two new divalent cation binding sites per myosin is found that have a lesser specificity and a lowered affinity for calcium. These sites are not present on intact scallop fibrils and are likely to be a modified form of the specific calcium binding sites. The weakened affinity and lesser specificity suggest that regulatory light chains either contribute directly to the ligand binding site or may indirectly stabilize its conformation. The heavy chain associated calcium binding sites are found only in molluscan myosins and are distinctly different from the divalent cation binding sites associated with the regulatory light chains (Bagshaw and Kendrick-Jones, 1979; Chantler and Szent-Györgyi, 1980). The regulatory light chains bind divalent cations (Werber *et al.*, 1972; Morimoto and Harrington, 1974; Kendrick-Jones *et al.*, 1976; Alexis and Gratzer, 1978; Bagshaw, 1977; Bagshaw and Kendrick-Jones, 1979; Chantler and Szent-Györgyi, 1978). This site, however, is not very specific for calcium, is readily recognizable both on isolated light chains or on myosins by manganese EPR spectroscopy and is not responsible for the calcium effects on regulation (Bagshaw, 1978; Bagshaw and Kendrick-Jones, 1979). Its occupancy by magnesium evidently stabilizes the binding of light chains to heavy chains (Kendrick-Jones *et al.*, 1976; Chantler and Szent-Györgyi, 1980).

Light chains stimulate the actin activated Mg-ATPase in the presence of calcium and are inhibitory in its absence (Fig. 4) (Chantler and Szent-Györgyi, 1980). The activating effect of light chains is species dependent,

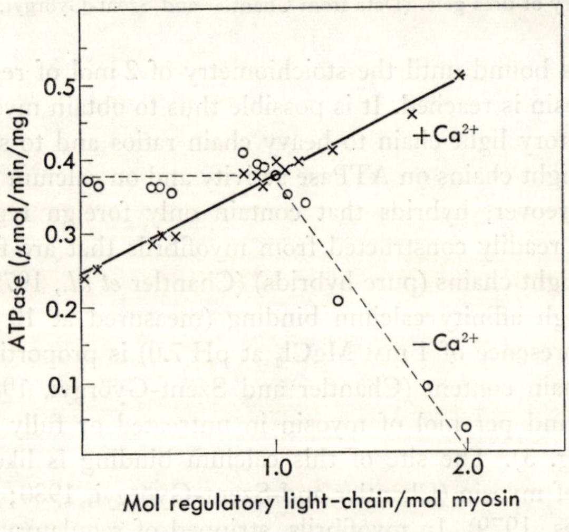

Fig. 4. Actin-activated Mg-ATPase ($\pm Ca^{2+}$) of *Aequipecten* myofibrils as a function of regulatory light chain content. ○ 0.1 mM EGTA; × 0.1 mM EGTA, 0.2 mM Ca^{2+} in addition to 25 mM NaCl and 1 mM MgCl at pH 7.5. Light chain ratios from densitometry of urea gels. (Data from Chantler and Szent-Györgyi, 1980).

and is considerably greater in Placopecten than in Aequipecten (Chantler and Szent-Györgyi, 1980). It is not clear what role if any this activation plays in regulation. It is unlikely to be very important, since the fibres develop about the same tension in the presence of calcium both with and without regulatory light chains (Simmons and Szent-Györgyi, 1980). The occupancy of the modified divalent cation binding sites of light chain denuded myofibrils by calcium somewhat lowers the ATPase activity.

The inhibitory action of regulatory light chains on ATPase activity is biphasic. Inhibition develops only at regulatory light chain contents between 1 to 2 mol per myosin molecule (Chantler and Szent-Györgyi, 1979) (Fig. 4). The results indicate that both regulatory light chains must be bound on myosin for inhibition, and that the two light chains are bound to myosin with a different affinity. The differences in affinity may signify that the two subfragment 1 (S-1) heads of myosin are different or may indicate a negative cooperativity of binding. Isolated S-1 preparations bind regulatory light chains with a uniformly high affinity and digestion of

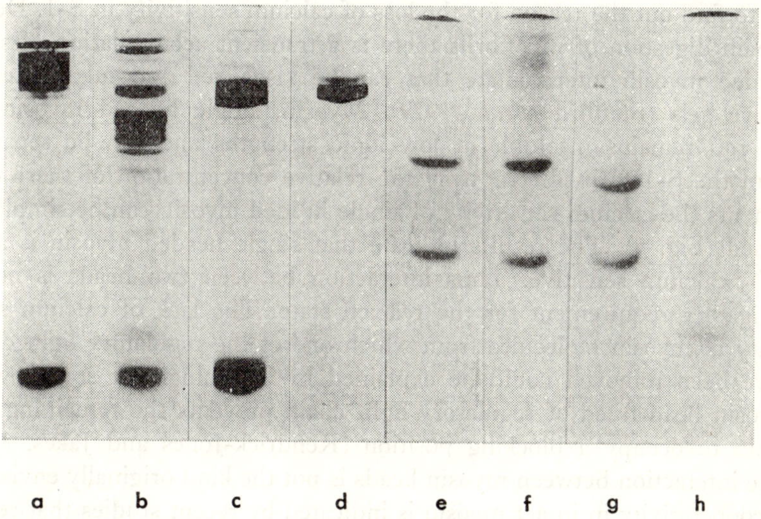

Fig. 5. Scallop myosin, heavy meromyosin and S-1 preparations. a, e, myosin; b, f, HMM; c, g, (Ca, Mg) S-1; d, h, (EDTA) S-1; a–d, 7 1/2% SDS acrylamide gels; e–h, 10% urea acrylamide gels. Preparations: myosin as in Chantler and Szent-Györgyi (1980); (Ca, Mg) S-1 and (EDTA) S-1 as in Stafford *et al.* (1979) from myosin; HMM according to Craig *et al.* (1980). Note that the regulatory light chain of (Ca, Mg) S-1 have a higher mobility and that although HMM is polydisperse on SDS gels, both of its light chains remained intact.

myofibrils from which one light chain has been removed at 0°C indicates that the preparative procedures did not select for S-1 having a higher affinity for light chains (Stafford *et al.*, 1979). We suggest that the differences in the binding of light chains result from a negative cooperativity between myosin heads.

S-1 preparations prepared in the presence of divalent cations [(CaMg)-S-1] retain both their regulatory light chains (slightly modified by the removal of 1,500-dalton piece) and essential light chains. S-1 prepared by papain in the presence of EDTA contains no regulatory light chains and only a 8,000-dalton fragment of the essential light chains [(EDTA)-S-1] (Fig. 5). Neither S-1 preparations or S-1 preparations containing intact regulatory light chains show any calcium sensitivity (Stafford *et al.*, 1979). In contrast heavy meromyosin preparations obtained by trypsin are calcium sensitive (Szent-Györgyi *et al.*, 1973; Craig *et al.*, 1980). Heavy meromyosin differs from S-1 in two respects: heavy meromyosin is a double headed species and it also retains the S-2 regions of myosin. It was therefore important to investigate the behaviour of single headed myosin to find out the reason for the loss of calcium sensitivity in S-1. During papain digestion of myofibrils there is a transient accumulation of single headed myosin intermediate that can be visualized and quantitated on native gels (Stafford *et al.*, 1979). Myosin, single headed myosin and the rod remains insoluble at low ionic strengths and can be separated from the S-1 formed. Knowing the relative concentration of these components the calcium sensitivity of single headed myosin can be simply obtained (Fig. 6). The results indicate that single headed myosin is about 75% calcium sensitive. Thus interaction between two heads is not an obligatory requirement for the relaxed state. The lack of calcium sensitivity in myosin molecules from which one of the regulatory light chains have been removed could be explained by assuming that somehow the myosin head freed of regulatory light chain prevents the remaining light chain to occupy a blocking position (Kendrick-Jones and Jakes, 1977). The interaction between myosin heads is not the kind originally envisaged. A cooperativity in intact myosin is indicated by recent studies that related calcium binding with the calcium dependence of the ATPase and showed that both myosin heads must bind calcium for activity (Chantler *et al.*, 1980). The experiments on single headed myosin at the same time point to the importance of the S-2 region, possibly the S-1–S-2 hinge, in maintaining the relaxed state. It is important to recall that regulatory light chains are elongated structures (Stafford and Szent-Györgyi, 1978; Mendelson

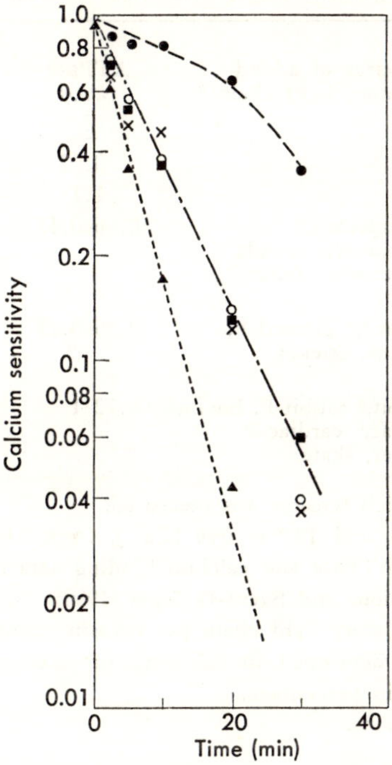

Fig. 6. Calcium sensitivity of single headed myosin. Time course of papain digestion of precipitated myosin. ▲ double headed myosin, estimated from densitometry of native gels; ■ myosin heavy chains from SDS gels; ○ S-1 heads not yet digested; × actin activated Mg-ATPase in the presence of Ca^{2+}; ● calcium sensitivity in the low salt insoluble fraction. Values normalized. If single headed myosin would not be calcium sensitive, calcium sensitivity should follow the curve of double headed myosin. (Data from Stafford *et al.*, 1979).

and Hartt, 1978; Alexis and Gratzer, 1978) with length approaching 100 Å and may easily span a considerable portion of the distance between the actin binding region and the S-1–S-2 hinge.

Fully denuded myofibrils recombine with 2 mol of the regulatory light chains of all the different myosins so far tested. The pure hybrids formed allows one to compare the ability of regulatory light chains to restore regulation (Chantler *et al.*, 1979; Sellers *et al.*, 1980). Regulatory light chains may be classed into two major groups, those that are inhibitory in the absence of calcium and those that are not. Inhibitory light chains may

TABLE I
Comparison of Muscle Regulatory Light Chains Hybridized with Placopecten Muscle

	Source of added regulatory light chains[b]	ATPase ratios	Ca binding in % of intact	Ca sensitivity of tension (20°C)[c]
Intact		0.02	100	100
Desensitized		1.18	4	0
Inhibitory LC's high calcium affinity	Aequipecten,[a] Placopecten spisula,[a] mercenaria,[a] loligo,[a] urechis	0.06–0.17	79–92	87–100
Inhibitory LC's low calcium affinity	Gizzard,[a] gizzard-P, limulus, cricket	0.06–0.13	12–47	100
Non functional LC's	Rabbit,[a] rabbit-P, bovine cardiac,[a] cardiac-P lobster, skate	0.72–1	0–13	0

[a] Denotes hybrids on which tension was measured.
Calcium binding at 2.1 and 10^{-6} M free [Ca^{2+}], 1 mM $MgCl_2$, 40 mM NaCl, 10 mM imidazole pH 7.0, 0°C. ATPase and calcium binding data from Sellers *et al.* (1980); tension data from Simmons and Szent-Györgyi (1980).
[b] More than 1.5 mol regulatory light chain per myosin recombined.
[c] Ratio of extra tension developed in calcium (compared with tension −Ca) to the total tension expressed in percentage.
LC, regulatory light chains.

differ in the manner their inhibitory function is reversed. Molluscan light chains (mercenaria, spisula, squid) and the light chain of the echuroid, Urechis, behave like scallop, restore high affinity calcium binding fully and are fully activated at relatively low calcium concentrations (10^{-6} M in the presence of 1 mM $MgCl_2$ at pH 7.5) (Table I) (Sellers *et al.*, 1980). It is of interest that none of the molluscan light chains are phosphorylated (Frearson *et al.*, 1976; Kendrick-Jones and Jakes, 1977) and resist phosphorylation by light chain kinases such as gizzard kinase (Sellers, personal communication). The presence of high affinity specific calcium binding sites on these myosins and the full reversal of the inhibition at relatively low calcium concentrations suggests that in these muscle calcium interacts directly with myosin to switch on the system. Tension data agree with the ATPase studies. Fibre bundles containing mercenaria, spisula, and loligo regulatory light chains regain their ability to relax (90–100% sensitivity) (Simmons and Szent-Györgyi, 1980).

Another subclass of the light chains that are inhibitory when hybridized with scallop myosin heavy chains, are the ones that require a considerably higher calcium concentration for ATPase activation (3×10^{-6}–2×10^{-5} M free calcium, in the presence of 1 mM $MgCl_2$ at pH 7.5). Correspondingly these hybrids bind calcium with a lower affinity. Regulatory light chains from cricket, limulus, platelet, and gizzard belong to this group (Sellers *et al.*, 1980). Most, possibly all of these light chains are phosphorylatable. Phosphorylation of platelet light chains (Adelstein and Conti, 1975) gizzard (Sobieszek, 1977; Aksoy *et al.*, 1977) limulus (Sellers, 1980) accompanied with a significant activation of the actin activated ATPase has been demonstrated. Phosphorylation of these light chains had no effect on the hybrids, even though phosphorylation was maintained during the ATPase studies. The results may simply mean that hybrids do not adequately reflect the *in vivo* behaviour of the myosins. The requirement for higher calcium concentrations may indicate that these light chains restore the heavy chain associated calcium binding sites only incompletely. Triggering of these myosins by calcium is likely to be an indirect effect (Sherry *et al.*, 1979; Sobieszek and Small, 1977) and a direct calcium binding by these myosins may not be a requirement for regulation.

The second major group of light chains are those that are not inhibitory and are not activated by calcium as pure hybrids. These light chains do not restore the specific high affinity calcium binding sites of scallop myosin. Light chains of this class originate from myosin that are not regulated in ATPase activity tests including those of vertebrate striated muscles, (such as rabbit back, chicken pectoralis, and skate muscle) bovine cardiac, and lobster tail muscle (Sellers *et al.*, 1980). One notes that these light chains are phosphorylatable (Perrie *et al.*, 1973; Frearson and Perry, 1975) but phosphorylation does not alter the behaviour of the hybrids. Tension studies and ATPase measurements agree and these light chains are unable to restore any calcium dependence of tension generation (Simmons and Szent-Györgyi, 1980). Frequently the ATPase activity of some of these light chain hybrids, particularly those of cardiac and rabbit DTNB light chains, have a lowered ATPase activity both in the absence and presence of calcium (one fourth to one half of the activity of desensitized myofibrils) but we have reason to suspect that this lowered activity is artefactual, possibly due to solubilization at the very low protein concentrations used in ATPase studies ($<0.01\%$ protein). In fact, tension generation does not seem to be reduced in the hybrids made with rabbit regulatory light chains. The inability of the regulatory light chains of vertebrate striated muscles to relax scallop myosin supports the interpretation that in these muscles myosin control is not operating.

The experimentally demonstrated components of myosin linked regulation are the regulatory light chains that interact with the myosin heads in a way that allows for both an "off" and an "on" state corresponding to rest and activity. However additional regions of the myosin molecule and additional components may also be involved in supporting the relaxed state. The possible importance of the S-2–S-1 hinge region is indicated by the calcium sensitivity of single headed myosin in contrast to the lack of sensitivity of S-1 preparations. In addition there is indirect evidence suggesting that the essential light chains may also be regulatory components. Antibodies, specific against essential light chains that do not cross react with regulatory light chains or with myosin heavy chains, selectively elevate the actin activated Mg-ATPase in the absence of calcium; at a ratio of about 1.3–1.5 mol antibody bound per mol of myosin head the ATPase rate in EGTA equals the rate in calcium (Wallimann and Szent-Györgyi, 1979). Interestingly, Fab' fragments are as effective in abolishing calcium sensitivity as intact antibodies, thus desensitization by the antibody is not due to cross linking of the myosin heads. Furthermore, some rabbits produced antibodies that although bound to essential light chains failed to abolish calcium sensitivity (Wallimann and Szent-Györgyi, 1979). To be effective antiessential light chain antibodies must be bound at a particular region of the essential light chain.

Myosin control thus appears to involve the rearrangement of both kind of light chains and a substantial stretch of the S-1 head, possibly extending from the actin combining sites to the S-1–S-2 hinge. The great length of the regulatory light chains (*ca.* 100 Å) would indicate that they alone or in conjunction with the essential light chain may straddle a considerable portion of the distance that separates the actin binding sites from the S-1–S-2 hinge. Calcium ions, the presence or the absence of regulatory light chains have only little effect on the spectroscopic properties of myosin or on the light chains. The lack of substantial change in absorbancy, circular dichroism or fluorescence is in sharp contrast with the large effects of calcium on parvalbumin or troponin C. The type of structural alteration in myosin and the light chain evidently is not a large scale conformational change but may be restricted to relatively few residues (Chantler and Szent-Györgyi, 1978). The movement, if it takes place, may be a "hinge-like" one rather than a general "swelling" of the molecule.

The subunit rearrangement of myosin is a structural problem and requires structural approaches. The techniques that are applied to soluble system have not been informative so far. In contrast effects of the removal of the regulatory light chain were readily seen by X-ray diffraction or by electron microscopy. Light chain removal resulted in a substantial increase in the 384 Å meridional reflection in fibre bundles of glycerinated scallop muscles with a moderate increase in the density of the 192 Å reflection (Vibert *et al.*, 1978). At present higher order reflections are not sufficiently resolved to give a precise choice for the possible reasons for these changes, such as altered cross-bridge angle or occupancy of actin sites. A possible explanation that the increased intensity of the 384 Å layer line arises from an increased number of cross-bridges myosin can form, once regulatory light chain has been removed, was made unlikely by the direct visualization in the electron microscope that both heads of a calcium sensitive heavy meromyosin (HMM) combines with the neighbouring monomers of the same strand of an actin filament (Craig *et al.*, 1980). In accordance muscle stiffness in rigor is not changed by the removal of regulatory light chains (Simmons and Szent-Györgyi, unpublished observation).

The appearance of arrowheads of decorated actin filaments depends on the regulatory light chains (Craig *et al.*, 1980). S-1 or HMM preparation that do not contain regulatory light chains give rise to "blunted" arrowhead pattern: the outlines of the wide portion of the arrowhead are convex. "Barbed" arrowheads are formed by S-1 and HMM preparations that contain their full complement of regulatory light chains. "Barbed" patterns have a concave appearance and are easily distinguished from "blunted" ones (Craig *et al.*, 1980). Rabbit S-1 and HMM also form "barbed" and "blunted" arrowheads depending on their regulatory light chain content. An interpretation of the difference between the two arrowhead pattern must await three dimensional reconstruction studies. Both X-ray and electron microscopy demonstrate that the rigor pattern depends on the regulatory light chains. So far these changes were restricted to conditions of rigor. It will be of considerable interest to see if regulatory light chain induced differences can also be observed in myosin preparation in the absence of actin. Hopefully, future structural studies will give more precise informations on how myosin control operates.

Acknowledgment

This research was supported by grants from the Public Health Service (AM 15963) and the Muscular Dystrophy Association.

REFERENCES

Adelstein, R. S. and Conti, M. A. (1975). *Nature* **256**, 597–598.

Aksoy, M. O., Williams, D., Sharkey, E. M., and Hartshorne, D. J. (1976). *Biochem. Biophys. Res. Commun.* **69**, 35–41.

Alexis, M. N. and Gratzer, W. B. (1978). *Biochemistry* **17**, 2319–2325.

Bagshaw, C. R. (1977). *Biochemistry* **14**, 59–67.

Bagshaw, C. and Kendrick-Jones, J. (1979). *J. Mol. Biol.* **130**, 317–336.

Chantler, P. D., Sellers, J., and Szent-Györgyi, A. G. (1979). *Biophys. J.* **25**, 72a.

Chantler, P. D., Sellers, J. R., and Szent-Györgyi, A. G. (1980) (in preparation).

Chantler, P. D. and Szent-Györgyi, A. G. (1978). *Biochemistry* **17**, 5440–5448.

Chantler, P. D. and Szent-Györgyi, A. G. (1980). *J. Mol. Biol.* (in press).

Craig, R., Szent-Györgyi, A. G., Beese, L., Flicker, P., Vibert, P., and Cohen, C. (1980). *J. Mol. Biol.* (in press).

Frearson, N., Focant, B.W.W., and Perry, S. V. (1976). *FEBS Lett.* **43**, 27–32.

Frearson, N. and Perry, S. V. (1975). *Biochem. J.* **151**, 99–107.

Kendrick-Jones, J. (1974). *Nature* **249**, 631–634.

Kendrick-Jones, J. and Jakes, R. (1977). In *International Symposium on Myocardial Failure* (Riecker, G., Weber, A., and Goodwin, H., eds.), pp. 28–48, Springer-Verlag, Berlin.

Kendrick-Jones, J., Lehman, W., and Szent-Györgyi, A. G. (1970). *J. Mol. Biol.* **54**, 313–326.

Kendrick-Jones, J., Szentkiralyi, E. M., and Szent-Györgyi, A. G. (1976). *J. Mol. Biol.* **104**, 747–775.

Kumagai, H., Ebashi, S., and Takeda, F. (1955). *Nature* **176**, 166.

Lehman, W. and Szent-Györgyi, A. G. (1975). *J. Gen. Physiol.* **66**, 1–30.

Marsh, B. B. (1951). *Nature* **167**, 1065–1066.

Mendelson, R. A. and Hartt, J. (1978). *Biophys. J.* **25**, 71a.

Moisescu, D. G. (1976). *Nature* **262**, 610–613.

Morimoto, R. and Harrington, W. F. (1974). *J. Mol. Biol.* **88**, 693–709.

Natori, R. (1954). *Jikeikai Med. J.* **1**, 119–126.

Perrie, W. T., Smiley, L. B., and Perry, S. V. (1973). *Biochem. J.* **135**, 151–164.

Sellers, J. R., Chantler, P. D., and Szent-Györgyi, A. G. (in preparation).

Sellers, J. R. (1980) (in preparation).

Sherry, J.M.F., Gorecka, A., Aksoy, M. O., Dabrowska, R., and Hartshorne, D. J. (1978). *Biochemistry* **17**, 4411–4418.

Simmons, R. M. and Szent-Györgyi, A. G. (1978). *Nature* **273**, 62–64.

Simmons, R. M. and Szent-Györgyi, A. G. (1980). *Nature* (in press).

Sobieszek, A. (1977). In *The Biochemistry of Smooth Muscle* (Stephens, N. L., ed.), pp. 413–443, Univ. Park Press, Baltimore.

Sobieszek, A. and Small, J. V. (1977). *J. Mol. Biol.* **112**, 559–576.

Stafford, W. F. and Szent-Györgyi, A. G. (1978). *Biochemistry* **17**, 607–614.

Stafford, W. F., Szentkiralyi, E. M., and Szent-Györgyi, A. G. (1979). *Biochemistry* **18**, 5273–5280.

Szent-Györgyi, A. G. (1975). *Biophys. J.* **15**, 707–723.

Szent-Györgyi, A. G., Szentkiralyi, E. M., and Kendrick-Jones, J. (1973). *J. Mol. Biol.* **74**, 179–203.

Vibert, P., Szent-Györgyi, A. G., Craig, R., Wray, J., and Cohen, C. (1978). *Nature* **273**, 64–66.

Wallimann, T. and Szent-Györgyi, A. G. (1979). *Biophys. J.* **25**, 72a.

Werber, M. M., Gaffin, S. L., and Oplatka, A. (1972). *J. Mechanochem. Cell. Mot.* **1**, 91–96.

Muscle Contraction: Its Regulatory Mechanisms (Ebashi, S. et al., eds.), pp. 391–399, Japan Sci. Soc. Press, Tokyo / Springer-Verlag, Berlin (1980)

30

Myosin-linked Calcium Regulation in Squid Muscle

KUNIHIKO KONNO, KEN-ICHI ARAI, AND SHIZUO WATANABE

Summary: Aiming at finding the primary role of light chains of myosin in calcium regulation, we studied the regulatory system of squid mantle muscle and that of scallop adductor muscle. The results thus obtained were as follows:

1. Squid myosin and scallop myosin were capable of binding 2 mol of calcium per mol of myosin with an apparent dissociation constant of approximately 3 μM.
2. At 10 μM free calcium, squid regulatory light chain (abbreviated as LC-2) and scallop EDTA-light chain, both in an isolated form, were able to bind only about 0.5 and 0.1 mol, respectively, of calcium per mol of light chain. Moreover, the amount of calcium bound to the isolated form of light chains was greatly reduced by magnesium.
3. Not 1 mol but 2 mol of EDTA-light chain were removed by mere washing with 10 mM EDTA at 0°C, resulting in a complete loss of calcium binding capacity. A full recovery of calcium binding capacity required recombination of not 1 mol but 2 mol of either EDTA-light chain or squid LC-2 with 1 mol of EDTA-washed (desensitized) scallop myosin.

4. Removal of 2 mol of EDTA-light chain from 1 mol of scallop myosin also resulted in a large decrease in the activities of actomyosin-superprecipitation and of actomyosin-ATPase. A full recovery of the lost activities also required recombination of 2 mol of either squid LC-2 or scallop EDTA-light chain with 1 mol of desensitized scallop myosin.
5. Mg-ATPase of squid myosin alone (with no actin) and that of scallop myosin alone were activated by very low concentrations of calcium, half the maximal activation occurring at approximately 3 μM. Mg-ATPase activity of squid myosin was much lower in the absence of calcium and much higher in the presence of calcium than was that of scallop myosin. A hybrid myosin reconstituted from squid LC-2 and scallop-desensitized myosin did not show a high ATPase activity in the presence of calcium but showed a very low ATPase activity in the absence of calcium.

In a previous report (Konno *et al.*, 1978), we showed that a myosin-linked regulatory system and an actin-linked regulatory system were both present in squid mantle muscle. The two systems were found to be different in the concentrations of calcium and strontium required for activation of actomyosin-ATPase, and the calcium requirement for activation of squid myosin B-ATPase was shown to be identical with that for activation of skeletal acto-squid myosin-ATPase. From this and some other findings, we suggested that myosin-linked regulation rather than actin-linked regulation is the one effectively operating in squid myosin B.

A further study has since been made on the myosin-linked regulatory system, and some of the results obtained will be presented in this report.

GENERAL PROPERTIES OF SQUID MYOSIN

1. *Calcium and Strontium Sensitivities of Actomyosin-ATPase*

As previously reported (Konno *et al.*, 1979), Mg-ATPase of skeletal acto-squid myosin and that of skeletal acto-scallop myosin were both activated by either calcium or strontium. The concentration of calcium or strontium required for the activation was the same for ATPases of both actomyosins, half the maximal activation occurring at 3 μM calcium or at 100 μM strontium.

2. *Calcium Binding Ability of Myosins*

Calcium binding was measured by using ^{45}Ca in an equilibrium dialysis method (Kuwayama *et al.*, 1979). As shown in Fig. 1, both squid and

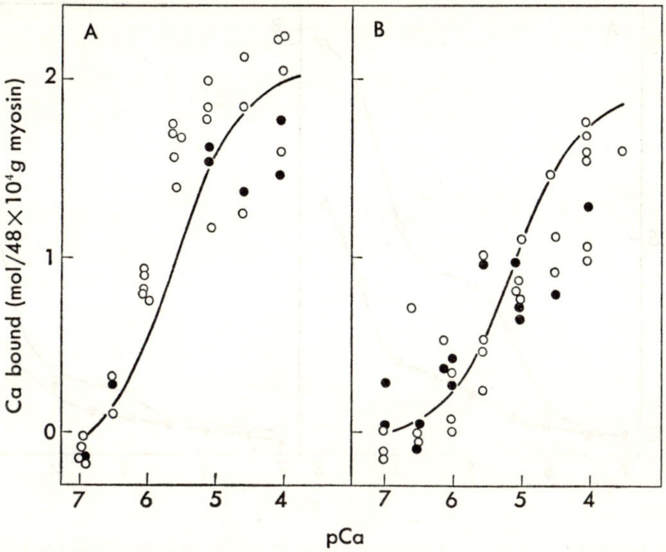

Fig. 1. Calcium binding of squid and scallop myosin. Calcium binding was measured by the dialysis equilibration method using ^{45}Ca in the medium of 0.3 M KCl, 20 mM His (pH 6.8) with (●) or without (○) 2.5 mM $MgCl_2$ at 20°C. Free calcium concentration was determined by using Ca-EGTA buffer. A: squid. B: scallop.

scallop myosins were capable of binding 2 mol of calcium per mol of myosin. Half the maximal binding occurred at 3 μM calcium with squid myosin and at 6 μM calcium with scallop myosin. Therefore, we conclude that the calcium sensitivity of acto-squid myosin ATPase and that of acto-scallop myosin-ATPase (Konno *et al.*, 1979) are closely interrelated with the calcium binding ability of myosins. It should be noted that calcium binding to myosin was not affected by the presence of 2.5 mM $MgCl_2$ (filled circles in Fig. 1).

3. *Calcium Binding Ability of Light Chains*

Calcium binding ability was also measured with regulatory light chains in an isolated form. As shown in Fig. 2, at 10 μM free calcium, the isolated form of squid regulatory light chain (LC-2) bound only about 0.5 mol of calcium per mol of light chain, and that of scallop EDTA-light chain only about 0.1 mol of calcium per mol of light chain. Moreover, calcium binding with light chains in the isolated form was sensitive to magnesium; 2.5 mM $MgCl_2$ caused a large decrease in the amount of calcium bound to light chains.

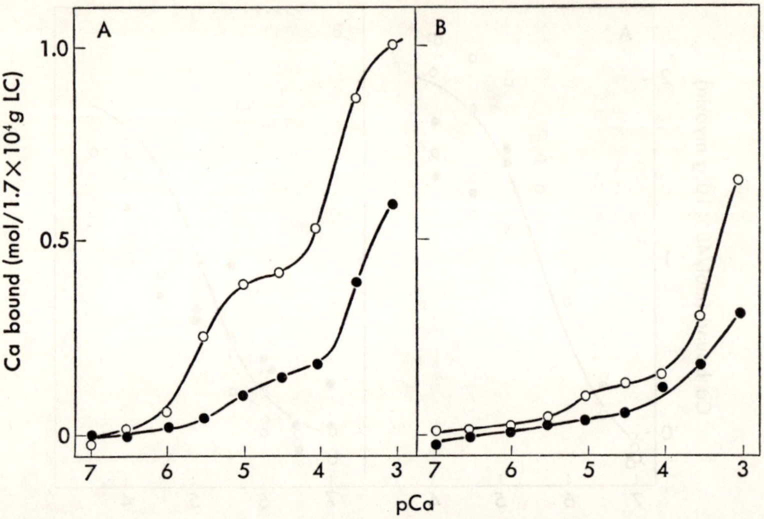

Fig. 2. Calcium binding of isolated regulatory light chains. Calcium binding of light chains were measured under the same conditions as described in Fig. 1, except that 0.1 M KCl was used instead of 0.3 M KCl. The symbols used were also the same as in Fig. 1. A: squid LC-2. B: scallop EDTA-LC.

4. *Calcium- and Strontium-induced Formation of Difference Spectra in UV Absorption of Light Chains*

Calcium and strontium were both found to induce formation of UV absorption difference spectra in the isolated form of squid LC-2 as well as in that of scallop EDTA-light chain. The concentrations of calcium and of strontium required for half the maximal saturation in the formation of difference spectra were 7 μM calcium and 630 μM strontium with squid light chain (measured at 293 nm), and 280 μM calcium and 2,800 μM strontium with scallop EDTA-light chain (measured at 288 nm).

The value of 7 μM calcium is sufficiently low to account for calcium sensitivity of the ATPase activity of acto-squid myosin (Konno *et al.*, 1979) and for the calcium binding of squid myosin (Fig. 1). On the other hand, the value of 280 μM calcium is too high to account for the calcium sensitivity or the calcium binding ability associated with scallop myosin.

DESENSITIZATION AND RESENSITIZATION

1. Regulatory Light Chain of Squid Myosin

As has been reported (Szent-Györgyi *et al.*, 1973), washing of scallop adductor myosin in EDTA resulted in the loss of calcium sensitivity of skeletal acto-scallop myosin-ATPase; the loss was restored on addition of scallop EDTA-light chain to EDTA-washed (desensitized) scallop myosin. Squid LC-2 was also shown to be capable of conferring calcium sensitivity on desensitized scallop myosin.

2. Calcium Binding Ability of Desensitization and Resensitization

Calcium binding was measured with desensitized scallop myosin and with resensitized myosin (Fig. 3). EDTA-wash of scallop myosin resulted in a complete loss of calcium binding capacity, and the lost capacity was fully recovered on addition of EDTA-light chain to EDTA-washed scallop myosin. Squid LC-2 was also capable of restoring the full capacity of calcium binding to desensitized scallop myosin. The restored capacity of

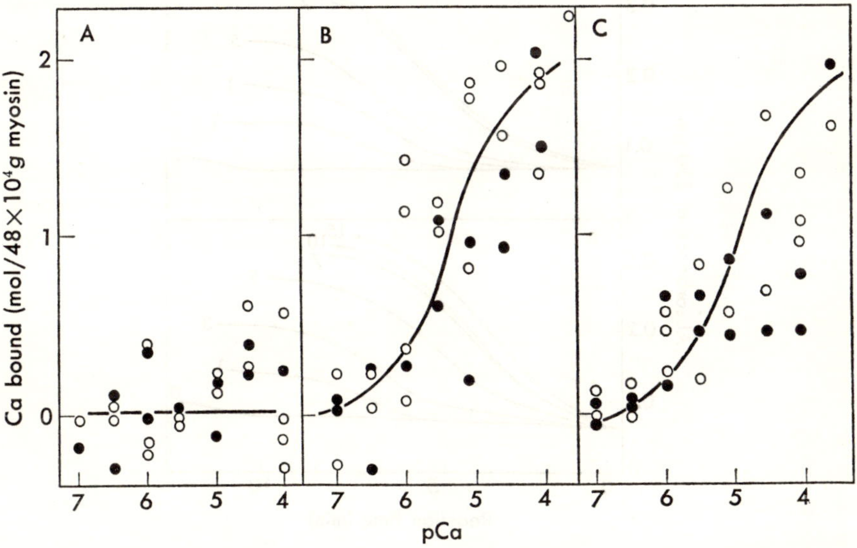

Fig. 3. Calcium binding of desensitized and resensitized scallop myosin. Calcium binding of scallop desensitized myosin and resensitized myosin with scallop EDTA-LC or squid LC-2 was measured under the same conditions as in Fig. 1. The symbols used are the same as in Fig. 1. A: desensitized myosin. B: resensitized myosin with squid LC-2. C: resensitized myosin with scallop EDTA-LC.

calcium binding was insensitive to magnesium; that is, it was not at all affected by the presence of 2.5 mM $MgCl_2$.

3. Removal and Recombination of Light Chains

Urea gel electrophoresis (Konno *et al.*, 1979) showed that EDTA-light chain of scallop myosin was removed not by half but nearly completely, and that 1 mol of desensitized scallop myosin was capable of recombining 2 mol of squid LC-2 or of scallop EDTA-light chain.

4. The Amount of Light Chain Required for Recombination and for Resensitization

As the amount of light chain added to desensitized myosin increased, the light chain content in myosin and the calcium sensitivity of actomyosin-ATPase were both increased linearly; both saturated at 7.5% of light chain bound per myosin. The value of 7.5% is equivalent to 2 mol of light chain per mol of myosin.

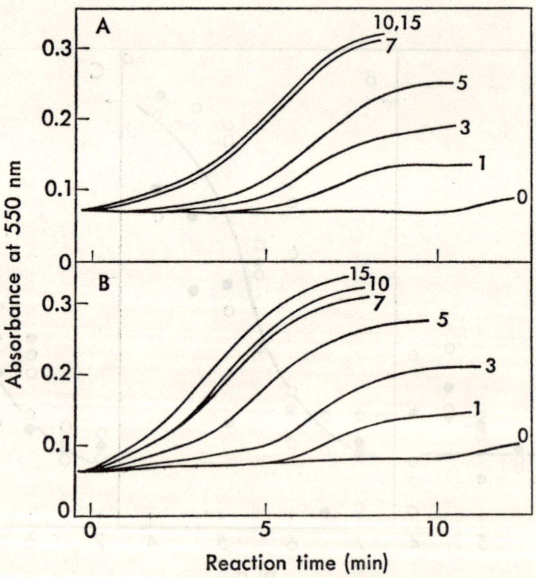

Fig. 4. Effect of light chain on the superprecipitation ability of acto-desensitized scallop myosin. Superprecipitation was measured in 0.03 M KCl, 20 mM Tris-maleate (pH 6.8), 2 mM $MgCl_2$, 300 μM ATP, and 0.1 mM $CaCl_2$ at 25°C. Numbers in figures indicate the amounts of light chain added to desensitized myosin (%, w/w). A: squid LC-2. B: scallop EDTA-LC.

5. Superprecipitation

Asada *et al.* (1979) have reported that the superprecipitation ability was lost in skeletal acto-desensitized clam myosin, and that the lost ability was restored by adding regulatory light chain (LC-1) of clam myosin. We observed, as shown in Fig. 4, that skeletal acto-desensitized scallop myosin also lost the superprecipitation ability even in the presence of calcium, and that the lost ability was restored by addition of either EDTA-light chain or squid LC-2. As the amount of light chains added increased, the recovery proceeded linearly, and saturated at approximately 7% of light chains added to desensitized myosin.

CALCIUM SENSITIVITY OF MYOSIN

1. Myosin ATPase

Asada *et al.* (1979) also reported that Mg-ATPase of clam foot myosin alone (with no actin) was as sensitive to calcium as that of skeletal acto-clam myosin. As shown in Fig. 5, we found that Mg-ATPase activity of either squid myosin or scallop myosin was also sensitive to calcium, half the maximal activation being obtained with 3 μM calcium. However, we noticed a) that the extent of calcium activation was very small with scallop myosin (2 times) as compared with that obtained with squid myosin (25

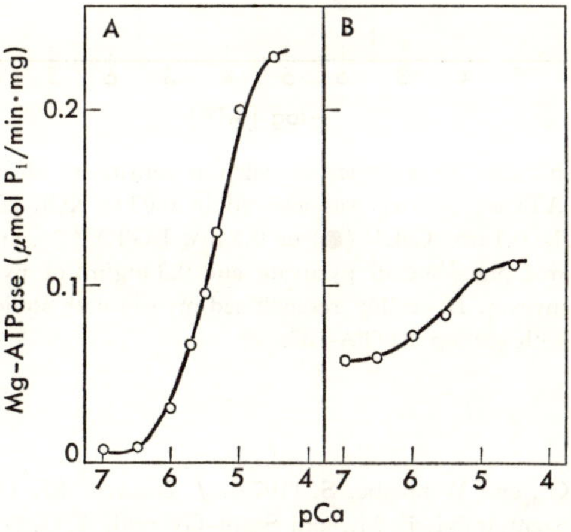

Fig. 5. Calcium sensitivity of Mg-ATPase of squid myosin or scallop myosin. ATPase activity was assayed in the medium of 0.03 M KCl, 20 mM Tris-maleate (pH 6.8), 2 mM $MgCl_2$, 1 mM ATP, and various concentrations of calcium (Ca-EGTA buffer system) at 25°C. A: squid myosin. B: scallop myosin.

times), and also b) that in the absence of calcium, the Mg-ATPase activity of scallop myosin was much higher than that of squid myosin.

2. Hybrid Myosin

We then asked of whether the difference observed above between squid and scallop myosin in the Mg-ATPase activity stemmed from the difference in the property of light chains. A hybrid myosin was prepared from squid LC-2 and scallop desensitized myosin, and a homologue myosin from scallop EDTA-light chain and scallop desensitized myosin. These were then compared for their Mg-ATPase activity. As shown in Fig. 6, squid LC-2 did not improve the activity in the presence of calcium but effectively reduced the activity in the absence of calcium, thus showing calcium sensitivity even at an ATP concentration as low as 1 μM.

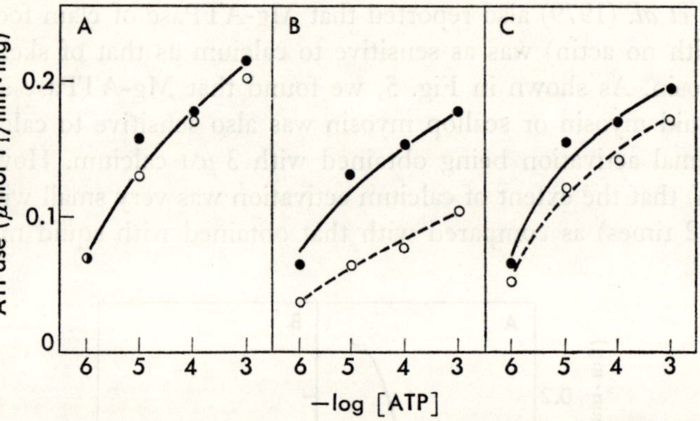

Fig. 6. ATP concentration dependence of calcium sensitivity of hybrid myosin and homologue myosin. ATPase activity was assayed in 0.03 M KCl, 20 mM Tris-maleate (pH 6.8) 2 mM $MgCl_2$, 0.1 mM $CaCl_2$ (●) or 0.5 mM EGTA (○), and various concentrations of ATP (1 mM phosphoenol pyruvate and 0.3 mg/ml of pyruvate kinase). A: scallop desensitized myosin. B: scallop resensitized myosin with squid LC-2. C: scallop resensitized myosin with scallop EDTA-LC.

REFERENCES

Asada, T., Ashiba, G., and Watanabe, S. (1979). *J. Biochem.* **85**, 1543–1546.

Kendrick-Jones, J., Szentkiralyi, E. M., and Szent-Györgyi, A. G. (1976). *J. Mol. Biol.* **104**, 747–775.

Konno, K. (1978). *J. Biochem.* **84**, 1431–1440.

Konno, K., Arai, K., and Watanabe, S. (1979). *J. Biochem.* **86**, 1639–1650.

Kuwayama, K. and Yagi, K. (1979). *J. Biochem.* **85**, 1245–1255.

Nishita, K., Ojima, T., and Watanabe, S. (1979). *J. Biochem.* **86**, 663–673.
Szent-Györgyi, A. G., Szentkiralyi, E. M., and Kendrick-Jones, J. (1973). *J. Mol. Biol.* **74**, 179–203.

Muscle Contraction: Its Regulatory Mechanisms (*Ebashi, S. et al., eds.*), *pp. 401–407,*
Japan Sci. Soc. Press, Tokyo / Springer-Verlag, Berlin (1980)

31

Regulatory Mechanisms in Squid Muscle

TAKAHIDE TSUCHIYA AND
JUICHIRO J. MATSUMOTO

Summary: The regulatory mechanism of the squid muscle was studied. The squid tropomyosin showed functions similar to those of carp tropomyosin when it cooperates with carp troponin in Ca^{2+} regulation. Ca^{2+}-sensitivity of squid muscle was also studied by desensitizing squid natural actomyosin and by cross combination tests between squid myosin and actin and between carp myosin and squid thin filament fraction. After the trypsin treatment, the squid natural actomyosin showed a decrease in Ca^{2+}-sensitivity. These results proved that squid muscle has an actin-linked regulatory system rather than a myosin-linked one. However, according to detailed studies on Ca^{2+}-sensitivity of synthetic actomyosin between squid actin and myosin and on the effect of Ca^{2+} on the electrophoretic mobility of myosin light chains, it is likely that the regulation of squid is conveyed not only by the actin-linked regulatory system but also by the myosin-linked one. These results suggest that, in squid muscle, there is a dual regulation system.

In search of the regulatory factors in the molluscan muscles, Kendrick-

Jones *et al.* (1970) found that there is no troponin in scallop muscle, and that a myosin light chain rather than troponin functions as the calcium regulation unit of the skeletal muscles. Based on their competition tests on the muscles of various animal origins, Lehman and Szent-Györgyi (1975) proposed that all the molluscan muscles have a myosin-linked regulatory system. The simultaneous presence of both actin- and myosin-linked regulatory systems has been found in other animals such as Arthropoda. Squid muscle has been included in the myosin-linked regulatory system group with scallop and other molluscan muscles. In a series of studies on the proteins of the obliquely striated muscle of the squid mantle, we found the presence of an actin-linked regulatory system along with the myosin-linked system (Tsuchiya *et al.*, 1978a). Thence, a troponin-like protein has been detected in squid myofibrils by SDS disc electrophoresis. These studies suggested modification of the conventional theory which excludes the actin-linked regulatory system from all the molluscan muscles. Subsequently, Konno (1978) reported that squid muscle has both actin- and myosin-linked regulatory systems. Lehman *et al.* (1978) revealed the existence of troponin-like protein accompanying actin in the scallop muscle, and they proposed the existence of dual regulation by both myosin-linked and actin-linked regulatory systems in the scallop muscle.

CONSTITUENT PROTEINS

As the basis of study, information so far available on the squid myofibrillar proteins will be reviewed first.

1. *Musculature*

The obliquely striated musculature of the squid mantle was illustrated electronmicroscopically by Kawaguti and Ikemoto in 1957.

2. *Actomyosin*

Matsumoto (1959) has described the characteristic properties of actomyosin in salting-in profiles and lability. Its ATPase activity was studied qualitatively by Matsumoto (1959), de Villafranca (1955), and Horie *et al.* (1975). Attempts to isolate myosin from squid were never successful, and thus studies were confined to actomyosin.

3. *Myosin*

At the beginning, myosin was difficult to isolate due to its lability during the extraction and isolation procedures. It was successfully isolated by inhibiting its decomposition by the addition of the soybean trypsin inhibitors, Trasylol and EDTA. The physicochemical properties of squid myosin were comparable with those of the skeletal muscle: molecular weight, 20×10^4 (heavy chain), 1.5 and 1.2×10^4 (light chains); sedimentation coefficient, 8.0 S; and intrinsic viscosity, 2.18 dl/g. Similarity was found also in the amino acid composition and in the propensity to form dumbell-shaped filaments. With respect to the ATPase activity, the squid myosin is similar to the shellfish myosins in the mode of dependence on pH and ionic strength, but dissimilar in the effects of EDTA and PCMB (Tsuchiya *et al.*, 1978a).

4. *Actin*

Squid actin is similar to skeletal and shellfish actins in molecular weight (4.3×10^4), sedimentation coefficient (3.3 S), intrinsic viscosity (0.12 dl/g), and amino acid composition. The periodicity of squid F-actin, 35–36 nm, corresponds to that of rabbit F-actin (Tsuchiya *et al.*, 1977).

5. *Paramyosin*

Squid paramyosin resembled the paramyosins of various shellfishes in molecular weight (10×10^4, $\times2$), sedimentation coefficient (3.5 S), intrinsic viscosity (2.03 dl/g), and amino acid composition (Tsuchiya *et al.*, 1980). The squid paramyosin formed paracrystals with spacings similar to those of other Mollusca. Paramyosin is present also in the octopus (Bailey, 1956).

6. *Tropomyosin*

Squid tropomyosin was first isolated by Yoshimura (1955) and Tsao (1956). Further study revealed that: squid tropomyosin is comparable to the skeletal tropomyosins in molecular weight (3.7 and 3.5×10^4), sedimentation coefficient (3.0 S), and intrinsic viscosity (0.31 dl/g). The amino acid composition is generally similar, but there are differences between squids and vertebrates in the content of some amino acids, such as, Pro, Trp, and Arg. Moreover, squid tropomyosin is less likely to form paracrystals than is skeletal tropomyosin (Tsuchiya *et al.*, 1980a).

7. *Protein Composition*

The squid muscle proteins include; sarcoplasmic proteins (12–20%), myofibrillar proteins (77–85%), and stroma (2–3%) (Matsumoto, 1958).

The squid myofibrils consist of; myosin (54%), actin (17%), paramyosin (14%), tropomyosin (8%), and other proteins (7%) (Horie *et al.*, 1975).

REGULATORY MECHANISM

Based on the above knowledge of the constitutional proteins, the regulatory mechanism of the squid muscle has been studied.

1. Preliminary Survey

1) To test the function of tropomyosin *per se*, squid tropomyosin was added to carp desensitized actomyosin, but no modification was found in the Ca^{2+}-sensitivity of ATPase activity and superprecipitation. 2) Squid tropomyosin exhibited a propensity to coprecipitate with carp actin. 3) With added squid tropomyosin, carp desensitized actomyosin and carp troponin recovered Ca^{2+}-sensitivity of ATPase activity and superprecipitation (Table I). 4) With respect to the possible myosin-linked regulation, the Ca^{2+}-sensitivity of the squid actomyosin was investigated, but no precise effect was found by the surplus addition of either carp myosin or squid actin (Tsuchiya *et al.*, 1978a). 5) Ca^{2+}-sensitivity of the natural actomyosin of squid did not decrease with the EDTA treatment. 6) The actomyosin synthetized from the myosin and actin of squid did not reveal any Ca^{2+}-sensitivity. These data did not provide any support for myosin-linked regulation.

TABLE I
Effect of Squid Tropomyosin on Ca^{2+}-sensitivity of Carp Desensitized Actomyosin

	ATPase activity[a]		Ca^{2+}-sensitivity
	0.1 mM $CaCl_2$	1 mM EGTA	(%)
Carp nAM	0.366	0.062	83
Carp desAM	0.400	0.396	1
Carp desAM+carp TM+carp TN	0.344	0.066	81
Carp desAM+squid TM+carp TN	0.234	0.120	49
Carp desAM+squid TM	0.256	0.236	8

[a] P_i μmol/mg·min.

2. Actin-linked Regulation

The results showed the existence of an actin-linked regulatory system in obliquely striated muscle of squid. 1) The Ca^{2+}-sensitivity of the squid

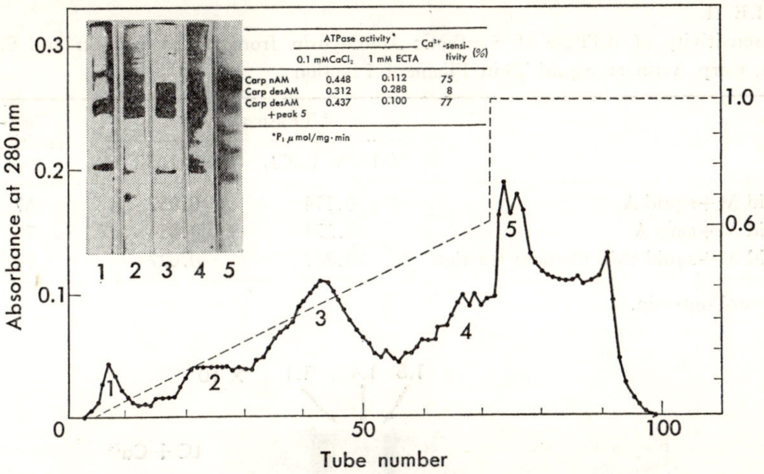

Fig. 1. DEAE-Sephadex chromatography of proteins obtained by washing. The inserted figures show the SDS disc electrophoretic patterns of each peak and the effect of peak 5 on Ca^{2+}-sensitivity of carp desensitized actomyosin.

natural actomyosin was lost by washing with low ionic strength solution. SDS disc electrophoresis of the washing showed a band of actin (4.3×10^4), tropomyosin (3.7 and 3.5×10^4), and other low molecular weight components (5.3, 3.4, 2.5, 1.9, and 1.7×10^4). 2) The Ca^{2+}-sensitivity of the mixture of carp myosin and the above washing fraction was restored to 48% (Tsuchiya *et al.*, 1978). 3) The proteins in the washing fraction were fractionated by anion-exchange chromatography (Fig. 1). The mixture of the low molecular chromatographed components and the squid tropomyosin restored the Ca^{2+}-sensitivity of the desensitized squid actomyosin to the level of the natural squid actomyosin. 4) By a mild trypsin treatment, the squid natural actomyosin showed a decrease in Ca^{2+}-sensitivity from 82% to 41%, as the skeletal natural actomyosin does. By this treatment, the low molecular components were decomposed the fastest.

3. *Myosin-linked Regulation*

1) When squid actin or carp actin was added to squid myosin, the Ca^{2+}-sensitivity of mixture was about 70%. Moreover, if the squid thin filament fraction was added to squid myosin, the Ca^{2+}-sensitivity increased to the level of squid natural actomyosin (Table II). 2) In the presence of Ca^{2+}, the lighter component of the myosin light chains showed an increased electrophoretic mobility, while the heavier component showed no

TABLE II
Ca^{2+}-sensitivity of ATPase of Synthetic Actomyosin from Squid Myosin and Squid Actin, Carp Actin or Squid Thin Filament Fraction

	ATPase activity[a]		Ca^{2+}-sensitivity (%)
	0.1 mM $CaCl_2$	1 mM EGTA	
Squid M+squid A	0.174	0.057	67
Squid M+carp A	0.224	0.054	76
Squid M+squid thin filament fraction	0.312	0.027	91

[a] P_i μmol/mg·min.

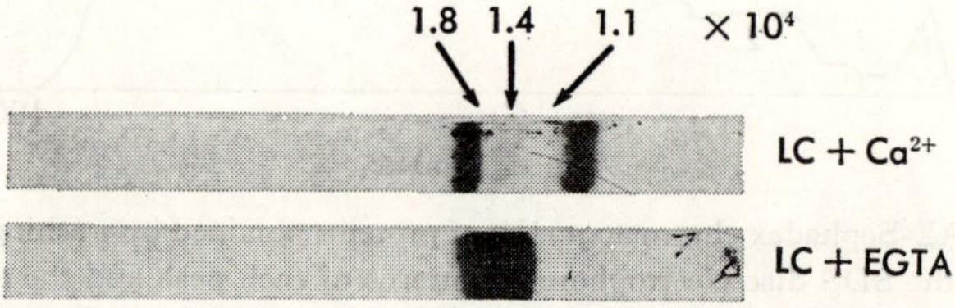

Fig. 2. Effect of Ca^{2+} on the electrophoresis of isolated squid myosin light chains.

change (Fig. 2). This fact suggests that the lighter one might be involved in Ca^{2+}-binding of squid myosin, like the skeletal troponin-C. These results suggest that, in the squid muscle, there are dual regulatory systems, both myosin-linked and actin-linked. This is in accord with the view of Konno (1978).

REFERENCES

Bailey, K. (1956). *Pubbl. Staz. Zool. Napoli* **24**, 96–108.

de Villafranca, G. W. (1955). *Biol. Bull.* **108**, 113–119.

Goldberg, A. and Lehman, W. (1978). *Biochem. J.* **171**, 413–418.

Horie, N., Tsuchiya, T., and Matsumoto, J. J. (1975). *Bull. Japan. Soc. Sci. Fish.* **41**, 1039–1045.

Kawaguti, S. and Ikemoto, N. (1957). *Biol. J. Okayama Univ.* **3**, 196–207.

Kendrick-Jones, J., Lehman, W., and Szent-Györgyi, A. G. (1970). *J. Mol. Biol.* **54**, 313–326.

Konno, K. (1978). *J. Biochem.* **84**, 1431–1440.

Lehman, W. and Szent-Györgyi, A. G. (1975). *J. Gen. Physiol.* **66**, 1–30.

Matsumoto, J. J. (1958). *Bull. Japan. Soc. Sci. Fish.* **24**, 133–139.

Matsumoto, J. J. (1959). *Bull. Tokai Reg. Fish. Res. Lab.* **23**, 51–63.

Tsao, T. C., Tan, P. H., and Pong, C. M. (1956). *Sci. Sinica* **5**, 91–111

Tsuchiya, T., Suzuki, H., and Matsumoto, J. J. (1977). *Bull. Japan. Soc. Sci. Fish.* **43**, 1233–1240.

Tsuchiya, T., Kaneko, T., and Matsumoto, J. J. (1978a). *J. Biochem.* **83**, 1191–1193.

Tsuchiya, T., Yamada, N., Mori, H., and Matsumoto, J. J. (1978b). *Bull. Japan. Soc. Sci. Fish.* **44**, 1039–1045.

Tsuchiya, T., Fukuhara, S., and Matsumoto, J. J. (1980). *Bull. Japan. Soc. Sci. Fish.* **46**, 197–200.

Tsuchiya, T., Sinohara, T., and Matsumoto, J. J. (1980a). *Bull. Japan. Soc. Sci. Fish.* **46** (in press).

Yoshimura, K. (1955). *Mem. Fac. Fish. Hokkaido Univ.* **3**, 159–176.

EXCITATION-CONTRACTION COUPLING

Muscle Contraction: Its Regulatory Mechanisms (Ebashi, S. et al., eds.), pp. 411–420, Japan Sci. Soc. Press, Tokyo / Springer-Verlag, Berlin (1980)

32

Voltage Dependence of Calcium Release in Vertebrate Skeletal Muscle Fibers

W. KNOX CHANDLER

Summary: During activation of contraction in frog skeletal muscle, Ca^{2+} is released into the sarcoplasm from the sarcoplasmic reticulum (SR). This release is normally controlled by the muscle membrane potential. The first part of the talk will be concerned with a description of intramembranous charge movement in muscle, a process which may underly this control. The second part of the talk will describe changes in intracellular Ca^{2+} as monitored by Arsenazo III, a Ca^{2+} sensitive dye. The latter experiments indicate that the Ca^{2+} release process is strongly voltage dependent.

The normal sequence of events leading to contraction in a vertebrate twitch fiber begins with action potential propagation along the surface membrane and into the interior of the cell by way of the T system (transverse tubular system). The change in potential difference across the walls of the T system results in intracellular calcium release from the sarcoplasmic reticulum (SR). This talk will be divided into two parts. The first part will be concerned with a description of small electrical currents which are seen during muscle activation. These appear to arise from rearrangements of charged groups confined to the surface or tubular membranes, possibly providing a mechanism for conveying information about tubular membrane potential to the adjacent SR. The second part of the talk will be devoted to some more recent experiments on measurements of changes in sarcoplasmic Ca^{2+} concentration using the metallochromic dye Arsenazo III. Both sets of experiments were carried out on frog twitch fibers.

CHARGE MOVEMENT EXPERIMENTS

The normal mechanism for contractile activation in twitch fibers is strongly dependent on the level of membrane potential across the transverse tubular system (Hodgkin and Horowicz, 1960). Such voltage regulation might arise from rearrangements of charged molecules within the membrane, along the lines suggested by Hodgkin and Huxley (1952) to account for voltage regulation of sodium and potassium channels in axon membrane. With this idea in mind, Dr. Schneider and I set out in 1971 to see whether we could detect, under voltage clamp, the small electrical currents which would necessarily be associated with these hypothetical intramembranous charge movements.

The first evidence came from voltage dependence of membrane capacitance measured just below mechanical threshold, such as that shown in Fig. 3. Encouraged by this, we then investigated the properties of the charge movement current over a wide range of potentials, from −80 to −20 mV (Schneider and Chandler, 1973). The strategy of the experiments was 1) to eliminate the relatively large voltage and time dependent sodium and potassium currents with tetrodotoxin (TTX) and tetraethylammonium ions (TEA), 2) to suppress mechanical movement with sucrose hypertonicity (so that we could use microelectrodes without damaging the muscle fiber), and 3) to subtract the linear capacitative component by taking the difference in membrane currents recorded during a 'test' voltage step and a 'control' voltage step. The test step was from −80 or −90 mV, near the normal resting potential, to a more positive potential at which activation occurs. The current would contain contributions from the linear capacitative current plus any 'activating' current. The voltage of the control step was equal in magnitude to the voltage of the test step but was superimposed on a conditioning hyperpolarization so that no 'activating' current would be obtained, leaving only the capacitative component. The

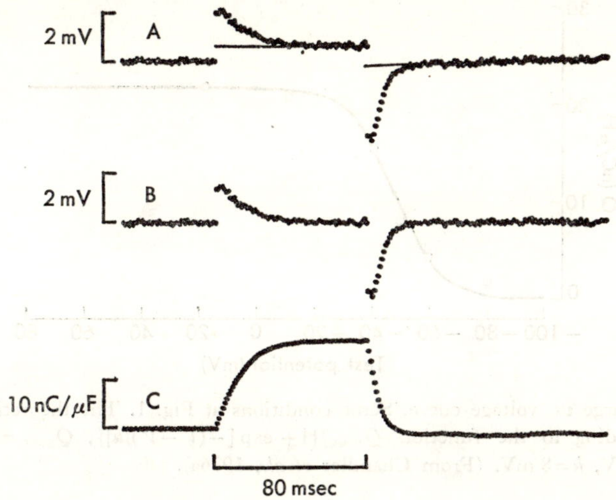

Fig. 1. Current record obtained using the three microelectrode voltage clamp technique. Records A and B show current, with 1 mV corresponding to 0.75 μA/μF. Record C shows the time integral of record B, normalized by fiber capacitance. The external solution contained 117.5 mM (TEA) Cl, 5 mM RbCl, 1.8 mM $CaCl_2$, 10^{-6} g/ml TTX, 1 mM Tris maleate pH 7.1 plus 467 mM sucrose; 1.4°C. Additional details in text. (From Chandler *et al.*, 1976a).

difference between the test and control current records, then, should consist mainly of 'activating' current and should therefore be free from contributions from either the linear membrane capacitance or the voltage dependent ionic channels which produce the action potential.

Figure 1 shows a current record of this type, obtained during a somewhat later study by Dr. Rakowski, Dr. Schneider, and myself. Record A shows the subtracted 'test' minus 'control' current record. The 'test' voltage step was from −80 to −30 mV and the 'control' step was from −130 to −80 mV. Record B was obtained by subtracting the ionic component, indicated by the straight line segments in record A; this component reflects a small nonlinearity in the current-voltage relation of the ionic channels which are not blocked by TTX, TEA or Rb.

The current in record B is transient and outward during depolarization and transient and inward on repolarization. The rates of decay of the 'on' and 'off' currents are different and, in general, are voltage dependent.

There are two other properties of these currents which are sufficiently important to require special mention. The first property is that the total

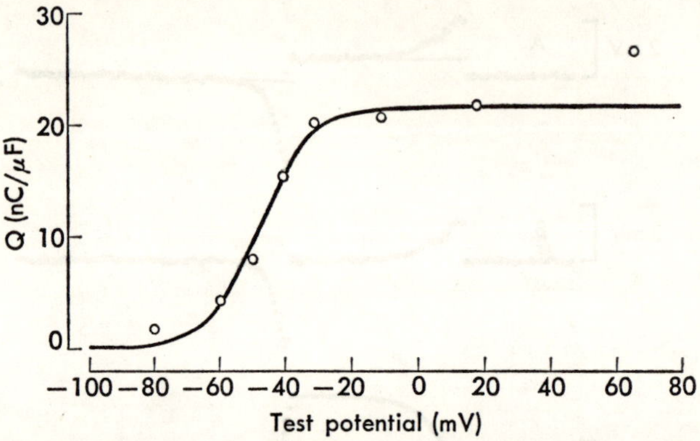

Fig. 2. Charge *vs.* voltage curve. Same conditions at Fig. 1. The theoretical curve is drawn according to the function $Q_{\max}/\{1+\exp[-(V-\bar{V})/k]\}$, $Q_{\max}=21.5$ nC/μF, $\bar{V}=-47.7$ mV, $k=8$ mV. (From Chandler *et al.*, 1976a).

amount of charge, given by the time integral of current, is the same during the 'on' and 'off' transients. This point is illustrated in Fig. 1C. The record shows the time integral of record 1B, beginning with the first point during the pulse. It is clear that the charge which moved during the step depolarization, 17 nC/μF as given by the final level in 1C, returned to the resting location on replolarization, as indicated by the final near zero level in 1C. This charge equality is preserved when the amount of charge is varied by changing either the voltage level of the pulse or the duration of the pulse (Schneider and Chandler, 1973; Adrian and Almers, 1976; Chandler *et al.*, 1976a).

The second important property of the charge movement is that the amount of charge when plotted against voltage is a saturating sigmoid function, Fig. 2. This curve is qualitatively similar to the curve relating tension against voltage, determined by Hodgkin and Horowicz (1960), except that the charge movement curve is considerably less steep.

The experiments in Figs. 1 and 2 were done in a hyperosmotic solution so that fiber movement would not dislodge the microelectrodes. It is also possible to obtain information about charge movement currents in isosmotic conditions by using voltages which do not reach mechanical threshold. Figure 3 shows fiber capacitance plotted against membrane potential. As the potential is made more positive than −100 mV, the capacitance increases exponentially with an *e*-fold change every 12.5 mV.

Although there is little evidence which directly links charge movement

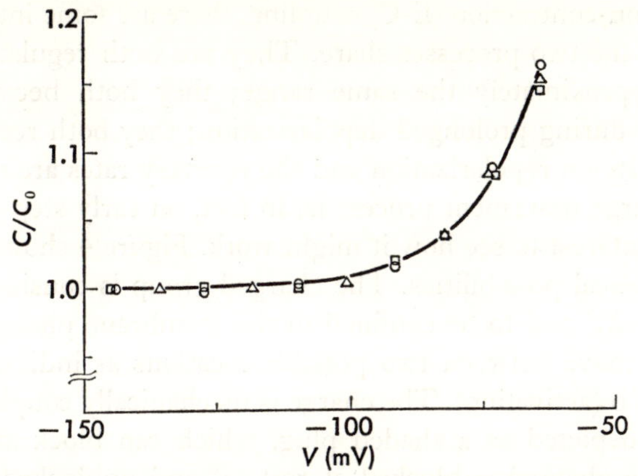

Fig. 3. Capacitance *vs.* voltage curve. Muscle bathed in solution of same composition as given in Fig. 1 except no sucrose was added; 1.7°C. The smooth curve is plotted according to an exponential function which increases *e*-fold every 12.5 mV. (From Schneider and Chandler, 1976).

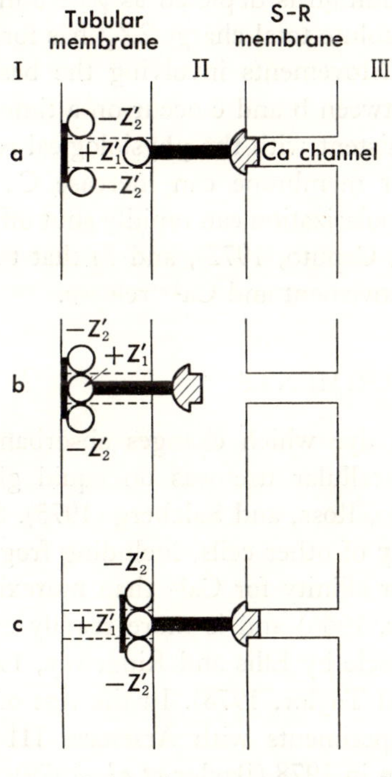

Fig. 4. Hypothetical diagram relating charge movement to SR calcium release. (From Chandler *et al.*, 1976b).

with excitation-contraction (E-C) coupling, there are some interesting properties which the two processes share. They are both regulated by voltage and over approximately the same range; they both become refractory (inactivated) during prolonged depolarization; they both recover from the refractory state on repolarization and the recovery rates are similar.

If the charge movement process is, in fact, an early step in E-C coupling, it is of interest to see how it might work. Figure 4 shows one of many very hypothetical possibilities. The charged group is considered to have a valence of $+Z_1'$ and to be confined to the membrane phase of the T system. It can move between two possible locations as indicated in parts a (resting) and b (activating). The charge is mechanically coupled to a molecular entity, depicted as a shaded plug, which can block an SR calcium channel. The channel is blocked at rest (a) and unblocked on activation (b). When unblocked, the channel allows Ca^{2+} to move from inside the lateral cisternae of the SR (III) into the sarcoplasm (II). Transitions between a and b can occur rapidly, on a time scale of milliseconds.

The refractory condition is depicted as part c in Fig. 4. In this case a basket complex of absolute total charge $2Z_2'$ has forced Z_1' into the closed position, $2Z_2' > Z_1'$. Movements involving the basket occur very slowly so that transitions between b and c occur on a time scale of seconds.

The model is consistent with the physiological results 1) that depolarization of the tubular membrane can increase Ca^{2+} flux across the SR membrane, 2) that repolarization can rapidly shut off Ca^{2+} release (Hodgkin and Horowicz, 1960; Caputo, 1972), and 3) that the refractory condition affects both charge movement and Ca^{2+} release.

ARSENAZO III EXPERIMENTS

Arsenazo III is a dye which changes absorbance in the presence of Ca^{2+}. The first intracellular use was on squid giant axons by Brown, Cohen, deWeer, Pinto, Ross, and Salzberg (1975). Since then the dye has been used in a variety of other cells, including frog muscle (Miledi *et al.*, 1977). It has a higher affinity for Ca^{2+} than murexide (used in muscle by Jöbsis and O'Connor, 1966) and is more readily available than aequorin (used in barnacle muscle by Ellis and Ridgeway, 1970 and in frog muscle by Blinks, Rudel, and Taylor, 1978). In the rest of the talk I should like to describe some experiments with Arsenazo III that Dr. Baylor, Dr. Marshall, and I began in 1978 (Baylor *et al.*, 1979).

In our experiments intact single fibers were dissected, mounted in an experimental chamber, and stretched to a sarcomere spacing of 3.8–4.3 μm in order to minimize movement artifacts. A 35–65 μm spot of light was positioned on the fiber and the light collected from the other side was measured with a photodiode. Dyes were injected iontophoretically using microelectrodes.

One of the difficulties encountered with using Arsenazo III as a Ca^{2+} indicator is that its color depends on H^{+} and Mg^{2+} as well as Ca^{2+}. It was therefore necessary to find out whether changes in concentrations (strictly speaking, activities) of these ions occur during muscle activation.

Phenol Red was injected to estimate internal pH. Resting fibers gave a value 6.9, similar to the value 7.0–7.1 determined by Bolton and Vaughan-Jones (1977) using glass pH sensitive microelectrodes. A small change, less than 0.01 pH unit, occurred following a single action potential. This change is sufficiently small that it can be ignored in analyzing the Arsenazo III Ca^{2+} signal. Changes in intracellular Mg^{2+} were similarly estimated using Arsenazo I and, again, the change during activity was negligibly small. We could now be confident that any relatively large Arsenazo III signal would not be due to changes in pH or Mg^{2+} (although the difference spectrum and apparent dissociation constant for Ca^{2+} are influenced by the steady level of pH and Mg^{2+}).

The upper set of traces in Fig. 5 shows optical records from a single fiber injected with Arsenazo III. To allow for changes in dye concentra-

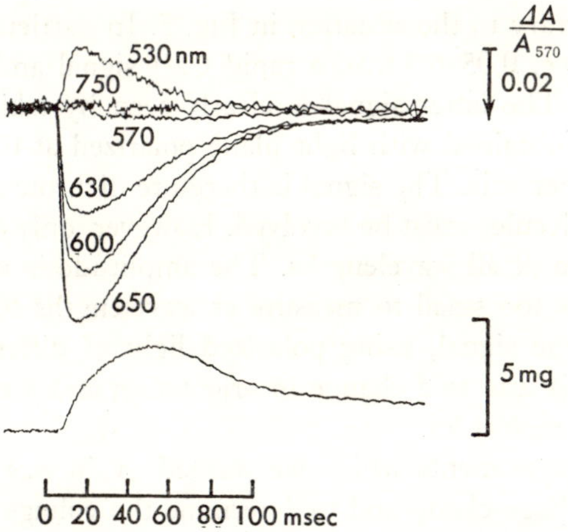

Fig. 5. Changes in absorbance at different wavelengths in a fiber injected with Arsenazo III, concentration 0.6–0.9 mM. Fiber immersed in Normal Ringer, 16°C, and stimulated at time 0 to give a twitch. (From Baylor *et al.*, 1979).

tion, the changes in absorbance ΔA have been normalized by A_{570}, the resting absorbance at the isosbestic point, 570 nm. The records are reasonably flat at 570 nm, the isosbestic point, and at 750 nm, a wavelength at which the dye does not absorb light. This indicates that the optical signal is unlikely to contain appreciable movement artifact.

Substantial absorbance signals are seen at 530, 600, 630, and 650 nm. These signals show the same essential waveform but scaled by different factors. This suggests, but doesn't prove, that the dye is reacting to a single underlying process. Since the relative amplitude of the signal *vs.* wavelength agrees qualitatively with cuvette determinations of a Ca^{2+} difference spectrum, and since pH and Mg^{2+} change little, if at all, it seems reasonable to conclude that the signal is due to an increase in ionized Ca^{2+} in the sarcoplasm.

The lower record in Fig. 5 shows tension recorded by a transducer attached to the end of the fiber. The fiber was highly stretched so there was no overlap of thick and thin filaments in the middle region, and presumably no tension production in that region. The small tension signal recorded in Fig. 5 most likely arose from the very ends of the fiber where some filament overlap remained (Huxley and Peachey, 1961). The Ca^{2+} signal clearly preceeds tension, although some caution must be applied in comparing an optical signal recorded in the middle of the fiber with tension recorded at the end.

In some circumstances signals at different wavelengths show different waveforms, contrary to the situation in Fig. 5. In particular, at lower concentrations of dye, 0.05–0.3 mM, a rapid Ca^{2+} signal and a second slower signal are seen. The latter signal is clearly seen by taking the difference between signals obtained with light plane polarized at 0° and at 90° with respect to the fiber axis. The signal is therefore dichroic and indicates that oriented dye molecules must be involved. However, only one basic dichroic waveform is seen at all wavelengths. The amplitude is maximal near 570 nm and becomes too small to measure at wavelengths 650 nm or greater. An analysis of the signal, using polarized light of different wavelengths, suggests that it is due to a change in orientation and a change in color of some of the dye molecules.

One of the experiments which we wanted to do was to measure Ca^{2+} signals under voltage clamp and to determine the voltage dependence near threshold. If Ca^{2+} release were regulated by the charge movement process in the simple manner represented in Fig. 4, then Ca^{2+} release should show

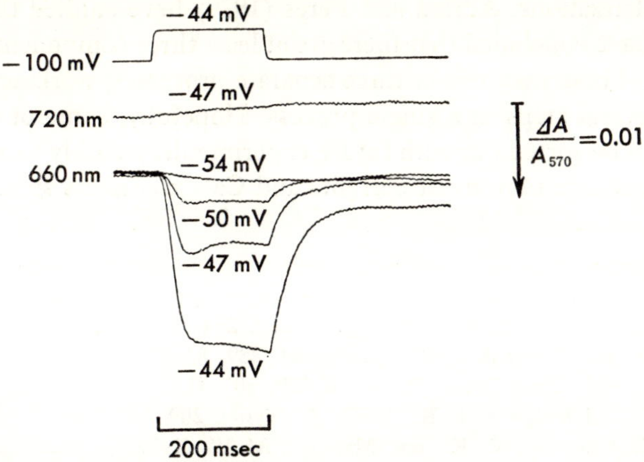

Fig. 6. Voltage-clamp experiment in a fiber injected with 0.8 mM Arsenazo III. D_2O Ringer containing TTX and TEA (instead of Na), 16°C. Similar results have also been obtained with H_2O. (From Baylor *et al.*, 1979).

the same voltage dependence as charge movement, roughly 12 mV in isosmotic solution as indicated in Fig. 3.

Figure 6 shows the experiment. The upper trace shows one of the records of voltage. The middle trace shows an optical record at 720 nm; this is a control for non dye-related movement artifacts since the dye does not absorb light at this wavelength. The bottom records illustrate changes in absorbance at 660 nm which arise from changes in sarcoplasmic free Ca^{2+}. Each small increase in the amplitude of the voltage step produces a marked increase in the Ca^{2+} signal, showing that Ca^{2+} release is strongly voltage dependent. An *e*-fold change occurs every 3–4 mV.

The experiment in Fig. 6 provides clear evidence that the simple scheme shown in Fig. 4 does not quantitatively fit the results. If the initial event in E-C coupling is, in fact, charge movement, some modification of the diagram is required. For example, the two state hypothesis (indicated by states 4a and 4b) may be too simple, the 1 : 1 relation between charge movement and Ca^{2+} release may be incorrect, or there may be cooperative interaction between charge groups.

Some important clues concerning how to modify the model may be provided by results obtained in Dr. Richard Adrian's laboratory. Adrian and Almers (Almers, 1975) showed that the decay of the charge movement current seen with depolarization is not a simple exponential, especially for small depolarizations. Adrian and Peres (1979) have studied this in more detail and have concluded that there are at least three components of charge movement. These may reflect three separate processes, working independently, or several steps in a single process. Hopefully, some of these complexities can be sorted out with future experiments, possibly using methods to measure charge movement currents and Ca^{2+} transients simultaneously.

REFERENCES

Adrian, R. H. and Almers, W. (1976). *J. Physiol.* **254**, 339–360.
Adrian, R. H. and Peres, A. (1979). *J. Physiol.* **289**, 83–97.
Almers, W. (1975). *Phil. Trans. R. Soc. B* **270**, 507–513.
Ashley, C. C. and Ridgway, E. B. (1970). *J. Physiol.* **209**, 105–130.
Baylor, S. M., Chandler, W. K., and Marshall, M. W. (1979). *J. Physiol.* **287**, 23–24P.
Blinks, J. R., Rüdel, R., and Taylor, S. R. (1978). *J. Physiol.* **277**, 291–323.
Bolton, T. B. and Vaughan-Jones, R. D. (1977). *J. Physiol.* **270**, 801–833.
Brown, J. E., Cohen, L. B., de Weer, P., Pinto, L. H., Ross, W. N., and Salzberg, B. M. (1975). *Biophys. J.* **15**, 1155–1160.
Caputo, C. (1972). *J. Physiol.* **223**, 483–505.
Chandler, W. K., Rakowski, R. F., and Schneider, M. F. (1976a). *J. Physiol.* **254**, 245–283.
Chandler, W. K., Rakowski, R. F., and Schneider, M. F. (1976b). *J. Physiol.* **254**, 285–316.
Jöbsis, F. F. and O'Connor, M. J. (1966). *Biochem. Biophys. Res. Commun.* **25**, 246–252.
Hodgkin, A. L. and Horowicz, P. (1960). *J. Physiol.* **153**, 386–403.
Hodgkin, A. L. and Huxley, A. F. (1952). *J. Physiol.* **117**, 500–544.
Huxley, A. F. and Peachey, L. D. (1961). *J. Physiol.* **156**, 150–165.
Miledi, R., Parker, I., and Schalow, G. (1977). *Proc. Roy. Soc. Lond. B* **198**, 201–210.
Schneider, M. F. and Chandler, W. K. (1973). *Nature* **242**, 244–246.
Schneider, M. F., and Chandler, W. K. (1976). *J. Gen. Physiol.* **67**, 125–163.

Muscle Contraction: Its Regulatory Mechanisms (*Ebashi, S. et al., eds.*), *pp. 421–433,*
Japan Sci. Soc. Press, Tokyo / Springer-Verlag, Berlin (*1980*)

33

The Sarcoplasmic Reticulum and Its Composition in Resting and in Contracting Muscle

A. P. SOMLYO, A. V. SOMLYO, H. GONZALEZ-SERRATOS, H. SHUMAN, AND G. McCLELLAN

Summary: A brief, historical review of structural and biochemical studies of the sarcoplasmic reticulum (SR)-T-tubule system is given, and, based on recently published and ongoing electron probe analytical studies of cryosections of rapidly frozen frog striated muscles, the following results and tentative conclusions are presented:

1. The Na and Cl concentrations in the terminal cisternae (TC) are not higher than in the cytoplasm and are unchanged during a 1.2-sec tetanus. The SR does *not* have an extracellular composition and, therefore, does not appear to be in ionic communication with the extracellular space.
2. The K content of the TC in resting muscle is somewhat higher than the cytoplasmic K concentration, presumably due to K-binding by calsequestrin.
3. In fatigued muscle, the Ca concentration of the TC is comparable to that found in normal muscle fibers and the T tubules are vacuolated. Fatigue, therefore, is not due to depletion of Ca from the TC, and may possibly be associated with failure of the T tubule action potential.
4. The Ca content of the TC in resting frog semitendinous muscle ac-

counts for 50–70% of fiber Ca, and approximately 60% (63 mmol/kg dry TC wt.) of it is released from the TC by the end of a 1.2-sec tetanus. This amount released is in good agreement with that required to saturate the combined calcium binding sites present in frog muscle on troponin and on parvalbumin.

5. The Mg and K content of the TC is increased after a 1.2-sec tetanus, but the increase in these elements (alone or together) is not sufficient to account for the total charge (126 mEq/kg dry TC) lost as the result of Ca release. Possible mechanisms (*e.g.*, proton movement) that may also contribute to charge neutrality are discussed, and it is cautioned that some of the changes in the TC, measured after a 1.2-sec tetanus, may be due to recovery processes associated with calcium pumping.

The triadic structure, and its two major components, the terminal cisternae (TC) of the sarcoplasmic reticulum and the T tubule (Porter and Palade, 1957; Peachey, 1965) are the major site of excitation-contraction coupling in skeletal muscle (see Ebashi and Endo, 1968; Ebashi, 1976 for review; Endo, 1977). The recognition of the sarcoplasmic reticulum (SR) as the particulate relaxing system by Kumagai and the Ebashis (1955), (Ebashi and Endo, 1968), the characterization of this system as an energy dependent calcium pump (Weber *et al.*, 1966; Weber, 1971; Hasselbach, 1978) and the structural studies (with radioautography and oxalate precipitation) that showed the presence of calcium in the region of the TC (Winegrad, 1968; Podolsky *et al.*, 1970), firmly established the SR as a site of active calcium sequestration. A. Weber (1966, 1971) subsequently suggested that only a fraction of the Ca content of the SR lumen is free and that a major portion of it is bound internally. The largest binding site is probably the intracisternal calcium binding protein, calsequestrin (MacLennan and Wong, 1971; Ikemoto *et al.*, 1974; MacLennan and Holland, 1975). The skinned fiber preparation of Natori (1954) and its modification (the "split fiber") by Endo (1977; Endo and Nakajima, 1973) have been particularly useful for the demonstration of the release of activator calcium from the SR (*e.g.*, Ford and Podolsky, 1972; Fabiato and Fabiato, 1977; Endo and Kitazawa, 1977; Stephenson, 1978; Moisescu and Thieleczek, 1978; Mobley, 1979).

The structural association between the T tubule and the SR at the triadic junction and the results of local stimulation experiments by A. F. Huxley and Taylor (1958) suggested that the triad is the site where an extracellular signal, later thought to be the regenerative action potential of the T tubule (Gonzalez-Serratos, 1966; Costantin, 1975), triggers the release of stored calcium from the TC. However, the basic mechanism of release and even the molecular structure of the connections between the two membrane components of the triad are, at best, uncertain. Large molecules (*e.g.*, ferritin) in the lumen of the T tubule do not enter the SR (Huxley, 1964; Page, 1964), and the periodic electron dense processes between the T tubule and the junctional SR membrane do not contain demonstrable large channels (see Franzini-Armstrong, 1975 for review). An electron lucent core, best visualized with the use of tannic acid, has been demonstrated in these bridges that span the 10–12 nm gap between the TC and the T tubule (A. V. Somlyo, 1979; Eisenberg and Gilai, 1979). The structure of the membrane bridges (Fig. 1) with electron lucent core has been interpreted as representing a splitting of the trilaminar membrane of, respectively, the T tubule and the TC *without* direct communication between the lumen of the two structures (A. V. Somlyo, 1979).

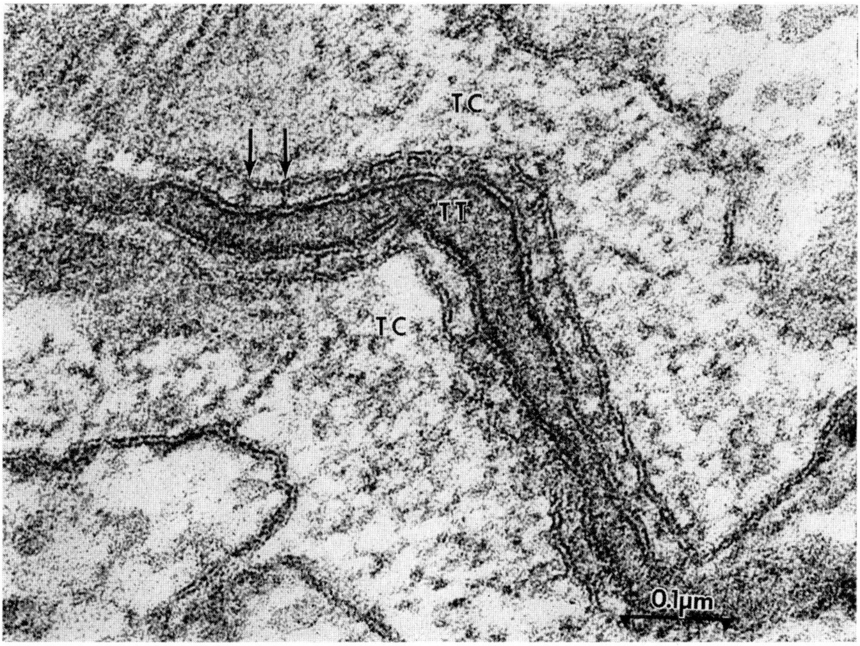

Fig. 1. Triad illustrating bridging structures (arrows) across the junctional gap. TC, terminal cisterna; TT, transverse tubule. The frog semtideninous muscle was fixed in glutaraldehyde, post fixed in osmium tetroxide followed by 2% tannic acid, stained en block with uranyl acetate.

It is probable, in view of the structural similarities, that the couplings between the junctional SR and the surface membrane in cardiac (Sommer, 1979) and in smooth (A. P. Somlyo *et al.*, 1976) muscle play a role similar to that of the triad in excitation-contraction (E-C) coupling.

The possibility of ionic, although not macromolecular, communication between the T tubule and SR lumen had been raised on the basis of Cl and Na flux studies interpreted as suggesting that the SR has an ionic composition that is identical to that of the extracellular fluid (*e.g.*, Rogus and Zierler, 1973; see A. V. Somlyo *et al.*, 1977 for review; Neville, 1979). Ultrastructural studies of muscles treated with hypertonic solutions were similarly interpreted. The absence of a large electrical capacitance expected of a membrane system having the surface of the SR, however, was difficult to reconcile with ionic communication between the latter and the extracellular space (Costantin, 1975), although the validity of this electrophysiological argument has been questioned on the basis of recent electrical circuit models of muscle (Mathias *et al.*, submitted).

Isolated SR preparations were not very suitable for determining the monovalent ion composition of the *in situ* SR or the mechanism of physiological calcium release (see also Endo, 1977), because the SR is highly permeable to physiological monovalent ions (Duggan and Martonosi, 1970; Meissner and McKinley, 1976; Tada *et al.*, 1978) and its K, Na, and Cl content is readily affected by isolation. Also, it has not been possible to unequivocally mimic "physiological" release in any of the isolated fractions. These considerations led us to the use of electron probe analysis as a means of determining the *in situ* composition of the SR in resting and, subsequently, in activated and in fatigued frog muscle (A. V. Somlyo *et al.*, 1977a, b; Gonzalez-Serratos *et al.*, 1978).

Electron probe analysis is based on the inelastic interactions of fast electrons with the specimen and the resultant generation of X-rays having energies characteristic of the elements (or more specifically of their radiative transitions) contained within the microvolume excited by the focused electron beam (see for review: Hall, 1971; Shuman *et al.*, 1976; Lechene, 1977; Hutchinson, 1979). The use of high brightness, field emission electron sources can provide enough signal with small probes for analysis at the relatively high spatial resolution of approximately 10 nm (A. P. Somlyo *et al.*, 1979). As an example, in Ca X-ray maps the two adjacent TC in a triad (Fig. 2) can be resolved in highly "favorable" test-objects, such as fibers heavily loaded with calcium (Fig. 2). A spatial resolution of 50 nm

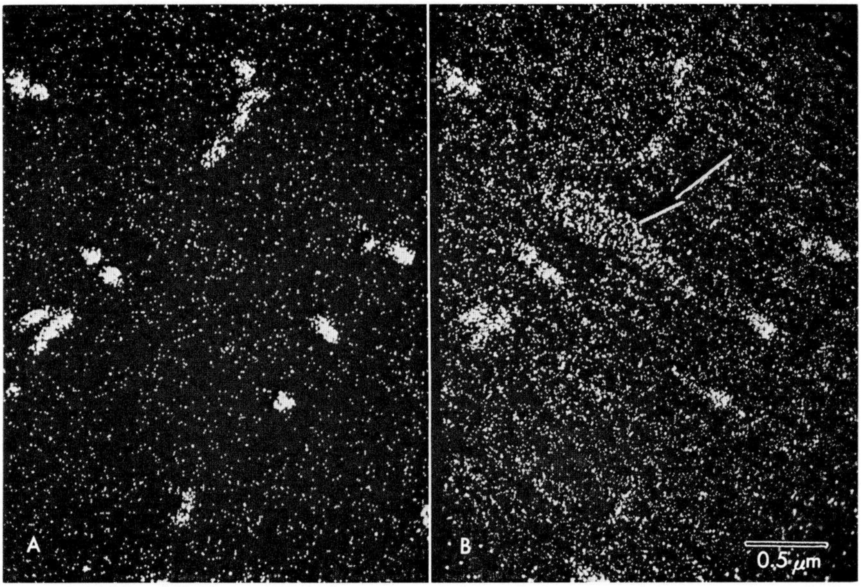

Fig. 2. X-ray maps (A, calcium and B, phosphorus) of cryo section of a frog muscle fiber that was "calcium loaded" due to damage during dissection. X-rays in the calcium energy window (centroid 3.69 kV) were collected without subtraction of the adjacent $K K_\beta$ peak. This and X-ray background contribute to the counts observed over regions of the fiber outside the SR. Maps obtained with field emission gun. The very high calcium content of the SR in this heavily calcium loaded muscle provides a "test object" to demonstrate the spatial resolution instrumentally obtainable with electron probe analysis. Note the presence of the mitochondrion (arrow) in the phosphorus, but not in the calcium X-ray map.

is routinely attainable with stationary probes in normal muscle (A. V. Somlyo *et al.*, 1977).

The preservation of diffusible elements in their *in situ* location requires rapid freezing techniques, estimated to freeze within 1–3 msec the 5–15 μm depth of tissue used for analysis. Rapid freezing retains the transmembrane ionic gradients (K, Na, Cl) of striated muscle measured by independent techniques (A. V. Somlyo *et al.*, 1977). In addition to the maintenance of cytoplasmic diffusible ion composition across the surface membrane, rapid freezing also preserves the extracellular monovalent ions present in the intracellular vacuoles (portions of the T system) of muscles that have been fatigued or treated with hypertonic solutions (A. V. Somlyo *et al.*, 1977; Gonzalez-Serratos *et al.*, 1978; Franzini-Armstrong *et al.*,

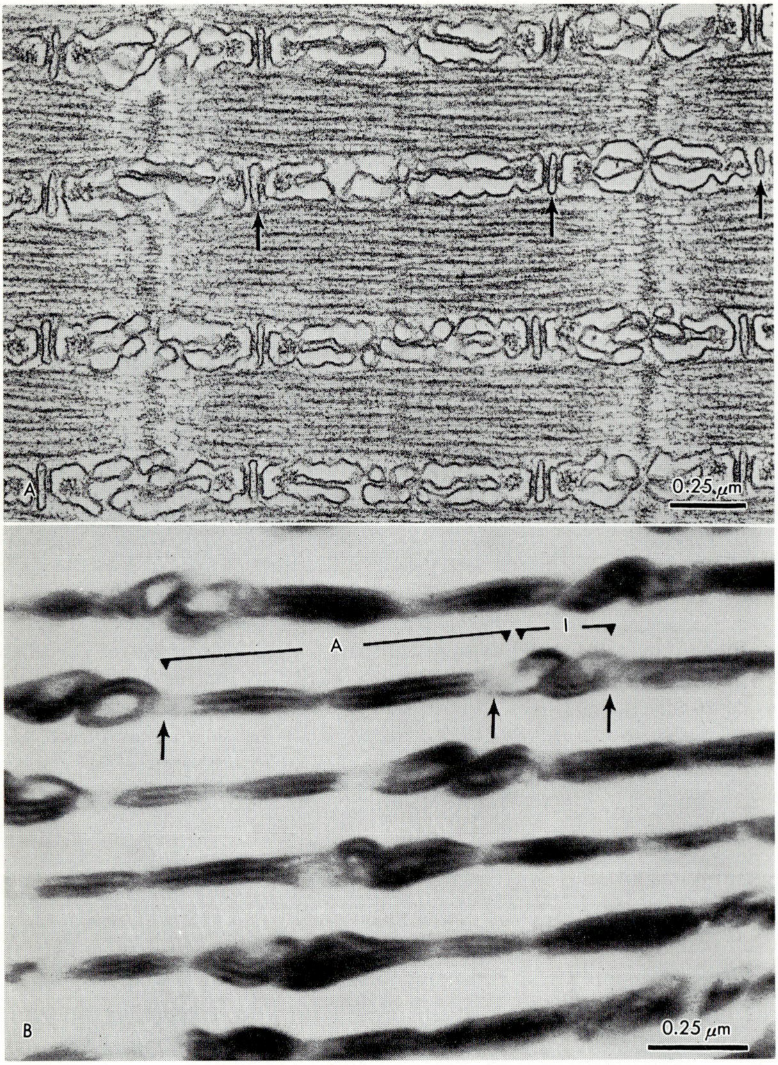

Fig. 3. A: longitudinal section from a toadfish swimbladder muscle. Two triads per A band. T tubules—arrows. Osmium fixation, plastic embedding. (Somlyo *et al.*, 1977). B: longitudinal (slightly oblique) cryo section of toadfish swimbladder muscle. Unfixed muscle, frozen in supercooled Freon 22, sectioned at $-130°C$, picked up dry onto Cu grid with C foil and dried at $<10^{-6}$ Torr. This dried section was stained with osmium vapor under vacuum to enhance contrast. Electron probe X-ray microanalyses were done on similar, but unstained material. T-tubule regions indicated by arrows. (Reprinted from Somlyo *et al.*, 1977b).

1978). These and similar findings in smooth muscle (A. P. Somlyo *et al.*, 1979) and in non-muscle cells (*e.g.*, Gupta and Hall, 1979) indicate that rapid freezing maintains the transmembrane ionic gradients present *in vivo.*

THE NA, CL, K, AND P CONTENT OF THE TERMINAL CISTERNAE IN RESTING MUSCLE

The monovalent ion composition of the SR has been determined with electron probe analysis in toadfish swimbladder and in frog skeletal muscle (A. V. Somlyo *et al.*, 1977a, b). The former is an extremely fast muscle that contains a very large (30% of fiber) volume of regularly organized SR (Fig. 3 and Fawcett and Revel, 1961; A. V. Somlyo *et al.*, 1977b).

The NaCl content of the SR is *not* higher than the respective cytoplasmic concentration in either toadfish swimbladder (Table I) or frog striated muscle (Table II and A. V. Somlyo *et al.*, 1977). Furthermore, the results with toadfish muscle unequivocally excluded the possibility of the SR discharging a (postulated) high NaCl content into the cytoplasm during freezing, because the latter process (with muscles bathed in 231 mM NaCl and having a 30% SR volume) would have resulted in a minimal cytoplasmic NaCl content of 277 mmol/kg dry wt. The actual measured cytoplasmic values are 30 (Na) and 47 (Cl) mmol/kg dry wt. These results of electron probe analysis clearly indicate that the SR of resting muscle does *not* have an "extracellular" composition and, therefore, is *not* in ionic communication with the extracellular space.

Other evidence *against* ionic communication of the SR with the extracellular space derives from ultrastructural studies of rapidly frozen muscle, showing that only the T tubules, rather than the SR, swell in muscles incubated in hypertonic solutions prior to freezing (A. V. Somlyo *et al.*,

TABLE I
Elemental Concentrations of Sarcoplasmic Reticulum and Cytoplasm of Toadfish Swimbladder Determined by Paired "Small Spot" Electron Probe Analysis (mmol/kg dry wt.±SEM (n=25 pairs))

Region analyzed	P	S	Cl	K	Ca	Mg	Na
SR	443±2.5	362±2.0	42±1.0	589±2.8	77±0.9	40±1.8	34±4.2
Cytoplasm (between SR)	275±2.0	335±2.0	47±1.1	510±2.7	4±0.8	47±1.9	30±4.3

(From Somlyo *et al.*, 1977).

1977; Franzini-Armstrong *et al.*, 1978). Similar results were subsequently also obtained with aldehyde-fixed preparations (Davey and O'Brien, 1978). Finally, recent solute flux studies on single frog muscle fibers also show the absence of an extracellular compartment that could be identified with the SR (Neville, 1979).

The K concentration is significantly higher in the TC than in the cytoplasm in both toadfish and in frog muscle. Assuming the free K concentrations in the SR lumen to be approximately 130 to 140 mM (the same as in the cytoplasm) and the free calcium concentration approximately 0.5–1 mM, with the relative affinity of calsequestrin for Ca being approximately 100-fold greater than for K (Ikemoto *et al.*, 1974; Ostwald and MacLennan, 1974), the concentrations of K and Ca bound to calsequestrin in resting muscle should be about the same. On this basis, and within the errors of the assumptions and measurements involved, the excess K in the SR (above cytoplasmic values) can probably be accounted for as being due to binding to calsequestrin.

The above conclusions regarding monovalent ion distribution, and the very low Ca-permeability of the resting SR, suggest that there is not a significant transmembrane potential across the SR membrane of resting muscle. This finding need not exclude the possibility that a change in the SR potential from near zero to some finite value may cause calcium release (Endo, 1977; Ebashi, 1979). The increase in the Ca-permeability of the SR in activated muscle, whatever its cause, would be expected to lead to an inside (SR) negative Ca-potential.

The phosphorus concentration is generally higher in the SR than in the cytoplasm, some of it possibly due to the presence of membrane phospholipids. The TC also appear to contain a significant proportion of diffusible phosphates, because the P content is reduced in the TC of split fibers bathed in artificial solutions (preliminary observations in collaboration with M. Endo).

CALCIUM CONTENT IN THE TERMINAL CISTERNAE AT REST AND ITS RELEASE DURING TETANUS

The high calcium content of the TC in resting frog semitendinous muscle (Table II) suggests that these structures contain most of the fiber calcium in resting muscle, considering a TC calcium of 108 mmol/kg dry wt. and a TC volume of 4–5% (Peachey, 1965; Mobley and Eisenberg, 1975). Deposits of Ca in the longitudinal tubules of the SR have thus far only been detected in muscle incubated in hypertonic solutions (A. V. Somlyo *et al.*, 1977).

The calcium content of the terminal cisternae is not significantly reduced in fatigued muscle (Gonzalez-Serratos *et al.*, 1978). This finding and the vacuolation of the T tubules observed in fatigued muscles suggested failure of the T tubule action potential as a possible mechanism of fatigue (Gonzalez-Serratos *et al.*, 1978).

The amount of calcium released (Table II) during a 1.2-sec tetanus was unexpectedly large, representing approximately 60% of the total calcium content of the resting TC. The amount released (about 1 mM in fiber H_2O) is greatly in excess of the amount (0.2 mM) that has to be bound to troponin (Weber and Herz, 1963; Fuchs and Briggs, 1968) for full activation of contraction, but in very good agreement with the total calcium required for binding to both troponin (0.2 mM) *and* to the soluble calcium binding protein parvalbumin (0.6–0.8 mM) that is present in approximately 0.3–0.4 mM concentration in frog muscle (Gosselin-Ray and Gerday, 1977). Parvalbumin has two calcium binding sites that can be saturated with calcium (probably in exchange for magnesium) with an affinity similar to that of troponin (Haiech *et al.*, 1979). The slower kinetics of calcium binding to parvalbumin than to troponin (Haiech *et al.*, 1979) may allow activation of the contractile system with a smaller amount of calcium released during a twitch than during more prolonged stimulation, although the total calcium release during a single twitch still remains to be determined. It should be emphasized that, unlike calcium sensitive indicators that show the release of free calcium (*e.g.*, Jöbsis and O'Connor, 1966; Ashley and Ridgeway, 1970; Miledi *et al.*, 1977; Blinks *et al.*, 1978; Scarpa *et al.*, 1978), electron probe analysis measures the total (free and bound) calcium. The relatively large amount of calcium released and bound to calcium binding proteins during a tetanus recalls the possibility suggested by Davies (1963), that calcium release and binding might contribute to that portion of heat production in frog muscle that is not accounted for by ATP hydrolysis.

THE NA, CL, K, AND MG CONTENT OF THE SARCOPLASMIC RETICULUM DURING TETANUS

Direct analysis of the TC in activated muscles is still at its beginnings. Therefore, in this communication, we can only present some intermediate results obtained in frog (*Rana pipiens*) semitendinous muscles that were

TABLE II
Elemental Composition of the Terminal Cisternae of Frog Muscle ($\bar{X}\pm$SD mmol/kg dry wt.)
Effect of tetanus (mEq): -126 Ca$+46$ K$+24$ Mg$=-56$ mEq.

	n	Na	Mg	P	S	Cl	K	Ca
Control	192	54±35	58±21	413±78	216±39	42±19	549±128	108±46
Tetanus	179	52±32	70±21[a]	427±69	212±55	42±20	595±109[a]	45±19[a]

[a] Statistically significant difference between tetanized and control.

frozen during tetanus (1.2 sec after initiation of stimulation). Detailed analysis of the experimental data (of individual paired muscles, tetanized and resting from the same frog) has not been completed yet, but the overall mean results (Table II) suggest the following:

1. The NaCl concentrations of the TC of resting and of tetanized muscle are not significantly different. Barring extremely rapid removal of these ions from the TC, this finding indicates the *absence* of significant ionic communication permitting downhill movement of NaCl through membrane bridges in the tetanized (as in the resting) muscle.
2. The Mg content of the TC in tetanized muscle is significantly increased. This finding implies, in view of the low resting permeability of the SR to Mg (Duggan and Martonosi, 1970; Tada *et al.*, 1978) an increase in this permeability (or transport) during tetanus. Mg-influx into the TC could take place through Ca-channels opened during activation, but may also reflect accumulation related not to Ca release, but to Ca pumping. Our results do not permit us to choose between these or other possibilities.
3. The K content of the TC of tetanized, compared to resting, muscle is also (slightly, but significantly) increased. This, in view of the high permeability of the SR to K, presumably reflects increased binding to calsequestrin, as a consequence of the reduction in the free Ca concentration in the TC.

The above findings would suggest that neither Mg (increased by 23 mEq/kg TC dry wt.) nor K (increased by 45 mEq/kg TC dry wt.) alone can account for the approximately 127 mEq/kg dry TC Ca released at the end of a 1.2-sec tetanus. Proton movements into the SR could conceivably take place to maintain electroneutrality, and protons are generated at the SR as the result of ATP hydrolysis by the SR pump. In fragmented SR, the sum of Ca^{2+} and Mg^{2+} and K^{+} (and probably H^{+}) bound was found to be constant (Carvalho and Leo, 1967).

The interpretation of our findings must be tempered by the fact that our analyses do not measure organic ions and show only the (averaged) end result of a relatively long (1.2 sec) tetanus. The possibility of muscle to muscle variations (dependent perhaps on the cytoplasmic $[K^+]$, $[Mg^{2+}]$ or pH) in the amount of Mg and K gained by the TC during contraction remains to be determined. It is also possible that at least some of the changes in TC composition (*e.g.*, Mg uptake), reflect recovery processes associated with calcium ATPase activity, rather than the fluxes of counterions compensating the loss of positive charge due to calcium release. Analyses of muscles frozen during the rising phase of contraction may throw further light on some of these questions.

Acknowledgment

This work was supported by grant HL15835 to the Pennsylvania Muscle Institute and by USPHS Grant 1 RO1-NS14206-01PHG to H.G.-S.

REFERENCES

Ashley, C. C. and Ridgway, E. B. (1970). *J. Physiol* **209**, 105–130.

Blinks, J. R., Rudel, R., and Taylor, S. R. (1978). *J. Physiol.* **277**, 291–323.

Carvalho, A. P. and Leo, B. (1967). *J. Gen. Physiol.* **50**, 1327–1352.

Costantin, L. L. (1975). *Prog. Biophys. Mol. Biol.* **29**, 197–224.

Davey, D. F. and O'Brien, G. M. (1978). *Aust. J. Exp. Biol.* **56**, 409–419.

Davies, R. E. (1963). *Nature* **199**, 1068–1074.

Duggan, P. B. and Martonosi, A. (1970). *J. Gen. Physiol.* **56**, 147–167.

Ebashi, S. (1976). *Annu. Rev. Physiol.* **38**, 293–313.

Ebashi, S. (1979). In *Advances in Pharmacology and Therapeutics*, vol. 3, *Ions-Cyclic Nucleotides-Cholinergy* (Stoclet, J. C., ed.), pp. 81–98, Pergamon Press, Oxford and New York.

Ebashi, S. and Endo, M. (1968). *Prog. Biophys. Mol. Biol.* **18**, 123–183.

Eisenberg, B. R. and Gilai, A. (1979). *J. Gen. Physiol.* **74**, 1–16.

Endo, M. (1977). *Physiol. Rev.* **57**, 71–108.

Endo, M. and Kitazawa, T. (1976). *Proc. Japan Acad.* **52**, 595–598.

Endo, M. and Nakajima, Y. (1973). *Nature New Biol.* **246**, 216–218.

Fabiato, A. and Fabiato, F. (1977). *Circ. Res.* **40**, 119–129.

Fawcett, D. W. and Revel, J. P. (1961). *J. Biophys. Biochem. Cytol.* **10**, 89–109.

Ford, L. E. and Podolsky, R. J. (1972). *J. Physiol.* **223**, 21–33.

Franzini-Armstrong, C. (1975). *Fed. Proc.* **34**, 1382–1389.

Franzini-Armstrong, Heuser, J. E., Reese, T. S., Somlyo, A. P., and Somlyo, A. V. (1978). *J. Physiol.* **283**, 133–140.

Fuchs, F. and Briggs, F. N. (1968). *J. Gen. Physiol.* **51**, 655–676.

Gonzalez-Serratos, H. (1966). *J. Physiol.* **185**, 20–21P.
Gonzalez-Serratos, H., Somlyo, A. V., McClellan, G., Shuman, H., Borrerro, L. M., and Somlyo, A. P. (1978). *Proc. Natl. Acad. Sci. U.S.* **75**, 1329–1333.
Gosselin-Rey, C. and Gerday, C. (1977). *Biochem. Biophys. Acta* **492**, 53–63.
Gupta, B. L. and Hall, T. A. (1979). *Fed. Proc.* **38**, 144–153.
Haiech, J., Derancourt, J., Pechere, J.-F., and Demaille, J. G. (1979). *Biochemistry* **18**, 2752–2758.
Hall, T. A. (1971). In *Physical Techniques in Biological Research* (Oster, G., ed.), vol. 1A, pp. 158–275, Academic Press, New York.
Hasselbach, W. (1978). *Biochim. Biophys. Acta* **515**, 23–53.
Hutchinson, T. E. (1979). *Int. Rev. Cytol.* **58**, 115–158.
Huxley, A. F. and Taylor, R. E. (1958). *J. Physiol.* **144**, 426–441.
Huxley, H. E. (1964). *Nature* **202**, 1067–1071.
Ikemoto, N., Nagy, B., Bhatnagar, G. M., and Gergely, J. (1974). *J. Biol. Chem.* **249**, 2357–2365.
Jöbsis, F. F. and O'Connor, M. J. (1966). *Biochem. Biophys. Res. Commun.* **25**, 246–252.
Kumagai, H., Ebashi, S., and Takeda, F. (1955). Nature **176**, 166.
Lechene, C. P. (1977). *Am. J. Physiol.* **232**, F391–F396.
MacLennan, D. H. and Holland, P. C. (1975). *Annu. Rev. Biophys. Bioeng.* **4**, 377–404.
MacLennan, D. H. and Wong, P.T.S. (1971). *Proc. Natl. Acad. Sci. U.S.* **68**, 1231–1235.
Mathias, R. T., Levis, R. A., and Eisenberg, R. S. *J. Gen. Physiol.* (submitted).
Meissner, G. and McKinley, D. (1976). *J. Membrane Biol.* **30**, 79–98.
Miledi, R., Parker, I., and Schalow, G. (1977). *Proc. Roy. Soc. Lond. B.* **198**, 201–210.
Mobley, B. A. (1979). *J. Membrane Biol.* **46**, 315–329.
Mobley, B. A. and Eisenberg, B. R. (1975). *J. Gen. Physiol.* **66**, 31–45.
Moisescu, D. G. and Thieleczek, R. (1978). *J. Physiol.* **275**, 241–262.
Natori, R. (1954). *Jikeikai Med. J.* **1**, 119–126.
Neville, M. C. (with an Appendix by R. T. Mathias) (1979). *J. Physiol.* **288**, 45–70.
Ostwald, T. J. and MacLennan, D. H. (1974). *J. Biol. Chem.* **249**, 974–979.
Page, S. G. (1964). *J. Physiol.* **175**, 10P.
Peachey, L. D. (1965). *J. Cell Biol.* **2**, 209–231.
Podolsky, R. J., Hall, T., and Hatchett, S. L. (1970). *J. Cell Biol.* **44**, 699–702.
Porter, K. R. and Palade, G. E. (1957). *J. Biophys. Biochem. Cytol.* **3**, 269–299.
Rogus, E. Zierler, K. L. (1973). *J. Physiol.* **233**, 227–270.
Scarpa, A., Brinley, F. J., Tiffert, T., and Dubyak, G. R. (1978). *Ann. N.Y. Acad. Sci.* **307**, 86–112.
Shuman, H., Somlyo, A. V., and Somlyo, A. P. (1976). *Ultramicroscopy* **1**, 317–339, 1976.
Somlyo, A. P., Somlyo, A. V., and Shuman, H. (1979). *J. Cell Biol.* **81**, 316–335.
Somlyo, A. P., Somlyo, A. V., Ashton, F. T., and Vallieres, J. (1976). In *Cell Motility*. Cold Spring Harbor Laboratory, **3**, 165–183.
Somlyo, A. P., Somlyo, A. V., Shuman, H., and Stewart, M. (1979). In *Scanning Electron Microscopy* (Johari, O., ed.), pp. 711–722, Scanning Electron Microscopy, Inc.
Somlyo, A. V. (1979). *J. Cell Biol.* **80**, 743–750.
Somlyo, A. V., Shuman, H., and Somlyo, A. P. (1977a). *J. Cell Biol.* **74**, 828–857.
Somlyo, A. V., Shuman, H., and Somlyo, A. P. (1977b). *Nature* **268**, 556–558.
Sommer, J. R. and Johnson, E. A. (1979). In *Handbook of Physiology: The Cardiovascular*

System (Berne, R. M., Sperelakis, N., and Geiger, S. R., eds.), pp. 113–186, Williams & Wilkins Co., New York.

Stephenson, E. W. (1978). *J. Gen. Physiol.* **71**, 411–430.

Tada, M., Yamamoto, T., and Tonomura, Y. (1978). *Physiol. Rev.* **58**, 1–79.

Weber, A. (1971). *J. Gen. Physiol.* **57**, 50–63.

Weber, A. and Herz, R. (1963). *J. Biol. Chem.* **238**, 599–605.

Weber, A., Herz, R., and Reiss, I. (1966). *Biochem. Z.* **345**, 329–369.

Winegrad, S. (1968). *J. Gen. Physiol.* **51**, 65–83.

Muscle Contraction: Its Regulatory Mechanisms (Ebashi, S. et al., eds.), pp. 435–446,
apan Sci. Soc. Press, Tokyo / Springer-Verlag, Berlin (1980)

34

On the Controlling Factors for Activation of Rapid Cooling Contracture

TOSHIO SAKAI

Summary: In frog skeletal muscle active tension was elicited by rapid cooling of a bath solution containing various concentrations of caffeine (rapid cooling contracture; RCC). The RCC was analyzed by the use of three parameters; 1) dose response relation for the RCC, 2) peak of active tension for the RCC, 3) rate of rise of active tension in the RCC. The experiments were carried out in Ca-depleted and potassium-rich solutions. As experimental materials, toe muscles from *Rana nigromaculata* were used. Thermal stimulation was constantly given by changing the temperature of the medium from 23° to 0°C within about 1 sec.

In Ca-depleted solution in which the extracellular magnesium ions were substituted for Ca^{2+}, all parameters were inhibited. When the T-tubular membrane was depolarized by potassium rich solutions, ranging between 2.5 mM and 40 mM, the peak tension of the RCC was inhibited in proportion to the extracellular potassium concentration. Inversely, the peak tension of RCC rose in proportion to the medium at above 50 mM $[K]_0$. Such a relation was observed from the point of view of the rising rate of tension and the dose-response relation for the RCC. The activation of

RCC which is generated by the rapid release of Ca^{2+} from the sarcoplasmic reticulum at a low temperature might be influenced by some functional changes of the T system through the T-tubular membrane to the terminal cisternae.

Since 1960, a series of studies have been made on rapid cooling contracture (RCC) with various muscle fibers (Sakai and Kurihara, 1974). In the case of skeletal muscle the presence of caffeine at a subthreshold concentration was prerequisite for activation of this contracture. In the case of smooth muscle, the presence of caffeine is not always necessary, the contracture being easily induced by thermal stimulation (Sakai and Iizuka, 1972; Kurihara, 1973; Kurihara *et al.*, 1974). In the frog ventricle, RCC can be observed without caffeine in the Na-depleted Ringer's solution that can inhibit the Na-Ca pump (Sakai and Kurihara, 1974; Kurihara and Sakai, 1978). Thus the tension of the RCC can be regulated by changing Na concentration in Ringer's solution. In guinea-pig papillary muscle, the frequency of electrical shocks before thermal stimulation can determine the state of the RCC: the higher the frequency, the stronger the tension becomes. In this muscle, the rapid cooling contracture is composed of two types of contraction, phasic and tonic. The phasic contraction is enhanced by an increase in electrical pulses (Kurihara and Sakai, 1978). Thus, in the myocardium, the amount of intracellular inflow Ca^{2+} determines the grade of tension of the phasic contraction. In the skeletal muscle, however, the sequestered-Ca amount in the sarcoplasmic reticulum (SR) seems to be directly involved in tension development, while extracellular Ca is not considered to play any positive role in it.

In skeletal muscle the graded responses are elicited by changing the concentration of caffeine and the mode of thermal stimulation. Recently, Yu, Hartt, and Podolsky (1979) have demonstrated by using X-ray analysis that the graded response is due to the cross-bridge number between two contractile filaments.

At the beginning of our study, we thought that the activation of the RCC would be accounted for as follows: caffeine and thermal stimulation would induce Ca liberation from the SR, while Ca-uptake would be inhibited by it. After our report (Sakai, 1965), Weber and Herz (1968), and Ogawa (1970) have demonstrated that the Ca-release from the SR is accelerated in the presence of caffeine at a low temperature. Also, acceleration of the Ca-release was reconfirmed in our laboratory. In fragmented sarcoplasmic reticulum (FSR) sufficiently loaded with Ca, the rate of

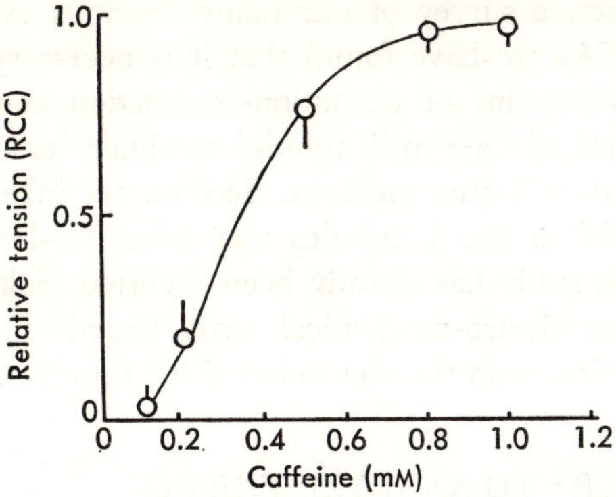

Fig. 1. Dose-response relation of active tension on RCC at various concentrations of caffeine. The sarcomere length of muscle was preserved at 2.4 μm. The active tension is compared with the maximum tension of RCC. ○ intact ($n=5$).

release of Ca is increased by rapid cooling and additionally accelerated in the presence of caffeine. Thus, it is assumed that rapidly released-Ca^{2+} from the surface membrane of the SR at a low temperature might play a triggering role in releasing Ca^{2+} from the SR.

In normal Ringer's solution, the threshold concentration of caffeine on the RCC is about 0.1 mM and the maximum tension of RCC is elicited at 0.8 or 1.0 mM caffeine (Fig. 1). Thus, the dose-response relation for the RCC shows an S-shaped curve at the resting sarcomere length. This dose-response relation seems to be fairly similar to the tension-pCa relation in skeletal skinned fibers (Endo, 1972). Regarding the dose-response relation for the RCC, when the skeletal muscle is bathed in Ringer's solution containing less permeant anions (NO_3^-, Br^-), it is shifted to low concentration of caffeine. Also, the dose-response curve is shifted to a high concentration of caffeine in normal Ringer's solution containing dantrolene-Na (Homma *et al.*, 1976).

It is well known that less-permeant anions have a close relation with excitation-contraction coupling (Hodgkin and Horowicz, 1960). They are also considered to affect the conductance of T-tubules membrane (Hodgkin and Horowicz, 1960; Hutter and Noble, 1960). We do not consider that these anions or drugs easily penetrate into the cell to act directly on the SR. If they did, some physiological changes in the process going on through T tubules and the triadic junction might indirectly exert some influence on the SR. On the other hand, sulphate anion, choline, and tetraethylammonium cations have no effect on the dose-reponse relation.

From the collective survey of our many previous experiments (Sakai and Kurihara, 1974), we have found that it is necessary to examine the RCC from the viewpoint of excitation-contraction coupling. First, we reduced the amount of Ca^{2+} in T tubules to obtain its action by washing the muscle with the Ca-free medium. Second, we investigated how the membrane potential of the T tubules may affect Ca-liberation from the SR—a part of this study has already been reported (Sakai and Kurihara, 1974). There is an electro-mechanical term "inactivation": we want to reconsider this notion from the viewpoint of SR function.

THE PARAMETERS TO ANALYZE THE RCC

At present, no estimation of transient myoplasmic free Ca on RCC has been carried out. So, in order to ascertain that free Ca is released from the SR, a tension curve of RCC has been analyzed as follows: 1) a maximum tension; 2) rate of rise of tension; 3) dose-response relation at various concentrations of caffeine. If tension development reflects the released Ca ions, the three parameters mentioned above might indicate the following: the value of the maximum tension is considered to reflect the amount of released-Ca in the myoplasm, and the rate of rise of tension indicates the liberation rate of Ca^{2+} from the SR. The dose-response curve enables us to observe the sensitivity of the SR to caffeine and changes in its characteristics.

Figure 2 represents a record of RCCs in a case in which the caffeine concentration was stepwisely elevated from an initial 0.8 mM to a final 1.4 mM. The peak tension of the RCC was obtained at the caffeine con-

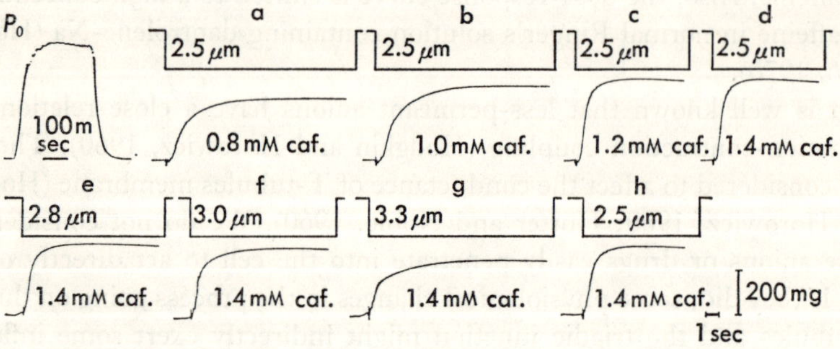

Fig. 2. Active tension of RCCs at various concentrations of caffeine and in the stretched state at 1.4 mM caffeine. Successive RCCs were carried out at 10-min intervals.

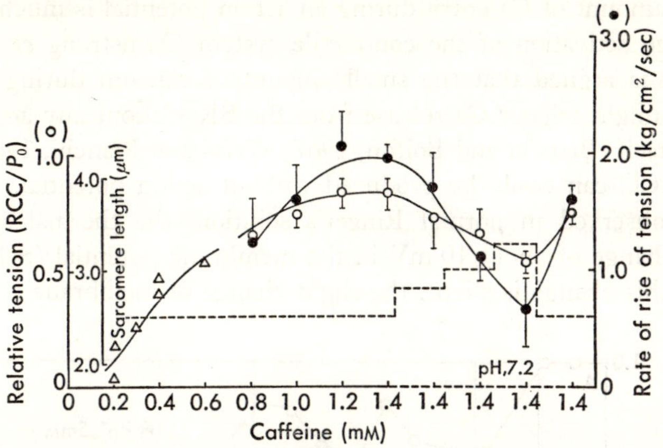

Fig. 3. Relation between relative tension and rate of rise of tension on RCC. The figure is partly composed of the results represented in Fig. 2. The active tension of RCCs was normalized in a relative tension as compared with the tetanic tension. Measurement of the rate of rise of tension was carried out by the procedure described in a previous paper. (Sakai, 1978).

centration of 1.0 mM in normal Ringer's solution, and, as seen in (c) and (d), the further elevation to 1.2 mM and 1.4 mM failed to elicit any change of the peak tension. But the rate of rise of tension became higher with increase in caffeine concentration. This is considered to mean that the maximal amount of Ca^{2+} may have been freed by thermal stimulation. In the next experiment, the average sarcomere length was stretched up to 2.8 μm and 3.3 μm and the same observations were repeated (Figs. 2 and 3). In this case, the peak tension and the rate of rise of tension was lowered with increase in the sarcomere length (Sakai, 1979).

The present experiments were carried out by use of toe muscle dissected from *Rana nigromaculata.* Thermal stimulation was constantly given by changing the temperature of the medium from 23°C to 0°C within about 1 sec. The sarcomere length before the RCC was adjusted by a manipulator and the first order of laser diffraction line given by the muscle using the equation: $d=n\lambda \sin \theta n$. The laser beam (He-Ne, 632.8 nm) was adjusted on the middle portion of the muscle and oriented perpendicular to its axis.

EXTRACELLULAR CALCIUM AND RCC

It has been reported that depolarization increases the Ca permeability of excitable membrane (Bianchi and Shanes, 1959; Curtis, 1966) and that the total amount of Ca entry during an action potential is much too small to produce activation of the contractile system (Armstrong *et al.*, 1972). Hence, it is argued that the small amount of calcium during an action potential might trigger Ca release from the SR without any action of the calcium itself (Bianchi and Bolton, 1967; Weiss and Bianchi, 1965).

The RCC can easily be produced without action potential. When the RCC is observed in normal Ringer's solution, the thermal stimulation elicits a change of about 10 mV in the membrane potential (Sakai, 1965). Though this change is slight, the slight change of membrane potential on

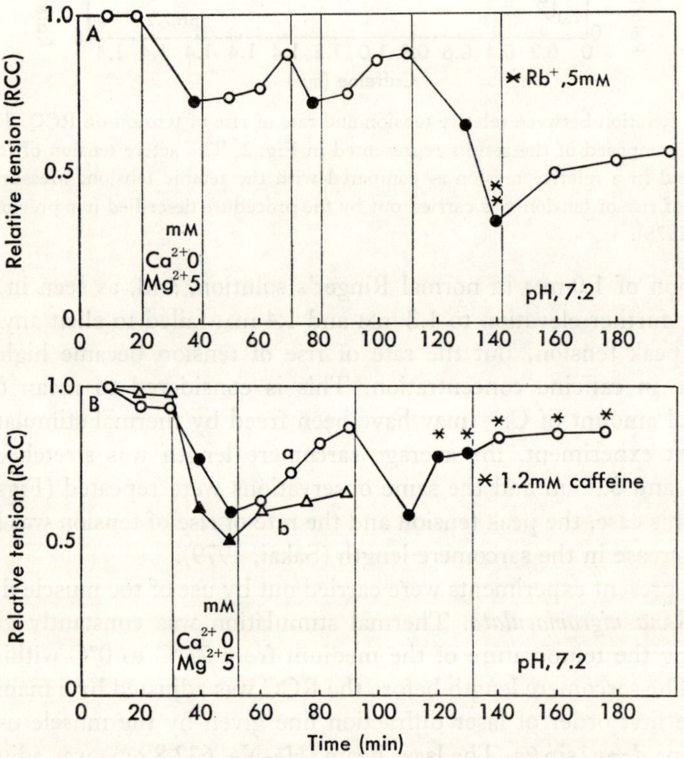

Fig. 4. Calcium-removal effect on active tension of RCC. The experiments were done in TEA modified Ringer's solution at 0.6 mM (B) and 1.2 mM (A) caffeine. Calcium removal effect was observed in Ca-free TEA Ringer's solution containing 5 mM Mg^{2+}. In the presence of Rb^+, potassium was omitted from the TEA modified Ringer's solution. The active tension of RCC in the presence of Rb^+ was inhibited and the reappearance of tension was reduced. In the experiment (B) 1 mM EGTA was added in the Ca-free TEA modified Ringer's solution in the absence of Rb^+.

the T tublules can not be denied. To substantiate this, the RCC was examined in the presence of tetrodotoxin (TTX), tetraethylammonium chloride (TEA), and Rb, which inhibit Na- and K-currents (Schneider and Chandler, 1973). But the result was identical to that in the normal Ringer's solution. This experiment was carried out in a solution containing 1.8 mM of Ca^{2+} besides TTX, TEA, and Rb. However, tension development in RCC was inhibited in the Ca-free TEA•Rb solution containing TTX, and it was thus found that the absence of Ca^{2+} was rather more effective. It was also observed that in the presence or absence of Rb, the

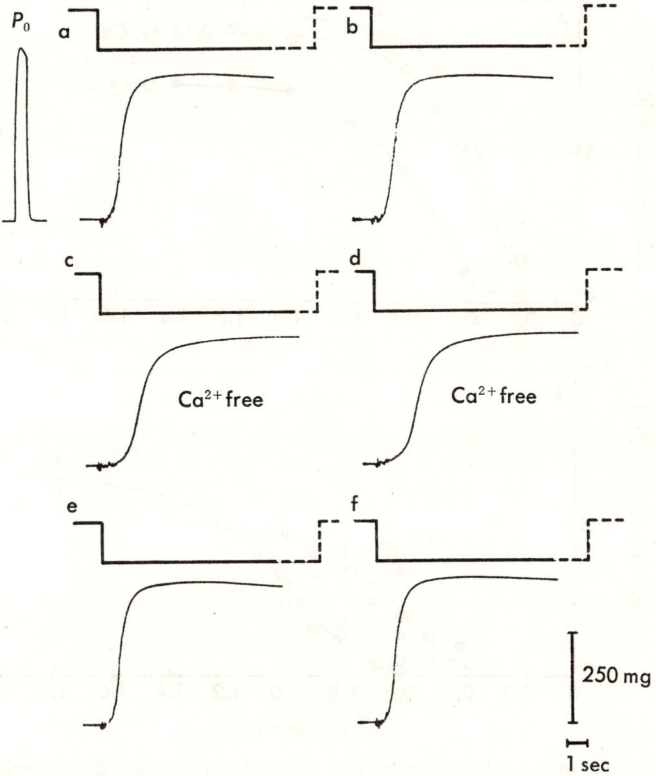

Fig. 5. Inhibition of rising rate of tension on RCC. The experiment was carried out in an inactive state after the muscle was equilibrated with 95 mM $[K]_0$ solution. In the Ca-free 95 mM $[K]_0$ solution, the equimolar Mg^{2+} was substituted for Ca^{2+}. In Ca-free solution a latency for activation of RCC was prolonged and both parameters, peak tension and rate of rise of tension, were reduced. The muscle completely recovered from these inhibitions about 10 min after the addition of calcium.

dose-response curve for RCC in a Ca-free TEA solution was shifted to the site of higher caffeine concentration (Fig. 6), and the peak tension of RCC was inhibited. Thus the threshold for Ca-liberation by thermal stimulation proved to be elevated. This can be interpreted as the result of depression, through the T system, of Ca-releasing mechanism of the SR. The figure represents the Ca-removal effect in the TEA modified Ringer's solution. Moreover it was confirmed that the removal of Ca^{2+} also depressed the dose-response curve. The removal of Ca^{2+} alone, without the addition of EGTA resulted at the depression of tension develop-

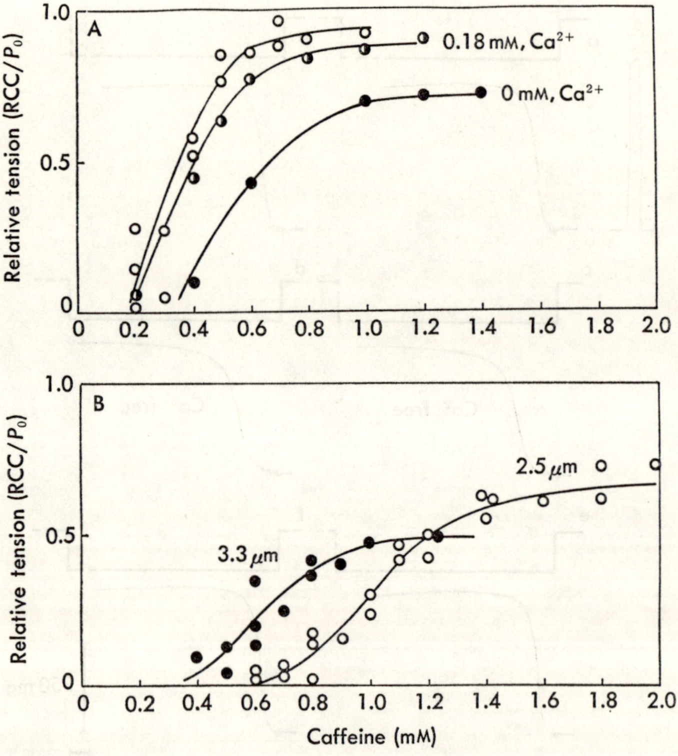

Fig. 6. Dose-response relation for RCC in Ca-depleted medium. Dose-response relation was normalized in relative tension as compared with the tetanic tension. The TEA modified Ringer's solution had the following composition (mM): TEA-chloride, 117; KCl, 2.5; $CaCl_2$, 1.8; Tris buffer, 5; pH 7.2 (adjusted by HCl); 3×10^{-2} mg/ml of d-tubocurarine was added to all solutions as per by Schneider and Chandler's paper (1973). A: ○ normal Ringer sol. (n=3); ◑ Ca depleted TEA Ringer (n=5); ● Ca removal TEA Ringer (n=5). B: Ca free, TEA-Rb Ringer TTX, 7×10^{7} g/ml.

ment in the RCC and decreased in the rate of the rise in tension. The depression grade of tension ranged between 30–50% depending on the caffeine concentration (Fig. 4). These phenomena, depression of tension and decrease in the rate of rise of tension, were also observed in the 95 mM-K solution (Fig. 5). The muscle was immersed in the Ca-free medium for 10–20 min, and if Ca^{2+} were removed from the T tubules, it would be possible to demonstrate that Ca^{2+} in T tubules is related to the development of RCC by some mechanism. Recently it has been suggested that the transverse tubular calcium contributes to excitation-contraction coupling by crossing the tubular membrane, and directly or indirectly enhancing release of Ca^{2+} from the terminal cisternae (J. N. Barrett and E. F. Barrett, 1978). Also, on the RCC the tubular Ca may contribute to the activation of Ca-releasing mechanism in the SR. Since the RCC is elicited without action potential as mentioned above, our results suggest that Ca^{2+} influx crossing through the T-tubules membrane does not directly trigger activation of contractile system. At present, we do not obtain an activator to elucidate the Ca-releasing mechanism on the SR. Therefore, it might be concluded that activation of RCC is induced by the mechanism used in the Ca-induced Ca release on the SR (Endo, 1977). There is a distinct possibility that this mechanism is modified with a characteristic change of the SR through the T system.

RELATION BETWEEN MEMBRANE POTENTIAL AND ACTIVATION OF RCC

If it is assumed, as above mentioned, that the tubular distributed calcium exerts its effect on the Ca-releasing mechanism through the T system, what change would be elicited in the development of tension in RCC, when the depolalization is induced in the tubular membrane? An initial experiment already has been reported in the muscle in which depolarization was induced by extracellular $[K]_0$ at various concentrations (Sakai and Kurihara, 1974).

The RCCs were observed by use of muscles which were depolarized at various concentrations of $[K]_0$ after treatment with the TEA-Rb modified Ringer's solution (Fig. 7). The peak tension at 30 mM $[K]_0$ seems to be inhibited compared with that in normal Ringer's solution. Relation between peak tension of RCC and the extracellular potassium concentration is shown in Fig. 8. It was shown that the peak tension of RCC is lowered with rise in $[K]_0$ concentration, reaching the minimum at 30–40 mM $[K]_0$.

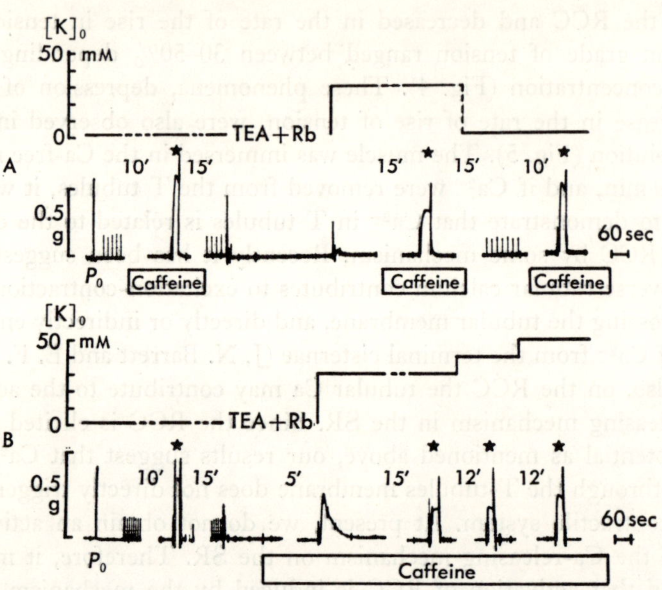

Fig. 7. Relation between extracellular potassium concentration and active tension of RCC. A: after the muscle was immersed in the Schneider and Chandler's solution for 10 min, the peak tension of RCC was observed in 30 mM $[K]_0$ medium. Compare with that in normal Ringer's solution. B: the same experiments were successively carried out in the medium at 30 mM, 40 mM, and 50 mM $[K]_0$. The K-rich medium was kept constant at $[K]_0 \times [Cl]_0 = 300\ \overline{mM}^2$ by Hodgkin and Horowicz's procedure. In both experiments 1.0 mM caffeine was used.

Simultaneously, we examined the dose-response relation for different $[K]_0$ concentrations to determine the maximal necessary caffeine concentration and characteristic change of the SR. This was found to be about 1.0 mM. It was subsequently confirmed that at this maximal caffeine concentration, the inhibitory effect of depolarization on the peak tension of RCC persisted. This inhibition resembles the inactivation reported by Hodgkin and Horowicz (1960). Thus the amount of Ca released from the SR has been shown to undergo inactivation. Why is Ca release from the SR inhibited? One explanation is that some activator may have been freed from the tubular membrane owing to depolarization, changing the function of the terminal cisternae to Ca. In other words, the activation of the Ca-release mechanism on the SR by thermal stimulation might be reduced, as a result attenuating the efficiency of the Ca-induced Ca release.

However, when $[K]_0$ concentration increases at above 50 mM, the ten-

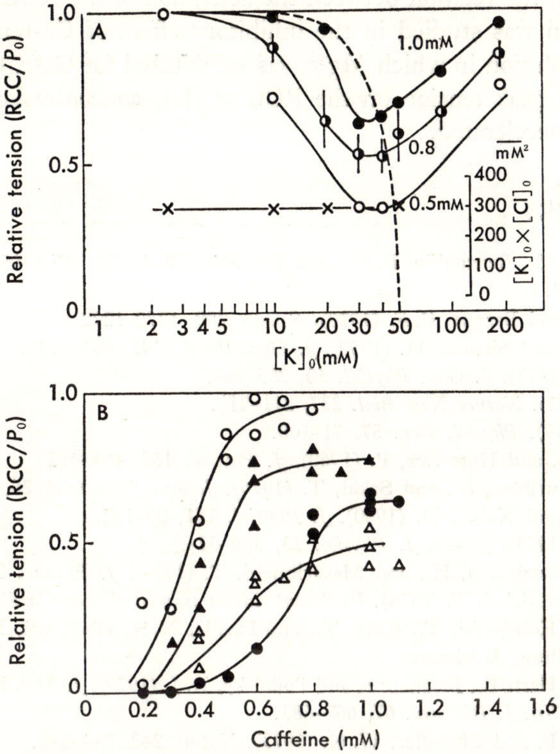

Fig. 8. Relation between peak tension and extracellular potassium concentration at various concentrations of caffeine. Relative tension of RCCs was normalized as compared with active tension in normal Ringer's solution. A: the symbols represent the average value of relative tension at 0.5 mM, 0.8 mM, and 1.0 mM caffeine. At 0.8 mM caffeine the average and the standard deviation were calculated from seven preparations. Another average value was calculated from five preparations. The dotted line shows the inactivation curve by Hodgkin and Horowicz (1960). B: the dose-response relation of RCC for the muscle equilibrated with various potassium rich media. ○ 2.5 mM $[K]_0$; ▲ 95 mM $[K]_0$; ● 50 mM $[K]_0$; △ 40 mM $[K]_0$

sion of the RCC begins to show an increase in reciprocal proportion with further fall of the resting potential. This phenomenon appears to contradict the above assumption on "inactivation." But if the functional connection between the T tubules and the lateral cisternae is tentatively presumed to have been broken in $[K]_0$ concentration of above 50 mM and to lose the controlling ability of the Ca release on the SR, the contradiction could be accounted for.

Moreover, the relation between the activation of the RCC and the $[K]_0$ concentration was studied in the inhibitory effect of Ca-removal. In the potassium solution in which Mg^{2+} was substituted for Ca^{2+}, the inhibitory effect of the peak tension on the RCC at $[K]_0$ concentration less 50 mM was further accelerated.

REFERENCES

Armstrong, C. M., Benzanilla, F. M., and Horowicz, P. (1972). *Biochim. Biophys. Acta* **267**, 605–608.

Barrett, J. N. and Barrett, E. F. (1978). *Science* **200**, 1270–1272.

Bianchi, C. P. and Shanes, M. (1959). *J. Gen. Physiol.* **42**, 803–815.

Curtis, B. A. (1973). *J. Gen. Physiol.* **50**, 255–267.

Endo, M. (1972). *Nature New Biol.* **237**, 211–213.

Endo, M. (1977). *Physiol. Rev.* **57**, 71–108.

Hodgkin, A. L. and Horowicz, P. (1960). *J. Physiol.* **153**, 404–412.

Homma, I., Kurihara, S., and Sakai, T. (1976). *Japan. J. Physiol.* **26**, 53–61.

Hutter, O. F. and Noble, D. (1960). *J. Physiol.* **151**, 89–102.

Kurihara, S. (1973). *Japan. J. Physiol.* **23**, 309–324.

Kurihara, S., Kuriyama, H., and Magaribuchi, T. (1974). *J. Physiol.* **238**, 413–426.

Kurihara, S. and Sakai, T. (1978). In *Recent Advances in Studies on Cardiac Structure and Metabolism* (Kobayashi, T., Sano, T., and Dhalla, N. S., eds.), vol. II, pp. 181–184, Univ. Park Press, Baltimore.

Leepo, C. Yu, Hartt E., Jacqueline, and Podolsky, R. J. (1979). *J. Mol. Biol.* **132**, 53–67.

Ogawa, Y. (1970). *J. Biochem.* **67**, 667–683.

Schneider, M. F. and Chandler, W. K. (1973). *Nature* **242**, 244–246.

Sakai, T. (1965). *Jikeikai Med. J.* **12**, 88–102.

Sakai, T. and Iizuka, T. (1972). *Japan. J. Physiol.* **22**, 135–145.

Sakai, T., Geffner, E. S., and Sandow, A. (1971). *Am. J. Physiol.* **220**, 712–717.

Sakai, T. and Kurihara, S. (1974). *Jikeikai. Med. J.* **21**, 47–88.

Sakai, T. and Kurihara, S. (1974). *Japan. J. Physiol.* **24**, 649–666.

Sakai, T. (1978). In *Cross-bridge Mechanism in Muscle Contraction* (Sugi, H. and Pollack, G. H., eds.), pp. 311–318, Univ. Tokyo Press, Tokyo.

Weber, A. and Herz, R. (1968). *J. Gen. Physiol.* **52**, 750–759.

Weiss, G. B. and Bianchi, C. P. (1965). *J. Cell. Comp. Physiol.* **65**, 385–392.

Muscle Contraction: Its Regulatory Mechanisms (Ebashi, S. et al., eds.), pp. 447–463, Japan Sci. Soc. Press, Tokyo / Springer-Verlag, Berlin (1980)

35

Different Features of Responses of the Sarcoplasmic Reticulum in Cardiac and Smooth Muscles

MAKOTO ENDO, TOSHIO KITAZAWA,
AND SHINOBU YAGI

Summary: Certain characteristics of the sarcoplasmic reticulum (SR) in cardiac and smooth muscle, which have not been observed in the skeletal muscle SR, were described. In cardiac fibers, removal of external Na ions enhanced the Ca-induced Ca release mechanism to evoke characteristic cyclic contractions, which were immediately suppressed by reintroduction of external Na ions. Caffeine contracture of living cardiac cells, probably a result of the enhancement of the Ca-induced Ca release, was also strongly inhibited by external Na ions. This inhibition was not the result of removal of the intracellular Ca ions through the Na-Ca exchange mechanism, but the Ca-releasing action of caffeine was somehow suppressed. The implication of these facts in connection with physiological excitation-contraction coupling was discussed. The presence of an SR-like intracellular Ca store in smooth muscle of the taenia caeci, from which caffeine caused Ca release, was demonstrated. After specific perforation of the surface membrane of smooth muscle cells with saponin, the function of the store was essentially preserved and showed very similar characteristics to those of the SR in skeletal muscle. Ca release by carbachol that had already been well described appeared to occur from the same store, since 'depletion' and 'reloading' of Ca from and to the store tested by carbachol was about the same as those tested by caffeine. After skinning, the Ca-releasing action of carbachol disappeared, while that of caffeine remained. The mechanism of the Ca-releasing action of carbachol was discussed.

It is well established that in skeletal muscle, the sarcoplasmic reticulum (SR) plays a central role in regulating Ca ion concentration in the cell during the contraction-relaxation cycle, although the precise mechanism is still unknown (Ebashi and Endo, 1968; Endo, 1977). The action potential of skeletal muscle fibers somehow triggers a release of Ca from the SR to cause contraction, and the SR then actively takes up Ca to cause relaxation. However, the role of the SR in cardiac and smooth muscles is much less clearly understood. Even the problem as to whether the Ca ions activating the contractile protein system physiologically come from the SR or not is not yet completely solved. We have been examining the properties of the SR of cardiac and smooth muscle by using intact and chemically skinned fibers, and found that the SR of these kinds of muscles have properties different from those of the skeletal muscle SR (Kitazawa, 1977; Yagi and Endo, 1980). Some of the differences, which we thought important, will briefly be described.

EFFECT OF EXTRACELLULAR NA IONS ON THE CA-INDUCED CA RELEASE IN THE CARDIAC SR

The most unexpected property of the cardiac SR found was the fact that its Ca-induced Ca release appears to be affected by extracellular Na ions. Suzuki (1962) showed that caffeine can cause contracture of the frog heart muscle only when Na ions in the extracellular medium were withdrawn. Chapman and Miller (1974) made a similar observation. Kimoto (1972) reported that the same is true in mammalian cardiac muscle, but a clear analysis of this Na effect has not so far been made.

Kitazawa (1977) dissected thin bundles of cardiac fibers, about 100 μm in diameter, from papillary muscles of the guinea-pig right ventricle, in which the extracellular environment of each cardiac cell could be altered rapidly, within several seconds, because of the short diffusion distance. This is important especially when responses to be observed are transient like caffeine contracture, since in thicker bundles, slow diffusion of the stimulating substance into the bundle would make response of individual cells in the bundle temporarily so dispersed that only small and quantitatively unreliable responses could be obtained.

Examples of the caffeine response of a thin bundle in an Na-free Li solution are shown in Fig. 1. Caffeine, if applied in such a high concentration as 25 mM, produced a large contracture in Li solution, the peak tension of which was much larger than twitch tension that had already been very much augmented in an Na-free condition. In fact, the contracture tension was almost the maximal tension of the preparation. In Fig. 1, it is also demonstrated that a very similar contracture could be produced even when external Ca ions were practically removed by applying a Ca free, EGTA-containing solution. This indicates that caffeine must have caused Ca release from an internal Ca store, most probably from the SR, and that in the SR an amount of Ca at least enough to produce maximal tension had been stored. It should also be noticed that unlike caffeine contractures in amphibian skeletal muscle fibers, the caffeine contracture of the cardiac fibers was transient and tension was reduced to a low level within a few minutes, as already extensively analyzed (Chapman and Léoty, 1976).

As shown in Fig. 2, if Na ions were progressively added in place of Li ions to the external solution, the duration of cardiac muscle caffeine contracture was gradually reduced until no response was finally obtained in a normal, 100% Na-containing solution. Thus, the contracture-producing action of caffeine was almost completely inhibited in an Na solution. This appears to be due to the specific inhibitory action of Na ions rather than to the enhancing action of Li ions, because, when Na ions are replaced by other cations, *i.e.*, Tris or K, similar contracture could also be produced by caffeine as reported by Suzuki (1962). In the case of K replacement, membrane potential change does not seem to be important, since as far as Na ions were absent, replacement of Li ions by K ions to various degrees did not affect the contracture size produced by a high concentration of caffeine at all. These results are quite different from those obtained by Chapman and Léoty (1976), in which external Na ions are reported not to affect caffeine contracture. The reason for this discrepancy is not known at present.

In the frog atria, caffeine contracture could be elicited after either the Na-withdrawal contracture or the K contracture had subsided, and Chapman and Miller (1974) interpreted this as indicating that an amount of Ca must be accumulated in the SR before caffeine becomes effective.

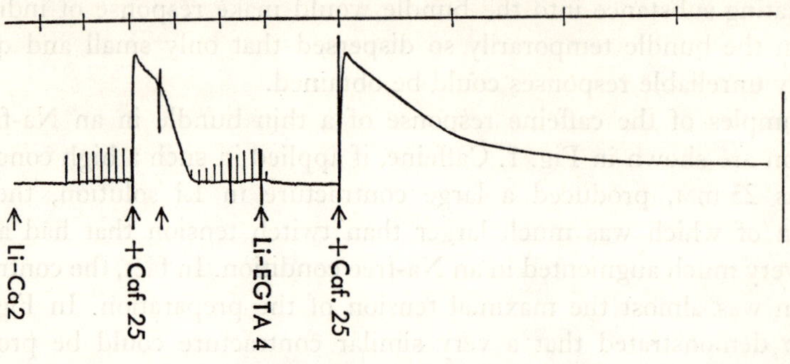

Fig. 1. Twitches and caffeine contractures of cardiac muscle in Li-solution. Solutions contained 150 mM LiCl, 2 mM KCl, 2 mM Mg-methanesulfonate, 5 mM HEPES (N-2-hydroxyethylpiperazine-N′-2-ethanesulfonic acid) neutralized with Tris-(hydroxymethyl)-aminomethane (pH 7.4), 5.6 mM glucose and either 2 mM Ca-methanesulfonate or 4 mM EGTA (ethyleneglycol-bis-(β-aminoethylether)-N-N′-tetraacetic acid). At the times indicated, 25 mM caffeine was applied. The first caffeine contracture was terminated by the removal of the drug at the arrow. Time, 1 min; tension calibration 20 mg.

However, such a 'Ca accumulation' was not necessary, at least in the case of guinea-pig papillary muscle, because even when Na-withdrawal was made in a Ca-free EGTA-containing solution, and, therefore, accompanied by no visible contracture, caffeine contracture of a similar size could still be obtained (Fig. 3B).

One of the conceivable mechanisms through which extracellular Na ions inhibit the caffeine contracture is an operation of the Na-Ca exchange system in the surface membrane as proposed by Chapman (1979). Even if caffeine normally releases Ca from the SR, a rapid enough exchange of intracellular Ca ions with extracellular Na ions might reduce the intracellular Ca ion concentration sufficiently to inhibit tension production. However, since in the absence of extracellular Na ions, caffeine releases Ca in an amount that produces maximal tension within a few seconds (Figs. 1 and 2), it is difficult to imagine that the Na-Ca excahnge system operates so rapidly that such an amount of Ca ions released could be mostly transported before it could produce any tension. The following facts are also against the theory of the Na-Ca exchange system as the main mechanism of the inhibitory action of the extracellular Na ions on caffeine contracture. According to the 'Na-Ca exchange theory,' since Ca release from the SR by caffeine is assumed to be unaltered in an Na solution, if caffeine is applied in a Ca-free EGTA-containing Na solution, much of the Ca should be transported out of the cells in an exchange with Na

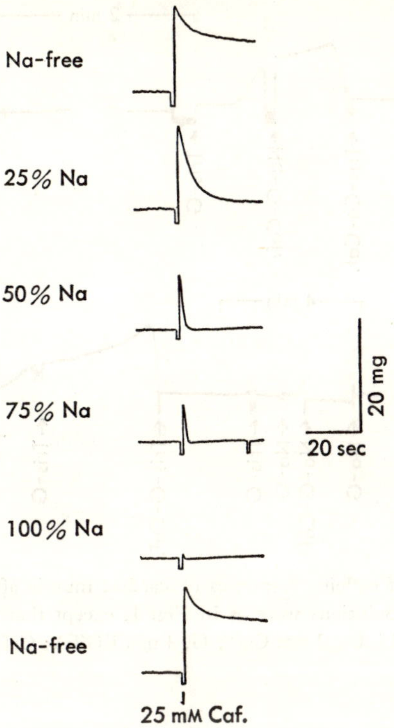

Fig. 2. Effect of Na ions on caffeine contracure of cardiac muscle. Solutions used were as in Fig. 1 (2 mM Ca), except that various parts of LiCl were replaced by NaCl.

ions, and the amount of Ca in the SR should be substantially reduced. However, as shown in Fig. 3B, after such a caffeine treatment in an EGTA-containing Na solution, which was not accompanied by an appreciable tension, if Na ions were replaced by Tris, a subsequent application of caffeine still produced large tension. Since Ca cannot re-enter the cell from the external Ca-free medium containing EGTA, the SR must have retained an appreciable amount of Ca in spite of the application of caffeine in the Na solution. An essentially similar experiment is shown in Fig. 3A. The time course of caffeine contracture in an EGTA-containing Tris solution was similar to that shown in Fig. 1, with slow spontaneous relaxation in the presence of the alkaloid. According to the 'Na-Ca exchange theory,' if a caffeine contracture was interrupted by introducing Na ions in place of Tris ions, it should have reduced the amount of Ca in the cell substantially, since caffeine-induced Ca release itself was assumed to be

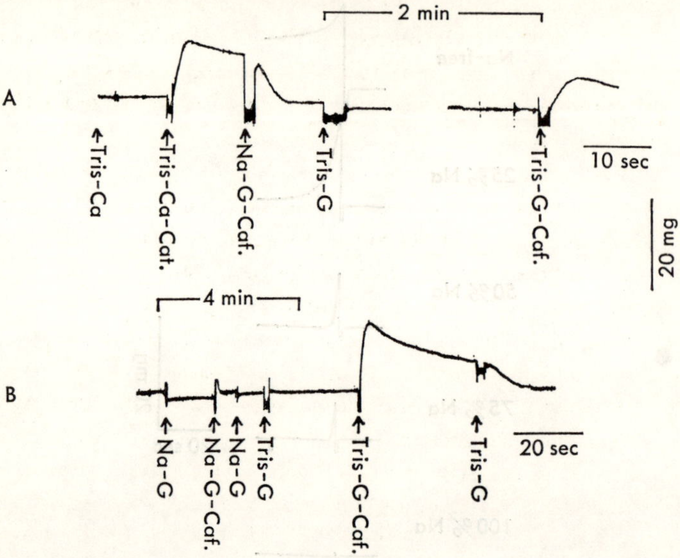

Fig. 3. Persistence of caffeine responses of cardiac muscle after caffeine treatments in Ca-free Na-solution. Solutions were as in Fig. 1, except that main cations were either Na or Tris instead of Li. Ca, 2 mM Ca^{2+}; G, 4 mM EGTA: Caf., 25 mM caffeine. Further explanation, see text.

unaffected. However, as shown in Fig. 3A, after some time in the EGTA-containing Na solution, Na removal and subsequent reapplication of caffeine produced a tension, the magnitude of which was about equivalent to that at the corresponding time without Na interruption. These results indicate that inhibition of caffeine contracture in an Na solution is not due to the removal of intracellular Ca ions through the Na-Ca exchange system but to an abolition of the Ca-releasing action of caffeine somehow brought about by the extracellular Na ions.

The depression of cardiac caffeine contracture by the extracellular Na ion does not appear to be due to a resulting increase in intracellular Na ion concentration. When fibers of such a thin cardiac bundle were skinned by a treatment with saponin (Endo and Kitazawa, 1978), replacement of most of K ion by Na ion affected neither the size of contracture produced by a high concentration of caffeine nor its time course.

Caffeine is considered to act directly on the SR to cause Ca release. Yet, the extracellular Na ion somehow inhibits this action of caffeine. At present the mechanism of this Na action is, although extremely interesting, completely unknown. The mechanism of action of caffeine is to enhance the Ca-induced Ca release in the SR (Endo, 1975b). Therefore, one can suppose that the Ca-induced Ca release mechanism itself might also be inhibited by the extracellular Na ion. Figure 4 shows that this is indeed the case. Figure 4A compares tension records at a high magnification of a thin cardiac bundle in normal Na-containing solution and in a Na-free K solution. It is clearly seen that the tension record in the K solution, which is a tail part of K contracture where most of tension had already subsided, showed small, more-or-less cyclically repeated contractions with irregular sizes. Membrane potential is not expected to be changing during these contractures. These cyclic contractions are similar to those produced by the Ca-induced Ca release mechanism (Endo *et al.*, 1970; Fabiato and Fabiato, 1972; Endo and Kitazawa, 1978), and in fact they were inhibited by procaine, an inhibitor of the Ca-induced Ca release (*cf.* Endo, 1977) (Fig. 4B). Similar cyclic contractions could be evoked when external Na ions were replaced by Li ions (Fig. 4C), so that depolarization is not essential for them, but the removal of Na appears to be responsible. As shown in Fig. 4C, reintroduction of Na ions in the external medium immediately stops the cyclic contractions. Thus, as well as caffeine-induced Ca release, the Ca-induced Ca release mechanism appears to be inhibited by extracellular Na ions. Again, in chemically skinned fibers Na ions did not inhibit cyclic contractions at all, suggesting that it is the extracellular but not intracellular Na ions that exert the action. Essentially the same cyclic contractions were reported by Glitsch and Pott (1975). They believed that the phenomenon occurs only under the condition when the intracellular Ca concentration was raised drastically. Although it probably enhances the cyclic contractions to raise their detectability, the increased cellular Ca content does not seem to be a necessary condition for the essential phenomenon. In fact, even in a Ca free EGTA-containing solution that precludes Ca accumulation, the removal of extracellular Na ions still induced similar cyclic contractions.

Ca-induced Ca release is more easily evoked in the cardiac SR than in the skeletal SR. Even in the cardiac SR, however, a relatively high concentration of Ca, 3×10^{-6} M, is required to cause Ca release in guinea-pig papillary muscle under a condition thought to be more or less physiological (Endo and Kitazawa, 1978), although a specifically lower figure was reported in the rat ventricle (Fabiato and Fabiato, 1977). If our above figure is true also under the really physiological condition in living cardiac cells, it appears rather difficult to believe that the Ca-induced Ca release

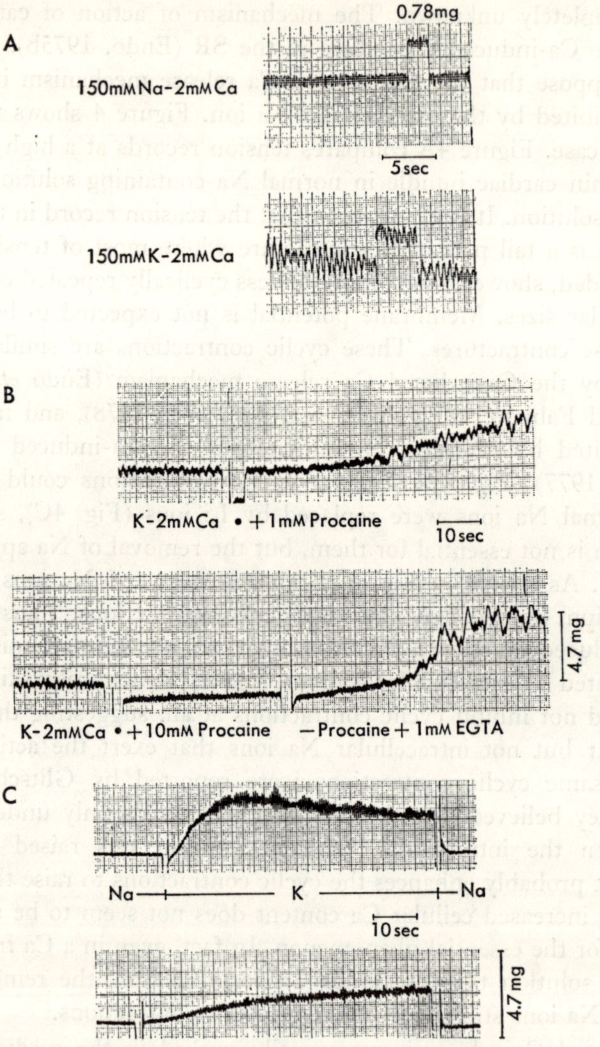

Fig. 4. Cyclic contractions of living cardiac cells in Na-free solutions. Solutions were the same as in Fig. 1 except that main cations were altered in some cases. A: demonstration of cyclic contractions in K-solution in comparison with quiescence in Na-solution. B: suppressing effect of procaine on cyclic contractions. C: records showing time courses of development and disappearance of cyclic contractions on Na-withdrawal and -reapplication.

mechanism plays a major role in physiological cardiac excitation-contraction coupling except in the case of rat ventricle. However, our present result indicates that there is at least a presently unknown mechanism controlling the Ca-induced Ca release in the cardiac SR, through which the extracellular Na ions exert the inhibitory action. Until this controlling mechanism is fully examined, conclusion about the physiological role played by the Ca-induced Ca release mechanism must be suspended. Thus, it is urgently necessary to elucidate the mechanism controlling the Ca-induced Ca release in the SR of cardiac cells and the mechanism of action of the extracellular Na ions on it.

THE SR-LIKE CA STORE IN SMOOTH MUSCLE AND ACTION OF THE AUTONOMIC AGONIST ON IT

The presence of the SR in smooth muscle cells has been electron-microscopically demonstrated (Somlyo *et al.*, 1971), but studies of its function in living cells have so far not been performed in a clear way. Caffeine that has a strong Ca-releasing action on the SR of striated muscle produces rather capricious responses in smooth muscle, sometimes contractile and sometimes relaxing, since it definitely has a relaxing action on smooth muscle as well, a well-known general property of methylxanthines. We have found a condition in which caffeine faithfully produces contracture due to Ca release from an internal store, most probably from the SR, in smooth muscle cells. A thin bundle of smooth muscle cells, 100–150 μm in diameter, was dissected from the guinea-pig taenia caeci, tied at each end with a single silk thread, and suspended in a trough through which solutions could be perfused rapidly. Isometric tension was recorded with a straingauge transducer (U-gauge, Shinkoh, Tokyo) on a pen recorder. To avoid possible changes in the membrane potential due to caffeine and other drugs, the preparation was kept almost completely depolarized by using a solution in which all the Na ions were replaced by K ions throughout the experiments. To avoid also the possibility that external Ca ions might enter into the cell and contribute to the contracture, application of caffeine was done in a Ca-free, EGTA (4 mM)-containing solution. Figure 5A shows an example of caffeine contracture in such a Ca-free, EGTA-containing K solution. Similar contractures were reproducibly obtained if pretreatments before the caffeine application filled certain requirements. As shown in the figure, caffeine contracture did not last long but spontaneous relaxation occurred in spite of the continued presence of the drug.

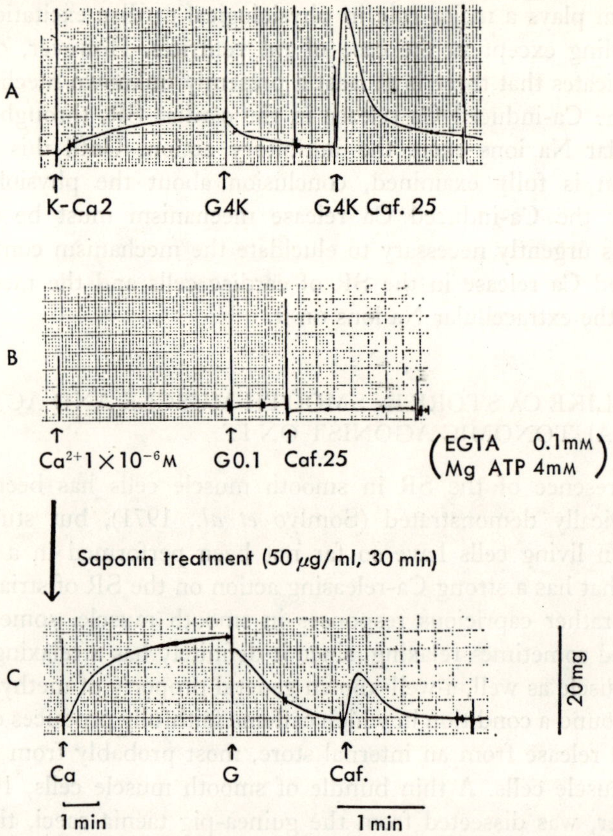

Fig. 5. Caffeine contracture of smooth muscle (taenia caeci) before (A, B) and after (C) saponin treatment. After a former exposure to caffeine, the preparation was pretreated with a solution containing either 2 mM Ca^{2+} (A) or 10^{-6} M Ca^{2+} (B, C) for 5 min before the next test exposure to 25 mM caffeine. Solutions used in A were essentially the same as in Fig. 1 except that the main cation was K. Solutions used in B and C were essentially the same as a relaxing solution described in Endo *et al.* (1977).

Reapplication of caffeine after the drug had been once washed out did not produce contractile responses any more. However, if the preparation was immersed in a Ca-containing solution for a certain period of time as shown in Fig. 5A, the responsiveness to caffeine recovered, as if once depleted Ca store was slowly refilled by Ca entry from the external medium. Of course, an alternative possibility is that, although Ca in the store is not depleted, the mechanism of caffeine-induced Ca release is somehow

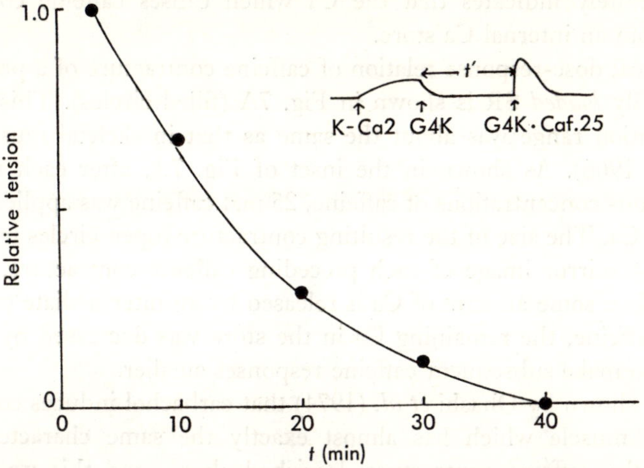

Fig. 6. Time course of *depletion* of Ca from the Ca store of smooth muscle in an EGTA-containing solution. The relative size of peak tension of caffeine contracture was plotted against the duration of immersion in a 4 mM EGTA-containing solution interposed between *loading* and 25 mM caffeine application (see inset).

inactivated by the continued presence of the drug, and recovery from the inactivation slowly occurs only in the presence of the external Ca ion. For the sake of brevity, however, we hereafter call the inactivation in the broad sense *depletion of Ca from the store*, and the recovery from the inactivation *loading of Ca to the store. Loading* of Ca was dependent both on external Ca ion concentration and on time of immersion in the Ca containing solution. In the presence of 2 mM Ca, the extent of *loading* gradually increases with time until saturation is reached at about 5 min. At least several tens of μM of Ca ion concentration appears to be necessary for appreciable *loading*. If, for example, the external Ca ion concentration was as low as 10^{-6} M, no *loading* occurred as shown in Fig. 5B. When a preparation, of which the Ca store had been fully *loaded*, was immersed in a Ca-free EGTA-containing solution, caffeine response gradually decreased with time (Fig. 6). Since the time for extracellular Ca ion to diffuse out of these thin preparations should be rather short, of the order of several tens of seconds at most, and also since the time course of the decline was practically independent on EGTA concentration up to 20 mM, the gradual decline in caffeine responses in Fig. 6 is due not to the removal of extracellular Ca ions, but to the *depletion* of Ca from the Ca store.

This definitely indicates that the Ca which causes caffeine contracture comes from an internal Ca store.

A typical dose-response relation of caffeine contracture of a preparation with a fully *loaded* SR is shown in Fig. 7A (filled circles). This effective concentration range was about the same as that in skeletal muscle fibers (Caputo, 1966). As shown in the inset of Fig. 7A, after each treatment with various concentrations of caffeine, 25 mM caffeine was applied without *reloading* Ca. The size of the resulting contracture (open circles in Fig. 7A) was just a mirror image of each preceding caffeine contracture. Thus, it is as if when some amount of Ca is released by an intermediate concentration of caffeine, the remaining Ca in the store was decreased by just that amount to make subsequent caffeine responses smaller.

It was shown by Ohashi *et al.* (1974) that carbachol induces contracture of taenia muscle which has almost exactly the same characteristics as those of the caffeine contracture described above, and this was recently confirmed and extended by Casteels and Raeymaekers (1979). Thus, carbachol also can cause a release of Ca from an internal Ca store of the taenia muscle. We have examined the problem as to whether the store from which carbachol releases Ca is the same as that on which caffeine acts or not. In the experiments shown in Fig. 7B, after a full *loading* of the store, various concentrations of carbachol were applied and then, after washing out, a sufficiently high concentration, in this case 50 mM, of caffeine was added. It is clear that whenever carbachol induced contracture, the subsequent response to caffeine decreased by about the appropriate amount to make a mirror image very similar to that shown in Fig. 7A. Thus, carbachol appears to *deplete* Ca from the store on which caffeine also acts. This conclusion is supported by the fact that both the time course of *loading* in various external Ca concentrations and that of *depletion* in an EGTA-containing solution tested by the carbachol response are about the same as those tested by the caffeine response.

As shown in Fig. 5C, when thin bundles of taenia were treated with saponin to obtain chemically skinned smooth muscle fibers (application of 10^{-6} M free Ca ions as in Fig. 5B now produces an appreciable tension), caffeine responses of a respectable size could still be obtained. Although the caffeine response in this figure is about 30% of that before skinning (Fig. 5A), we have obtained responses after skinning of a size up to 60% of those before saponin treatment. Considering the fact that Ca released by caffeine may be partly diffused out of the cells without acting on the contractile proteins in the case of skinned fibers, we tend to conclude

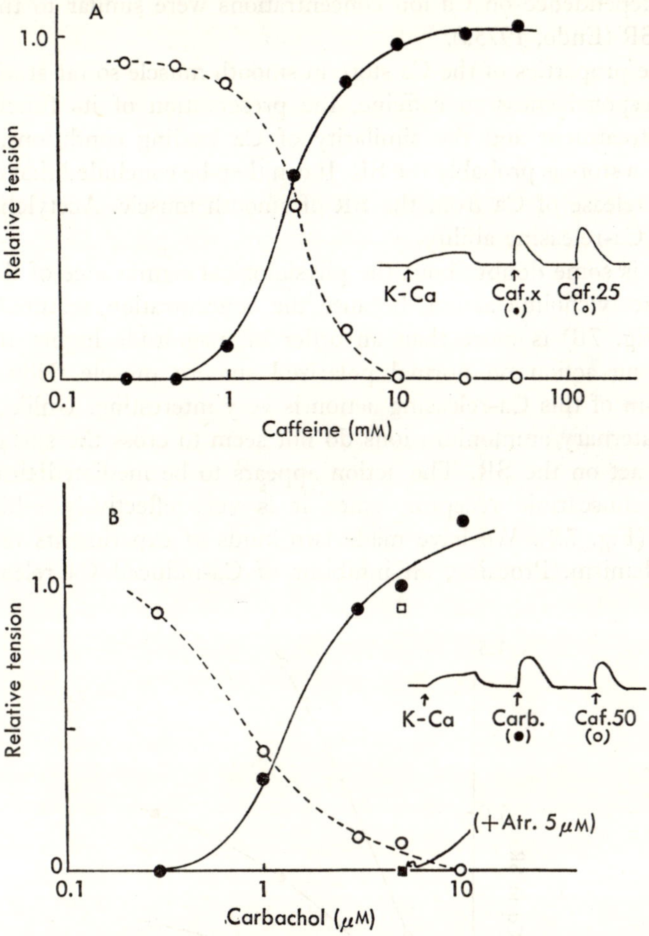

Fig. 7. Dose-response relations (●) of caffeine contracture (A) and carbachol contracture (B) and resulting *depletion* of Ca store (○). In B, a complete inhibition of the carbachol response by 5 μM atropine is also demonstrated (■, □). For experimental procedures, see inset and text.

that the function of the Ca store sensitive to caffeine is mostly preserved after saponin treatment; however, as naturally expected, the condition for *loading* (in skinned fibers, *loading* is real loading), is completely different before and after saponin treatment. After skinning, a free Ca ion concentration on the order of 10^{-6} M is enough to load appreciably (Figs. 5C and 8). The time course of loading of the Ca store in skinned smooth muscle and its dependence on Ca ion concentrations were similar to that in the skeletal SR (Endo, 1975a).

All the properties of the Ca store in smooth muscle so far studied, such as the responsiveness to caffeine, the preservation of its function after saponin treatment and the similarity of Ca loading conditions, indicate that the Ca store is probably the SR. It can then be concluded that carbachol causes a release of Ca from the SR of smooth muscle. Acetylcholine has a similar Ca-releasing ability.

There is some doubt about the physiological significance of this carbachol or acetylcholine action, because the concentration required for this action (Fig. 7B) is more than an order of magnitude higher than their depolarizing action on normal polarized smooth muscle. However, the mechanism of this Ca-releasing action is very interesting. Unlike caffeine, these quaternary ammonium ions do not seem to cross the surface membrane to act on the SR. The action appears to be mediated through the ordinary muscarinic receptor, since it is very effectively inhibited by atropine (Fig. 7B). We have made two kinds of experiments relevant to this mechanism. Procaine, an inhibitor of Ca-induced Ca release, fairly

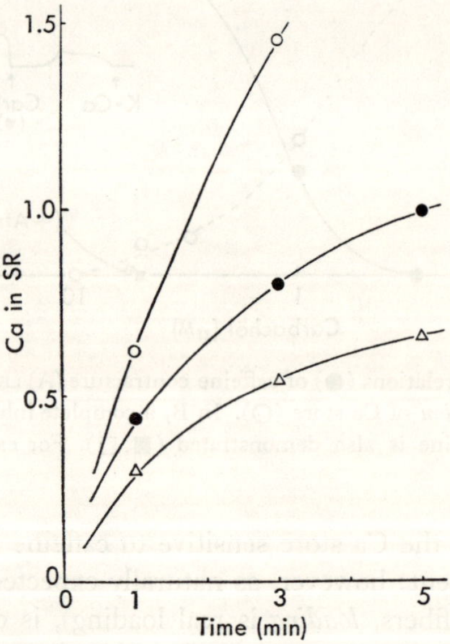

Fig. 8. Time course of loading of Ca to the SR-like Ca store under various concentrations of Ca^{2+}. Experimental procedures were essentially the same as shown in Fig. 5C. ○ Ca^{2+} 1×10^{-5} M; ● Ca^{2+} 1×10^{-6} M; △ Ca^{2+} 5×10^{-7} M; EGTA 0.1 mM.

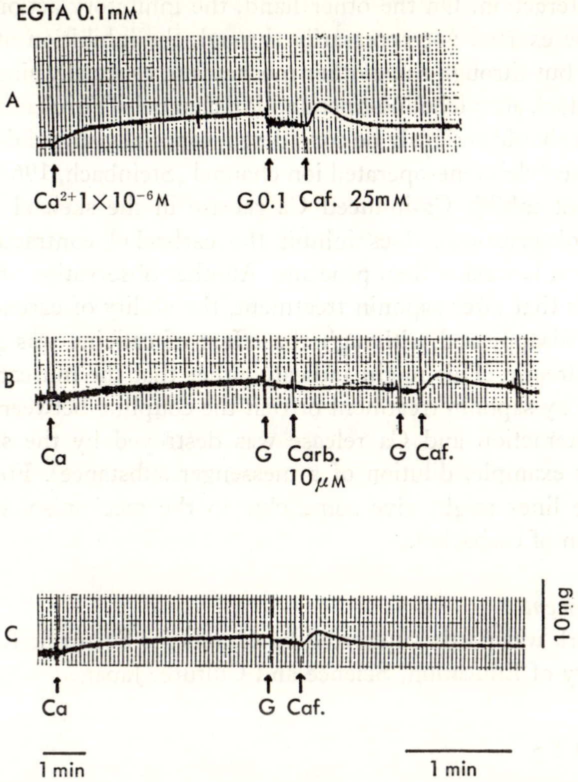

Fig. 9. The absence of carbachol responses in saponin-treated smooth muscle. The preparation was checked to give normal carbachol responses, and then treated with 50 μg/ml saponin for 30 min. A and C show that caffeine responses were preserved in this preparation as that in Fig. 5C. B, however, shows that the carbachol response could not be evoked any more and that Ca remained in the SR after the carbachol treatment to give a normal caffeine response.

strongly inhibited carbachol-induced Ca release. Therefore, it is possible that carbachol, for example, increases the Ca ion concentration in the vicinity of the SR and then the Ca-induced Ca release mechanism comes into operation. Although carbachol is known to open ion channels that pass Ca ions, since Ca ions are almost absent in the external EGTA-containing solution, no Ca influx is expected during the action of carbachol. However, it is conceivable that some Ca ions bound to, for example, the inner surface of the plasma membrane, are released as a result of the agonist-receptor interaction. On the other hand, the inhibitory action of procaine may well be exerted not necessarily through the inhibition of Ca-induced Ca release, but through some other mechanism, especially since procaine is known to block acetylcholine-operated ion channels (Furukawa, 1957; Ruff, 1977). This should seriously be considered, especially since lidocaine, which blocks the acetylcholine-operated ion channel (Steinbach, 1968; Ruff, 1977) but does not inhibit Ca-induced Ca release in the skeletal SR (our unpublished observation), does inhibit the carbachol contracture although the inhibition is weaker than procaine. Another observation of ours, shown in Fig. 9, is that after saponin treatment, the ability of carbachol to cause Ca release disappeared, although the effect of caffeine was preserved, as described already. This might indicate either that the muscarinic receptor was altered by saponin treatment or that the coupling between the agonist-receptor interaction and Ca release was destroyed by the skinning procedure (for example, dilution of a messenger substance). Further studies along these lines might give some clue to the mechanism of this interesting action of carbachol.

Acknowledgment

This work was partly supported by Grants for Scientific Research from the Ministry of Education, Science and Culture, Japan.

REFERENCES

Caputo, C. (1966). *J. Gen. Physiol.* **50**, 129–139.

Casteels, R. and Raeymaekers, L. (1979). *J. Physiol.* **294**, 51–68.

Chapman, R. A. (1979). *Prog. Biophys. Mol. Biol.* **35**, 1–52.

Chapman, R. A. and Léoty, C. (1976). *J. Physiol.* **256**, 287–314.

Chapman, R. A. and Miller, D. J. (1974). *J. Physiol.* **242**, 589–613.

Ebashi, S. and Endo, M. (1968). *Prog. Biophys. Mol. Biol.* **18**, 123–183.

Endo, M. (1975a). *Proc. Japan Acad.* **51**, 467–472.

Endo, M. (1975b). *Proc. Japan Acad.* **51**, 479–484.

Endo, M. (1977). *Physiol. Rev.* **57**, 71–108.

Endo, M. and Kitazawa, T. (1978). In *Biophysical Aspects of Cardiac Muscle* (Morad, M., ed.), pp. 307–327, Academic Press, New York.

Endo, M., Kitazawa, T., Yagi, S., Iino, M., and Kakuta, Y. (1977). In *Excitation-Contraction Coupling in Smooth Muscle* (Casteels, R., Godfraind, T., and Rüegg, J. C., eds.), pp. 199–209, Elsevier/North-Holland, Amsterdam.

Endo, M., Tanaka, M., and Ogawa, Y. (1970). *Nature* **228**, 34–36.

Fabiato, A. and Fabiato, F. (1972). *Circ. Res.* **31**, 293–307.

Fabiato, A. and Fabiato, F. (1977). *Circ. Res.* **40**, 119–129.

Furukawa, T. (1957). *Japan. J. Physiol.* **7**, 199–212.

Glitsch, H. G. and Pott, L. (1975). *Pflügers Arch.* **358**, 11–25.
Kimoto, Y. (1972). *Japan. J. Physiol.* **22**, 225–238.
Kitazawa, T. (1977). *J. Mol. Cell. Cardiol.* **9**, Suppl. 48.
Ohashi, H., Takewaki, T., and Okada, T. (1974). *Japan. J. Pharmacol.* **24**, 601–611.
Ruff, R. L. (1977). *J. Physiol.* **264**, 89–124.
Somlyo, A. P., Devine, C. E., Somlyo, A. V., and North, S. R. (1971). *J. Cell Biol.* **51**, 722–741.
Steinbach, A. B. (1968). *J. Gen. Physiol.* **52**, 144–161.
Suzuki, K. (1962). *Japan. J. Physiol.* **12**, 186–199.
Yagi, S. and Endo, M. (1980). *Folia Pharmacol. Japon.* **76**, 43 p.

*Muscle Contraction: Its Regulatory Mechanisms (Ebashi, S. et al., eds.), pp. 465–473,
Japan Sci. Soc. Press, Tokyo / Springer-Verlag, Berlin (1980)*

36

Contractile Activation and Calcium Movement in Mammalian Cardiac Muscles

TOSHIO KITAZAWA

Summary: The problem of whether or not the alterations in twitch tension of mammalian cardiac muscle under varied conditions are due to the changes in the amount of Ca in the intracellular Ca store has been re-examined using very thin bundles dissected from guinea-pig papillary muscle.

Half-time of twitch responses of the thin bundles after a sudden decrease in extracellular Ca concentrations was about 1–2 sec, which lies within the range of theoretical values calculated by assuming simple diffusion models. As expected from simple diffusion models, the same half-time was obtained for twitch decrease and increase when Ca concentration was decreased or increased; no half-time dependence on the magnitude of steps in Ca concentration or on the stimulus frequencies was observed. Half-time for the Ca effect was the same order of magnitude as that for the effects of transition metal ions, K or Na ions. The effects of altered rate of stimulus on twitch contraction could develop latently at almost the same rate even under conditions in which the magnitude of the Ca inward current was very small and Ca accumulation in the sarcoplasmic reticulum (SR) could not be expected. The time course of development of adrenaline effects was also the same under the conditions in which the amount of Ca influx was negligible and Ca accumulation in the intracellular stores would not occur. In a Na-free solution 30 mM of caffeine produced maximum tension response in the absence of external Ca ion regardless of the conditions in which the bundle had just been equilibrated. This suggests that the SR even under normal conditions has an amount of Ca which is enough to produce the maximum contraction of cardiac muscle.

It is well known that contraction of cardiac muscle is regulated by intracellular Ca ions as in skeletal muscle (Allen and Blinks, 1978). Unlike the skeletal muscle, however, twitch responses of cardiac muscle are greatly altered when extracellular Ca ion concentrations, stimulus frequencies or other conditions are changed. The mechanism of these alterations in mammalian cardiac muscle is generally assumed to be due to changes in the amount of Ca in the intracellular Ca store, probably the sarcoplasmic reticulum (SR), which results in more or less proportional changes in the amount of Ca released by each action potential (Morad and Goldman, 1973; Fozzard, 1977; Katz, 1977). This generally believed assumption has been re-examined in this study by the use of thin bundles of cardiac muscle. In these thin bundles, unlike ordinary thick cardiac preparations, diffusion through extracellular space was rapid enough to be distinguished from supposedly slow processes of accumulation or depletion of Ca in the SR.

Small bundles of fibers 75–120 μm in diameter and 1–1.5 mm in length were dissected from papillary muscles of guinea-pig right ventricles. Isometric tension was measured with a strain-gauge transducer at 25°C. Composition of the normal external solution was 150 mM NaCl, 4 mM KCl, 2 mM Ca-methanesulfonate, 2 mM Mg-methanesulfonate, 5.6 mM glucose, 5 mM N-2-hydroxyethyl-piperazine-N′-2-ethane-sulfonic acid (pH 7.40).

INOTROPISM INDUCED BY CHANGE IN $[\mathrm{Ca}]_0$

Figure 1 compares the time course of changes in the isometric twitch tension of whole papillary muscle and that of small bundles, produced by a sudden alteration in extracellular Ca concentration, $[\mathrm{Ca}]_0$. As shown in the figure, the half-time, $t^1/_2$, of the effect of $[\mathrm{Ca}]_0$ decrease was more than 30 sec in the whole muscle, while it was only a few seconds in the small bundle. The latter $t^1/_2$ appears too fast to be the time for accumulation or depletion of Ca in the SR induced by the changes in $[\mathrm{Ca}]_0$. Instead,

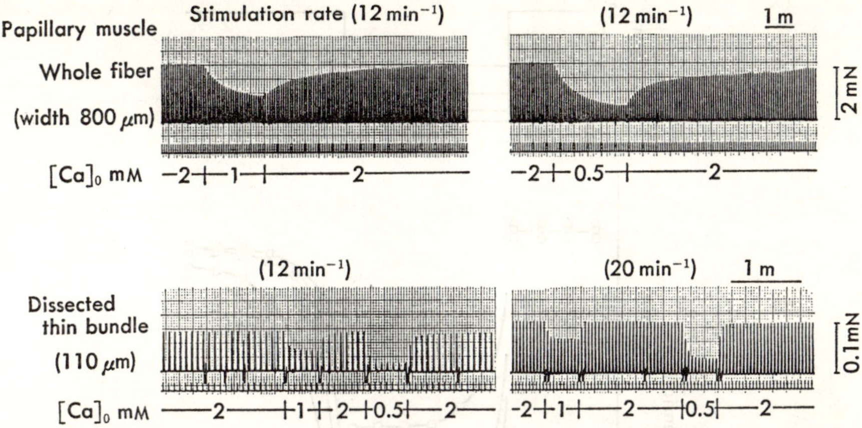

Fig. 1. Comparison between the time courses of the change in isometric twitch tension of whole papillary muscle and that of a small dissected bundle. Solutions were exchanged very rapidly at the points indicated in the figure, at a flow speed of about 5 ml/sec through the trough with the volume of about 0.9 ml. In the experiments using the whole muscle, continuous flow was done besides the points for exchange of the solution. The fibers were stretched to their optimal length in the whole muscle and to the average sarcomere length of 2.4 μm at rest in the small bundle. They were directly stimulated through external platinum electrodes. Temperature 25°C.

it most probably indicates the changes in $[Ca]_0$ at the cardiac cell surface as a result of the simple diffusion of Ca ions through the extracellular space. The following lines of evidence support this idea. 1) As in Fig. 1, most of the good preparations give $t\,^1/_2$ of about 1–2 sec. These values lie within the range of theoretical values that were calculated by assuming reasonable models and simple diffusion of Ca ions through the extracellular space between cardiac cells. 2) Since the relation between $[Ca]_0$ and steady twitch tension is almost linear under the conditions of Fig. 1, it is expected from the simple diffusion theory that $t\,^1/_2$ with a sudden decrease in $[Ca]_0$ should be the same as that with a sudden increase in $[Ca]_0$, which was in fact the case (see the case of 1 mM $[Ca]_0$ in Fig. 2). 3) Similarly, as expected from the simple diffusion theory, almost the same $t\,^1/_2$ was always obtained for twitch decline when $[Ca]_0$ was changed from 2 mM to various lower concentrations (Fig. 2).

When the normal $[Ca]_0$ solution was reintroduced after the bundles were soaked in solutions of $[Ca]_0$ lower than 0.5 mM for a certain period of time, twitch tension during the recovery phase in the absence of Mg ion temporarily became much larger than the steady tension in normal solu-

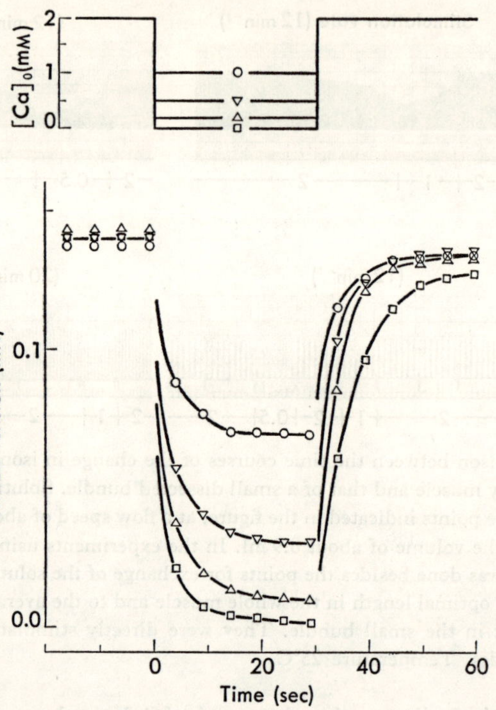

Fig. 2. Changes in isometric twitch tension produced by changes in $[Ca]_0$. The duration in lower $[Ca]_0$ solution was 30 sec; ○ 1 mM; ▽ 0.5 mM; △ 0.2 mM; □ no added Ca. Stimulus rate 12 min^{-1}. The abscissa shows the time after $[Ca]_0$ decrease.

tion. Sometimes contractures even occurred and twitches disappeared. This potentiation after a treatment with low $[Ca]_0$ solution was usually not observed in the presence of 2 mM $[Mg]_0$, while the time course of twitch decline associated with decreasing $[Ca]_0$ was not influenced by Mg ions. As a result, the recovery time course in the presence of 2 mM $[Mg]_0$, when $[Ca]_0$ was changed from a level lower than 0.5 mM to 2 mM, was slower than that in its absence, and furthermore a slower component was observed clearly (see the case of 0 mM $[Ca]_0$ in Fig. 2). The half-time of the fast component of twitch recovery in Mg-containing media was independent of $[Ca]_0$ immediately before reintroduction of 2 mM $[Ca]_0$. 4) If the time course of twitch responses is solely determined by diffusion, it should not be altered by the stimulus rate. This was also the case with stimulus frequencies of 6 to 30 min^{-1}, if the nonlinearity between $[Ca]_0$ and twitch tension at high stimulus frequencies was taken into account.

5) Again, from the simple diffusion theory, $t^1/_2$ for the Ca effect should be the same order of magnitude as that for effects of other kinds of ions that are considered to exert their actions from outside the cardiac cells. Indeed, an addition of 1 mM Mn or other transition metal ions to the normal solution depressed the twitch tension with a similar time course to that associated with twitch decline produced by a decrease in $[Ca]_0$, although, unlike $[Ca]_0$ decrease, the rapid twitch decline induced by transition metals was followed by a slower decrease in twitch tension. Magnesium ions had the same effect, but of much smaller extent. Similarly, when $[K]_0$ was changed from 4 mM to 12 or 16 mM, the twitch tension was reduced, probably because of rapid inactivation of pNa and pCa due to depolarization, at almost the same rate as that in the tension change produced by changes in $[Ca]_0$. In addition to the rapid decline phase induced by an increase in $[K]_0$, a slower phase was observed like that in the addition of transition metals. When Na concentration of the external solution was suddenly changed, for example increased from 75 to 112.5 mM $[Na]_0$, the twitch contraction was reduced also, at the same rate as that produced by the change in $[Ca]_0$. The steady level of twitch tension was linearly related to the ratio of $[Ca]_0/[Na]_0^2$.

STAIRCASE PHENOMENA

As is well known, a sudden increase in the rate of stimulus causes ventricular twitch tension to increase gradually until a new plateau is reached (Koch-Weser and Blinks, 1963). Voltage-clamp experiments revealed that slow inward current was unchanged during the gradual increase in twitch tension (Beeler and Reuter, 1970). Therefore, it has been assumed that the positive staircase is caused by an increased filling of the intracellular Ca store(s), probably as a result of an increase in the amount of Ca influx per min.

However, the following results did not support the assumption. For example, on increasing the rate of stimulus from 12 to 60 min^{-1}, the twitch tension increased initially rapidly and then slowly to reach a new steady state about 3 min later (Fig. 3). When $[Ca]_0$ was decreased to 0.2 mM just before the rate of stimulus was increased from 12 to 60 min^{-1}, the twitch tension was very small in spite of the increased rate of stimulus, and hence no Ca accumulation could be expected. After 3 min incubation in the low $[Ca]_0$ solution, $[Ca]_0$ returned to 2 mM. The time course of twitch recovery in this case was exactly as rapid as that following sudden $[Ca]_0$ change

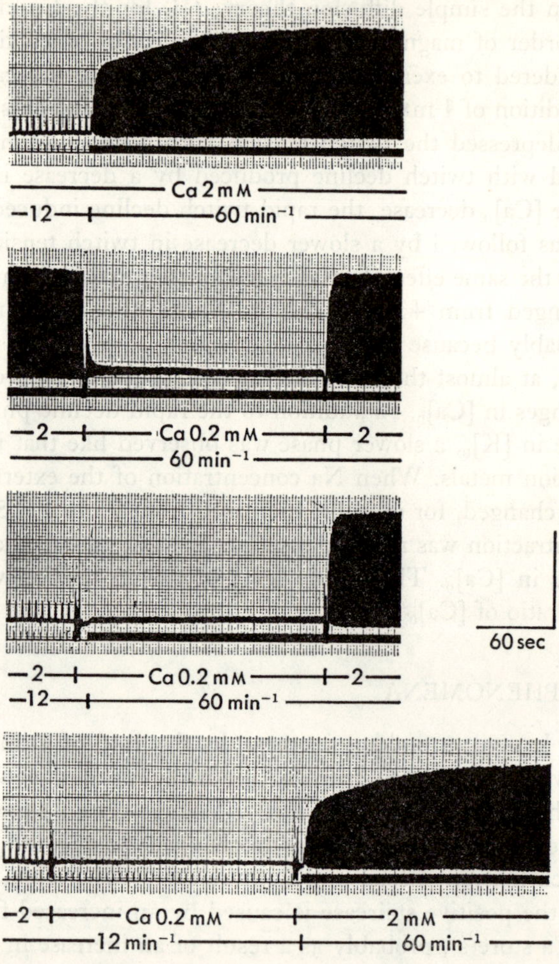

Fig. 3. Ascending staircase induced by acceleration of the stimulus rate from 12 to 60 min^{-1} and the effect of changes in $[Ca]_0$ on it. $[Ca]_0$ and stimulus rate as indicated. Bottom trace is the signal of the stimulus.

during the steady state at the higher rate of stimulus (Fig. 3). Qualitatively similar results were obtained, instead of 0.2 mM $[Ca]_0$, in the presence of 1 mM EGTA with no added Ca, which lowered the free Ca ion concentration to less than 10^{-7} M.

Conversely, a decrease in the rate of stimulus caused the potentiated twitch tension to decrease gradually until a new lower level of twitch height was reached. When $[Ca]_0$ was increased from 2 to 5 mM simultaneously with a reduction in the stimulus rate from 30 to 5 min^{-1}, the cardiac muscle developed large twitch tensions in spite of the much lower rate of stimulus. After 5 min of incubation in the high $[Ca]_0$ solution, $[Ca]_0$ was reduced to normal concentration and twitch tension decreased suddenly to the normal steady level at the lower stimulus rate, without any gradual slow reduction phase expected on the basis of changes in the filling of intracellular Ca stores.

From these results it is certain that the effects of an altered rate of stimulus on twitch contraction could develop at almost the same rate independent of the magnitude of Ca inward current.

POSITIVE INOTROPISMS CAUSED BY CATECHOLAMINES

Catecholamines such as adrenaline, noradrenaline, and isoprenaline have a potentiating effect on twitch tension in cardiac muscle. As shown in Fig. 4 (the case of 2 mM $[Ca]_0$), the inotropism due to adrenaline at 25°C usually consists of at least 3 phases—rapid positive, intermediate negative and slow positive which have hardly been observed in whole papillary muscle. Niedergerke and Page (1977) showed that in small frog heart atrial trabeculae the initial rapid increase was mediated by adrenergic α-receptors. However, in guinea-pig papillary muscle all three components were similarly developed by noradrenaline and by isoprenaline, completely inhibited by a β-blocker, propranolol, and not in the least affected by an α-blocker, phentolamine. These suggest that all three phases of adrenaline effect in the present experiments were mediated by adrenergic β-receptors.

It has been shown by voltage-clamp experiments of mammalian cardiac muscle that the electrical effects of adrenaline—that is, an increase in Ca inward currents—precedes an increase in contractile tension (Reuter, 1974). Therefore, it has been assumed that the increased Ca inward current as a result of adrenaline effect was to increase the amount of Ca stored in SR gradually, which resulted in an increase in amounts of the Ca ions released from the SR, causing the tension to rise. This possibility, however, seems again to be excluded by the following experiments with thin cardiac bundles.

As in Fig. 4, 3 μM adrenaline was suddenly applied and simultaneously $[Ca]_0$ was reduced from 2 to 0.2 mM or to zero. After the certain duration in low $[Ca]_0$ solution, $[Ca]_0$ was returned to the normal concentration in the presence of adrenaline. The strength of the first twitch elicited 3–4

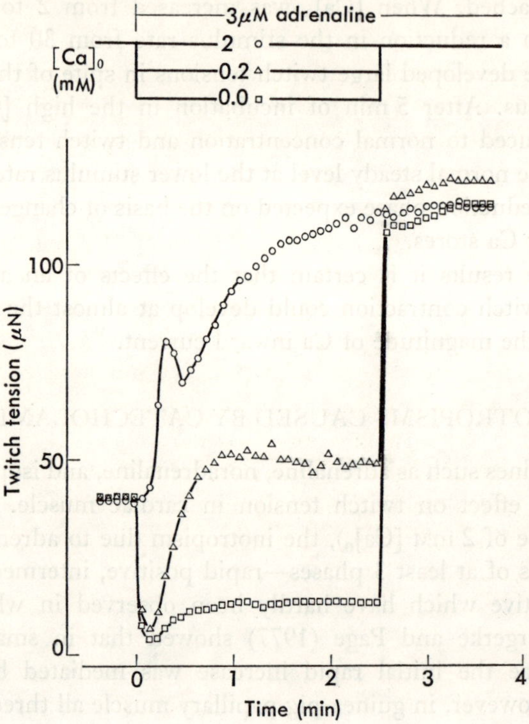

Fig. 4. Time course of development of adrenaline effects in conditions of varying $[Ca]_0$. Changes in $[Ca]_0$ from 2 to 0.2 mM or to zero were simultaneously made at the time of addition of 3 μM adrenaline. Stimulus rate 12 min^{-1}

sec after restoration of the original $[Ca]_0$ was already almost the same as that at the equivalent moment in the control where $[Ca]_0$ was kept at 2 mM throughout (Fig. 4). Similar results were obtained even in the presence of 1 mM EGTA with no added Ca. Therefore, these results suggest that the progress of the effect of adrenaline is not dependent on the amount of Ca inflow and its accumulation in the intracellular stores even in the guinea-pig heart, as had already been shown in the frog heart (Niedergerke and Page, 1977).

PRESENCE OF SUFFICIENT CALCIUM IN SARCOPLASMIC RETICULUM

In a Na-free solution a high concentration of caffeine, such as 30 mM, which is known to release most of the Ca stored in the SR in cardiac skinned fibers, produced a large contracture of the intact cardiac muscle even in the absence of external Ca ion (Endo and Kitazawa, 1978). The peak tension obtained during caffeine contracture was always much larger than the twitch and was the maximum tension that the cardiac bundle can develop. All the fibers in the cardiac bundle were then skinned by treating with 50 μg/ml saponin for 30 min (Endo and Kitazawa, 1978). The maximum tension developed by this skinned fiber by 10^{-4} M free Ca ions in the presence of 4 mM Mg-ATP was only slightly larger than that of caffeine contracture, before saponin treatment. This slight difference may be caused by incomplete synchronization of caffeine contracture due to the spontaneous rapid relaxation following peak tension. The magnitude of caffeine contracture was independent of the conditions in which the bundle had just been equilibrated: in other words, the same caffeine response was obtained at the peak of positive or negative inotropisms. These results seem to indicate that the SR, even under normal conditions, contains enough Ca to produce the maximum contraction of cardiac muscle.

Acknowledgment

This study was supported in part by research grants awarded to Professor M. Endo by the Ministry of Education, Science and Culture of Japan.

REFERENCES

Allen, D. G. and Blinks, J. R. (1978). *Nature* **273**, 509–513.

Beeler Jr., G. W. and Reuter, H. (1970). *J. Physiol.* **207**, 211–229.

Endo, M. and Kitazawa, T. (1978). In *Biophysical Aspects of Cardiac Muscle* (Morad, M., ed.), pp. 307–327, Academic Press, New York.

Fozzard, H. A. (1977). *Annu. Rev. Physiol.* **39**, 201–220.

Katz, A. M. (1977). In *Physiology of the Heart*, pp. 137–195, Raven Press, New York.

Koch-Weser, J. and Blinks, J. R. (1963). *Pharmacol. Rev.* **15**, 601–652.

Morad, M. and Goldman, Y. (1973). *Prog. Biophys. Mol. Biol.* **27**, 259–316.

Niedergerke, R. and Page, S. (1977). *Proc. Roy. Soc. Lond. B.* **197**, 333–362.

Reuter, H. (1974). *J. Physiol.* **242**, 429–451.

Muscle Contraction: Its Regulatory Mechanisms (Ebashi, S. et al., eds.), pp. 475–481,
Japan Sci. Soc. Press, Tokyo / Springer-Verlag, Berlin (1980)

37

Sarcomere Disorder in Skinned Fibers during Contraction

YOSHIKI UMAZUME AND
TOSHITADA YOSHIOKA

Summary: It is well known that the intensity drop of optical diffraction lines of muscle fibers occurs during isometric contraction. As a possible cause of this intensity drop, the occurrence of small random movements of A and/or I band positions has been proposed.

In order to *visualize* the disorder of the sarcomere structure, the dispersions of Z-to-Z and M-to-M distances in skinned fibers at rest, during contraction and in rigor have been examined by means of electron microscopy combined with optical diffractometry. Disordering not only in interfibrillar but also in intrafibrillar Z and M line registers was observed on electron micrographs taken from sections of the contracting fiber. The excess dispersions of both Z-to-Z and M-to-M distances compared with those at rest were estimated to be 0.2 μm (during contraction) or 0.1 μm (in rigor), both of which are comparable to the previous suggestion from optical diffractometry.

Although preliminary, the present method seems to provide useful information about dynamic/static disorder in the sarcomere structure at various conditions.

During the plateau of a complete tetanus of a single skeletal muscle fiber, the intensity of the first-order laser diffraction line drops significantly (Kawai and Kuntz, 1973). Fujime (1975) observed the intensity drops of all observable lines during electrophoretical injection of Ca ions into skinned fibers. He suggested that the cause of this intensity drop was small random fluctuations in A and/or I band (in other words, M and/or Z line) positions along the fiber axis. We tried in this study to visualize the occurrence of dynamic and/or static disordering of the sarcomere structure in various conditions by means of electron microscopy. The dispersions of Z-to-Z and M-to-M distances in Natori's skinned fibers at rest, during contraction and in rigor will be discussed.

MATERIALS AND METHODS

Semitendinosus frog (*Rana catesbeiana*) muscle was used. Skinned fibers were prepared in liquid paraffin oil (Natori, 1954). The skinned fiber segment (5–10 mm long) was mounted horizontally in a bath between a force transducer and a fixed hook. To remove oil from the surface of the fiber segment, a relaxing solution containing 0.5% Triton X-100 was flushed briefly.

The following solutions were used: rigor solution; 100 mM KCl, 10 mM morpholinopropanesulfonic acid (MOPS) at pH 7.0, 4 mM EGTA, and 5 mM $MgCl_2$. Contracting solution; composition of rigor solution *plus* 4 mM $CaCl_2$ and 5 mM ATP. Relaxing solution; composition of rigor solution *plus* 5 mM ATP.

He-Ne laser diffraction patterns were projected on a white screen (Umazume and Fujime, 1975) and photographed by a camera with a telescope to minimize parallax.

For fixation, the bathing solution was drained from the bath and the same solution (but containing 1% glutaraldehyde) was poured, keeping the level of the solution constant, and left standing for 30 min. Little or no differences were observed in diffraction patterns (both in line spacing and intensity) and tension during fixation, suggesting that no serious artifacts were introduced. Then the central portion of the fiber segment, where the diffraction patterns had been recorded, was embedded in an Epon mixture and sectioned. A JEM 100C electron microscope was used at 80 kV and at 5,000×.

We randomly picked up more than 100 sarcomeres on the enlarged photographs for each condition and made histograms of neighboring

Z-to-Z and M-to-M distances. Accuracy of measurements was ±0.25 mm on photographs, which corresponds to ±0.015 μm in the myofibrillar scale. The observed width or dispersion σ_{obs} of the histogram will be given by

$$\sigma^2_{obs} = \sigma^2_{dyn} + \sigma^2_{stat} \tag{1}$$

where σ_{stat} and σ_{dyn} are the dispersions at rest (static) and during contraction (dynamic).

RESULTS AND DISCUSSION

Sarcomere lengths measured in this experiment from diffraction angles and electron micrographs are summarized in Table I.

Figure 1a shows an example of the histograms of Z-to-Z and M-to-M distances in preparations at rest. The dispersions at both distances are considered to come partly from intrinsic dispersions in the intact fiber and partly from artifacts introduced during mechanical skinning. The dispersions at rest are the controls.

Next, we tried to examine these dispersions in skinned fibers during contraction. In spite of many trials, the diffraction lines of fully activated skinned fibers at room temperature disappeared. This is probably due to a loss of parallel elastic components of skinned fibers. Then we tried a

TABLE I
Sarcomere Lengths Measured from Diffraction Angles and Electron Micrographs
L_d, $L_{d,f}$, sarcomere lengths measured from peak-to-peak diffraction angles of the first-order lines before and after fixation, respectively. Mean±SD are given. SD≡σ_{obs} in Eq. (1).

Condition	Fiber reference number	L_d (μm)	$L_{b,f}$ (μm)	Z-to-Z (μm)	M-to-M (μm)
Relax	8311	2.30	2.29	2.31±0.05	2.30±0.05
Relax	9012	2.27	2.27	2.36±0.07	2.36±0.07
Relax	8312	3.02	3.02	2.82±0.05	2.82±0.04
Relax	9014	3.65	3.65	3.60±0.22	3.57±0.25
Relax	9041	4.41	4.50	4.30±0.21	—[a]
Relax	9011	4.45	4.54	4.82±0.29	4.84±0.24
Contract	2222	2.77	2.77	2.80±0.20	2.78±0.18
Rigor	9042	2.48	—[b]	2.50±0.07	2.49±0.06
Rigor	9043	2.88	2.79	2.68±0.11	2.66±0.11

[a] Unmeasurable due to vague contrast of M lines.
[b] Observation missing.

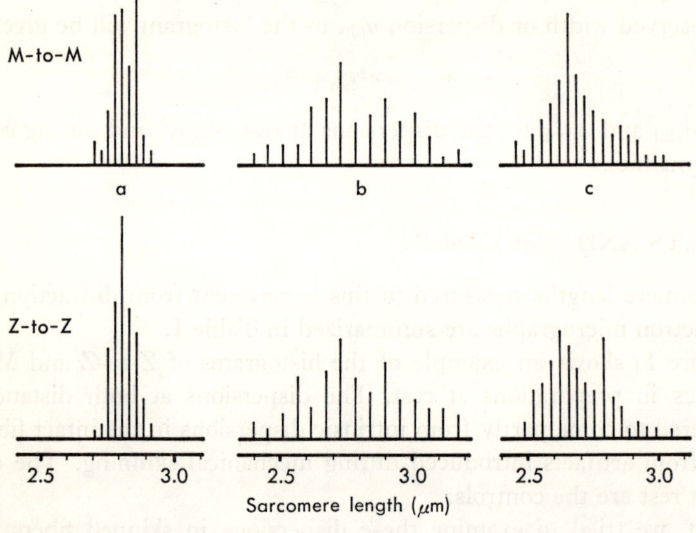

Fig. 1. Some examples of histograms of M-to-M and Z-to-Z distances. a, at rest; b, during contraction; c, in rigor.

slow activation technique (Iwazumi and Pollack, 1978). Whenever we got a fully activated preparation showing clear diffraction lines with lower intensity, the diffraction pattern completely returned in the relaxing solution to that before activation. (We actually had several such preparations in *winter*.) So one of these was fixed by glutaraldehyde. (Later we tried the slow activation again, but no good preparation has so far been obtained in *summer*.) Figure 2 shows the densitometer traces of the first order lines before contraction, during maximum contraction and after fixation. During contraction, the intensity of the diffraction lines decreased, keeping the asymmetry of intensity in right and left lines. This fact held also for the case of rigor preparations mentioned below. This asymmetry of intensity may come from the populations of myofibrils with different tilt angles (Rüdel and Zite-Ferenczy, 1979). However, the intensity decrease during contraction is not due mainly to the change in the populations, so far as our present cases are concerned. Figure 3 shows the electron micrographs of the above preparation. Disordering not only in the interfibrillar but also in the intrafibrillar Z and M line registers can be seen. Figure 1b shows the histograms of the Z-to-Z and M-to-M distances for the above

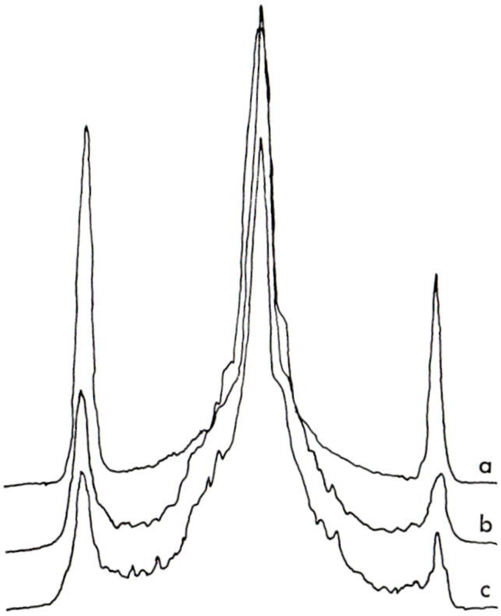

Fig. 2. Densitometer traces of laser diffraction patterns. a, at rest; b, during contraction; c, after fixation.

Fig. 3. An example of electron micrographs from the specimen fixed during contraction. Bar: 2 μm.

preparation. Both dispersions during contraction are much greater than those at rest. This result does not necessarily contradict Fujime's suggestion, because the present result was taken from fully activated skinned fiber, and the sarcomere length (L) was not sufficiently close to 3.0 μm. In fully activated muscle, all the scatterers are considered to move randomly, as dynamic studies have suggested (Bonner and Carlson, 1975; Fujime and Umazume, 1978). It is noted that the calculated dispersion σ_{dyn} from Eq. (1), *i.e.*, 0.19 μm or 7% of the sarcomere length, is very close to that previously suggested (Fujime, 1975).

It is well known that, even in the absence of free Ca ions, the skinned fiber does contract at around 10^{-5} M Mg-ATP due to the release of substrate inhibition. When the bathing solution is changed from the relaxing solution to the rigor one, Mg-ATP in the myofibrillar space will diffuse out. During this process, the skinned fiber will transiently contract and go into the rigor state. Thus, the contracting state is expected to be frozen by fixation for electron microscopy. On the basis of this idea we used preparations in rigor as a model of those during contraction. Figure 1c shows an example of the histograms of the Z-to-Z and M-to-M distances in a preparation in rigor. Both dispersions are definitely larger than those at rest, but smaller than those during contraction. We must add here the following: the difference in the dispersions (or SD's in Table I) in two preparations in rigor is considered to come from the fact that the skinned fiber went into the rigor state slowly in the preparation giving larger dispersions, but quickly in the other preparation. We think that the former preparation is closer to the preparation during contraction. The dispersion σ_{dyn} is about 0.1 μm or 4% of the sarcomere length.

The intensities I_1 and I_2 of the first- and second-order diffraction lines, respectively, have relations $I_1 \gtrsim I_2$ for $L \lesssim 6$ μm and $I_1 \simeq 0$ at $L \simeq 8$ μm (Umazume and Fujime, 1975). To explain these relations in terms of diffraction theory (Fujime and Yoshino, 1978), large *static* disordering of the sarcomere structure must be assumed (Fujime, private communication). The large dispersions of the Z-to-Z and M-to-M distances in extremely stretched fibers were actually observed (see Table I).

The fixation procedure with 1% glutaraldehyde presented here seems to freeze not only static but also dynamic disordering faithfully. Although this study is a preliminary one, we hope that our method will provide useful information on the sarcomere structure in various conditions.

Acknowledgment

We would like to thank Professor Reiji Natori for his warmful guidance since our student days.

REFERENCES

Bonner, R. F. and Carlson, F. D. (1975). *J. Gen. Physiol.* **65**, 555–581.
Fujime, S. (1975). *Biochim. Biophys. Acta* **379**, 227–238.
Fujime, S. and Umazume, Y. (1978). Abstr. 6th Int. Biophys. Congr., Kyoto, VIII-7-(J).
Fujime, S. and Yoshino, S. (1978). *Biophys. Chem.* **8**, 305–315.
Iwazumi, T. and Pollack, G. H. (1978). Abstr. 6th Int. Biophys. Congr., Kyoto, VIII-8-(J).
Kawai, M. and Kuntz, I. D. (1973). *Biophys. J.* **13**, 857–876.
Natori, R. (1954). *Jikeikai Med. J.* **1**, 18–28.
Rüdel, R. and Zite-Ferenczy, F. (1979). *J. Physiol.* **290**, 317–330.
Umazume, Y. and Fujime, S. (1975). *Biophys. J.* **15**, 163–180.

CYTOSKELETON

Muscle Contraction: Its Regulatory Mechanisms (Ebashi, S. et al., eds.), pp. 485–496, apan Sci. Soc. Press, Tokyo / Springer-Verlag, Berlin (1980)

38

Elastic Structure of Connectin in Muscle

KOSCAK MARUYAMA

Summary: The structure and function of connectin, elastic protein in muscle, are summarized. Connectin extracted with 1% SDS from salt- and water-insoluble stroma consisted of a high molecular weight (HMW) component and a 43,000-dalton component. Both components had similar amino acid compositions. Immunofluorescent technique using antiserum against HMW connectin revealed that connectin is located mainly in A-I junctions, and slightly in the I band. In myosin-removed myofibrils, fluorescent antiserum stained A-I junction regions and I bands. It appears that connectin plays a role in keeping the A band at the center of a sarcomere. Judging from the similarity of the localization in myofibrils and also of the amino acid composition, it is suggested that "titin" isolated from an SDS extract of myofibrils (Wang *et al.*, 1979) is identical with HMW component of connectin. The elastic net structure of connectin is discussed in relation to Porter's microtravecular lattice. The transmission of tension produced in myofibrils to tendon is summarized on the basis of information available.

I

Early in 1975, Teru Hayashi and I were working together at the Department of Biophysics, Kyoto University. The research project was to test whether Oplatka's exciting findings were reproducible or not. According to Oplatka (Borejido *et al.*, 1974), myosin-extracted "ghost myofibrils" could be contracted on adding heavy meromyosin in a Mg-ATP solution. I could not confirm this remarkable result, although a slow and slight shortening was observed.*

When I saw under a microscope a ghost myofibril from which myosin and actin had been removed, the remaining Z lines did not fall into parts; as water flows, lines of Z lines moving just like the neighboring Z lines were connected by some transparent material.** There must have been something between the Z lines! Therefore, I treated the ghost myofibril with various reagents. In dilute alkali, Z lines swelled and became unrecognizable. However, on neutralization, a very thin fibrous material reappeared. Then, a number of attempts were made to obtain "ghost material" from muscle stroma.

II

Over the past 20 years an annual meeting on biological movements has been continuously held in Japan under the leadership of Professors S. Ebashi, Y. Tonomura, F. Oosawa, and N. Kamiya. At each meeting Professor Reiji Natori always delivered a speech on his recent work using Natori's skinned fibers. Often he emphasized the existence of an "elastic internal membrane" which he first postulated 25 years ago (Natori, 1954). The elastic membrane could be digested away by collagenase treatment.***

When I went to Tokyo early in August 1975, I called up Professor Natori and told him about my observations. He urged me to come to his laboratory to do some experiments with skinned fibers.

There was a conversation between us before the work.

---

* Later, Shimizu's group confirmed Oplatka's finding (Mitsuku *et al.*, 1979). After hearing the exact conditions for the experiment directly from Dr. Oplatka at Kyoto in 1978, I repeated the experiments but without success.

** Later, it was pointed out by Sir Andrew Huxley that H. E. Huxley and the late Jean Hanson (1954) already described this fact 25 years ago.

*** It turned out that commercial samples of collagenase were contaminated with other proteolytic enzymes (*cf.* Maruyama *et al.*, 1977a).

Natori: "In summer time, frog muscle is not in good condition, and therefore I have never worked in summer. But this is a special case. I will do anything for you."

Maruyama: "If possible, I'd like to see whether an alikali-treated skinned fiber is still elastic or not."

Natori: "It's easy to do."

Professor Natori showed me that a ghost fiber, treated with 0.1 N NaOH and neutralized, behaved like an elastic string (Maruyama *et al.*, 1976a).

The next time I visited him, he was busy as chairman of the faculty of the Jikei Medical School, because President I. Higuchi suddenly died. Professor Natori was soon elected as his successor. It was really impressive that President Natori continued our joint work while he was frequently interrupted by telephone calls and by visitors.

III

Dr. Saburo Matsubara, then in Professor Natori's laboratory, did further work in close contact with myself. He performed a number of physiological studies along the lines President Natori and I had started (Matsubara and Maruyama, 1977).

When skinned fiber of frog skeletal muscle was stretched beyond the resting sarcomere length, tension developed. On removal of myosin and actin, there was a drop in tension development at relatively shorter sarcomere length, but practically no change was observed at extreme stretch. On further treatment with dilute alkali, ghost fibers still developed tension on extreme stretch, although considerably lowered. Matsubara measured the temperature dependence of passive tension development of ghost muscle fibers treated with 1 N NaOH and 2% SDS, and concluded that the ghost fibers had a rubber-like property.

After glycerinated muscle fibers were treated with 1 N NaOH and 2% SDS at room temperature, the ghost fibers which still generated tension on stretch were washed with water and subjected to amino acid analysis. The results showed that the remaining protein had an amino acid composition very similar to that of isolated connectin (Maruyama *et al.*, 1977a). Judging from the small amount of hydroxyproline, it was suggested that contamination with collagen was at most only 5%. This result clearly indicated that the protein material responsible for passive tension generation was largely connectin, and not collagen, an extracellular connective tissue protein, covering the cell membrane of the muscle fiber.

IV

The KI-insoluble material from muscle could be largely solubilized in 1% SDS at pH 7.0, when heated for a few minutes at 100°C. The still insoluble residue was almost all insoluble collagen. However, unless mechanically dispersed, clumps of connectin cannot easily be dissolved in hot SDS. This was the case with "SDS-insoluble" connectin (*cf.* Maruyama *et al.*, 1977a).

The SDS-solubilized connectin shows two main bands on SDS gel electrophoresis: an almost immovable one and the 43,000-dalton component. The two components can be separated by a gel filtration using Bio-gel A 50 m chromatography.

Amino acid compositions of the two components are very similar except for the glycine content (Table I). It should be mentioned that the amino acid compositions are similar to those of actin and eu-actinin (Kuroda and Masaki, 1979). They differ in methionine content from actin and in

TABLE I
Amino Acid Composition of Connectin and Titin

	Titin	Connectin		
		A	B	C
Asp	95	92	93	96
Thr	75	64	66	62
Ser	69	70	67	63
Glu	116	116	118	130
Pro	74	60	67	63
Gly	71	110	76	90
Ala	62	80	75	81
Cys/2	11	7	6	2
Val	85	69	78	63
Met	10	15	16	24
Ile	59	57	56	54
Leu	67	76	76	74
Tyr	30	29	30	30
Phe	27	31	29	29
Lys	82	56	79	66
His	15	22	18	17
Arg	55	46	50	56

A, 43,000-dalton component of connectin; B, high molecular component of connectin; C, connectin (high molecular component) (Data from King and Kurth, 1979).

alanine content from eu-actinin. It is possible that the 43,000-dalton component is the monomer of high molecular weight (HMW) connectin. It should be noted that this 43,000-dalton component resisted extraction with 3 M KI for a few weeks at 5°C. In other words, the SDS extract showed the same pattern as the usual one on SDS gel electrophoresis.

The connectin could be only partially solubilized by 6 M urea or by 6 M guanidine-HCl. However, in the presence of 2-mercaptoethanol (0.1 M), it became completely soluble by 6 M guanidine-HCl, as was first pointed out by King and Kurth (1979). After dialysis against 6 M urea, connectin was still soluble in 6 M urea, although 6 M urea in the presence of 0.1 M 2-mercaptoethanol poorly solubilized KI-extracted muscle residues.

King and Kurth (1979) were able to separate HMW connectin using DEAE-Sepharose CL-6B chromatography. The amino acid composition was quite similar to that of HMW connectin prepared in our laboratory (Table I). King's connectin showed doublet bands in SDS gel electrophores is, and King suggested that the molecular weight was approximately 700,000. Judging from a similar mobility of tetramer of myosin heavy chains, we have estimated the molecular weight to be about 800,000. King and Kurth (1980) observed the doublet bands of connectin in an SDS extract of intact myofibrils.

Quite recently, Wang *et al.* (1979) isolated a HMW (1,000,000 daltons) material from an SDS extract of total myofibrils of rabbit or chicken skeletal mucle. They named the high molecular protein "titin." We have repeated Wang's isolation procedure and obtained titin by gel filtration. The amino acid composition of titin we determined is also listed in Table I. The amino acid composition of titin is almost identical with that of HMW connectin. In the preparation procedures, titin was prepared directly from total myofibrils, and connectin was isolated from insoluble muscle stroma. Therefore, it is possible that both preparations contained some contaminants.

It was observed that titin bands easily disappeared by a very mild treatment with trypsin. Connectin is also very sensitive to trypsin in intact myofibrils (Maruyama *et al.*, 1977a; Maruyama and Yamamoto, 1979). It is to be noted that titin band was present in SDS-solubilized skinned fibers of frog skeletal muscle and disappeared after a mild trypsin treatment (Natori, Umazume, Kimura, and Maruyama, unpublished).

Thus it is very likely that connectin (HMW component) is the same as titin, as already suggested by the work of King and Kurth (1979).

V

We thought that the HMW connectin could not be a single peptide. Although we do not deny the role of SS bonds in the formation of very insoluble material, the molecular weight of connectin is still as high as one million even in the presence of excess SH reagents (King and Kurth, 1979; Wang *et al.*, 1979).

As is well known in collagen and elastin, intermolecular cross-linking of a covalent nature may result in the formation of polymers of an apparently huge molecular weight (*cf.* Tanzer, 1973; Bailey *et al.*, 1974).

Therefore, we made efforts in isolating such lysine derivatives from hydrolyzates of connectin. We isolated a candidate in a pure form. However, it turned out to be lysinoalanine which was artificially formed from lysine and cysteine residues of connectin during alkaline treatment. Next, ^{3}H-labelled borohydride was incorporated into the protein to reduce dihydro-form of lysine-derivatives as in collagen (Fujii *et al.*, 1978; Kimura *et al.*, 1979). After hydrolysis, the lysinonorleucine peak was identified as compared with that from human fetal skin collagen.

Unfortunately, we have not been successful in isolating lysinonorleucine from hydrolyzates of connectin, very probably due to its smaller content. It is still possible that there are other cross-linking amino acids such as glutamyl lysine.

VI

In 1955, Hanson and Huxley described the possibility of the existence of an "S-filament" connecting actin filaments from the both sides of the Z lines in a sarcomere, to explain the fact that "ghost" myofibrils maintain mechanical continuity. Later, Sjöstrand (1962) claimed that very thin filaments connect the edges of myosin filaments to Z lines and called these "gap filaments." G. Hoyle's group extended Sjöstrand's work and described "T filaments," about 2.5 nm in diameter, connecting neighboring Z lines (Hoyle, 1967; Hoyle *et al.*, 1968). Guba *et al.* (1968) also emphasized the role of such thin filaments and assumed that actin-like protein, "fibrillen," constitutes the thin filaments.

More recently, Locker and his associates (1975, 1976), revived the concept of "gap filaments" in order to explain the structural changes in extremely stretched muscle fibers. Dos Remedios and Gilmour (1978) showed the presence of very thin filaments extending from a Z line to an adjacent Z line in a sarcomere, after extraction of myosin and actin. This was confirmed by Takahashi (1979), who showed that such thin filaments disappeared during *post mortem* changes.

When frog skeletal muscle was extracted with Hasselbach-Schneider solution followed by 1 N alkali treatment, the Z lines of the myofibrils deteriorated, but there were nets of very thin filaments between the adjacent Z lines (Maruyama *et al.*, 1977a). This was more clearly observed in KI-, SDS-treated frog heart myofibrils (Toyoda and Maruyama, 1978). The nets consisted of very thin filaments, approximately 2 nm in diameter. The nets were seen in thin sections cut in any direction relative to the longitudinal axis of a myofibril, suggesting that the nets are three-dimensional and present within a whole myofibril. This was rather striking to me, because I had thought that the connectin nets covered a myofibril like a sock. We showed later that filaments thinner than actin filaments could be noticed in intact myofibrils (Maruyama *et al.*, 1980). This had already been shown by Hoyle (1967) and also by Franzini-Armstrong (1970). In the A band region, it is rather difficult to show the thinner filaments because of the presence of cross-bridges.

It is thought that the nets of thinner filaments are comprised of connectin: purified connectin was shown to be entanglements of very thin filaments, 2 nm in diameter (Maruyama *et al.*, 1977, 1980). In order to see such a structure, it is necessary to disperse water-washed connectin clumps by prolonged treatment with 1% Triton-X 100 (Maruyama *et al.*, 1980).

VII

We have repeatedly obtained antisera against HMW connectin: connectin isolated from chicken skeletal muscle was injected to rabbits. Unfortunately, we can't do a conventional immunodiffusion test, since connectin is not soluble in the usual salt solutions. The fluorescent dye-labelled antiserum stained a clump of connectin. The antiserum did not react with myosin and actin.

The fluorescent antiserum strongly stained A bands and I bands were faintly stained.* In the A band, the A-I junction was most fluorescent. When myosin was extracted away by 0.6 M KCl containing 1 mM ATP,

* We reported that anti-connectin also stained Z band (Maruyama *et al.*, 1980). However, later, the anti-connectin preparations did not stain Z lines. It is thought that previous connectin preparations were contaminated with some Z line proteins.

the A-I junction region was still stained and the I band was weakly stained. The gap between the A-I junctions in a sarcomere was very faintly fluorescent.

At present, it is rather difficult to interpret these localizations. It might be the case that connectin is abundant around the A-I junction; it may serve to maintain the position of the A band at the center of a sarcomere. If so, this is in good agreement with Locker's view (*cf.* Locker and Fleet, 1976; Locker and Daines, 1980).

Recently, Wang *et al.* (1979) have shown the localization of "titin" in a sarcomere, using an immunofluorescence technique. According to them, the localization of titin is similar to that of connectin, except that titin is also located at the Z lines and the M line. The latter might be due to a contamination of anti-M protein (Masaki and Takaiti, 1969), which often contaminates any antiserum against myofibrillar proteins.

VIII

What is the nature of connectin? We do not yet know all the details.

First of all, its apparent molecular weight is so huge—800,000—that it is extremely difficult to characterize connectin physicochemically. Second, connectin has not yet been obtained in nondenaturating agents. Connectin can be solubilized either in 1% SDS or in 6 M guanidine chloride (subsequently in 6 M urea), and it can be isolated by gel filtration. It is certain that such isolated connectin is already denatured.

In this connection, it is worth mentioning that connectin nets in intact myofibrils are very sensitive to tryptic action, but they become insensitive to it as "purification" proceeds. Even in the state of KI-extracted residues, they are barely digested by trypsin at neural pH or pepsin at pH 2. Connectin filaments may be tightly tangled so that proteolytic enzymes cannot attack.

As for the net structure of connectin, we have some evidence for the structure (section VII). In this connection, the "microtrabecular lattice" structure in cells recently described by Wolosewick and Porter (1979) is very interesting. According to Porter, most of the cytoplasmic structures are linked together through the filamentous lattice in a three-dimensional way. The lattice consists of very thin filaments, approximately 3 nm in diameter. Is the connectin net a specialized form of Porter's microtrabecular lattice? This is a subject for further study.

IX

Connectin-like proteins have been obtained from the cortical layers of sea-urchin eggs (Maruyama *et al.*, 1977b), erythrocyte cell membranes (Maruyama *et al.*, 1977b), leucocytes (Tatumi and Maruyama, unpublished data), and *Physarum* plasmodium (Ozaki and Maruyama, 1980) (Table II).

The observations so far have strongly suggested that connectin-like protein is concentrated at the cytoplasmic surface of cell membranes (Maruyama *et al.*, 1977b). It has also been shown that anti-connectin stains isolated sarcolemma from rat skeletal muscle (Maruyama *et al.*, 1977a). If so, connectin (like protein) forms nets at the cytoplasmic surface of the cell membrane which play an important role in the change as well as the

TABLE II
Amino Acid Compositions of Connectin-like Proteins from Various Sources

	Rabbit skeletal muscle[a]	Leucocytes[b]	Erythrocytes[a]	Sea urchin egg[c]	Plasmodium[d]	Carp muscle[e]
Asp	95	90	71	101	100	94
Thr	61	53	55	57	64	58
Ser	61	66	60	67	61	51
Glu	130	98	107	104	116	125
Pro	63	50	66	57	62	51
Gly	89	104	80	83	70	70
Ala	82	72	85	80	84	78
Cys/2	2	15	2	2	4	10
Val	62	66	77	69	60	70
Met	24	20	27	27	18	27
Ile	53	48	54	56	49	60
Leu	73	97	129	89	88	80
Tyr	30	33	26	33	29	33
Phe	29	43	52	48	42	31
Lys	66	41	40	60	74	78
His	16	25	19	22	24	20
Arg	52	78	50	52	49	51
						Try (12)

[a] Maruyama *et al.* (1977b).
[b] Maruyama and Tatsumi, unpublished data.
[c] Maruyama *et al.* (1976b).
[d] Ozaki and Maruyama (1980).
[e] Kimura *et al.* (1978).

maintenance of cell shape. Here the connectin nets may be identical to, or closely related to, the microtrabecular lattice.

X

Finally, I should like to briefly mention how tension developed in myofibrils of vertebrate skeletal muscle by the sliding mechanism is transmitted to tendons so as to move bones.

At the myotendinous junction, thin filaments identified as actin (Maruyama and Shimada, 1978) attach to the thickened cell membrane. There are very thin "connecting filaments" between muscle cell membrane and basement membrane to which collagen fiber binds (Nakao, 1976). Tension developed in the sarcomere must be transmitted in the following way: sarcomere → actin filament → thickened cytolplasmic surface of cell membrane → connecting filament → basement membrane → collagen fiber → bone.

What is the role of connectin in the process of tension transmission of vertebrate skeletal muscle? One possibility is that dense connectin nets are present in the cytoplasmic surface of the muscle cell membrane to which actin filaments attach. The chemical nature of Nakao's connecting filaments is entirely obscure. It is of some interest that glycoproteins (reticulin) of the basement membrane* have amino acid compositions similar to that of connectin (Pras and Glynn, 1973; *cf.* Maruyama, 1976). In this connection, it should be mentioned that the amino acid composition of connectin resembles that of fibronectin, a HMW cell surface protein, which is involved in the attachment of collagen fibers to cell membranes (*cf.* Ruoslahti *et al.*, 1979).

The connectin nets in myofibrils are not likely to be directly involved in physiological tension transmission. In an extreme case, they may be involved: a glycerinated myofibril contracts into a small mass on addition of Mg-ATP, but after a mild trypsin treatment, it falls apart into several pieces with Mg-ATP, where most of the connectin was digested away (Maruyama and Yamamoto, 1979).

Under the physiological conditions of contraction and relaxation, the connectin nets may be in a slackened state. Physiologically, the connectin nets are safety-guarded to keep the mechanical continuity of each sarcomere in a whole myofibril. Here it is noteworthy that 10 nm filaments consisting

* It consists of collagen (Type IV)and glycoproteins in about equal amounts.

of desmin (skeletin) exist around the periphery of each Z line and connect neighboring myofibrils at the Z lines (Lazarides *et al.*, 1979).

Acknowledgments

I am most indebted to President Reiji Natori of Jikei University for his warm encouragements and for his continued cooperation during the present work. I take this opportunity to acknowledge Emeritus Professor Hiroshi Kumagai for his kind help given to me during my difficult days.

I am very much grateful for the helpful assistance of my present collaborators: Ms. S. Kimura, Dr. K. Ohashi, Mr. K. Ozaki, Mr. M. Tsuneoka, Mr. M. Kimura, and Mr. Y. Kuwano.

This work was supported by grants from the Ministry of Education, Culture and Science, the Ministry of Health and Welfare, the Yamada Science Foundation, and the Muscular Dystrophy Association.

REFERENCES

Bailey, A. J., Robin, S. P., and Balian, G. (1974). *Nature* **251**, 105–109.

Borejdo, J., Gadasi, H., and Oplatka, A. (1974). *FEBS Lett.* **45**, 55–58.

Dos Remedios, C. and Gilmour, D. (1978). *J. Biochem.* **84**, 235–238.

Franzini-Armstrong, C. (1970). *Tissue Cell* **2**, 327–338.

Fujii, K., Kimura, S., and Maruyama, K. (1978). *Biochem. Biophys. Res. Commun.* **81**, 1248–1253.

Guba, F., Harsänyl, V., and Vajda, E. (1968). *Acta Biochim. Biophys.* **3**, 353–365.

Hanson, J. and Huxley, H. E. (1955). *Symp. Soc. Exp. Biol.* **9**, 228–264.

Hoyle, G. (1967). *Am. Zool.* **7**, 435–449.

Hoyle, G., McNeil, A., Walcott, B., and Selverston, A. (1968). *Symp. Biol. Hung.* **8**, 34–41.

Kimura, S., Akaboshi, Y., and Kubota, M. (1978). *J. Biochem.* **83**, 321–323.

Kimura, S., Fujii, K., Kubota, M., and Maruyama, K. (1979). *Bull. Japan. Soc. Sci. Fish.* **45**, 241–243.

King, N. L. and Kurth, L. (1980). In *Fibrous Proteins* (Parry, D.A.D., ed.), vol. II, pp. 57–66, Academic Press, New York.

Kuroda, M. and Masaki, T. (1979). Abstr. 6th Int. Biophys. Congr., Kyoto.

Lazarides, E., Hubbard, B. D., and Granger, B. L. (1979). In *Cell Motility: Molecules and Organization* (Hatano, S., Ishikawa, H., and Sato, H., eds.), pp. 521–543, Univ. Tokyo Press, Tokyo.

Locker, R. H. and Leet, N. G. (1975). *J. Ultrastruct. Res.* **52**, 64–75.

Locker, R. H. and Leet, N. G. (1976). *J. Ultrastruct. Res.* **56**, 31–38.

Locker, R. H. and Daines, C. J. (1980). In *Fibrous Proteins* (Parry, D.A.D., ed.), vol. II, pp. 43–56, Academic Press, New York.

Maruyama, K. (1976). *J. Biochem.* **80**, 405–407.

Maruyama, K., Natori, R., and Nonomura, Y. (1976a). *Nature* **262**, 58–59.

Maruyama, K., Mabuchi, I., Ohashi, K., and Matsubara, S. (1976b). *Biochim. Biophys. Acta* **446**, 321–324.

Maruyama, K., Matsubara, S., Natori, R., Nonomura, Y., Kimura, S., Ohashi, K., Murakami, F., Handa, S., and Eguchi, G. (1977a). *J. Biochem.* **82**, 317–337.

Maruyama, K., Murakami, F., and Ohashi, K. (1977b). *J. Biochem.* **82**, 339–345.

Maruyama, K., Kimura, S., Kuroda, M., and Handa, S. (1977c). *J. Biochem.* **82**, 347–351.

Maruyama, K. and Shimada, Y. (1978). *Tissue Cell* **10**, 741–748.

Maruyama, K. and Yamamoto, K. (1979). In *Cross-bridge Mechanism in Muscle Contraction* (Sugi, H. and Pollack, G. H., eds.), pp. 319–328, Univ. Tokyo Press, Tokyo.

Maruyama, K., Kimura, S., Toyota, N., and Ohashi, K. (1980). In *Fibrous Proteins* (Parry, D.A.D., ed.), vol. II, pp. 33–41, Academic Press, New York.

Masaki, T. and Takaiti, O. (1974). *J. Biochem.* **75**, 367–380.

Matsubara, S. and Maruyama, K. (1977). *Japan. J. Physiol.* **27**, 589–600.

Mitsuku, M., Yamada, T., and Shimizu, H. (1979). *J. Biochem.* **85**, 559–565.

Nakao, T. (1976). *Cell Tissue Res.* **166**, 241–254.

Natori, R. (1954). *Jikeikai Med. J.* **1**, 18–28.

Ozaki, K. and Maruyama, K. (1980). *J. Biochem.* in press.

Pras, M. and Glynn, L. E. (1973). *Br. J. Exp. Path.* **54**, 449–456.

Ruoslahti, E., Hayman, E. G., Kuusela, P., Shively, J. E., and Engvall, E. (1979). *J. Biol. Chem.* **254**, 6054–6059.

Sjöstrand, F. S. (1962). *J. Ultrastruct. Res.* **7**, 225–246.

Takahashi, K. and Saito, K. (1979). *J. Biochem.* **85**, 1539–1542.

Tanzer, M. L. (1973). *Science* **180**, 501–566.

Toyoda, N. and Maruyama, K. (1978). *J. Biochem.* **84**, 239–241.

Wang, K., McCure, J., and Tu, A. (1979). *Proc. Natl. Acad. Sci. U.S.* **76**, 3698–3702.

Wolosewick, J. J. and Porter, K. R. (1979). *J. Cell Biol.* **82**, 114–139.

Muscle Contraction: Its Regulatory Mechanisms (Ebashi, S. *et al.*, eds.), pp. 497–505,
Japan Sci. Soc. Press, Tokyo / Springer-Verlag, Berlin (1980)

39

A New Structural Protein Located in the Z Line of Chicken Skeletal Muscle

KAZUYO OHASHI AND KOSCAK MARUYAMA

Summary: Myofibrils of chicken skeletal muscle, from which myosin, actin and other proteins had been removed, was extracted with 0.6 M KI solution for more than 24 hr. From the extract, a structural protein located in the Z line was purified by the $(NH_4)_2SO_4$ precipitation and by repeated DEAE-cellulose column chromatography. This protein showed a chain weight of about 55,000 daltons, estimated from the SDS polyacrylamide gel electrophoretic mobility. By the indirect immunofluorescent technique, the antiserum against the 55,000-dalton protein exclusively stained the Z lines of a chicken myofibril. This 55,000-dalton protein differed from α-actinin, eu-actinin and desmin in amino acid composition, and was distinguished from α-actinin and desmin immunologicaly. This 55,000-dalton protein formed lattice structures *in vitro* which were similar to those of the Z lines *in situ*, in addition to membrane-like structures. The conditions under which these lattice structures were formed, have not yet been specified. It is not yet clear whether the 55,000-dalton protein can bind with actin and α-actinin or not.

Electronmicroscopic images of the Z lines of vertebrate skeletal muscle have been frequently reported, and it is now well known that the vertebrate Z line have a lattice structure. Several modeles of the Z line have been proposed (Knappeis and Carlsen, 1962; Kelly, 1967; Landon, 1970; Franzini-Armstrong, 1973; Ullrick *et al.*, 1977).

On the other hand, several articles on the protein composition and the protein components of the vertebrate Z line have been published (Saide *et al.*, 1974; Muguruma *et al.*, 1978; Suzuki *et al.*, 1978). α-Actinin, a well-established Z line component, is supposed not to form the lattice structure of Z line (Masaki *et al.*, 1967, 1969; Stromer *et al.*, 1972). The role of eu-actinin, which is also localized at the Z line, is not clear yet (Sugita *et al.*, 1974; Kuroda *et al.*, 1978). Desmin (skeletin), the 10-nm filament protein, has been shown to be localized around the peripheral region of the vertebrate Z line and appears to link neighboring Z lines (Small and Sobieszek, 1978; Lazarides and Hubbard, 1976; Lazarides and Granger, 1978). However, there is no information available about the component that constitutes the lattice structure of vertebrate Z lines.

In this article, we report a new protein component of the Z line of

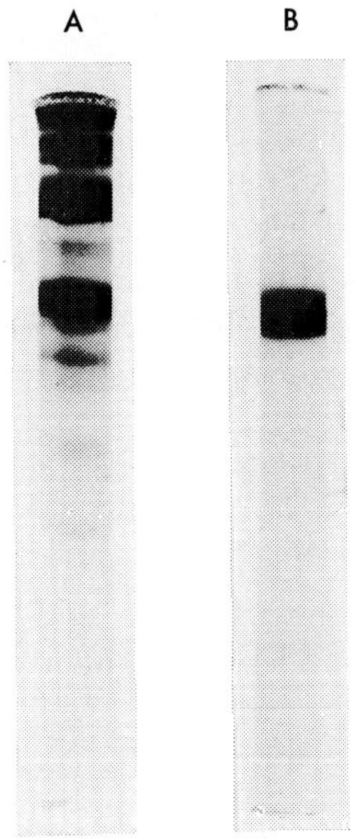

Fig. 1. SDS polyacrylamide gel electrophoretic patterns of the 55,000-dalton protein. A: crude preparation before column chromatography. B: column purified 55,000-dalton protein.

chicken skeletal muscle which can form the lattice structure *in vitro* (Ohashi and Maruyama, 1979).

PURIFICATION OF PROTEIN

Myofibrils of chicken skeletal muscle were extracted with 0.6 M KI solution, after they had been treated with Hasselbach-Schneider's solution. In this extract, several minor protein components of the myofibril such as α-actinin (Ebashi and Ebashi, 1965) and I-protein (Ohashi *et al.*, 1977; Maruyama *et al.*, 1977) were found, in addition to actin, tropomyosin and troponin. α-Actinin was partialy extracted with this solution, although the Z lines are relatively preserved during the extraction with 0.6 M KI solution at the level of light microscopy. The residue of this extraction was further extracted with 0.6 M KI solution for more than 24 hr at 4°C. After being dialyzed against 1 mM $NaHCO_3$, the extract was precipitated

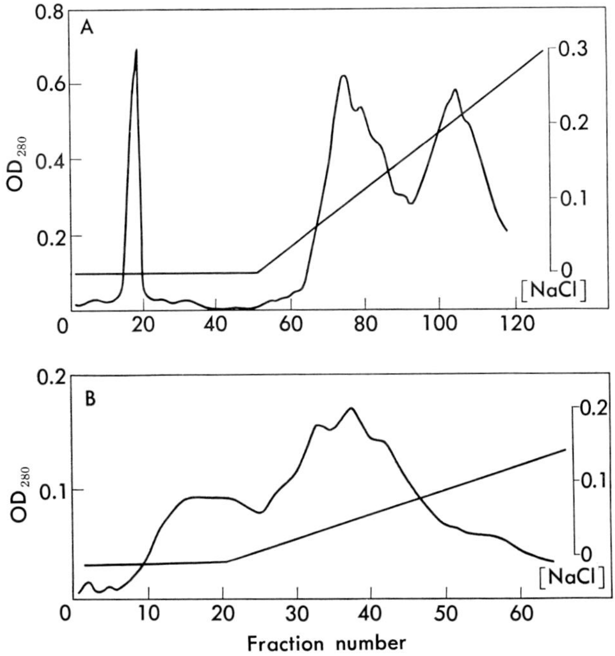

Fig. 2. Elution profils of DEAE-cellulose column chromatography (see text). A: first chromatography of crude the 55,000-dalton protein. B: rechromatography of the partialy purified 55,000-dalton protein by first chromatography.

with $(NH_4)_2SO_4$ between 20 and 40% saturation. This precipitate (Fig. 1A) was loaded onto a DEAE-cellulose column after being dialyzed against 20 mM Tris-HCl buffer (pH 7.5). Elution was carried out with a linear concentration gradient of NaCl, from 0 to 0.3 M. Fractions eluted between 0.05 and 0.1 M NaCl were collected and rechromatographed on the same column under the concentration gradient of NaCl, from 0 to 0.15 M. The elution profiles are shown in Fig. 2. The first peak eluted around 0.03 M NaCl contained one protein whose chain weight was estimated to be about 55,000 daltons from the SDS polyacrylamide gel electrophoretic pattern (Fig. 1B). The yield was very low: about 1 mg from 1 kg muscle. This low yield was perhaps due to the partialy extraction of this protein with 0.6 M KI solution, and also due to precipitation during dialysis.

COMPARISON WITH KNOWN PROTEINS

Antiserum against the 55,000-dalton protein was obtained by injecting the protein into a rabbit. By the indirect immunofluorescent technique (Kawamura, 1969), this antiserum clearly stained the Z line of myofibrils

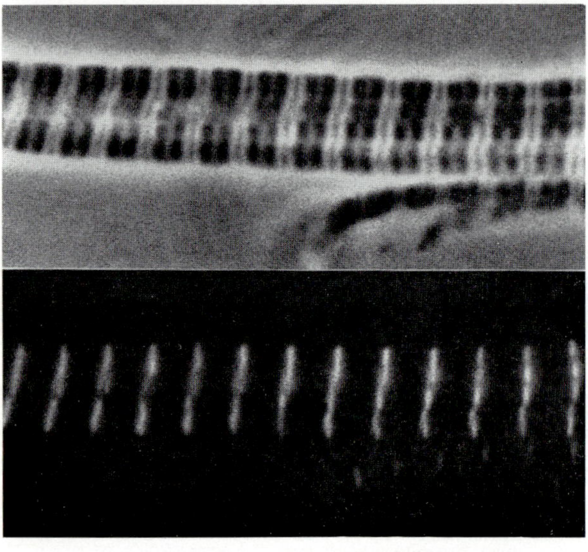

Fig. 3. Indirect immunofluorescent staining of chicken myofibrils. Upper: phase-contrast microscopic image. Lower: fluorescent microscopic image. Bar: 10 μ

TABLE I
Amino Acid Compositions of the 55,000-Dalton Protein and Desmin (Skeletin)

	55,000-Dalton protein	Skeletal desmin[a]	Gizzard skeletin[b]	Gizzard desmin[a]
Asp	102	97	86	88
Thr	51	56	57	57
Ser	57	71	60	74
Glu	116	176	199	197
Pro	58	36	18	23
Gly	63	64	40	60
Ala	49	92	93	86
Cys/2	5	3	3	5
Val	55	60	52	50
Met	19	18	26	25
Ile	64	41	44	42
Leu	117	94	108	99
Tyr	37	28	29	30
Phe	52	30	28	27
Lys	76	51	49	47
His	23	14	16	16
Arg	56	77	93	87

[a] J. M. O'Shea *et al.* (1979).
[b] J. V. Small and A. Sobieszek (1977).

of chicken skeletal muscle (Fig. 3). This fluorescence disappeared by absorbing the antiserum on the 55,000-dalton protein.

Desmin also has a molecular weight of 55,000 daltons and is localized at around the Z line (Lazarides and Granger, 1978). However, the 55,000-dalton protein distinctly differed from desmin in its solubility in salt solution. And, as is shown in Table I, amino acid composition of the 55,000-dalton protein largely differs from that of desmin in the contents of Glu, Pro, Ala, Phe, and Lys (Small and Sobieszek, 1977; O'Shea *et al.*, 1979). The 55,000-dalton protein has relatively large amounts of Asp, Glu, and Leu. The 55,000-dalton protein was also distinguished from desmin immunologicaly. Immunodiffusion was carried out on the 1% agarose gel containing 0.1% SDS and 2.0 mM 2-mercaptethanol. Antiserum against the 55,000-dalton protein formed a precipitine line with the 55,000-dalton protein in 1% SDS but did not react with desmin solubilized by 1% SDS solution (Fig. 4). α-Actinin and eu-actinin (chain weight 95,000 and 43,000, respectively) are different from the 55,000-dalton protein in chain weight

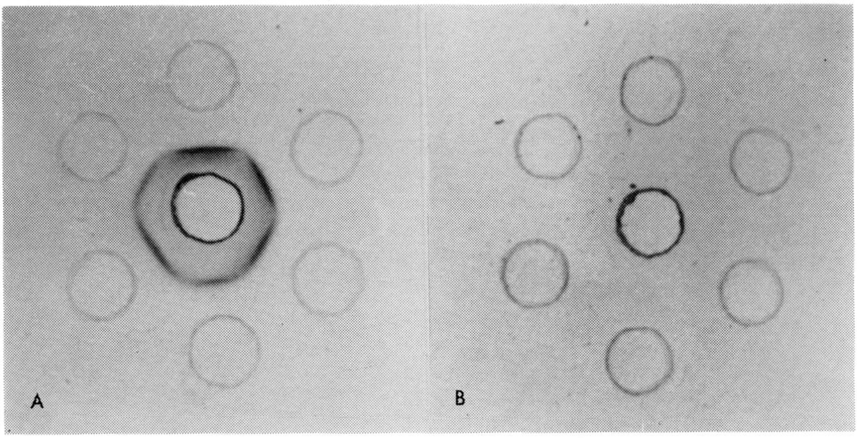

Fig. 4. Immunodiffusion of antiserum against the 55,000-dalton protein with the 55,000-dalton protein and desmin in the 1% agarose gel containing 0.1% SDS and 2 mM 2-mercaptoethanol. Central wells: antiserum. Surrounding wells: A: the 55,000-dalton protein at a concentration of 0.4 mg/ml in 1% SDS solution was diluted twice and placed clockwise from the top. B: desmin at a concentration of 0.6 mg/ml solubilized by 1% SDS solution was treated in the same way as in A.

and amino acid composition (Masaki *et al.*, 1967; Sugita *et al.*, 1974; Kuroda and Masaki, 1978). Furthermore, the antiserum against α-actinin did not react with the 55,000-dalton protein and the antiserum against 55,000-dalton protein did not react with α-actinin. So the 55,000-dalton protein appeared not to be a proteolytic product of α-actinin. This protein was also distinguished from all other previously characterized structural proteins of myofibrils.

LATTICE STRUCTURE

Fine precipitate was formed in the solution of the 55,000-dalton protein when the solution was stored for a few days at 4°C in 1 mM $NaHCO_3$. This precipitate was negatively stained with 1% uranyl acetate and observed under an electron microscope. Most of the precipitate formed a "membrane-like structure" (Fig. 5A) and amorphous aggregate. In some places, however, lattice structures of various sizes were seen which seemed to be made up of very thin filament. Typical structures are shown in Fig. 5B. At least two types of lattice structure were observed. The periodicity of the lattice was about 8 nm. It has been fairly well established that the

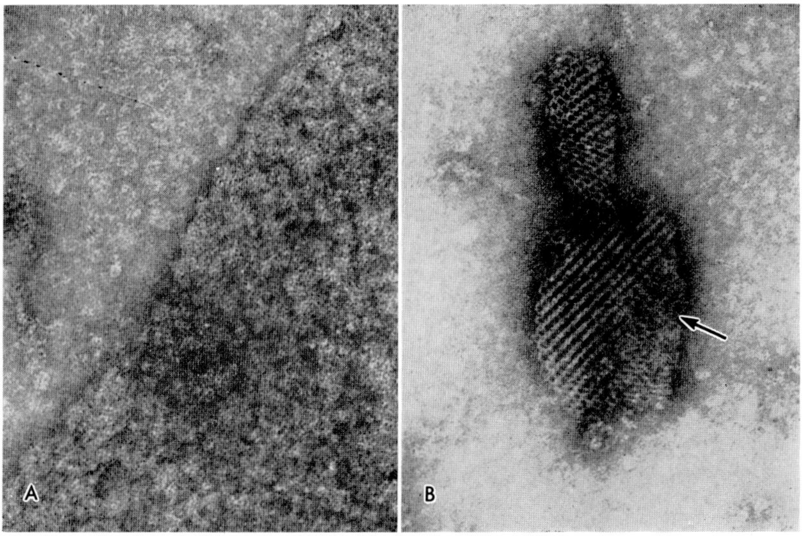

Fig. 5. An electron micrograph of a negatively stained aggregate of the 55,000-dalton protein. A: membrane-like structure. B: lattice structure. Magnification ×14,400.

vertebrate Z line presents three fundamental patterns: 1) a large-square lattice with a periodicity of about 22 nm, 2) a small-square lattice with a periodicity of about 11 nm, and 3) a basket-weave pattern with a spacing of about 22 nm (Ullrick *et al.*, 1977). The appearance of the lattice of the 55,000-dalton protein resembled large-square lattice of the vertebrate Z line. However, they differ in periodicity. The periodicity is rather similar to that of small-square lattice. From these similarities, it is expected that the 55,000-dalton protein forms the lattice structure of the Z line.

Optimal conditions under which the 55,000-dalton protein forms the lattice structure *in vitro* has not been specified yet. There is a possibility that the lattice structure is due to contamination of the 55,000-dalton protein preparation. However, as is shown in Fig. 5B (arrow), there is a "membrane-like structure" part in the lattice structure formed *in vitro.* Therefore, the 55,000-dalton protein is thought to form the lattice structure.

FURTHER PROBLEMS

As the 55,000-dalton protein is localized at the Z line, it is expected to bind with actin and α-actinin. However, binding experiments with the

55,000-dalton protein and F-actin in the presence of 0.1 M KCl, 10 mM Tris-HCl (pH 7.5) showed that the 55,000-dalton protein does not bind appreciably with F-actin. Using affinity chromatography of the α-actinin-immobilized Sepharose 4B column, the conditions under which the 55,000-dalton protein binds with α-actinin were not found. The purified 55,000-dalton protein may have lost its binding capacity with actin and α-actinin during the long extraction. Or, another protein may be needed in order for the 55,000-dalton protein to bind with actin and α-actinin. Saide and Ullrick (1974) found that in isolated Z-discs of insect muscle several proteins are present which, on the basis of the moelcular weight, are not tropomyosin, actin or α-actinin. But they did not report the existence of 55,000-dalton protein. Muguruma *et al.* (1978) reported several proteins solubilized by 1 mM $CaCl_2$ from the I-Z-I segments of rabbit skeletal muscle. Suzuki *et al.* (1978) showed that proteins, released from rabbit myofibrils by CAF, reconstituted the Z lines. In each case, the existence of the 55,000-dalton protein was unclear. Evidently, more information is needed to elucidate the protein composition of the vertebrate Z line.

Three remaining questions must be answered: 1) Under what conditions, does the 55,000-dalton protein form a lattice structure? 2) Does the 55,000-dalton protein really construct the lattice structure of the vertebrate Z line or not? 3) Does the 55,000-dalton protein can bind with actin, α-actinin, eu-actinin, and some other proteins?

Acknowledgment

This work was supported by grants from the Ministry of Education, Science and Culture (Japan), the Ministry of Health and Welfare (Japan), the Yamada Science Foundation and the Muscular Dystrophy Association.

REFERENCES

Ebashi, S. and Ebashi, F. (1965). *J. Biochem.* **58**, 7–12.

Franzini-Armstrong, C. (1973). *J. Cell Biol.* **58**, 630–642.

Kawamura, A. (1969). *Fluorescent Antibody Techniques and Their Applications*, Univ. Tokyo Press, Tokyo.

Kelly, D. E. (1967). *J. Cell Biol.* **34**, 837–840.

Knappeis, G. G. and Carlsen, F. (1962). *J. Cell Biol.* **13**, 323–335.

Kuroda, M. and Masaki, T. (1978). Abstr. 6th Int. Biophys. Congr., Kyoto, VII-10-(1).

Landon, D. N. (1970). *J. Cell Sci.* **6**, 257–276.

Lazarides, E. and Granger, B. L. (1978). *Proc. Natl. Acad. Sci. U.S.* **75**, 3683–3687.

Lazarides, E. and Hubbard, B. D. (1976). *Proc. Natl. Acad. Sci. U.S.* **73**, 4344–4348.

Maruyama, K., Ishii, T., Kimura, S., Kuroda, M., Ohashi, K., and Muramatsu, S. (1977). *J. Biochem.* **81**, 215–232.

Maruyama, K., Kunitomo, S., Kimura, S., and Ohashi, K. (1977). *J. Biochem.* **81**, 243–247.

Masaki, T., Endo, M., and Ebashi, S. (1967). *J. Biochem.* **62**, 630–632.

Masaki, T. and Takaiti, O. (1969). *J. Biochem.* **66**, 637–643.

Muguruma, M., Yamada, M., and Fukazawa, T. (1978). *Biochim. Biophys. Acta* **532**, 71–80.

Ohashi, K., Kimura, S., Deguchi, K., and Maruyama, K. (1977). *J. Biochem.* **81**, 233–236.

Ohashi, K. and Maruyama, K. (1979). *J. Biochem.* **85**, 1103–1105.

Ohashi, K., Masaki, T., and Maruyama, K. (1977). *J. Biochem.* **81**, 237–242.

O'Shea, J. M., Robson, R. M., Huiatt, T. W., Hartzer, M. K., and Stromer, M. H. (1979). *Biochem. Biophys. Res. Commun.* **89**, 972–980.

Saide, J. D. and Ullrick, W. C. (1974). *J. Mol. Biol.* **87**, 671–683.

Small, J. V. and Sobieszek, A. (1977). *J. Cell Sci.* **23**, 243–268.

Stromer, M. H. and Goll, D. E. (1972). *J. Mol. Biol.* **67**, 489–494.

Sugita, H., Masaki, T., and Ebashi, S. (1974). *J. Biochem.* **75**, 671–673.

Suzuki, A., Saito, M., Iwai, H., and Nonami, Y. (1978). *Agric. Biol. Chem.* **42**, 2117–2122.

Ullrick, W. C., Toselli, P. A., Saide, J. D., and Phear, W.P.C. (1977). *J. Mol. Biol.* **115**, 61–74.

Muscle Contraction: Its Regulatory Mechanisms (*Ebashi, S. et al., eds.*), *pp. 507–514*,
Japan Sci. Soc. Press, Tokyo / Springer-Verlag, Berlin (1980)

40

On the 42,000-Dalton Proteins in Vertebrate Skeletal Muscle: Actin and Eu-actinin

MASAAKI KURODA AND TOMOH MASAKI

Summary: Proteins were extracted from the myosin-removed myofibrils (called I-Z-I segments in this paper) of chicken skeletal muscle with 2 mM Tris-HCl, pH 6.5, 2 mM Tris-HCl, pH 8.0 and 0.6 M KI solution, successively. The 42,000-dalton components of the extracts were purified by the use of SDS- polyacrylamide gel electrophoresis of the preparative scale and the isoelectric points were compared using the isoelectric focusing technique. The presence of a 42,000-dalton component whose isoelectric point was slightly more acidic than that of skeletal muscle actin was detected from the pH 6.5-extract. This protein was designated eu-actinin.

Eu-actinin was purified from the pH 6.5-extract by column chromatography on Sepharose 4B and Sephadex G-200. The chain molecular weight of eu-actinin measured on SDS-PAGE was 42,000, identical to that of actin. The results of analytical chromatography on Sephadex G-200 indicated that eu-actinin exists as a dimer in a solvent of low ionic strength near neutral pH. The intrinsic viscosity of eu-actinin determined in the presence of 2 mM Tris-HCl, pH 8.0, was 6.4 ml/g. Antibody to eu-actinin was raised in a goat. Monospecific IgG was prepared by the use of affinity chromatography. Direct immunofluorescence microscopy of myofibrils showed that eu-actinin localized at the Z line.

Antiserum against chicken skeletal muscle actin was also raised in a goat and made monospecific by affinity chromatography. The anti-actin IgG formed precipitine lines with actin from chicken gizzard, chicken heart, and rabbit skeletal muscle as well as with actin from chicken skeletal muscle. The antibody did not interact with eu-actinin.

Thus, it is concluded that there exists a new protein, eu-actinin, whose molecular weight is identical to that of actin, as the Z-line constituent.

Although the molecular organization of actin filaments of skeletal muscle (I-filament) has become much clearer, the interaction of actin filaments with the Z line of myofibrils remains an unsolved problem. We examined the extractability of proteins from the myosin-removed residues of myofibrils in detail with special regard to the 42,000-dalton components. Isoelectric focusing studies of 42,000-dalton components purified from the I-Z-I segments* indicated that there exist two kinds of molecular species which differed in isoelectric points. The isoelectric point of one of the two species coincidated with that of skeletal muscle actin, *i.e.*, α-actin (Rubenstein and Spudich, 1977). The other species (designated eu-actinin) had a slightly more acidic isoelectric point. Antibody prepared against purified eu-actinin did not interact with actin. The results of direct immunofluorescence microscopy indicated that eu-actinin is the Z-line component.

Some physico-chemical properties and the interaction of eu-actinin with actin will be discussed in this report.

RESULTS AND DISCUSSION

G-F transformation is one of the most characteristic properties of the actin molecule (Oosawa and Higashi, 1967). When placed in a solvent of low ionic strength, actin filaments depolymerize into the monomer. However, depolymerizability of actin at low ionic strength depends greatly on pH. Measurement of flow-birefringence indicated that the rate of depolymerization of isolated thin filaments was very small at pH 6.5.

On the other hand, when I-Z-I segments were repeatedly extracted with 2 mM Tris-HCl at pH 6.5, a considerable amount of protein was

* It refers to the myosin-free residues of myofibrils prepared by the repeated treatment of myofibrils with Hasselbach-Schneider's solution.

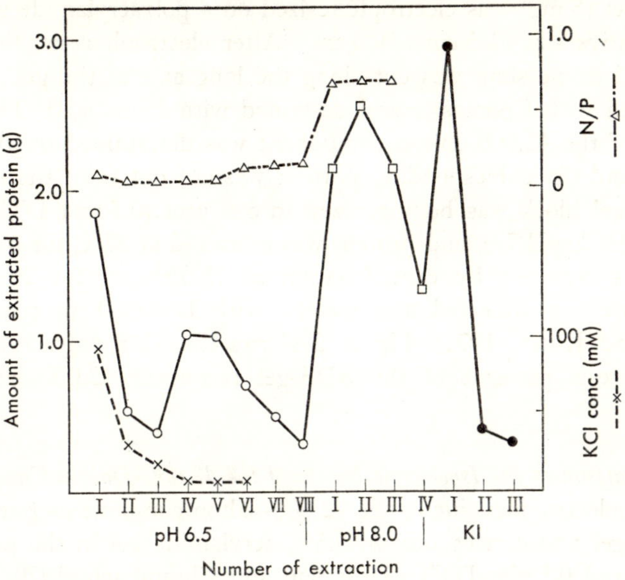

Fig. 1. Relationship between the amount of protein extracted and the number of extractions. The I-Z-I segments (21.5 g) were extracted repeatedly with 2 mM Tris-HCl (pH 6.5), 2 mM Tris-HCl (pH 8.0), and 0.6 M KI solution by means of resuspension and centrifugation. Each extraction was carried out for 30 min at 4°C. The amount of protein liberated in the supernatant fraction is shown in the ordinate. × KCl concentration of the suspension; △ molar ratio of adenine nucleotides to protein.

found to be liberated in the supernatant fraction (Fig. 1). The results of SDS-polyacrylamide gel electrophoresis indicated that the 42,000-dalton component comprised 70–80% of the extracted proteins. The I-Z-I segments were further extracted with 2 mM Tris-HCl, pH 8.0 and 0.6 M KI solution (containing 0.6 M KI, 10 mM sodium thiosulfate, 0.5 mM dithiothreitol (DTT) and 10 mM Tris-HCl, pH 7.0). Twenty-eight percents of the proteins of the initial I-Z-I segments were liberated by pH 6.5-extraction, 48% by the pH 8.0- and 17% by the KI-extraction. SDS-polyacrylamide gel electrophoresis showed that the 42,000-dalton components were the major constituents of these extracts.

1. Purification of the 42,000-Dalton Components by SDS-polyacrylamide Gel Electrophoresis

The 42,000-dalton components of the extracts were purified by the use of SDS-polyacrylamide gel electrophoresis (Weber and Osborn, 1969).

The extract (5 mg) was electrophoresized on a polyacrylamide gel (7.5%) of preparative size (1.5 cm × 10.0 cm). After electrophoresis, the gel was frozen and a fine strip was cut along the long axis of the gel. The strip was stained with Coomassie and destained with acetic acid. The relative mobility of the 42,000-dalton component was determined on the stained gel strip and the corresponding portion was cut out from the un-stained gel. The gel block was homogenized in 8 M urea, 0.5 mM DTT, and 10 mM Tris-HCl, pH 7.0, and protein was extracted at 37°C for 6 hr. Acrylamide was removed by centrifugation at 15,000 rpm for 20 min. The supernatant was collected and treated with Dowex-1 to remove SDS (Weber and Kuter, 1971). Figure 2 shows the SDS-polyacrylamide gel electrophoresis patterns of the SDS-gel recovered 42,000-dalton components.

2. *Comparison of the Isoelectric Points of the 42,000-Dalton Components*

The isoelectric focusing of the 42,000-dalton components purified from the SDS-gel was carried out on 3.5% acrylamide gel in the presence of 8 M urea and 0.5 mM DTT (Finlayson and Chrambach, 1970). The isoelectric point of 42,000-dalton components recovered from the pH 8.0-extract and KI-extract was found to be identical to that of skeletal muscle actin. On the other hand, 42,000-dalton components from the pH 6.5-extract contained another molecular species whose isoelectric component

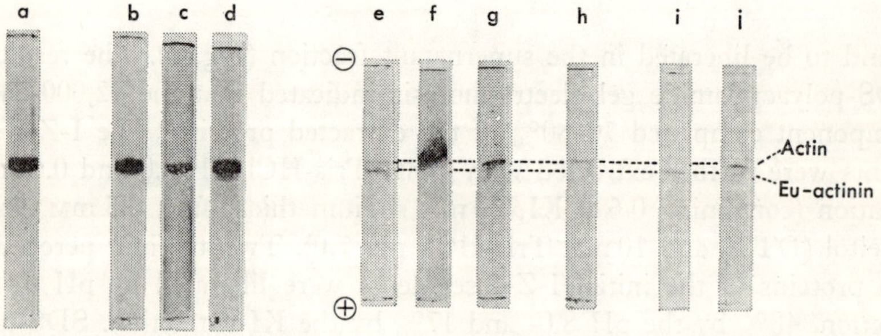

Fig. 2. SDS-polyacrylamide gel electrophoresis and isoelectric focusing patterns. a, SDS-polyacrylamide gel pattern of purified eu-actinin; b, SDS-polyacrylamide gel pattern of SDS-gel purified 42,000-dalton component from pH 6.5-extract; c, from pH 8.0-extract; d, from KI-extract; e, Isoelectric focusing pattern of 42,000-dalton component from the pH 6.5-extract; f, from the pH 8.0-extract; g, from the KI-extract; h, isoelectric focusing pattern of pure actin; i, isoelectric focusing pattern of purified eu-actinin; j, isoelectric focusing pattern of the mixture of actin and eu-actinin.

was more acidic than that of actin (Fig. 2). This component was designated eu-actinin because of its close similarity in molecular weight to actin.

3. Purification of Eu-actinin

Eu-actinin was purified out from the supernatant-IV of the pH 6.5-extraction. The supernatant-IV was fractionated by means of ammonium sulfate fractionation. The precipitates obtained between 8–13% ammonium sulfate saturation were collected and dialyzed against 0.3 mM DTT and 10 mM Tris-HCl, pH 8.0 (called Buffer A). The crude preparation was chromatographed on Sepharose 4B column (26 cm × 80 cm) equilibrated with Buffer A. Two distinct peaks, P-1 and P-2, were separated. P-1 was eluted at the position of the void volume of the column. The retarded peak, P-2, was collected and rechromatographed on Sephadex G-200 equilibrated with 0.5 mM DTT and 10 mM Tris-HCl, pH 6.7. The top portion and the latter half of the main peak were collected and used as a pure preparation of eu-actinin.

4. Some Physicochemical Properties of Eu-actinin

The isoelectric point of eu-actinin measured in the presence of 8 M urea was slightly more acidic than that of actin (Fig. 2i, j). The chain molecular weight determined on SDS-polyacrylamide gel was 42,000, identical to that of actin. Judging from the results of analytical chromatography on Sephadex G-200, eu-actinin exists as a dimer in a solvent of low ionic strength at neutral pH. The intrinsic viscosity of eu-actinin in 2 mM Tris-HCl (pH 8.0) was 6.4 ml/g, indicating considerable solvation or dissymmetry of the molecule. The addition of DTT did not affect the intrinsic viscosity. The dimer of eu-actinin dissociated into its monomer in the presence of both DTT and 6 M guanidine-HCl. The addition of KCl to eu-actinin caused the aggregation of eu-actinin. As far as we have investigated, the difference in amino acid composition from actin was rather small. Significant differences were detected in nonpolar residues. The cysteine content of eu-actinin (3 mol/42,000 *g*) was smaller by two residues than that of actin.

5. Immunological Studies of Eu-actinin

Antiserum against eu-actinin was raised in a goat by multiple subcutaneous injections of purified eu-actinin (5 mg) admixed with complete Freund's adjuvant. Monthly booster injections (2 mg each) were

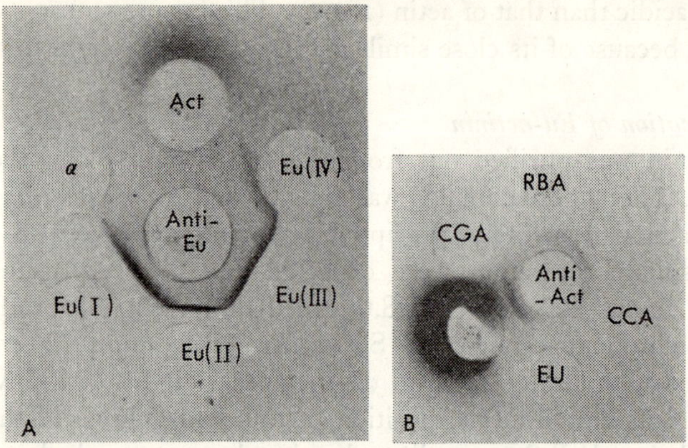

Fig. 3. Ouchterloney's test of anti-eu-actinin and anti-actin antibody. Conditions: 1% agarose, 0.1 M NaCl and 10 mM Tris-HCl, pH 7.0. A: specificity of anti-eu-actinin IgG. Anti-eu, anti-eu-actinin IgG; Act, actin (1.0 mg/ml); α, α-actinin (1.0 mg/ml); Eu (I), eu-actinin (4.0 mg/ml); Eu (II), eu-actinin (2.0 mg/ml); Eu (III), eu-actinin (1.0 mg/ml); Eu (IV), eu-actinin (0.5 mg/ml). B: Specificity of anti-actin IgG. CBA, chicken breast muscle actin; RBA, rabbit skeletal muscle actin; CGA, actin from chicken gizzard; DA, denatured actin (chicken breast muscle); Eu, eu-actinin. Concentration of antigens: 1 mg/ml.

adminstered over 6 months. Specific antibody to eu-actinin was prepared by the use of affinity chromatography. Purified eu-actinin (20 mg) was coupled with 5 ml of Sepharose 4B by the CNBr activation method (Kato and Anfinsen, 1969). The coupling efficiency was 86.4%. Antiserum was applied to the affinity column (1.0 cm×10.0 cm), and antibody was eluted stepwise with (I) 1 M $MgCl_2$ and (II) 4 M $MgCl_2$ (Mains and Eipper, 1976). Antibody was found to be eluted with 4 M $MgCl_2$. After desalting by Sephadex G-25 column chromatography, the antibody was condensed by the use of ammonium sulfate. The purity of the antibody was confirmed by the use of SDS-polyacryamide gel electrophoresis. Judging by the immunoelectrophoresis of the antibody, the purified antibody was found to be composed of IgG.

The double immunodiffusion test on 1% agarose (Ouchterloney, 1968) showed that anti-eu-actinin IgG formed a precipitine line only with eu-actinin. Interaction with neither actin nor α-actinin was observed (Fig. 3A). Direct immunofluorescence microscopy of chicken myofibrils stained with fluorescence isothiocyanate (FITC)-conjugated anti-eu-actinin IgG revealed that eu-actinin localized at the Z line of myofibrils (Fig. 4a).

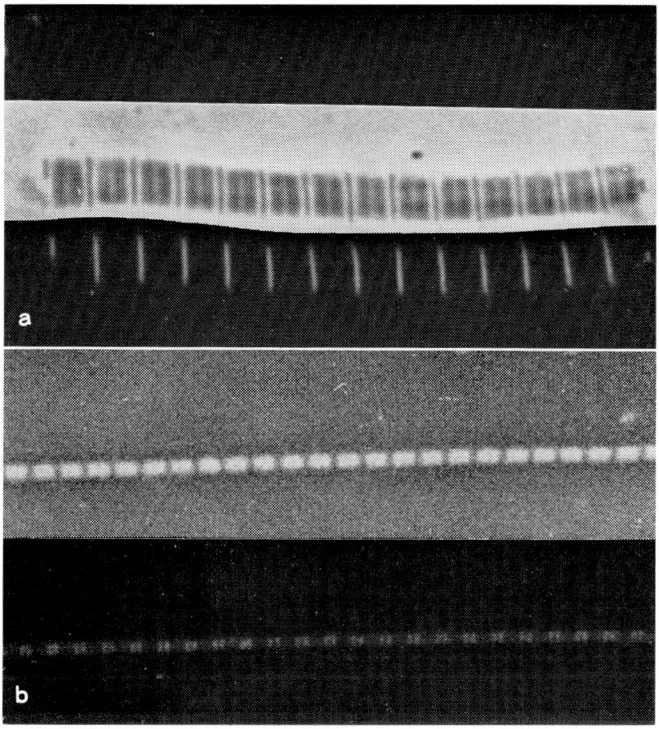

Fig. 4. Immunofluorescence microscopy of myofibrils. A: direct immunofluorescence microscopy of chicken skeletal muscle using the FITC-conjugated antibody to eu-actinin. B: FITC staining of chicken skeletal myofibrils using anti-actin IgG.

6. *Antibody to Skeletal Muscle Actin*

Antiserum against chicken breast muscle actin was raised in another goat. The immunization schedule and the purification procedure of mono-specific antibody were the same as described above. Denatured actin purified from the pH 8.0-extracts was used as antigen. Actin purified from the acetone-dried powder of chicken breast muscle was coupled with the affinity column. The specificity of anti-actin IgG was examined by the use of Ouchterloney's immunodiffusion test (Fig. 3B). The anti-actin antibody was shown to form precipitine lines with actin from rabbit skeletal muscle, chicken gizzard muscle, and chicken cardiac muscle as well as with actin from the chicken skeletal muscle. The antibody did not react with eu-actinin. The results of indirect immunofluorescence microscopy indicated that anti-actin antibody binds to the I bands of myofibrils. It is to be noted that the entire length of the I-filament was not stained unless fully stretched myofibrils or myofibrils freed from myosin were used (Fig. 4b). When F-actin treated with anti-actin antibody was examined under the electron microscope by the negative staining method, F-actin decorated with the antibody was observed. Where the ratio of antibody to actin was low, the typical Y-shaped IgG molecules attached to F-actin were recognized. Judging from these observations, it is very likely that the reason the entire I-filament was not stained in the relaxed myofibrils at relax is the steric hindrance of the head portions of myosin molecules to antibody binding.

7. *Immunological Comparsion of the Extracts from I-Z-I Segments*

The immunological differences in the extracts from I-Z-I segments were examined using anti-eu-actinin and anti-actin IgG. It was found that anti-eu-actinin antibody interacted only with the pH 6.5-extracts, while antibody to actin interacted with all the extracts. Thus, it is concluded that eu-actinin was selectively extracted at pH 6.5.

Acknowledgment

The authors thank Mr. T. Tanaka of the Tokyo Metropolitan Institute of Medical Science (TIMS) for his technical assistance, and Dr. H. Yoshizawa of TIMS for his advice on affinity column chromatography.

REFERENCES

Finlayson, G. R. and Chrambach, A. (1970). *Anal. Biochem.* **40**, 292–311.

Kato, J. and Anfinsen, C. B. (1969). *J. Biol. Chem.* **244**, 5849–5855.

Kawamura, K. (1969). *Fluorescent Antibody Techniques and Their Applications*, Univ. Tokyo Press, Tokyo.

Mains, R. F. and Eipper, B. A. (1976). *J. Biol. Chem.* **251**, 4115–4120.

Oosawa, F. and Higashi, S. (1969). *Prog. Theor. Biol.* **1**, 79–164.

Ouchterloney, O. (1968). *Handbook of Immunodiffusion and Immunoelectrophoresis*, 215 pp., Ann Arbor Publ., Ann Arbor, Michigan.

Rubenstein, P. A. and Spudich, J. (1977). *Proc. Natl. Acad. Sci. U.S.* **74**, 120–123.

Tuboi, K. K. (1968). *Biochim. Biophys. Acta* **160**, 420–434.

Weber, K. and Kuter, D. J. (1971). *J. Biol. Chem.* **246**, 4505–4517.

Weber, K. and Osborn, M. (1969). *J. Biol. Chem.* **244**, 4406–4412.

Muscle Contraction: Its Regulatory Mechanisms (Ebashi, S. et al., eds.), pp. 515–525,
Japan Sci. Soc. Press, Tokyo / Springer-Verlag, Berlin (1980)

41

Difference in Protein Constituents among Various Chicken Skeletal Muscles

TAMIO HIRABAYASHI

Summary: Protein components of glycerinated chicken skeletal muscles were investigated by SDS-polyacrylamide gel electrophoresis in order to determine how much difference there is in the protein constituents among chicken skeletal muscles. From the difference detected, chicken skeletal muscles were roughly classified into 4 groups: group A (mainly white muscles), group B (muscles with no particular characteristics in protein constituents), group C (seemingly intermediate between group A and group D), and group D (mainly red muscles). M. pectoralis major, M. caudoilio-femoralis, M. biventer cervicalis, and M. gastrocunemius (interior) can be listed as four representatives of chicken skeletal muscle. However, un-like rabbit skeletal muscle, chicken muscles differ in protein constituents from muscle to muscle.

Skeletal muscle differentiation has been one of the main subjects in developmental biology. Muscle has a simple function and definite protein components, and the biochemical and biophysical characteristics of the protein are already well known. Although muscle tissue seems to be rather homogeneous as a whole in comparison with other organs, heterogeneity in several respects has been reported. Classification of fast-twitch and slow-twitch muscles is generally accepted (Barany, 1967; Peachey, 1968; Hess, 1970; Close, 1972). However, it is not so simple to categorize all muscles into fast-twitch white muscle and slow-twitch red muscle (Barnard *et al.*, 1971). Fast-twitch muscle can be divided into two types (IIA and IIB) and is revealed to be interconvertible in response to training or inactivity (Andersen and Henriksson, 1977). Furthermore, in slow-twitch muscle (soleus) some amount of fast-twitch muscle-type component has been found (Amphelett and Perry, 1975). On the protein components of muscles, much information has been accumulated (Perry, 1974). Myosins in vertebrate fast-twitch and slow-twitch muscles are structurally distinct (Sarker *et al.*, 1971; Rowey and Risby, 1971; Jean *et al.*, 1973), as has also been acertained by immunological techniques (Gröschel-Stewart and Doniach, 1969; Masaki, 1974; Arndt and Pepe, 1975; Rubinstein *et al.*, 1977; Lutz *et al.*, 1978). The fact that myosin exists as several isozymic forms (Hoh *et al.*, 1976; Holt and Lowey, 1977; Hoh, 1978; Weeds, 1978) and as two polymorphic forms even in fast-twitch muscle (Weeds *et al.*, 1975; Gauthier and Lowey, 1977) causes us much concern about fetal myosin. It has been controversial whether embryonic myosin is identical with adult fast-twitch muscle myosin (Pelloni-Müeller *et al.*, 1979; Rubinstein and Kelly, 1978) or a mixture of first-twitch and slow-twitch muscle myosins (Gauthier *et al.*, 1978). Huszar (1972) and Sròter *et al.* (1975) reported that fetuses had a distinct fetal form of myosin differing in primary structure from adult myosins. The most conclusive evidence for the existence of fetal myosin was presented recently by Hoh (1979) and by Hoh and Yoeh (1979). According to their experiments, fetal slow-twitch muscle myosin is predominantly slow-twitch muscle-type myosin which appears to contain fast-twitch muscle myosin light chain in addition to slow-twitch muscle myosin light chain, and fetal myosin heavy chain is distinct from fast-twitch and slow-twitch muscle myosin in the two-dimensional peptide map.

Skeletal muscle actin is homogeneous protein. Although β- and γ-actins are found in culture muscle cells, they are considered molecules in non-muscle tissue (Whalen *et al.*, 1976).

Tropomyosin is notable for its polymorphic forms (Hirabayashi and Hayashi, 1970; Cummins and Perry, 1973). This protein is a mixture of two different polypeptides (α and β) (Cummins and Perry, 1973), and the existence of still more kinds of peptides is presumed from the results obtained by immunological techniques (Hayashi *et al.*, 1977).

Heterogeneity of troponin-I in skeletal muscles was first reported by Syska *et al.* (1974), and it was reported that troponin-I polymorphism was changed by cross-innervation (Amphlett *et al.*, 1975) and constant during development (Amphlett *et al.*, 1976). Likewise, the change in tropomyosins and myosins in development (Roy *et al.*, 1979a) and during transformation (Roy *et al.*, 1979b) presented a dynamic aspect of protein synthesis in skeletal muscle.

As far as the developmental aspect is concerned, chicken skeletal muscle is a suitable material to use since we can easily obtain it at any stage of development. A large part of our knowledge on muscle protein, however, has come from study with rabbit muscles. Sometimes there is a slight difference in muscle protein characteristics between rabbit and chicken, and this can cause discrepancies in results. The definition of muscle differentiation can also vary from person to person depending on their research interests. When quality and quantity of muscle proteins are used as one feature of muscle differentiation, it is an essential prerequisite for us to know how much variation there is in protein constituents among chicken skeletal muscles and which muscle in the chicken body is a typical or representative one in terms of protein composition. Such basic information on chicken skeletal muscles is rather scarce as compared with that on mammalian muscles. Therefore, it is the purpose of this work to present a basic biochemical description of as many chicken skeletal muscles as possible in order to determine which muscle is the representative one in its protein composition and which one is most suitable for the study of muscle differentiation.

MATERIALS AND METHODS

Fresh chicken muscles were removed, cut into 3 to 5 mm-thick slices, and immersed in cold 50% glycerol for 12 hr. After the glycerol was replaced with a fresh batch, the muscles were stored at −20°C for longer than 2 weeks before use. Samples from five hens and two cocks were used in order to find individual variation in protein constituents. The muscles used are listed in Table I. The numbers assigned to them in Table I will be used again in describing the results.

In order to find as many protein components as possible, slab-type electrophoresis on SDS-polyacrylamide gel was carried out with serial dilution of muscle homogenate. All the muscles listed in Table I were firstly analyzed using a phosphate buffer system (Weber and Osborn, 1969),

TABLE I
Chicken Skeletal Muscles Used

1. M. pectoralis major	23. M. longus colli anticus
2. M. pectoralis minor	24. M. peroneus superficialis
3. M. biceps brachii	25. M. levator coccygis
4. M. infrascapularis	26. M. flexor perforatus digiti II
5. M. rhomboideus superficialis	27. M. ambiens
6. M. infraspinatus	28. M. ilio-tibialis (anterior)
7. M. coraco-brachialis	29. M. extensor digitorum communis
8. M. latissimus dorsalis posterior	30. M. sternohyoideus
9. M. extensor propatagialis longus	31. M. femori-tibialis medius
10. M. triceps brachii	32. M. complexus
11. M. scalenus	33. M. femori-tibialis externus
12. M. tensor metapatagialis	34. M. cutaneus dorsalis
13. M. ilio-trochantricus anticus	35. M. semimembranosus
14. M. rhomboideus profundus	36. M. flexor perforatus digiti III
15. M. obliquus internus abdominis	37. M. biventer cervicalis
16. M. femori-tibialis internus	38. M. spinalis
17. M. ilio-tibialis (posterior)	39. M. pubi-ischio-femoralis
18. M. gastrocunemius (exterior)	40. M. flexor perforatus digiti IV
19. M. obliquus externus abdominis	41. M. latissimus dorsalis anterior
20. M. sartorius	42. M. ilio-trochantricus posticus
21. M. tibialis anticus	43. M. gastrocunemius (interior)
22. M. caudo-ilio-femoralis	

which was reasonably reliable in determining the molecular weight of protein, and then using a Tris-glycine buffer system (Laemmli, 1970) which was found to be quite potent for separating proteins with close mobilities.

RESULTS AND DISCUSSION

When all the electrophoretic patterns obtained by the phosphate buffer system were surveyed, two remarkable features were noted in the patterns. One was that the relative heaviness (or amount) of band 6 and band 7 varied from muscle to muscle (Fig. 1), and the other was that band 14 and band 19 were main components in M. pectoralis major (Fig. 1A) and other muscles but were very faint or almost absent in M. latissimus dorsalis anterior (Fig. 1B) and others. A serious disadvantage of this buffer system is that it gives broad bands just beneath the actin band, where several important components of myofibrils are supposed to be located, so that identification of any band around there is almost impossible. On the other hand, the Tris-glycine buffer system gave sharp bands all along the gels

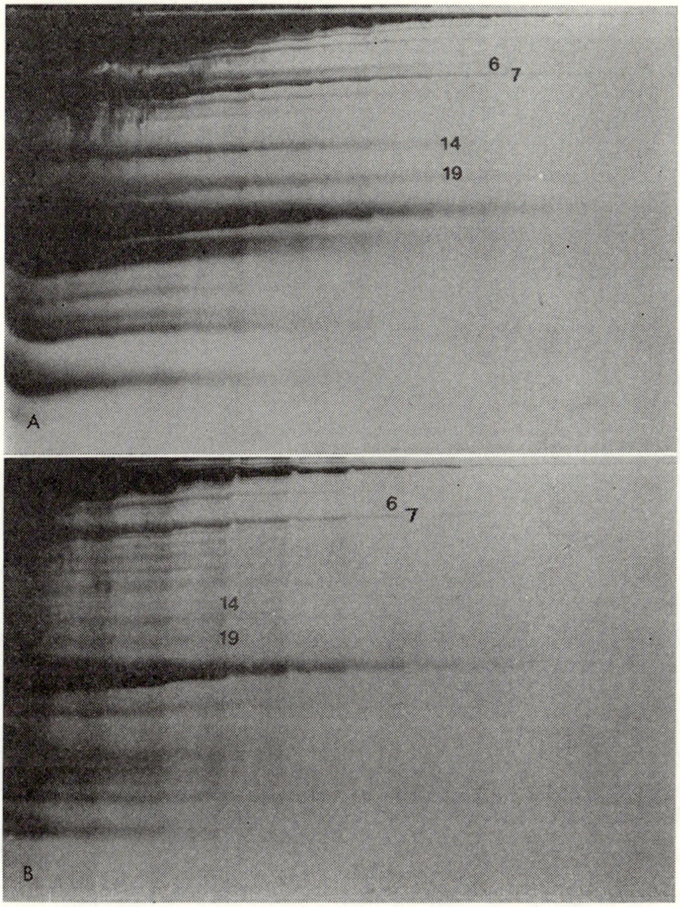

Fig. 1. SDS-polyacrylamide gel electrophoresis in phosphate buffer A: M. pectoralis major (♀). B: M. latissimus dorsalis anterior. For numbers, see Fig. 4. (Weber and Osborn, 1969).

(Fig. 2, see also Fig. 4). We could count 7 bands (band 20–band 26) beneath the actin band, and five sharp bands (band 1–band 5) beneath the myosin heavy chain band. To make correlations between the patterns in the phosphate buffer system and the Tris-glycine buffer system, two-dimensional electrophoresis was carried out, first in phosphate buffer horizontally and then in Tris-glycine buffer vertically. As shown in Fig. 3, many bands correspond to a broad band in the phosphate buffer system.

Again, all the muscles in Table I were analyzed by the electrophoresis

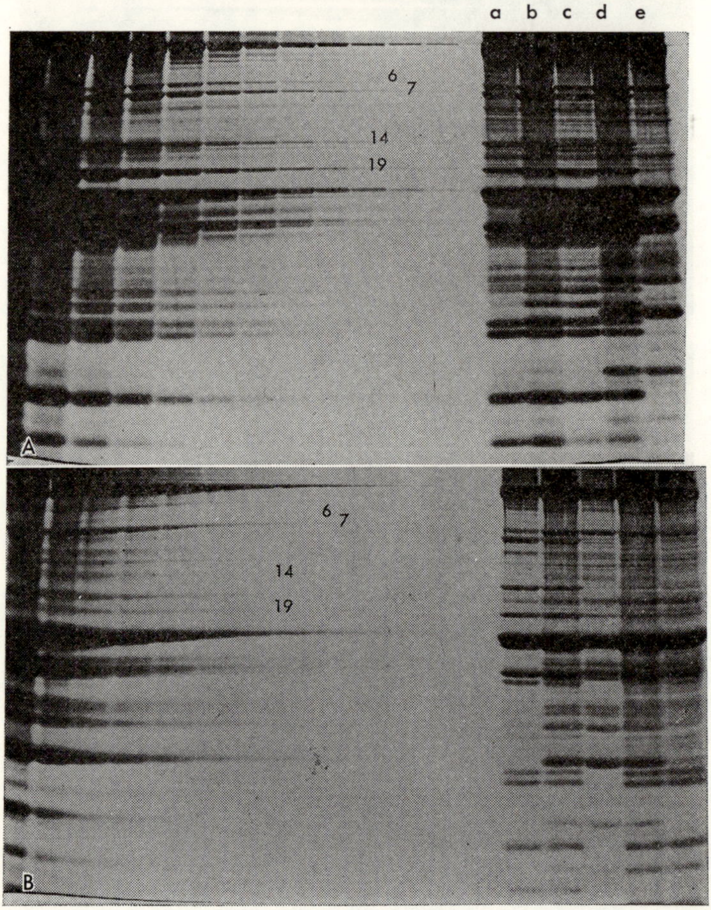

Fig. 2. SDS-polyacrylamide gel electrophoresis in Tris-glycine buffer. A: M. pectoralis major (♂). B: M. latissimus dorsalis anterior. a, M. pectoralis major (♀); e, M. caudo-iliofemoralis; upper c, M. pectoralis major (♂); lower c, M. latissimus dorsalis anterior; b, mixture of a and c; d, mixture of c and e. For numbers, see Fig. 4. (Laemmli, 1970).

in Tris-glycine buffer. In order to clarify the correspondence of bands between muscles, serially diluted muscle homogenates were charged on the gel with two reference muscle homogenates, M. pectoralis major (a in Fig. 2) and M. caudo-ilio-femoralis (e in Fig. 2). Mixture of each muscle homogenate and the reference muscle homogenate (b or d in Fig. 2) was coelectrophoresed on the gel side by side. By this method we could find a clear correspondence between the bands of muscle proteins.

Fig. 3. Two-dimensional electrophoresis. SDS-phosphate buffer, horizontally; SDS-Tris-glycine buffer, vertically. Sample, M. pectoralis major (♀).

As in the electrophoretic patterns in the phosphate buffer system, the relative heaviness (or amount) of band 6 and band 7 and the heaviness of band 14 and band 19 varied depending on the muscle source. Since band 6 and band 7 were located close together, and band 6 persisted in all muscles tested, the apparent relative heaviness of the two bands was used as a base for arranging the electrophoretic patterns of muscles in order. All muscles are lined, being based on this rather rough standard and shown in Table I starting from the higher value of band 6/band 7 (No. 1, M. pectoralis major) to the lowest (No. 43, M. gastrocunemius (interior)). Each pattern is presented in the same order in Fig. 4, and all the bands (band 1–band 41) observed are summarized at the end of Fig. 4.

Beneath the myosin heavy chain band, band 1 was found in muscles Nos. 1, 2, 5, 25, 29, 32, 33, 36, and 37. This band was visible only when a diluted sample of muscle homogenate was used; it probably is not an isoenzyme of myosin heavy chain (although it could be observed at low sample concentration) because purified chicken myosin did not show such a satellite band. In addition to band 1, four other bands (band 2–band 5) appeared beneath the myosin heavy chain band in most muscles.

When judged roughly from the patterns, the total concentration of band 6+band 7 seems to be constant in various muscles, although quantitative analysis is necessary before a final conclusion can be drawn. Band 14 and

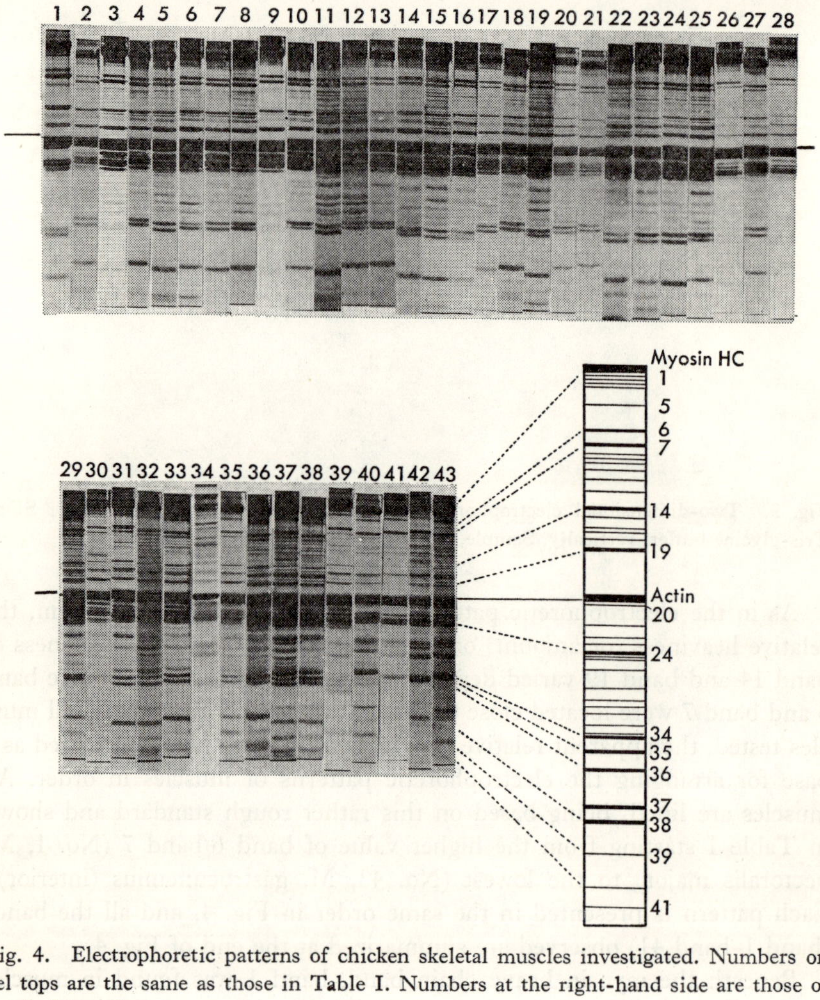

Fig. 4. Electrophoretic patterns of chicken skeletal muscles investigated. Numbers on gel tops are the same as those in Table I. Numbers at the right-hand side are those of bands detected.

band 19 made a good contrast to band 6 in that the two bands were almost absent in muscle No. 43 while band 6 is a main component in the muscle. Band 20 was found in muscles Nos. 1–8, and the largest amount was in M. pectoralis major (No. 1). It is very interesting that all the muscles containing the protein of band 20 are wing, dorsal or pectoral muscles, and no muscle in the legs contains the proteins. A large amount of band 21 protein was in muscles Nos. 3, 8, 9, 19, 25, and 29. This protein was not included in M. pectoralis major from a hen, but was included in that from a cock. The distribution of this protein in Table I and in M. pectoralis major from a cock suggests that the amount of the protein may be variable depending on the physiological conditions of the animals. Band 24 (α-tropomyosin) is present in all muscle except M. gastrocunemius (interior) (No. 43), in which band 22 (β-tropomyosin) is present instead. Band 34 and band 37 appear together in muscles Nos. 40–43, just as band 35, band 36, and band 38 appear as a set in muscles Nos. 1–30. The muscles from No. 31 to No. 39 showed an intermediate pattern with respect to these bands.

Although the electrophoretic patterns of the muscles vary significantly, we can roughly classify the patterns (except for that of M. cutaneus dorsalis (No. 34) into four groups: group A to which muscles from No. 1 to No. 10 belong; group B to which muscles from No. 11 to No. 30 belong; group C to which muscles No. 31 to No. 39 (except No. 34) belong; group D to which muscles from No. 40 to No. 43 belong. Group A contains mainly white muscles, and the representative is M. pectoralis major (No. 1). Group B contains many muscles with no distinctive feature in their composition. The pattern of M. Caudo-ilio-femoralis (No. 22) is one example in this group. Group C seems to have all the components of the muscles, combining the protein components of group A and group D, and the representative is M. biventer cervicalis (No. 37). Group D is the group of mainly red muscles, and the representative is M. gastrocunemius (interior) (No. 43).

Can we conclude that chicken skeletal muscles contain almost the same protein constituents or that the muscles contain rather different protein complements each other? The latter may be preferable because 7 skeletal muscles of the rabbit, prepared in the same way as in this experiment, showed two distinctive types of electrophoretic pattern and an intermediate one, so that the correspondence of the bands between the two types could easily be made (data not shown) to contrast with the chicken skeletal muscles.

Group A contains M. pectoralis major and others. The muscle has no band 21 or band 22, which were found in M. biceps brachii (No. 3), M. latissimus dorsalis posterior (No. 8) and M. triceps brachii (No. 10). The fact that M. pectoralis major differs in protein constituents from other white muscles in group A and even from M. pectoralis minor must be kept in mind when we use M. pectoralis major in the study of muscle differentiation. The muscle is only one example of chicken skeletal muscle.

The classification of muscles in this paper is very tentative, because of the existence of many kinds of intermediate types. In order to classify them in neat order and to study in detail the appearance of each component in cell differentiation, a reliable technique enabling us to analyze many protein constituents quantitatively must be developed.

REFERENCES

Amphelett, G. W. and Perry, S. V. (1975). *Nature* **257**, 602–604.

Amphelett, G. W., Perry, S. V., Syska, H., Brown, M. D., and Vrbova, G. (1975). *Nature* **257**, 602–604.

Amphelett, G. W., Syska, H., and Perry, S. V. (1976). *FEBS Lett.* **43**, 22–26.

Anderson, P. and Henriksson, J. (1977). *Acta Physiol. Scand.* **99**, 123–125.

Arndt, I. and Pepe, F. A. (1975). *J. Histochem. Cytochem.* **23**, 159–169.

Barany, M. (1967). *J. Gen. Physiol.* **50**, 197–218.

Barnard, R. J., Edgerton, V. R., Furukawa, T., and Peter, J. B. (1971). *Am. J. Physiol.* **220**, 410–414.

Close, R. (1972). *Physiol. Rev.* **52**, 129–197.

Cummins, P. and Perry, S. V. (1973). *Biochem. J.* **133**, 765–777.

Gauthier, G. F. and Lowey, S. (1977). *J. Cell Biol.* **74**, 760–779.

Gauthier, G. F., Lowey, S., and Hobbs, A. W. (1972). *Nature* **274**, 25–29.

Gröschel-Stewart, U. and Doniach, D. (1969). *Immunology* **17**, 981–994.

Hayashi, J., Ishimoda, T., and Hirabayashi, T. (1977). *J. Biochem.* **81**, 1487–1495.

Hess, A. (1970). *Physiol. Rev.* **50**, 40–62.

Hirabayashi, T. and Hayashi, Y. (1970). *Dev. Growth Diff.* **12**, 151–178.

Hoh, J.F.Y. (1978). *FEBS Lett.* **90**, 297–300.

Hoh, J.F.Y. (1979). *FEBS Lett.* **98**, 267–270.

Hoh, J.F.Y., McGrath, P. A., and White, R. I. (1976). *Biochem. J.* **157**, 87–95.

Hoh, J.F.Y. and Yeoh, G.P.S. (1979). *Nature* **280**, 321–323.

Holt, J. C. and Lowey, S. (1977). *Biochemistry* **14**, 4398–4402.

Huszar, G. (1972). *Nature New Biol.* **240**, 260–264.

Jean, D. H., Guth, L., and Albers, R. W. (1973). *Exp. Neurol.* **38**, 458–471.

Laemmli, U. K. (1970). *Nature* **227**, 680–685.

Lowey, S. and Risby, D. (1971). *Nature* **234**, 81–85.

Lutz, H., Ermini, M., and Jenny, E. (1978). *Histochemistry* **57**, 223–235.

Masaki, T. (1974). *Biochem. J.* **76**, 441–449.

Peachey, L. D. (1968). *Annu. Rev. Physiol.* **30**, 401–440.

Pelloni-Müeller, G., Ermini, M., and Jenny, E. (1976). *FEBS Lett.* **67**, 68–74.

Perry, S. V. (1974). In *Exploratory Concepts in Muscle* (Milhorat, A. T., ed.), vol. 2, pp. 319–328, Excerpta Medica, Amsterdam.

Roy, R. K., Mabuchi, K., Sarkar, S., Mis, C., and Srèter, F. A. (1979). *Biochem. Biophys. Res. Commun.* **89**, 181–187.

Roy, R. K., Srèter, F. A., and Sarkar, S. (1979). *Dev. Biol.* **69**, 15–30.

Rubinstein, N. A. and Kelly, A. M. (1978). *Dev. Biol.* **62**, 473–485.

Rubinstein, N. A., Pepe, F. A., and Holtzer, H. (1977). *Proc. Natl. Acad. Sci. U.S.* **74**, 4524–4527.

Sarkar, S., Srèter, F. A., and Gergely, J. (1971). *Proc. Natl. Acad. Sci. U.S.* **68**, 946–950.

Srèter, F. A., Balint, M., and Gergely, J. (1975). *Dev. Biol.* **46**, 317–325.

Syska, H., Perry, S. V., and Trayer, I. P. (1974). *FEBS Lett.* **40**, 253–257.

Weber, K. and Osborn, M. (1969). *J. Biol. Chem.* **244**, 4406–4412.

Weeds, A. (1978). *Nature* **274**, 417–418.

Weeds, A. G., Hall, R., and Spurway, M.C.S. (1975). *FEBS Lett.* **49**, 320–324.

Whalen, R. G., Butler-Browne, G. S., and Gros, F. (1976). *Proc. Natl. Acad. Sci. U.S.* **73**, 2018–2022.

Muscle Contraction: Its Regulatory Mechanisms (Ebashi, S. et al., eds.), pp. 527–540
Japan Sci. Soc. Press, Tokyo / Springer-Verlag, Berlin (1980

42

The Morphogenesis of Myofibrils from the Microtrabecular Lattice as Observed in Cultured Myoblasts

KEITH R. PORTER AND KAREN ANDERSON

Recent studies of whole cultured cells by high voltage microscopy and stereo images have provided convincing evidence that the cytoplasmic ground substance (CGS) is highly structured and in no sense a homogeneous cytosol. Besides the well-known cytoskeletal components (the microtubules, microfilaments and intermediate (10 nm) filaments) the CGS comprises a lattice of slender strands measuring 5–10 nm in diameter and 50–100 nm in length. These have been called microtrabeculae and the whole structure the microtrabecular lattice (MTL). In connecting and supporting the various cytoplasmic components, including those of the cytoskeleton, and in linking these to one another and to the cytoplasmic cortex, it parallels in certain respects the relationship between connective tissue and the components of other tissues such as blood vessels, muscles, nerves, skeletal elements, and epidermis. This analogy should not, however, be given more than limited attention.

Actually, the MTL is the matrix in which the more precisely characterized structures of the cytoplasm are contained. Its structure effectively divides the CGS into two phases, one protein-rich and the other a structure-free, water-rich phase. The latter would seem to provide a medium for the rapid diffusion of low-molecular-weight metabolites whereas the former appears to be a complex polymer that gives the cytoplasm its viscoelastic properties. The material comprising this structured phase covers the surfaces of the endoplasmic reticulum (ER) cisternae, the polysomes and microtubules. It extends as well to the cytoplasmic surface of the cell membrane where it includes (is co-extensive with) the cytoplasmic cortex. There is evidence, derived in part from studies on chromatophores and their pigment motion, to support the belief that the MTL is highly dynamic and that the contraction and elongation of its trabeculae provides the motive force for the several motions displayed as well as to regard the MTL, and the cytoplast it comprises, as representing the fundamental form-determining component of the cells. It is obviously distortable through, *e.g.*, the assembly of microtubules, but has a form it prefers. This is dependent for its maintenance on microtrabeculae and metabolic energy.

Since, in terms of volume, the MTL represents a major part of the cytoplasm and since it is the matrix and the continuous phase in which all other structures are suspended, it would seem to be the source of information that controls the non-random distribution of organelles and systems. These, in most instances, are products of cell differentiation and include ER and Golgi, as well as myofibrils, keratin fibers, *etc.* For the purposes of observing the relationship between a differentiating structure and the MTL, we have chosen the myofibril. It is non-randomly distributed in the cell and in its assembly it goes through some of the same sequences as stress fibers. These are most certainly co-extensive with the MTL and seem to result from a parallel clustering of elongated trabeculae. The questions we have chosen to ask in this preliminary study are the following:

a. Do myofibrils and Z bands emerge from the lattice of the undifferentiated cell and, if so, through what sequences?
b. Is the synthesis of muscle proteins required for myofibril assembly random or non-random?
c. Is there a residual lattice around the myofibrils after their differentiation?

There are numerous other questions one might ask and in time answer. But, for a beginning and present purposes, these seem sufficient.

MATERIALS AND METHODS

The cells are obtained from hearts of chick embryos after 7 days of incubation according to the method of DeHaan (1967). This involves the use of a disaggregation medium containing trypsin 0.05%. From the suspension, a population of cells rich in myoblasts settles out and attaches to formvarcoated gold grids and by the end of 24 hr many of the myoblasts are beating spontaneously. After 3 or 4 days, the percentage of beating cells increases. At this time, the cells are fixed with glutaraldehyde and processed according to procedures previously described (Wolosewick and Porter, 1979). The location of interesting cells on finder grids, identified for beating and fibril content, was noted by light microscopy previous to fixation and this makes it relatively easy to find the same cells in the high voltage microscope.

OBSERVATIONS

The myoblasts that first appear as spontaneously contracting cells in the cultures contain varying numbers of myofibrils. These in turn vary in their differentiation, some showing only a few distinct sarcomeres and others many. In most instances the sarcomeres showing distinct A, I, and Z bands are proximal to the cell center while the less completely formed are distal. These latter then are the most valuable for examination if one is interested in studying the phenomena of myofibril assembly (Fig. 1). The orientation of the fibrils is always non-random and parallel to one another and to the long axis of the cell.

In these preliminary studies we have been bothered by two concerns. One stems from the observation that some of the fibrils in the myoblasts are non-striated and could be in the process of disassembly. And the other is the distinct possibility that fixation, by the methods used, is too slow to avoid distortions of fine structure arising from contractions that occur in the presence of glutaraldehyde and before fixation is complete. It is evident from recent experiments that the very rapid immobilization and subsequent fixation that one can achieve with the techniques of freeze substitution will eliminate the latter concern and provide images in which all the myofibrils will show striations characteristic of either the relaxed or contracted morphology.

The CGS, in regions of myoblasts showing no evidence of myofibril formation is represented by a lattice of very slender trabeculae (Fig. 2).

These contain numerous polysomes, a feature of the cultural myoblast that repeats what has been reported from thin sections of developing muscle fixed *in situ* (Warren and Porter, 1969; Toselli and Pepe, 1968; Kelly, 1969). In places, the microtrabeculae of the random lattice elongate apparently to participate in the organization of slender stress fibrils. The two types of MTL are at very least confluent (Fig. 2). The membrane-limited components of these regions are mostly tubular in shape and without any patterns of organization that one can recognize. They probably represent precursors of what will eventually become the sarcoplasmic reticulum (SR) or the transverse tubular system. The cytoplasmic surfaces of these cisternae is apparently covered by lattice material and numerous microtrabeculae attach to these surfaces to join other surfaces including that of the plasmalemma.

Regions of contracting myoblasts, obviously involved in the assembly of myofibrils, show three major components: 1) membranous elements similar to the SR or transverse tubules, mostly associated with Z bands, 2) bundles of parallel filaments constituting various stages in the assembly of myofibrils and 3) elements of the surrounding lattice with contained polysomes (Fig. 3). These latter seem to be more numerous in the lattice in and around the myofibrils than in the lattice more peripheral to the fibrils.

The membrane-limited structures are preferentially associated with the Z bands. Though their outline is irregular, their width generally matches that expected of the Z band (50 nm). They are oriented transversely to the long axis of the fibril and show very few extensions or branches with other orientations. In other words they have the position of the so-called T tubules (Figs. 3 and 4).

The thin, presumably actin filaments, are clearly continuous with the microtrabeculae of the random lattice. Usually they are confined in their location to a position between the Z bands and seem to have one end attached to the relatively amorphous material which is the Z band. The thick myosin filaments appear when an identifiable A band appears (Fig. 3). They can be recognized by their size among the thin filaments and seem also to retain MTL attachments.

In regions where the fibril is more completely formed and an I band is evident the distribution of polysomes is most concentrated in microtrabeculae closely associated with the A band. Noticeably fewer reside over the I and Z bands. Presumably these large polysomes are actively synthesizing myosin for assembly of the thick filaments. In sarcomeres at an earlier stage in differentiation the polysome distribution is not so confined, but extends over the entire length (top sarcomere in Fig. 3). It is assumed that here actin is being assembled into F-actin as the microtrabeculae containing F-actin elongate to become part of the muscle sarcomere.

Except for the evidence of filament emergence from the MTL and the non-random distribution of polysomes which stereo images provide, these studies have not added greatly to our knowledge of fibrillogenesis.

The early appearance of the fibrous Z band material in association with the cell cortex, the early emergence of thin filaments and subsequently thick filaments, and the eventual organization of these into the paracrystalline patterns of mature fibers are all features of fibrillogenesis that have been reported from other systems (Warren and Porter, 1969; Toselli and Pepe, 1968; Kelly, 1969). The stereo images of these whole cells show the Z band as associated with the cell cortex next to the culture substrate. These may traverse substantial spaces between fibrils or parts of fibrils without associated thin filaments (Figs. 1 and 3).

DISCUSSION

This limited study of myofibril differentiation in cultured chick cardiac myoblasts was made possible by high voltage microscopy of whole cells and the capacity of stereo images to give information on 3-D fine structure. It is a beginning only and current investigations are exploring the potentials of selective extraction and immunocytochemistry for revealing additional details of the phenomena associated with fibrillogenesis. From these preliminary studies we have learned that the sarcoplasm of the embryonic myoblast comprises a finely divided lattice of microtrabeculae similar to that in other cells and that the filamentous components of the myofibril, including those of the Z band, are essentially local differentiations within this lattice. The sequence of conformational changes in the MTL that terminates in filamentous components of the myofibril is essentially the same as that which appears to take place in the formation of stress fibers, *i.e.*, elements of the lattice (microtrabeculae) elongate and fuse to become the long actin-rich trabeculae which, in parallel array, make up the fully formed stress fiber (Porter, 1978). Continuity with the lattice at the margins of the fiber persists. These phenomena, in muscle fibril assembly, are confined to the spaces between successive Z bands. Despite this difference there seems adequate reason to equate the two and therefore to derive the filamentous elements of the fibril from the local synthesis and assembly of, for example, G-actin into F-actin. This, according to our interpretation, is an intratrabecular event.

The origin of the Z band is less readily described from the observations available. The "skein like" filamentous material which Kelly reports as making up the Z bands appears in these stereo images as a condensed arrangement of short and thick trabeculae. These obviously accommodate less water in their composition and scatter electrons more intensively to provide a darker image in the micrographs. Again continuity with the surrounding MTL makes this structure also a derivative of the undifferentiated lattice. Presumably α-actinin is synthesized locally in special trabeculae at the edge of the Z band and the fusion of these trabeculae into the organized Z band contributes to its lateral growth. Just how the bands come to be positioned at equally spaced intervals within the myoblast is not immediately evident, although it must be that the information for this resides within the MTL. The basis for this structural information is obscure, to say the least, and relative questions are not likely to be answered without extensive experimental studies. A start may have been made with cytochalasin B. When cultured cells (rat or chick endothelial) are exposed to this extraordinary drug they arborize and the MTL takes on a very coarse configuration (Fonte *et al.*, 1978; Porter and Anderson, unpublished observations). Returned to normal culture media without cytochalasin B, the cells quickly resume their morphology including that of the microtrabecular lattice. Cultured myocytes from explants of chick embryos similarly exposed to cytochalasin B fail to form normal myofibrils or even the normal alignment of Z bands (Sachs *et al.*, 1974). Obviously cytochalasin B affects a change in the organization of the normal phenotype and simultaneously the morphology of the MTL. An extension of these studies is certainly indicated.

Another question among the many to emerge from these phenomena is whether the specific proteins required for fibril assembly are synthesized and distributed randomly or whether they are formed non-randomly and in close spatial relation to the fibril being assembled. Again, there is a shortage of useful observation. It will be difficult, though not impossible, to quantitate the distribution of polysomes per unit volume of cytoplasm. Thus far only a visual judgment has been made and that says that polysomes are more numerous over the A band when presumably myosin is being synthesized. Since, as mentioned earlier, those polysomes are "contained" in the MTL, the products of synthesis become part of trabeculae that are presumptive thick filaments. This is the current hypothesis. For a few reasons it makes good sense, for it means that the cell is conservative in its programmed synthesis of specific proteins; they are structure-bound, non-diffusible and made close to where they are used. This as a possibility has been suggested by Penman and his colleagues (Lenk *et al.*, 1977). It is supported by some preliminary immunocytochemistry (P-A-P method) plus high voltage microscopy on PtK_2 cells (DeMey *et al.*, 1980). In these preparations the test for tubulin distribution shows it to be concentrated in the MTL around the microtubules and not randomly distributed throughout the cytoplast. Perhaps one should not find the implied non-randomness in the distribution of polysomes so remarkable since we have known for a long time that ribosomes are concentrated on what is known as the rough endoplasmic reticulum. Since this differentiation (the ER) is contained in the trabeculae associated with the ER, the polysomes are presumably transported to their place on the ER cisternae *via* the MTL. These concepts will not be immediately accepted because the image of the cytoplast as a structured unit is relatively new.

The role of the microtrabecular lattice in the fully differentiated myocyte is difficult to assess. That it persists is evident in stereo views of thick sections. One is tempted on structural bases to equate the lattice with the network of filamentous material in micrographs of Toyoda and Maruyama (1978), but whether, on the strength of such meager evidence, one should suggest that the MTL in the myocyte comprises the cell's content of connectin (Maruyama *et al.*, 1977b) is doubtful. Connectin is thought to be responsible for the "passive elasticity" of muscle (Maruyama *et al.*, 1977a) and in this capacity may function in the return of the myocyte to rest length. Much more work on the MTL in muscle will be required to test this possibility.

Acknowledgments

This study is supported by a grant from the Muscular Dystrophy Association of America. It follows on some work initiated in collaboration with Judy Fleming and George Wray (Porter *et al.*, 1979) and will be part of a longer report to be published when the study has reached maturity. The authors are grateful to Joyce Albersheim for helping with the manuscript and to George Wray for assistance with the microscopy.

REFERENCES

De Haan, R. L. (1967). *Dev. Biol.* **16**, 216–249.

De Mey, J., Wolosewick, J. J., De Brabander, M., Geuens, G., Joniau, M., and Porter, K. R. (1980). In *Cell Movement and Neoplasia* (De Brabander,M. *et al.*, eds.), pp. 21–28, Pergamon Press, Oxford and New York.

Fonte, V. G., Anderson, K. L., Wolosewick, J. J., and Porter, K. R. (1978). *J. Cell Biol.* **79**, 74a.

Kelly, D. E. (1969). *Anat. Rec.* **163**, 403–426.

Lenk, R., Ransom, L., Kaufman, Y, and Penman, S. (1977). *Cell* **10**, 67–78.

Maruyama, K., Kimura, S., Kuroda, M., and Handa, S. (1977a). *J. Biochem.* **82**, 347–350.

Maruyama, K., Matsubara, S., Natori, R., Nonomura, Y., Kimura, S., Ohashi, K., Murakami, F., Handa, S., and Eguchi, G. (1977b). *J. Biochem.* **82**, 317–337.

Porter, K. R. (1978). In *Ninth International Congress on Electron Microscopy* (Sturgess, J. M., ed.), vol. III, pp. 627–639, Imperial Press, Ltd. Mississauga, Ontario.

Porter, K. R., Fleming, J., and Wray, G. (1979). *J. Cell Biol.* **83**, 46a.

Sachs, H. G., McDonald, T. F., and Springer, M. (1974). *J. Cell Sci.* **14**, 163–185.

Toselli, P. A. and Pepe, F. A. (1968). *J. Cell Biol.* **37**, 462–481.

Toyoda, N. and Maruyama, K. (1978). *J. Biochem.* **84**, 239–241.

Warren, R. H. and Porter, K. R. (1969). *Am. J. Anat.* **124**, 1–30.

Wolosewick, J. J. and Porter, K. R. (1979). *J. Cell Biol.* **82**, 114–139.

EXPLANATION OF FIGURES

Fig. 1. Margin of chick cardiac myoblast including differentiating myofibrils. This high voltage micrograph of the margin of a cultured myoblast is included to help the reader better relate the stereo images in Figs. 2, 3, and 4 to the whole cell. The most marginal myofibril, as it extends toward the bottom of the micrograph, is interpreted as depicting early stages in fibril assembly. Mitochondria (M) and Z bands (Z) are easily identified. Where the cell margin spreads toward the lower left, the cytoplast is very thin and it is possible to recognize fragments of the ER, or in this instance the SR. In this same region some protofilaments (Pf) extend away from nearby Z bands (*) into the surrounding sarcoplasm, as branches from the myofibril. These can be seen in Fig. 2 (Mf) and can be recognized there as local changes in the MTL. Additional myofibrils are evident deeper in the cell at the upper right. For some reason, possibly the thickness of the cells or fibrils, the sarcomere structure in these is less distinct. It is particularly interesting that in the marginal fibril the Z line is evident crossing expanses of sarcoplasm where filament formation is scarcely, if at all, evident. The areas outlined are shown at higher magnifications in Figs. 2, 3, and 4. Magnification ×10,000.

Fig. 2. Enlargement of area 2 on Fig. 1. This micrograph was selected to illustrate the structure of the sarcoplasm peripheral to the differentiating myofibrils. The morphology of the MTL is similar to that in other cells except in places as at SF where a stress fiber is shown and at Mf which is an extension of the myofibril. In both instances these appear as local differentiations of the MTL. The microtrabeculae, instead of being slender and randomly arranged, are thick, long and arranged in parallel bundles. Polysomes are suspended in the MTL as previously described (samples in circles). The region included in the micrograph contains numerous irregular profiles of SR. The surfaces of these are co-extensive with the trabeculae of the MTL. Magnification ×30,000.

Fig. 3. Enlargement of area 3 on Fig. 1. This is a stereo pair of micrographs taken with the high voltage microscope. It shows in three-dimensions the fine structure of the sarcoplasm associated with myofibril assembly in the margin of a cardiac myoblast from a chick embryo heart. The three Z lines (Z) evident in the field help to identify the limits of the myofibril. Within these limits the filamentous subunits (actin and myosin) are fairly discrete (*); outside the limits and to the left of the fibril there are long filaments which are co-extensive with the microtrabeculae of the lattice and which may have been destined to become part of the fibril at a later time in the differentiation of the cell. There is ample evidence in this and similar images of other myoblasts that the filaments (actin and myosin) develop as local differentiations of the lattice. |——| indicates roughly the limits of the A band. The distribution of polysomes (in circles) in the field portrayed is not uniform; there are more, as a rule, in the lattice elements within and over the myofibril than there are in the sarcoplasmic ground substance where fibrils are not being formed. Finally, the micrograph depicts a close association of membrane-limited saccules with the Z band (arrows). Whether these membranous structures are to be equated with invaginations of the plasmalemma (T system) or derivatives of the SR cannot be determined from the available images of this phenomenon. Magnification ×30,000.

Fig. 4. Enlargement of area 4 on Fig. 1. This stereo pair is selected to show the filament composition of a sarcomere of a developing myofibril. The Z lines (Z) mark the limits of the sarcomere. There appears to be a mixture of thick (arrows) and thin filaments (arrowheads). The organization is in such an early phase that the two are not easily separated and the I bands are not sharply defined. Within the limits of the technique it is possible, however, to discern the two and to observe their continuity with the microtrabeculae of the surrounding lattice. As is common in these preparations, there are numerous polysomes (circles) associated with forming elements of the fibril. The fine structure of the Z lines is difficult to decipher, but in other preparations it comes through as a compaction of relatively short and thick trabeculae which are extraordinarily dense to electrons. Thus the Z band, as a differentiation of the cell cortex and adjacent ground substance is not different in its basic structure from that that prevails elsewhere in the cytoplast. Magnification ×37,500.

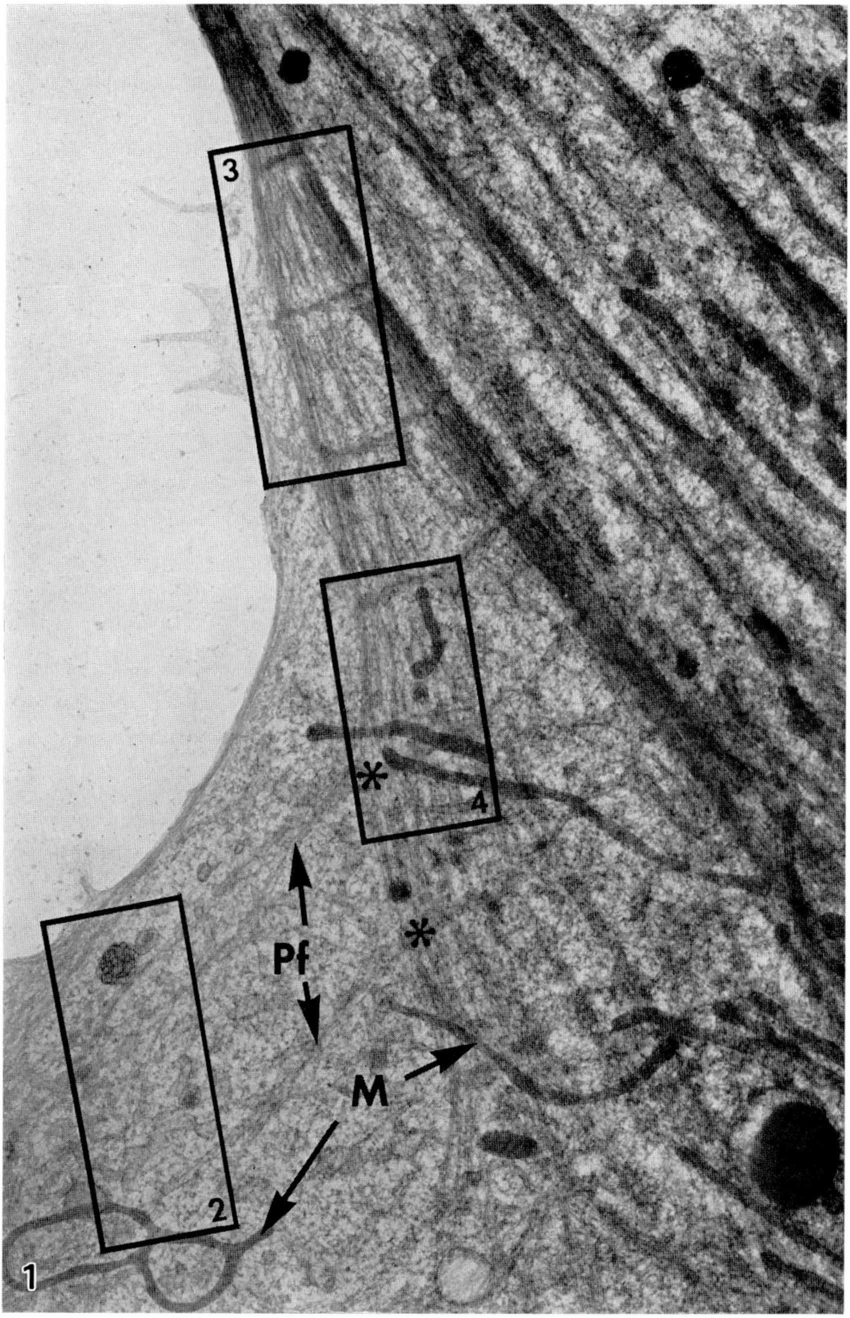
3
4
*
*
Pf
M
2
1

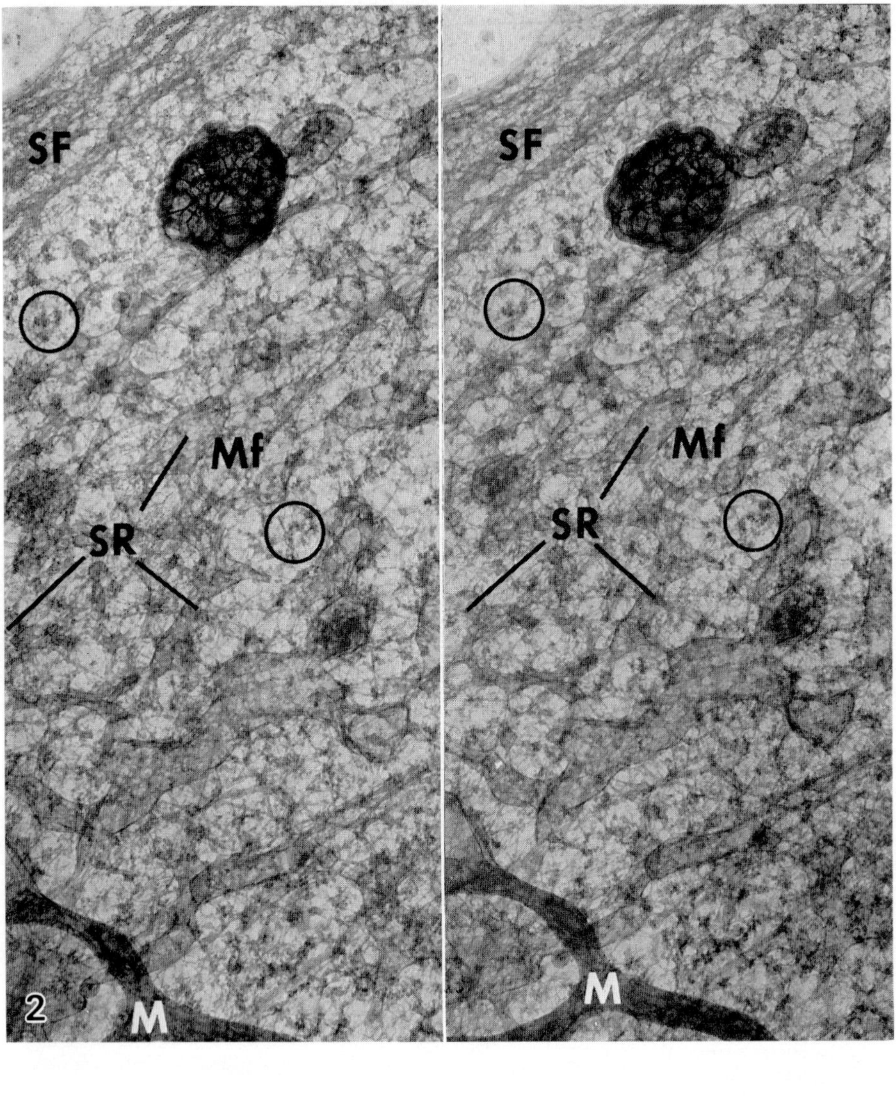
SF
Mf
SR
M
2
SF
Mf
SR
M

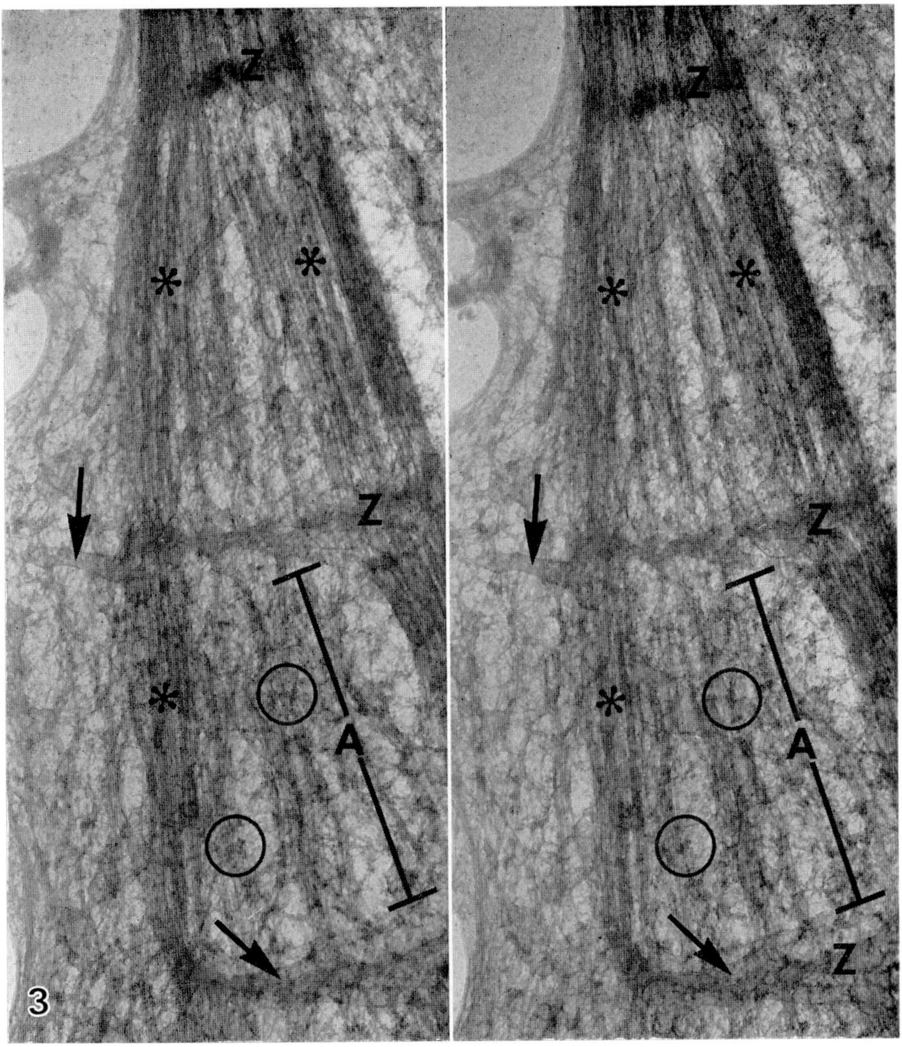
Z
Z
A
3
Z
Z
A
Z

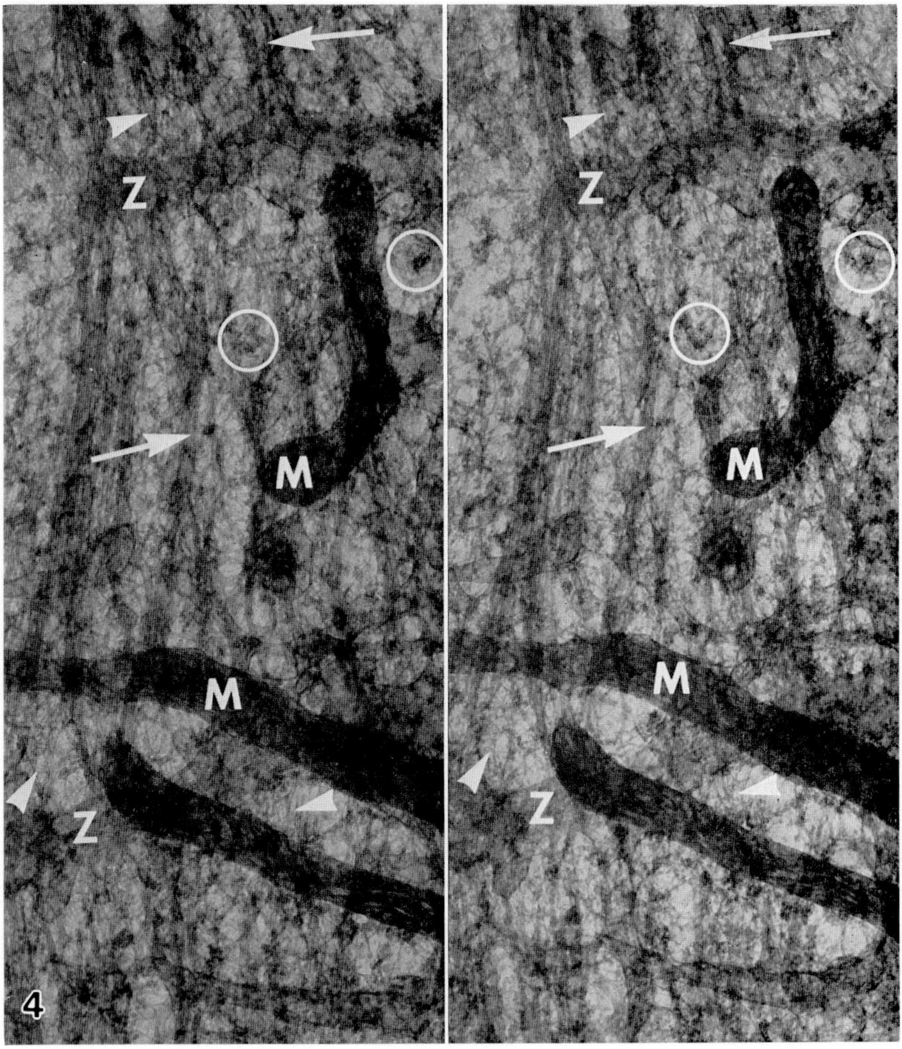
Z
M
M
Z
4

Concluding Remarks

The honor of being in this position in the program does not suit me well, for I am not an expert in the central subject, and in this assemblage I feel rather like a candle among lasers. However, I am not without thoughts concerning relations between the muscle research community of Japan and that of what we still quaintly call "the West."

These relations may be said to have begun in earnest in 1957. Japan had not truly recovered from the ravages of the war, but in muscle research the two leaders being honored here had already accomplished a great deal.

Professor Natori had fought hard to modernize Japan's physiology establishment, particularly to make it "think chemical." He was to continue a brilliant career in the service of his country and his University. In 1977 Japan was to reward him by making him Vice Chief of its Science Council, and in 1975 Jikei University was to reward him with its Presidency. But these honors were belated because by 1957 he had already organized his famous discussion groups and had strongly stimulated a young wave of Japanese muscle researchers. And of course already in 1949 he had invented a little fiber without a skin to impede biochemists. Many "westerners" became acquainted with his preparation in 1957, but its adoption worldwide had to wait on a Marco Polo effort by Richard Podolsky several years later.

Professor Kumagai was one of those rare individuals with both great research promise and great leadership quality. But I surmise that the war and the post-war period did not seem to him to be the time to pursue personal goals. Knowing that what his country needed was reconstruction, he gave himself wholeheartedly to the protection and encouragement of Japan's future pharmacologists and muscle researchers. Aided by his position as Professor in Tokyo University and President of Japan's Medical Society, he identified and fostered a remarkable group of today's leaders, including Endo, Nonomura, Ohtsuki, Masaki, and of course the Young Fellow to whom we owe knowledge of the SR pump and the Ca^{2+}-sensor, as well as this memorable symposium.

In 1957 the vision, the goals, and the friendship that Kumagai and Natori shared brought them together in an International Conference designed to open up relations with the West—Kumagai as the organizer, Natori as a star participant. That conference was successful beyond all expectations; in the decade that followed postdoctoral Fellows from Japan went in droves to England, Germany, and America, and many senior westerners started coming here as lecturers (and tourists). I think it is appropriate to thank Professors Kumagai and Natori for the enormously beneficial result of these intensified relations.

Anyone with sense would stop on this up-beat note, but I want to go on to express a concern. If I may venture a little psychoanalysis, I think that most of the young Japanese who attended the 1957 Conference sincerely thought of themselves as amateurs, eager to learn from their western teachers. In some (not all!) cases there was a realistic basis for this student-teacher attitude, and it certainly was a pleasant fairy tale for us westerners to swallow. Perhaps it was even reasonable for this unsymmetrical psychology to persist throughout the first decade following the Conference. But what about now, late in the second decade?

All who have recently changed dollars or pounds into yen are painfully aware of a new fiscal balance between Japan and the West. There has been a similar equalization in scientific achievement. In my area, the physical biochemistry of contraction, for example, it is clear that some of the finest laboratories in the world are in this country. Yet it is still the West that is accommodating Japanese postdoctoral fellows. Why do not the senior Japanese scientists prevail on their government to fund fellowships for western postdoctorates to come here? And it is still largely Japan that invites senior westerners to come here. Why are senior Japanese researchers with extended accomplishments so rarely invited to our western symposia? If there is some sense in these questions perhaps the many of influence in this audience will reflect on them and eventually respond. I think that such a re-ordering would keep the faith with the grand beginning that Kumagai and Natori made.

October 30, 1979 Manuel F. MORALES

Mr. Chairman, Sir Andrew Huxley, distinguished speakers from abroad, my dear Japanese colleagues, young and old, ladies and gentlemen.

It was a great pleasure to attend the Symposium on Regulatory Mechanism of Muscle Contraction, held in the Fuji-View Hotel surrounded by the beautiful scenery around Mt. Fuji, in company with a number of first class muscle scientists from abroad.

I had the pleasure of hearing the opening lecture by Sir Andrew Huxley, the Nobel Prize winner, and five sessions on various aspects of muscle contraction including 25 lectures and 15 poster sessions. So I was able to obtain up to date knowledge of muscle contraction which has made remarkable progress during the past twenty years.

My pleasure was doubled because these papers were delivered in celebration of the 25th or 30th anniversary of the skinned fiber invented by Professor Natori and of my 75th birthday.

Professor Natori has beautifully reviewed the past and present of Natori's fiber and Professor Morales has given us warm-hearted words. I am afraid that I do not deserve such high praise. These three days have been certainly the best days in my life.

Now I would like to present one slide. This shows you the participants in the Conference on Muscle Contraction in 1957 in Tokyo. In my talk at its closing session, I stated, "It is my great joy to have many young colleagues from both Japan and abroad, who will, I am sure, take leadership of muscle chemistry of tomorrow."

Attending this Symposium, I am convinced that my prediction was true. I am so delighted with this. I never felt happy with my longevity until today.

I very much hope that you will be the bridge—or corss-bridge—between the generations of today and tomorrow. I also hope that you will promote muscle research on a global scale, and eventually contribute to human welfare.

My sincere thanks are due to the organizers of this successful Symposium and to the Fujiwara Foundation of Science and the Japan Society for the Promotion of Science.

I would be very delighted if you could stay in Japan for a while, enjoying the beautiful autumn scenery, and take pleasant memories to your countries with you.

Finally I would like to extend my sincere thanks to all of you from the bottom of my heart.

October 30, 1979 Hiroshi KUMAGAI

Subject Index